Free Electron Lasers

ETTORE MAJORANA INTERNATIONAL SCIENCE SERIES
Series Editor:
Antonino Zichichi
European Physical Society
Geneva, Switzerland

(PHYSICAL SCIENCES)

Recent volumes in the series:

Volume 7 **UNIFICATION OF THE FUNDAMENTAL PARTICLE INTERACTIONS I**
Edited by Sergio Ferrara, John Ellis, and Peter van Nieuwenhuizen

Volume 8 **CURRENT ISSUES IN QUANTUM LOGIC**
Edited by Enrico G. Beltrametti and Bas C. van Fraassen

Volume 9 **ENERGY DEMAND AND EFFICIENT USE**
Edited by Fernando Amman and Richard Wilson

Volume 10 **INTERACTING BOSE-FERMI SYSTEMS IN NUCLEI**
Edited by F. Iachello

Volume 11 **THE SOLUTION OF THE INVERSE PROBLEM IN GEOPHYSICAL INTERPRETATION**
Edited by R. Cassinis

Volume 12 **NEUTRINO PHYSICS AND ASTROPHYSICS**
Edited by Ettore Fiorini

Volume 13 **UNCONVENTIONAL APPROACHES TO FUSION**
Edited by B. Brunelli and G. G. Leotta

Volume 14 **MINIATURIZATION OF HIGH-ENERGY PHYSICS DETECTORS**
Edited by A. Stefanini

Volume 15 **UNIFICATION OF THE FUNDAMENTAL PARTICLE INTERACTIONS II**
Edited by John Ellis and Sergio Ferrara

Volume 16 **THE SEARCH FOR CHARM, BEAUTY, AND TRUTH AT HIGH ENERGIES**
Edited by G. Bellini and S. C. C. Ting

Volume 17 **PHYSICS AT LEAR WITH LOW-ENERGY COOLED ANTIPROTONS**
Edited by Ugo Gastaldi and Robert Klapisch

Volume 18 **FREE ELECTRON LASERS**
Edited by S. Martellucci and Arthur N. Chester

A Continuation Order Plan is available for this series. A continuation order will bring delivery of each new volume immediately upon publication. Volumes are billed only upon actual shipment. For further information please contact the publisher.

Free Electron Lasers

Edited by

S. Martellucci
CNEN–Euratom
Frascati, Italy

and

Arthur N. Chester
Hughes Aircraft Company
El Segundo, California

Plenum Press • New York and London

Library of Congress Cataloging in Publication Data

Course on Physics and Technology of Free Electron Lasers of the International School of Quantum Electronics (7th: 1980: Erice, Italy)
Free electron lasers.

(Ettore Majorana international science series. Physical sciences; v. 18)
"Proceedings of the Seventh Course on Physics and Technology of Free Electron Lasers of the International School of Quantum Electronics, held August 17–29, 1980, in Erice, Sicily, Italy"—T.p. verso.
Bibliography: p.
Includes index.
1. Free electron lasers—Congresses. I. Martellucci, S. II. Chester, A. N. III. Title. IV. Series.

TA1673.C68 1980 621.36′6 83-13648
ISBN-13: 978-1-4613-3753-9 e-ISBN-13: 978-1-4613-3751-5
DOI: 10.1007/978-1-4613-3751-5

Proceedings of the Seventh Course on Physics and Technology of Free Electron Lasers of the International School of Quantum Electronics, held August 17–29, 1980, in Erice, Sicily, Italy

Softcover reprint of the hardcover 1st edition 1983
A Division of Plenum Publishing Corporation
233 Spring Street, New York, N.Y. 10013

PREFACE

The volume contains the proceedings of the 7th Course on Physics and Technology of Free Electron Lasers of the International School of Quantum Electronics, which was held in Erice (Italy) from 17 to 29 August 1980, under the auspices of the "Ettore Majorana" Centre for Scientific Culture.

The level of this Course was much closer to a workshop than to a school, and "Advances in Free Electron Lasers" might have been an appropriate title. Many of the world's leading scientists in the field (among them, the inventor of FEL, J.M.J. Madey) were brought together to review the accomplishments of FEL experiments, as well various trends in FEL theory.

In editing this material we did not modify the original manuscripts except to assist in uniformity of style. The papers. are presented without reference to the chronology of the Course but in the following topical arrangement:

A. "Fundamentals of free electron lasers," a group of tutorial papers;

B. "Free electron lasers operating in the Compton regime," where theories and experiments of FELs based on Compton scattering are reviewed;

C. "Free electron lasers operating in the Raman regime," a discussion of FELs based on Raman scattering;

D. "Optical klystrons," where the possibility of this class of FEL is discussed from a theoretical viewpoint;

E. "Accelerator and magnet technologies related to FEL design," where the technical state of the art is discussed in relation to proposed new FEL schemes; and,

F. "Applications of FELs," (It is unfortunate that among the various informal contributions given at the Panel and Round Tables, only the application of FELs to photochemistry is reported here).

Several reasons have prevented the inclusion of the contributions by some busy authors (C.D. Cantrell, Y. Farge and G.J. Yevick). A. Szöke's lectures were based on a paper already submitted for publication to another publisher and have not been included in these proceedings. The unexpected absence of Soviet scientists led the Course to be dominated by Western lectures.

Before concluding, we acknowledge the invaluable help of Mrs. M. Fiorini for much of the organization of the Course. Our thanks are also due to the organizations who sponsored the School, especially the E. Majorana Centre for Scientific Culture, whose financial support made this Course possible. The editing of these proceedings has been carried out at the Frascati CNEN Research Center, under the scientific supervision of Dr. Alberto Renieri. It has also been a pleasure to work with R.H. Andrews of Plenum Press in the preparation of this volume.

Directors of the Int. School of Quantum Electronics:

A.N. Chester
Hughes Aircraft Company
El Segundo, CA (USA)

S. Martellucci
Facoltá di Ingegneria
Universitá di Napoli (Italy)

December 16, 1980

CONTENTS

FUNDAMENTALS OF FREE ELECTRON LASERS

FREE ELECTRON LASERS OPERATING IN THE COMPTON REGIME

A UNIFIED LINEAR FORMULATION AND THE OPERATING PARAMETERS OF CERENKOV-SMITH-PURCELL, BREMSSTRAHLUNG AND COMPTON SCATTERING FREE ELECTRON LASERS*

A. Gover
Tel-Aviv University, Faculty of Engineering
Israel and Jaycor, Alexandria, Virginia 22304

P. Sprangle
N.R.L., Plasma Phys. Division
Washington, D.C. 20375

Abstract

This article discusses in a comparative way the main operating parameters of various free electron lasers, providing a useful tool for laser design, and for comparative evaluation of the various lasers. A general formulation is presented for the excitation of electromagnetic waves in any FEL structure. It is shown that any linear response theory for the electron beam current results for all FELs similar gain-dispersion relation which differs only by a single coupling parameter κ. The different gain regimes which are common to all FELs are delineated. We find the small signal gain in all the gain regimes (warm and cold beam, low or high gain, single electron, collective or strong coupling interaction). The laser gain parameter, radiation extraction efficiency, maximum power generation and spectral width are given and compared in the various kinds of FELs and gain regimes. The maximum power generation of all FELs (except Compton-Raman scattering) is shown to be limited by an interaction region width parameter. This parameter and consequently the laser power is larger relativistic limit by a factor $\sim \gamma_0$ in all bremsstrahlung FELS in comparison to Cerenkov-Smith-Purcell FELs.

*This research is supported in part by the Air Force Office of Scientific Research under grant No. AFOSR - 80 - 0073.

1. INTRODUCTION

We presently have a number of detailed theoretical analyses of free electron lasers (FEL) of various kinds: magnetic bremsstrahlung [1-10] electrostatic bremsstrahlung [11,12] stimulated Compton-Raman scattering [13-16] and Cerenkov-Smith-Purcell [17-21]. However, it would be desirable at this point to have a simple unified model which describes simultaneously all the kinds of FELs, allowing easy comparison among the various lasers and providing simple expressions for the various operating parameters required for laser design.

As is shown in the next sections, such a unified analysis is possible because the different kinds of FELs all satisfy to a good approximation similar dispersion and gain relations. The origin of the similarity of the various FELs is that they all involve longitudinal coupling between single electrons or electron plasma waves (space charge waves) and an electromagnetic wave. This coupling is carried out via different mechanisms in the different FELs discussed.

The qualitative distinction among the different kinds of FEL mechanisms was discussed in detail in Ref. [17]. The basic difference between bremsstrahlung FELs and Cerenkov-Smith-Purcell FELs is that in the first case a periodic (magnetostatic or electrostatic) force operates on the electron beam allowing phase matching (synchronism) with the electromagnetic wave by providing to the single electron or the space charge waves negative crystal momentum — k_o:

$$k_o \equiv \frac{2\pi}{L} \tag{1}$$

where L is the periodicity of the periodic field. On the other hand, in the Cerenkov-Smith-Purcell FELs the synchronism is obtained by increasing the momentum (wave number) of the electromagnetic wave in a slow wave structure (a periodic waveguide or a dielectric waveguide).

The stimulated Compton-Raman scattering problem is very similar to that of the bremsstrahlung FEL, except that instead of a static periodic force, the electron beam is modulated by an intense electromagnetic (pump) wave which propagates in a counter direction to the electron beam, and facilitates coupling between the electron plasma waves and a forward going scattered wave of higher frequency. The bremsstrahlung FEL is sometimes regarded as a special case of Compton-Raman scattering with zero frequency pump.

Figure 1 illustrates schematically the general structure of all the FELs discussed in the present article. They are all composed of an electron beam of uniform cross section which propagates at an average velocity v_{oz} through an electromagnetic waveguide and parallel to its axis (z direction). The crossed areas symbolically represent the source of interaction agent (pump) which allows the interaction between the electromagnetic wave and the electron beam. This can be in different FELs coil windings, periodic magnets, periodic electrodes, a helix, corrugated walls, dielectric walls etc. The figure shows a schematic laser amplifier structure in which the input radiative power P(o) is amplified along an interaction length ℓ, producing an output power $P(\ell) > P(o)$. A free electron laser oscillator structure will consist of similar elements but will need in addition also means for a feedback mechanism (for example a Fabri-Perot resonator).

In the next section the radiation condition (wavelength of radiation) is derived for all FELs. In Section 3 we introduce a formulation for calculating the excitation of an electromagnetic wave in a waveguide (which may be axially periodic or uniform) by a current wave passing through it. This formulation is used in Section 4 to derive the common dispersion gain relation of all FELs which is then normalized in Section 5. The different gain regimes, which evolve from the dispersion relation (and apply to all FELs) are delineated in Section 6. Expressions for the maximum gain, efficiency, power and spectral width of the various FELs are derived in Sections 7 to 10 and discussed in a comparative way.

Along the whole article we kept a full relativistic analysis. In most of the recent work on FELs the electron beam is highly relativistic. Therefore we gave the extreme relativistic limits of the relevant expressions (given in brackets). But in some FEL structures like the Smith-Purcell experiment [22], the Orotron [23,24] the Ledatron [25] or the Ubitron [26], the electron beam is non-relativistic or moderately relativistic. For this reason the derivation was made for all FELs in a general way without using

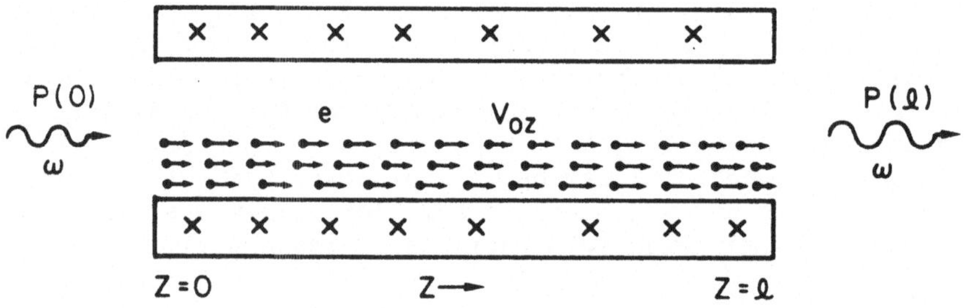

Figure 1. A schematic representation of all kinds of FEL amplifier structures.

the somewhat simplifying assumption that the beam is highly relativistic.

We use the M.K.S. unit system throughout this article.

2. THE RADIATION CONDITION

The radiation condition of the various free electron lasers can be derived from the kinematics of the interaction scheme, without requiring involved analysis.

In all FEL structures, shown schematically in Figure 1, an electron beam propagates inside the FEL structure together with a waveguided electromagnetic wave along with same direction (z axis). A necessary condition for interaction is close synchronism (phase matching) between the interacting electro,agnetic wave and electron (plasma) waves. When such "near synchronism" is obtained, energy can be transferred from the electron beam to the radiation field (amplification) or vice versa (electron acceleration). This "synchronism" condition results in the radiation relation which determines the wavelengths at which amplification should be expected.

Figure 2 describes schematically the interaction schemes of the various free electron lasers discussed. The approximate wave numbers of the interacting waves or wave components (space harmonics) are liated in the first two columns of Table 1. Equating the wave numbers (phase matching or momentum conservation) yields the radiation condition (third column in Table 1). The expressions in brackets eorrespond to the highly relativistic limit $\gamma_{oz} >> 1$ (γ_{oz} is defined later in Eq. (7)).

In Table 1 and Figure 1 k_{zo} represents the wave number of the interacting electromagnetic wave which is in general a waveguide mode for which $k_{zo} < k \equiv \omega/c$. For a plane wave propagating in the z direction, or for a low order mode in a wide waveguide, one has $k_{oz} = k = \omega/c$. In a dielectric waveguide (Cerenkov FEL) one may have $k_{zo} > k$. v_o is the mean velocity of the electron beam which is propagating in the z direction. In the case of the transverse pump FELs (two last rows in Table 1) the electron beam has at each point also transverse velocity and the parameter which is used in the synchronism condition is v_{oz} the longitudinal component of the average beam velocity. For the longitudinal FELs (first three rows in Table 1) $v_{oz} = v_o$ ($\beta_{oz} = \beta_o$).

The dispersion curve of the electron waves is represented symbolically in Figure 1 by the thick line of slope $\omega/k_z = v_{oz}$. A single electron which is forced to oscillate at frequency ω and at the same time propagates in the z direction with velocity v_z is generating a travelling current wave with wave number ω/v_z. In the case of a warm electron beam ther is some spread in the wave numbers ω/v_z due to the variance v_{zth} in the velocities. When the beam is cold enough the eigenmodes of the beam plasma will be the longitudinal space charge waves with wave number ($\omega \pm \omega_p'/v_{oz}$. Where ω_p'

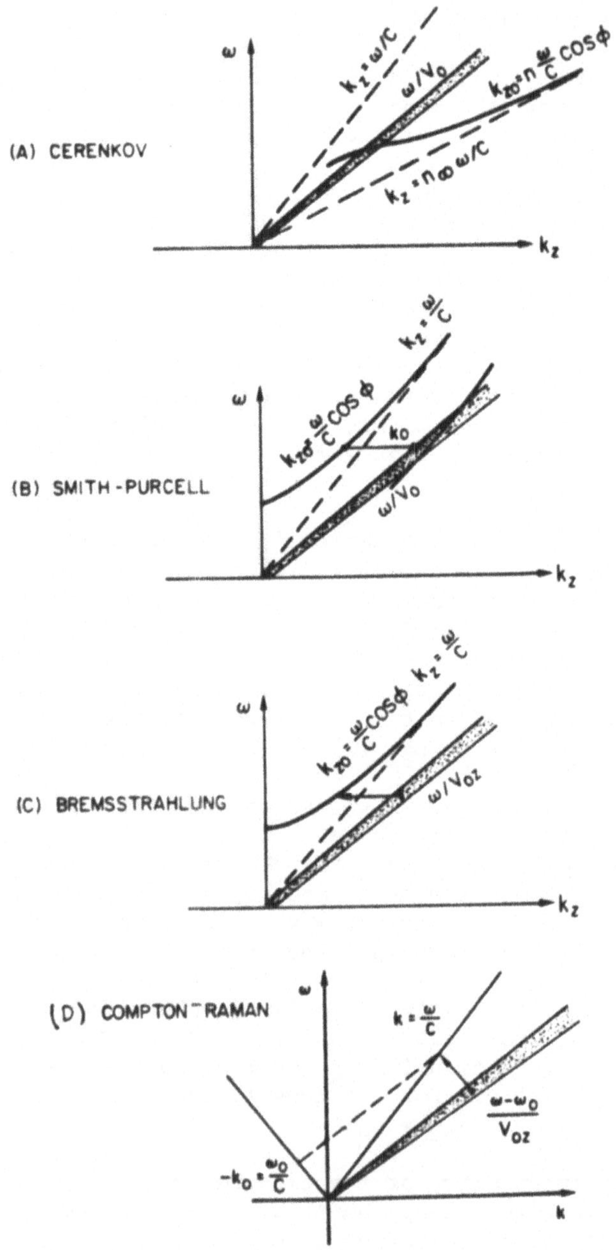

Figure 2. Dispersion diagrams of the interacting waves illustrating the synchronism (phase matching) condition for the difference FEL schemes.

is the modified plasma frequency of the electron beam. Either case is represented approximately by the thick dotted line of slope v_{oz} assuming $v_{zth} << v_{oz}$ and $\omega_p' << \omega$.

The dispersion curve of the electromagnetic wave is in the case of Cerenkov FEL (Figure 2a) the well known dispersion curve of a dielectric waveguide mode which at high enough frequency tends to a slope (phase velocity) c/n (n is the high frequency index of refraction). If the beam which is progated in close proximity to the dielectric waveguide, has a mean velocity $v_{oz} > c/n$, the synchronism between the electron beam and an electromagnetic mode may be possible around the curves crossing region:

$$\frac{\omega}{v_{oz}} \simeq k_{zo} \tag{2}$$

This condition is equivalent to a condition for synchronism between the wave phase velocity and the beam velocity $\omega/k_{oz} \simeq v_{oz}$. The Radiation condition in Table 1 row 1 is derived from (2) using the definition of the "mode zig-zag angle" ϕ

$$\cos\phi \equiv \frac{k_{oz}}{n(\lambda)\,k} \tag{3}$$

where $k \equiv \omega/c$.

TABLE 1. THE WAVE NUMBERS OF THE SYNCHRONOUS COMPONENTS OF THE INTERACTING WAVES AND THE RADIATION CONDITION FOR VARIOUS FELS

FEL	EM Component Wave Number	Plasma Component Wave Number	Radiation Condition
Cerenkov	k_{zo}	$\frac{\omega}{v_o}$	$n^{-1}(\lambda) = \beta_o \cos\phi$
Smith Purcell	$k_{zo} + k_o$	$\frac{\omega}{v_o}$	$\frac{\lambda}{L} \simeq \beta_o^{-1} - \cos\phi$
Longitudinal Bremsstrahlung	k_{zo}	$\frac{\omega}{v_o} - k_o$	$\frac{\lambda}{L} \simeq \beta_o^{-1} - \cos\phi$
Transverse Bremmstrahlung	k_{zo}	$\frac{\omega}{v_{oz}} - k_o$	$\frac{\lambda}{L} \simeq \beta_{oz}^{-1} - 1 = \frac{1}{(1+\beta_{oz})\,\beta_{oz}\,\gamma_{oz}^2} \left[\simeq \frac{1}{2\gamma_{oz}^2}\right]$
Compton-Raman	k	$\frac{\omega - \omega_o}{v_{oz}} - k_o$.	$\frac{\lambda}{\lambda o} = \frac{1}{(1+\beta_{oz})^2\,\gamma_{oz}^2} \left[\simeq \frac{1}{4\gamma_{oz}^2}\right]$

In the case of the Smith-Purcell FEL (Figure 1b) the phase matching condition between the electron beam waves and the electromagnetic mode is obtained by adding "crystal momentum k_o" to the mode wave number k_{zo}. To be more concrete we may say that in the periodic waveguide, of which the Smith-Purcell FEL is composed, the electromagnetic eigenmodes are Floquet-Bloch modes and the smooth waveguide dispersion diagram is modified into a Brillouin diagram with period k_o. (Which for clarity is only partly drawn as a curve section parallel to the smooth waveguide dispersion curve and displaced by k_o to the right.) The synchronism condition is obtained when the dispersion curve of the first order space harmonic (k_{zo} + k_o) matches the curve of the electron beam wave:

$$\frac{\omega}{v_{oz}} \simeq k_{zo} + k_o \tag{4}$$

This can also be interpreted in the velocity space as a synchronism condition between the wave phase velocity and the beam velocity: $\omega/(k_{zo} + k_o) = v_{oz}$. The radiation condition of the Smith-Purcell effect [22] is derived in Table 1 from (4) using the definition for the "mode zig-zag angle" ϕ:

$$\cos \phi \equiv \frac{k_{zo}}{k} \tag{5}$$

The scheme for the bremsstrahlung FEL (Figure 1c) is similar to that of the Smith-Purcell FEL and results the same synchronism condition (4) and consequently the same radiation condition (Table 1). The difference is that in this case a periodic magnetic or electric force operates on the electron beam and endows its waves with a negative crystal momentum $-k_o$. Thus the electromagnetic waveguide mode interacts with the -1 order space harmonic of the electron beam wave.

In the case of the most familiar transverse bremsstrahlung FEL (magnetic or electric), the interaction can take place with a transverse electromagnetic (TE) wave for which $k_{zo} = k$ ($\phi = 0$). Hence the radiation condition for this case which is given in Table 1 (row 4) is derived by simply substituting $\phi = 0$ in the previous result (row 3). In this case and also in the transverse electrostatic and Compton-Raman FELs the electron beam has a transverse component of the average beam velocity due to the transverse force applied by the pump, and $v_{oz} \neq v_o$. We defined for these cases

$$\beta_{oz} \equiv \frac{v_{oz}}{c} \tag{6}$$

$$\gamma_{oz} \equiv (1 - \beta_{oz}^2)^{1/2} \tag{7}$$

In the case of Compton-Raman scattering (Figure 1d) the periodic electrostatic or magnetostatic pump of the transverse bremsstrahlung FEL is replaced by an intense TEM wave with frequency ω_o and wave number $k_o = 2\pi/\lambda_o$) propagating in a counter direction to the electron beam. The combined effect of the scattering and scattered electromagnetic waves generates in the beam an electron current wave which oscillates with the difference frequency $\omega - \omega_o$ and propagates with wave number $(\omega - \omega_o)/v_{oz} - k_o$. In wave number space the phase matching condition between the electron wave and the amplified electromagnetic wave is:

$$\frac{\omega - \omega_o}{v_{oz}} - k_o \simeq k \tag{8}$$

which results in the radiation condition of Table 1 (row 5).

The phase matching condition (8) can be also explained in a different way. The combined effect of the scattering and scattered waves generates in a parametric process a current with frequency $\omega - \omega_o$ and wave number $k - k_o$. The scattering process is resonant when the phase velocity of the current is synchronous with the electron beam

$$\frac{\omega - \omega_o}{k + k_o} \simeq v_{oz} \tag{9}$$

This is of course equivalent to (8).

3. ELECTROMAGNETIC MODES EXCITATION IN A WAVEGUIDE STRUCTURE

The "kinematics" of the various FEL prccesses was sufficient to derive the radiation conditions (Table 1) for the difference FELs. In order to derive the gain expressions of FELs we have to investigate the "dynamics" of the interaction between the electromagnetic wave and the electron beam in the various FEL schemes.

The equations which govern the interaction are Maxwell equations on one hand and the electron equations on the other hand. The electron equations result in the time varying current wave which is induced on the electron beam by the electromagnetic wave traversing the FEL waveguide. The Maxwell equations (with current and charge sources terms) govern the excitation of electromagnetic waves in the FEL waveguide by the same current wave. The self consistent solution of the Maxwell and electron sets of equations would result in the dispersion equation of the FEL excitation, which is created by the coupling between the electromagnetic wave and the electrons. The solution of this dispersion equation would result in the growth rate (gain) of the electromagnetic wave.

In this section we concentrate on the excitation of the electromagnetic waves in the FEL waveguide by a time varying current. Following [27,20,21] we will simplify the Maxwell equations for the signal wave:

$$\underline{\nabla} \times \underline{E} = i\omega\mu_o \underline{H} \qquad (10)$$

$$\underline{\nabla} \times \underline{H} = -i\omega\varepsilon_o \underline{E} + \underline{J} \qquad (11)$$

$$\underline{\nabla} \cdot \underline{E} = \rho/\varepsilon_o \qquad (12)$$

$$\underline{\nabla} \cdot \underline{H} = 0 \qquad (13)$$

by expanding the fields in the waveguide in terms of its eigenmodes. This will allow to reduce the three dimensional Maxwell equations into equations with one variable (the axial z coordinate). We will keep the analysis general enough to include also the case of Smith-Purcell FEL and TWT amplifier, and therefore will assume in general that the waveguide is periodic. In all other FELs the waveguide is axially uniform.

To derive the equations for the excitation of electromagnetic waves in a periodic waveguide, we make use of the Lorentz theorem and the Floquet theorem. The Lorentz theorem states the following relation for any two sets of fields and currents ($\underline{E}_1$, $\underline{H}_1$, $\underline{J}_1$ and $\underline{E}_2$, $\underline{H}_2$, $\underline{J}_2$) which satisfy the Maxwell equations:

$$\underline{\nabla} \cdot (\underline{E}_1 \times \underline{H}_2 - \underline{E}_2 \times \underline{H}_1) = \underline{J}_1 \cdot \underline{E}_2 - \underline{J}_2 \cdot \underline{E}_1 \tag{14}$$

or in integral form

$$\iint_A (\underline{E}_1 \times \underline{H}_2 - \underline{E}_2 \times \underline{H}_1) \cdot \underline{ds} = \iint_V\!\!\int (\underline{J}_1 - \underline{E}_2 - \underline{J}_2 - \underline{E}_1)\, dV \tag{15}$$

This theorem can be proved straightforwardly by writing (11) for ($\underline{E}_1$, $\underline{H}_1$, $\underline{J}_1$) and scallar multiply it by $\underline{E}_2$, then writing the same equation for ($\underline{E}_2$, $\underline{H}_2$, $\underline{J}_2$) and scallar multiply by $\underline{E}_1$. Substructing the two resulting equations results in (14).

The Floquet theorem (which is derived from symmetry considerations) states that the eigenmodes of an axially periodic waveguide with period L can always be written in the form

$$\underline{E}_n (x, y, z) = \underline{\mathcal{E}}_n (x, y, z)^{i\gamma_n z} \tag{16}$$

where $\underline{\mathcal{E}}_n$ is a periodic function in z

$$\underline{\mathcal{E}}_n (x, y, z + L) = \underline{\mathcal{E}}_n (x, y, z) \tag{17}$$

The propagation parameter γ_n can be in general complex:

$$\gamma_n = k_{zn} + i\, \alpha_n \tag{18}$$

and the wave number of a wave going to the -z direction is defined by

$$\gamma_{-n} = -\gamma_n \tag{19}$$

In order to ortho-normalize the Floquet modes in the empty periodic waveguide ($\underline{J} = 0$), we plug (16) in the Lorentz equation (15) for two different modes: n and n', and perform the surface integration over a cylindrical surface with planar bases perpendicular to the waveguide axis which are one period apart ($z = z_1$ and $z = z_1 + L$) and with side walls on which the electromagnetic fields vanish (these can be the waveguide walls in the case that the walls are made of an ideal conductor, or a transversely far away surface where the fields of the waveguide modes decay). Making use of the periodicity of $\underline{\mathscr{E}}$ (17) we get

$$\iint_{-\infty}^{\infty} dxdy\,[\underline{\mathscr{E}}_n(x,y,z_1)\times\underline{\mathscr{H}}_{n'}(x,y,z_1)-\underline{\mathscr{E}}_{n'}(x,y,z_1)\times\underline{\mathscr{H}}_n(x,y,z_1)]$$

$$(z=z_1)\qquad [e^{i(\gamma_n+\gamma_{n'})(z_1+L)}-e^{i(\gamma_n+\gamma_{n'})z_1}] = 0 \tag{20}$$

Noting that Floquet modes wave numbers k_{zn} are defined module $k_0 = 2\pi/L$ we find from (20) that the different Floquet modes are orthogonal to each other and the ortho-normal relation of these modes can be written as:

$$\iint_{-\infty}^{\infty} dxdy\,[\underline{\mathscr{E}}_n(x,y,z)\times\underline{\mathscr{H}}_{n'}(x,y,z)-\underline{\mathscr{E}}_{n'}(x,y,z)\times\underline{\mathscr{H}}_n(x,y,z)]$$

$$= N_n(z)\,\delta_{n,-n'} \tag{21}$$

where

$$N_n(z) = \iint_{-\infty}^{\infty}(\underline{\mathscr{E}}_n\times\underline{\mathscr{H}}_{-n}-\underline{\mathscr{E}}_{-n}\times\underline{\mathscr{H}}_n)\,dxdy \tag{22}$$

is the norm of the nth order mode.

For many kinds of lossless periodic waveguides[27,21] and certainly for any uniform waveguide, the mode norm can be written in terms of the mode power

$$N_n = -4P_n \tag{23}$$

$$P_n = \frac{1}{2} \text{Re} \int_{-\infty}^{\infty}\int \underline{\mathscr{E}}_n \times \underline{\mathscr{H}}_n^* \, dxdy \tag{24}$$

Considering now a waveguide which contains a current source $\underline{J}(x, y, z)$ of frequency ω. The electromagnetic field which is generated and propagated in the waveguide can always be expanded in terms of the waveguide modes. For each z coordinate value we expand the transverse field of the total electromagnetic fields in terms of the transverse fields of the waveguide modes

$$\underline{E}_t = \sum_n [a_n(z)\, \underline{\mathscr{E}}_{nt}(x, y, z) + a_{-n}(z)\, \underline{\mathscr{E}}_{nt}(x, y, z)] \tag{25}$$

$$\underline{H}_t = \sum_n [a_n(z)\, \underline{\mathscr{H}}_{nt}(x, y, z) + a_{-n}(z)\, \underline{\mathscr{H}}_{-nt}(x, y, z)] \tag{26}$$

Substitution of (25,26) in the Maxwell equations (10-13) produces the equations for the axial field components and the mode amplitudes $a_n(z)$:

$$E_z = \sum_n [a_n(z)\, \mathscr{E}_{nz}(x, y, z) + a_{-n}(z)\, \mathscr{E}_{-nz}(x, y, z)] + \frac{1}{i\omega\varepsilon_0} J_z(x,y,z) \tag{27}$$

$$H_z = \sum_n [a_n(z)\, \mathscr{E}_{nz}(x, y, z) + a_{-n}(z)\, \mathscr{E}_{-nz}(x, y, z)] \tag{28}$$

$$\frac{da_n}{dz} - i\gamma_n a_n = \frac{1}{N_n} \int_{-\infty}^{\infty}\int \underline{J} \cdot \underline{\mathscr{E}}_n \, dxdy \tag{29}$$

$$\frac{da_{-n}}{dz} + i\gamma_n a_{-n} = -\frac{1}{N_n} \iint \underline{J} \cdot \underline{\mathscr{E}}_n \, dxdy \tag{30}$$

When $\underline{J} = 0$ (empty waveguide) the straightforward solution of (29,30) is

$$a_n(z) = a_n(0)\, e^{i\gamma_n z} \tag{31}$$

and Eqs (25-28) state that the total fields in the waveguide is a linear combination of the waveguide modes (16). For a finite current ($\underline{J} \neq 0$) we see from (29.30) that the amplitude of certain modes which overlap well (phase match) the current wave J(x, y, z) may grow due to the driving term on the right hand of the equations. In addition, the axial component of the current wave produces the longitudinal space charge field (second term in Eq. (27)) which adds up to the solenoidal field of the waveguide modes, This space charge field component, which was produced by the Poisson equation (12), guarantees that out formulation takes into account collective space charge effects.

In most FELs discussed here the waveguide is uniform, in which case (16) and (25- 30) still apply with

$$\underline{\mathscr{E}}_n\,(x, y, z) = \underline{\mathscr{E}}_n\,(x, y); \qquad \underline{\mathscr{H}}_n\,(x, y, z) = \underline{\mathscr{H}}_n\,(x, y) \tag{32}$$

For the uniform waveguide the derivation of (25-30) is somewhat shorter and given in standard textbooks[50]. The formulation used here applies also to periodic waveguides, and thus will allow us to include Smith-Purcell FELs (or T.W.T. amplifiers) in the proceeding generalized analysis.

For simplicity we will assume from now on that there is only a single mode excited in the waveguide and we will drop its mode index n. The waveguide will be assumed lossless ($\alpha_n = 0$). The equations for the electromagnetic wave excitation then simplify into

$$\underline{E}_t\,(x, y, z) = a(z)\,\underline{\mathscr{E}}_t\,(x, y, z) \tag{33}$$

$$E_z\,(x, y, z) = a(z)\,\mathscr{E}_z\,(x, y, z) + \frac{1}{i\omega\varepsilon_0}\,J_z\,(x, y, z) \tag{34}$$

$$\underline{H}\,(x, y, z) = a(z)\,\underline{\mathscr{H}}\,(x, y, z) \tag{35}$$

$$\frac{da(z)}{dz} - ik_{z0}a(z) = \frac{1}{4P}\iint_{-\infty}^{\infty} \underline{J}\,(x, y, z) \cdot \underline{\mathscr{E}}^*\,(x, y, z)\,dxdy \tag{36}$$

Equations (33-36) are a significant simplification of the Maxwell equations (10-13), since they form a simple set of algebraic and first order differential equations in a single variable z. These equations can be solved when the current $\underline{J}$ is given in terms of the electromagnetic fields.

In the limit $\underline{J} = 0$ (empty waveguide) Eqs. (33,34,36) result in

$$\underline{E}(x, y, z) = a(0)\, \underline{\mathscr{E}}(x, y, z)\, e^{ik_{z0}z} \tag{37}$$

This is the electromagnetic eigen mode which is interacting with the electron current wave.

4. FORMAL DERIVATION OF THE DISPERSION RELATION

In Section 3 we derived the equations which govern the excitation of a single electromagnetic mode in a waveguide by an electron current wave. To complete the analysis of the interaction between the electromagnetic wave and the electron beam we have to derive an expression for the current wave which is induced by the electromagnetic wave in the electron beam so that we can substitute it in (34,36) and solve the set of equations self consistently. In the present analysis we are interested in the linear current response only (i.e., first order in terms of the fields of the amplified electromagnetic wave).

In order to calucate the linear current response one has to solve the electron equations of a given electron beam to first order in the electromagnetic fields of the amplified wave. In terms of the scattering magnetostatic or electrostatic fields (in magnetic and electrostatic FELs respectively) or the scattering electromagnetic field (in Compton-Raman FELs), solution of the electron equations may be required to higher order. Different kinds of electron equations can be used to derive the linear response functions. They range through single electron force equations, moment (fluid) plasma equations, Vlasov equation, or even quantum mechanical equations. All of these different models were used in the different analyses of FELs[1-21]. It is beyond the scope of the present article to drive the linear response functions of the different FELs. We rather will derive in this section the gain dispersion relations of the different FELs assuming that the linear response relation of each FEL is known in terms of a general susceptibility tensor which could be derived from any set of electron equations. However, we later will specify the analysis to susceptibility functions which were derived in the existant literature for different FELs using the Vlasov equation.

The Cerenkov FEL

This FEL is perhaps the easiest to analize since it allows phase matching (synchronism) between an electromagnetic wave and the electron current waves (Figure 2a) in a structure which is axially uniform for both the electromagnetic wave and the electrons. In this case we can use (32) for the electromagnetic modes field.

Since we need to solve equations which are linear in a single variable z, and we want to include initial conditions, it is most natural to apply a Laplace transform in the spatial coordinate z:

$$\overline{f}(s) = \mathscr{L}\{f(z)\} = \int_0^\infty e^{-sz} f(z)dz \qquad (38)$$

When applied to (33-36) it results in

$$\bar{\underline{E}}_t\ (x, y, s) = \bar{a}(s)\ \underline{\mathscr{E}}_t\ (x, y) \tag{39}$$

$$\bar{E}_z\ (x, y, s) = \bar{a}(s)\ \mathscr{E}_z\ (x, y) + \frac{1}{i\omega\varepsilon_o}\ \bar{J}_z\ (x, y, s) \tag{40}$$

$$\bar{H}_z\ (x, y, s) = \bar{a}(s)\ \underline{\mathscr{H}}(x, y) \tag{41}$$

$$(s-ik_{zo})\ \bar{a}(s) - a(o) = \int_{-\infty}^{\infty}\!\!\int \bar{\underline{J}}\ (x, y, s) \cdot \underline{\mathscr{E}}^*\ (x, y)\ d\,x\,dy \tag{42}$$

The most general linear response relation between the induced electron current and the amplified electromagnetic wave fields in an axially uniform structure is:

$$\bar{\underline{J}}(x,y,\omega,s) = -i\omega\ \underline{\underline{\chi}}^E(x,y,\omega,s) \cdot \bar{\underline{E}}(x,y,\omega,s)\ -i\omega\underline{\underline{\chi}}^B(x,y,\omega,s) \cdot c\bar{\underline{B}}(x,y,\omega,s) \tag{43}$$

where in general $\underline{\underline{\chi}}^E$ and $\underline{\underline{\chi}}^\infty$ are 3 x 3 susceptibility tensors. The general relation (43) is indpendent of the particular model assumed in solving the electron equations. It is implying vanishing initial conditions for the current (the possibility of prebundhing the electron beam is not considered here).

Equations (39-43) constitute a set of algebraic equations which can be straightforwardly solved. Substitution of (39-41) in (43) results in

$$\bar{J}_z = -\frac{i\omega}{\varepsilon_{33}} \sum_{j=1}^{3} \left[\chi^E_{3j}\ \mathscr{E}_j + \chi^B_{3j}\ c\mathscr{H}_j\right] \bar{a}(s) \tag{44}$$

$$\bar{J}_k = -\, i\omega \sum_{j=1}^{3} \left[\chi^E_{kj}\ \mathscr{E}_j + \chi^B_{kj}\ c\mathscr{H}_j\right] \bar{a}(s)$$

$$-\frac{1}{\varepsilon_{33}}\frac{\chi^E_{k3}}{\varepsilon_o} \sum_{j=1}^{3} \left[\chi^E_{3j}\ \mathscr{E}_i + \chi^B_{3j}\ c\mathscr{H}_j\right] \bar{a}(s) \qquad (k = 1,2) \tag{45}$$

where

$$\varepsilon_{33}^{(\omega,s)} = 1 + \chi_{33}^{(\omega,s)}/\varepsilon_o \tag{46}$$

and

$$\underline{\mathscr{B}} = \mu_o \mathscr{H}.$$

Substitution of (44,45) in (42) enables us to express $\bar{a}(s)$ explicitly in terms of a(o) and in principle allows us to complete the analysis by inverse-Laplace transforming this expression. Ignoring for the moment the possible dependence of ε_{33} (46) on the transverse coordinates x, y (due to transverse variation of the electron beam density), we can take $1/\varepsilon_{33}$ out of the integral sign in (42) and obtain the following expression for the Laplace transformed mode amplitude:

$$\bar{a}(s) = \frac{\varepsilon_{33}\ (\omega,s)}{(s - ik_{zo})\ \varepsilon_{33}\ (\varepsilon,s) - iK}\, a(o) \tag{47}$$

$$K = \frac{\omega}{4P}\Bigg[\int_{-\infty}^{\infty} (\underline{\mathscr{E}} \cdot \underline{\underline{\chi}}^E \cdot \underline{\mathscr{E}} + \underline{\mathscr{E}} \cdot \underline{\underline{\chi}}^B \cdot c\underline{\mathscr{B}})\, d\,x\, dy$$

$$+ \frac{1}{\varepsilon_o}\int_{-\infty}^{\infty} \chi_{33}^E\, \underline{\mathscr{E}} \cdot (\underline{\underline{\chi}}^E \cdot \underline{\mathscr{E}})_3\, (\underline{\underline{\chi}}^{E^T} \cdot \underline{\mathscr{E}})_3\, d\,x\, dy$$

$$+ \frac{1}{\varepsilon_o}\int_{-\infty}^{\infty} \chi_{33}^E\, \underline{\mathscr{E}} \cdot (\underline{\underline{\chi}}^B\, c\mathscr{B}) - (\underline{\underline{\chi}}^B \cdot c\underline{\mathscr{B}})_3\, (\chi^{E^T} \cdot \underline{\mathscr{E}})_3\, d\,x\, dy\Bigg] \tag{48}$$

where the superscript T indicates a transposed tensor matrix. The derivation of (47,48) applies for an electron beam with finite cross section as long as the electron beam density is uniform across the beam, so that there is no dependence of ε_{33} (Eq. (46)) on the transverse coordinates x, y within the beam cross section. If this is not the case, the applicability of (47,48) is limited to regimes where collective effects are negligible and $\varepsilon_{33} \approx 1$.

In practice the exact but cumbersome form of (48) can be sufficiently reduced due to the vanishing of many terms. Using the

Vlasov equation model to calculate the susceptibility tensor, it is possible to show[43], that the two last terms in the right hand side of (48) are vanishing near the synchronism condition, and that in the susceptibility tensors $\underline{\underline{\chi}}^E$, $\underline{\underline{\chi}}^B$ the only significant term is the longitudinal-longitudinal response term $\chi^E_{33}(\omega,s)$ (the other terms are identically zero or negligible). Hence (47,48) reduce into

$$\bar{a}(s) = \frac{1 + \chi^E_{33}(\omega,s)/\varepsilon_o}{(s - ik_{zo})\,[1 + \chi^E_{33}(\omega,s)/\varepsilon_o] - i\kappa\chi^E_{33}(\omega,s)/\varepsilon_o}\, a(o) \tag{49}$$

where the coupling coefficient κ is defined by

$$\kappa = \frac{1/2\,\sqrt{\varepsilon_o/\mu_o}\displaystyle\int_{-\infty}^{\infty}\!\!\int |\,_z(x,\,y)|^2\,d\,x\,dy}{P} \tag{50}$$

The Smith-Purcell FEL

In Smith-Purcell FELs and TWT amplifiers an electron beam is passed through a periodic waveguide (a waveguide with corrugated walls, a helix etc.). The electromagnetic mode excitation equations are then given by (33-36) where the fields $\underline{\mathscr{E}}$ (x, y, z), $\underline{\mathscr{H}}$(x, y, z) are periodic functions in a z due to the Floquet theorem (17). Consequently we can expand the fields of the electromagnetic mode into a Fourier series:

$$\underline{\mathscr{E}}\,(x,\,y,\,z) = \sum_{m=-\infty}^{\infty} \underline{\mathscr{E}}^{(m)}(x,\,y)\,e^{imk_o z} \tag{51}$$

$$\underline{\mathscr{B}}\,(x,\,y,\,z) = \sum_{m=-\infty}^{\infty} \underline{\mathscr{B}}^{(m)}(x,\,y)\,e^{imk_o z} \tag{52}$$

Laplace transforming (33-36) with (51,52), results in

$$\underline{\bar{E}}_t(x,y,s) = \sum_{m=-\infty}^{\infty} \bar{a}\,(s - mik_o)\, \underline{\mathscr{E}}_t^{(m)}(x,y) \tag{53}$$

$$\bar{E}_z(x,y,z) = \sum_{m=-\infty}^{\infty} \bar{a}\,(s - mik_o)\, \mathscr{E}_z^{(m)}(x,y) + \bar{J}_z(x,y,s) \tag{54}$$

$$\underline{\bar{B}}(x,y,s) = \sum_{m=-\infty}^{\infty} \bar{a}\,(s - mik_o)\, \underline{\mathscr{B}}^{(m)}(x,y) \tag{55}$$

$$(s - ik_{zo})\,\bar{a}(s) - a(o) = -\frac{1}{4P} \sum_{m=-\infty}^{\infty} \int\!\!\int_{-\infty}^{\infty} \underline{J}(x,y,s + mik_o) \cdot \underline{\mathscr{E}}^{(m)*}(x,y)\, d\,x\, dy \tag{56}$$

For the electrons the interaction region is axially uniform and therefore the linear susceptibility relation (43) is applicable. We substitute (53-55) in (43) and obtain

$$\bar{J}_z(s) = -\frac{i\omega}{\varepsilon_{33}} \sum_{m=-\infty}^{\infty} \sum_{j=1}^{3} \left[\chi_{3j}^{E}(\omega,s)\, \mathscr{E}_j^{(m)} + \chi_{kj}^{B}(\omega,s)\, c\mathscr{B}_j^{(m)} \right] \bar{a}\,(s - mik_o) \tag{57}$$

$$\bar{J}_k(s) = -i\omega \sum_{m=-\infty}^{\infty} \left\{ \sum_{j=1}^{3} \chi^E_{kj}(\omega,s)\, \mathcal{E}_j^{(m)} + \left[\chi^B_{kj}(\omega,s)\, c\beta_j^{(m)}\right] - \frac{1}{\varepsilon_{33}} \frac{\chi^E_{k3}}{\varepsilon_o} \sum_{j=1}^{3} \left[\chi^E_{3j}\, \mathcal{E}_j + \chi^B_{3j}\, c\beta_j\right] \right\} \bar{a}(s-mik_o) \quad (k=1,2) \tag{58}$$

where ε_{33} is again given by (46).

Substitution of (57), (58) in (56) results in a set of equations for $\bar{a}(s-mik_o)$. In order to solve explicitly these equations we must assume some approximations. We are looking for an electromagnetic-like solution ($s \cong ik_{zo}$) and neglect $\bar{a}(s-mik_o)$ ($m \neq o$). We assume that phase matching takes place through the first order space harmonic (Eq. 4, Fig. 2.b), and therefore the only significant current excitation is due to the susceptibility term $\chi^E(\omega,s+ik_o)$ (where s k_{zo}). We also assume that the electron density is uniform across the beam or that space charge effects are negligible, so that we can neglect the dependence of ε_{33} on the transverse coordinates (x,y) and take $1/\varepsilon_{33}(\omega,s+ik_o)$ out of the integral sign in (56). In complete analogy to the Cerenkov FEL case these assumptions result in

$$\bar{a}(s) = \frac{\varepsilon_{33}(\omega,s+ik_o)}{(s-ik_{zo})\,\varepsilon_{33}(\omega,s+ik_o) - iK}\, a(o) \tag{59}$$

where K is given still by (48) with $\underline{\underline{\chi}} = \underline{\underline{\chi}}(\omega,s+ik_o)$. In practice most of the terms in (48) can be shown to vanish near synchronizm or to be negligible[43], and in analogy to (49,50) we have for the Smith-Purcell FEL (TWT)

$$\bar{a}(s) = \frac{1 + {}_{33}(\omega,s+ik_o)/\varepsilon_o}{(s-ik_{zo})\,[1+\chi_{33}(\omega,s+k_o)/\varepsilon_o] - i\,\kappa\,\chi_{33}(\omega,s+ik_o)/\varepsilon_o}\, a(o) \tag{60}$$

$$= \frac{1/2\sqrt{\varepsilon_o/\mu_o}\iint_{Ae} \left|\mathcal{E}_z^{(1)}(x,y)\right|^2 dxdy}{P} \tag{61}$$

where $\mathcal{E}_z^{(1)}(x,y)$ is the axial field of the first order space harmonic of the electromagnetic mode in the periodic waveguide (51)

$$\mathcal{E}_z^{(1)}\ (x,y) = \frac{1}{L}\int_0^L e^{-ik_o z}\mathcal{E}_z(x,y,z)dz \tag{62}$$

Thus the Smith-Purcell FEL mechanism is completely analogous to that of the Cerenkov FEL except that the Interaction is carried out via a wave component of the electromagnetic mode (for example the first order space harmonic). With proper design of the periodic structure the amplitude of this space harmonic can be made relatively large.

Bremsstrahlung FELs

In bremsstrahlung FELs a periodic force (electrostatic or magnetostatic) operates on the electrons in the electron beam. In an axially periodic structure, the most general linear response relation is

$$\bar{J}_k(x,y,\omega,s) = -i\omega\sum_{m=-\infty}^{\infty}\sum_{j=1}^{3}[\chi_{kj}^E(x,y,\omega,s,s+mik_o)\,\bar{E}_j(x,y,\omega,s+mik_o)$$

$$+\chi_{kj}^B\ (x,y,\omega,s,s+mik_o)c\bar{B}_j\ (x,y,\omega,s+mik_o)] \tag{63}$$

The electromagnetic waveguide is usually axially uniform and therefore the electromagnetic mode field satisfies (32) and the excitation equations are (39-42).

Substitution of (39-41) in the z component of (63) results in an infinite set of equations which relates the space harmonics of the current $\bar{J}_k(s+mik_o)$ to the electromagnetic wave amplitude $\bar{a}(s+mik_o)$. Again, in order to allow an explicit solution of the set of equations we must assume some approximations. We are looking for the electromagnetic-like solution ($s\cong ik_{zo}$) and neglect $\bar{a}(s+mik_o)(m\neq o)$. We assume that phase matching takes place through the -1 order space harmonic of the current wave (Eq. 4, Fig. 2c) and therefore the only significant harmonics of the current wave are $\bar{J}_z(s)$ and $J_z(s+ik_o)$ ($s\cong ik_{zo}\cong i\ (\frac{\omega}{v_{oz}}-k_o)$ and $s+ik_o\ \ i\frac{\omega}{v_{oz}}$ correspond to the -1 order and zero order space harmonics of the current wave). Neglecting all other space harmonics we solve for these two components

$$\bar{J}_z(s) = -\frac{i\omega}{\varepsilon(\omega,s,ik_o)} \Bigg\{\Bigg[1 + \frac{1}{\varepsilon_o} \chi_{33}(\omega,s + ik_o, s + ik_o) \sum_{j=1}^{3}$$

$$\Bigg[\chi_{3j}^{E} (\omega,s,s) \mathcal{E}_j + \chi_{3j}^{B} (\omega,s,s) c \beta_j\Bigg]$$

$$- \frac{1}{\varepsilon_o} \chi_{33}^{E}(\omega,s,s+ik_o) \sum_{j=1}^{3} [\chi_{3j}^{E}(\omega,s + ik_o.s) \mathcal{E}_i$$

$$+ \chi_{3j}^{B}(\omega,s + ik_o,s) c \beta_j\Bigg]\Bigg\} \bar{a}(s) \tag{64}$$

$$\bar{J}_z(s+ik_o) = - \frac{i\omega}{\varepsilon(\omega,s,ik_o)} \Bigg\{\Bigg[1 + \frac{1}{\varepsilon_o} \chi_{33}^{E}(\omega,s,s)\Bigg] \sum_{j=1}^{3}$$

$$\Bigg[\chi_{3j}^{E}(\omega,s,s) \mathcal{E} \sum_{j=1}^{3} {}_{3j}^{E}(\omega,s+ik_o,s)$$

$$+ \chi_{3j}^{B}(\omega,s+ik_o,s) c \beta_j]$$

$$- \frac{1}{\varepsilon_o} \chi_{33}^{E}(\omega,s+ik_o,s) \sum_{j=1}^{3} [{}_{33}^{E}(\omega,s,s) \mathcal{E}_j$$

$$+ \Bigg[\chi_{3j}^{B} (\omega,s,s) c \beta_j\Bigg]\Bigg\} \bar{a}(s) \tag{65}$$

$$\varepsilon(\omega,s,ik_o) = \Bigg[1 + \frac{1}{\varepsilon_o} {}_{33}^{E}(\omega,s,s)\Bigg] \Bigg[1 + \chi\frac{1}{\varepsilon_o} {}_{33}^{E}(\omega,s + ik_o,s + ik_o)\Bigg]$$

$$- \frac{1}{\varepsilon_o} 2 \chi_{33}^{E}(\omega,s,s + ik_o) \chi_{33}^{E}(\omega,s + ik_o,s) \tag{66}$$

The transverse components of the current (63) are then found explicitly in terms of $\overline{a}(s)$

$$
\begin{aligned}
\overline{J}_k(s) = - i\omega \sum_{j=1}^{3} \Bigg\{ & [\chi^E_{kj}(\omega,s,s)\,\mathcal{E}_j + \chi^B_{kj}(\omega,s,s)\,c\,\beta_j] \\
& - \frac{1}{\varepsilon(\omega,s,ik_o)} \, \frac{1}{\varepsilon_o} \Bigg({}^{E}_{k3}(\omega,s,s)\Big[1 + \frac{1}{\varepsilon_o}\chi^E_{33}(\omega,s+ik_o,s+ik_o)\Big] \\
& - \frac{1}{\varepsilon_o}\, 2\, \chi^E_{k3}(\omega,s,s)\, \chi^E_{33}(\omega,s+ik_o,s) \Bigg) [\chi^E_{3j}(\omega,s,s)\mathcal{E}_j \\
& + \chi^E_{3j}(\omega,s,s)\, c\beta_j] \\
& - \frac{1}{\varepsilon(\omega,s,ik_o)} \Bigg(\frac{1}{\varepsilon_o}\chi^E_{k3}(\omega,s,s+ik_o)\,\Big[1 + \frac{1}{\varepsilon_o}\chi^E_{33}(\omega,s,s)\Big] \\
& - \frac{1}{\varepsilon_o^2}\chi^E_{k3}(\omega,s,s)\, \chi^E_{33}(\omega,s,s+ik_o) \Bigg) \\
& [\chi^E_{3j}(\omega,s+ik_o,s)\mathcal{E}_j + \chi^B_{3j}(\omega,s+ik_o,s)\, c\beta_j] \Bigg\} \overline{a}(s) \quad (k=1,2)
\end{aligned}
\tag{67}
$$

We can now substitute (64,67) in (42) and solve for $\overline{a}(s)$ explicitly. The resulting expression can be simplified if $1/\varepsilon(\omega,s,ik_o)$ can be taken out of the integral sign in Eq. 42. The condition for doing this is that the electron beam density and the periodic static fields will be both transversely uniform across the electron beam, or alternatively that space charge effects are negligible ($\varepsilon \approx 1$). The expression for $\overline{a}(s)$ is then given by

$$
\overline{a}(s) = \frac{\varepsilon(\omega,s,ik_o)}{(s - ik_{zo})\varepsilon(\omega,s,ik_o) - iK}\, a(o) \tag{68}
$$

where

$$
\begin{aligned}
K = \frac{\omega}{4P}\iint_{\infty}^{\infty}\Big\{\Big(&\underline{\mathcal{E}}^* \cdot \underline{\underline{\chi}}^E(\omega,s,s)\underline{\mathcal{E}}. + \underline{\mathcal{E}}^* \underline{\underline{\chi}}^B(\omega,s,s) \cdot c\beta\Big)\\
&+ \frac{1}{\varepsilon_o}[\chi^E_{33}(\omega,s,s)\underline{\mathcal{E}}^* \cdot \underline{\underline{\chi}}^E(\omega,s,s)\underline{\mathcal{E}}. - (\underline{\underline{\chi}}^{E^T}(\omega,s,s) \cdot \underline{\mathcal{E}}^*)_3\\
&(\underline{\underline{\chi}}^{E^T}(\omega,s,s) \cdot \mathcal{E})_3] + \frac{1}{\varepsilon_o}[\chi^E_{33}(\omega,s,s)\underline{\mathcal{E}}^* \cdot \underline{\underline{\chi}}^B(\omega,s,s) \cdot \underline{\beta}c\\
&- (\underline{\underline{\chi}}^{E^T}(\omega,s,s) \cdot \underline{\mathcal{E}}^*)_3 (\underline{\underline{\chi}}^B(\omega,s,s) \cdot \underline{\beta}c)_3]\\
&+ \frac{1}{\varepsilon_o}[\chi^E_{33}(\omega,s+ik_o,s+ik_o)\underline{\mathcal{E}}^* \cdot \underline{\underline{\chi}}^{E^T}(\omega,s,s) \cdot \underline{\mathcal{E}}\\
&- (\underline{\underline{\chi}}^{E^T}(\omega,s,s+ik_o) \cdot \underline{\mathcal{E}}^*)_3 (\underline{\underline{\chi}}^E(\omega,s+ik_o,s) \cdot \underline{\mathcal{E}})_3]\\
&+ \frac{1}{\varepsilon_o}[\chi^E_{33}(\omega,s+ik_o,s+ik_o)\underline{\mathcal{E}}^* \cdot \underline{\underline{\chi}}^B(\omega,s,s) \cdot c\underline{\beta}\\
&- (\underline{\underline{\chi}}^{E^T}(\omega,s,s+ik_o) \cdot \underline{\mathcal{E}}^*)_3 (\underline{\underline{\chi}}^B(\omega,s+ik_o,s) \cdot c\underline{\beta}_3]\\
&- \frac{1}{\varepsilon_o^2}\chi^E_{33}(\omega,s,s+ik_o)[\chi^E_{33}(\omega,s+ik_o,s)\underline{\mathcal{E}}^* \cdot \underline{\underline{\chi}}^E(\omega,s,s) \cdot \underline{\mathcal{E}}\\
&- (\underline{\underline{\chi}}^{E^T}(\omega,s,s) \cdot \underline{\underline{\mathcal{E}}}^*)_3 (\underline{\underline{\chi}}^E(\omega,s+ik_o,s) \cdot \underline{\mathcal{E}})]\\
&- \frac{1}{\varepsilon_o^2}\chi^E_{33}(\omega,s,s+ik_o)[\chi^E_{33}(\omega,s+ik_o,s)\underline{\mathcal{E}}^* \cdot \underline{\underline{\chi}}^B(\omega,s,s) \cdot c\underline{\beta}\\
&- (\underline{\underline{\chi}}^{E^T}(\omega,s,s) \cdot \underline{\mathcal{E}}^*)_3 (\underline{\underline{\chi}}^B(\omega,s+ik_o,s) \cdot c\underline{\beta}]
\end{aligned}
$$

$$- \frac{1}{\varepsilon_0^2} (\underline{\underline{\chi}}^{E^T}(\omega,s,s+ik_0) \cdot \underline{\mathcal{E}}^*)_3 \; [\underline{\underline{\chi}}_{33}^E(\omega,s,s)(\underline{\underline{\chi}}^E(\omega,s+ik_0,s) \cdot \underline{\mathcal{E}})_3$$

$$- \chi_{33}^E(\omega,s+ik_0,s) \; (\underline{\underline{\chi}}^E(\omega,s,s) \cdot \underline{\mathcal{E}})_3]$$

$$- \frac{1}{\varepsilon_0^2}(\underline{\underline{\chi}}^{E^T}(\omega,s,s+ik_0) \cdot \underline{\mathcal{E}}^*)_3 \; [\underline{\underline{\chi}}_{33}^E(\omega,s,s) \; (\underline{\underline{\chi}}^B(\omega,s+ik_0,s) \cdot c\underline{\beta})_3$$

$$- \chi_{33}^E(\omega,s+ik_0,s)(\underline{\underline{\chi}}^B(\omega,s,s). \; c\underline{\beta})_3] \Bigg\} dxdy \qquad (69)$$

5. NORMALIZATION OF THE DISPERSION GAIN RELATION

The formulation of Section 4 is general enough to describe all kinds of FELS discussed in all operating regimes. The expressions for the (Laplace transformed) amplitude of the electromagnetic wave in the FEL waveguide which were derived in the different cases are all similar and are given explicitly in terms of the linear susceptibility tensor terms of the electron beam in the FEL structures. These linear susceptibility terms can be derived in principle using any model for the electron equations (single particle force equation, fluid plasma equations, Vlasov equation or quantum mechanical equations). The formulation is general enough to permit transverse variation of the amplified electromagnetic mode field across the beam (off axis wave propagation), finite cross section of the electron beam, and within certain limitations also transverse variation of the electron beam density and the pump (wiggler) field intensity (in the cases of bremsstrahlung FELs).

Calculation of the linear response susceptibility tensor terms for each FEL configuration is out of the scope of the present article. We are presently content with showing that the different FELs satisfy similar dispersion-gain relations and identify the specific susceptibility functions and coupling coefficients of the various FELs by comparing our general dispersion-gain relations to the results calculated before in the literature for specific FELs. The comparison is made to analyses based on the solution of Vlasov equation, in order to obtain general enough results which apply also in the warm beam regime.

It is found that in all FELs considered the expression for the Laplace transformed amplitude of the electromagnetic wave (the dispersion-gain relation) is

$$\bar{a}(s) = \frac{1 + \chi(\omega, s+ik_o)/\varepsilon_o}{(s-ik_{zo})[1 + \chi(\omega, s+ik_o)/\varepsilon_o] - i\,\kappa\,\chi(\omega, s+ik_o)/\varepsilon_o}\, a(o) \tag{70}$$

where

$$\chi(\omega, s+ik_o) = (1 + \alpha^2)\,\chi_p(\omega, s+ik_o) \tag{71}$$

$\chi_p\ (\omega,s)$ is the well known longitudinal plasma susceptibility of an electron beam "plasma" propagating in the z direction

$$\chi_p(\omega s) = -\, i\,\frac{e^2}{\omega} \iiint_{-\infty}^{\infty} \frac{\partial\, g^{(o)}(p_x,p_y,p_z)/\partial p_z}{s - i\omega/v_z}\, dp_x, dp_y, dp_z \qquad (72)$$

where $g^{(o)}$ is an electron distribution function centered around $(p_{ox},p_{oy},p_{oz}) = (0,0,p_{oz})$, and $v_z = p_z/(\gamma m)$ is the electrons longitudinal velocity component. The coupling coefficient κ is listed for the different FELs in Table 2. The parameter α is zero for all FELs considered except the longitudinal electrostatic bremsstrahlung FEL.

TABLE 2. THE COUPLING PARAMETER k OF VARIOUS FELs (The highly relativistic limit expressions are given in brackets).

FEL	
Cerenkov	$\dfrac{\frac{1}{2}\sqrt{\epsilon/\mu}\iint_{A_e}\lvert\mathcal{E}^{\kappa}_z(x,y)\rvert^2 dxdy}{P}\,\dfrac{\Pi}{\lambda}$
Smith-Purcell	$\dfrac{\frac{1}{2}\sqrt{\epsilon/\mu}\iint_{A_e}\lvert\mathcal{E}^{(1)}_z(x,y)\rvert^2 dxdy}{P}\,\dfrac{\Pi}{\lambda}$
Long. Elec.	$\alpha^2\dfrac{\frac{1}{2}\sqrt{\epsilon/\mu}\iint_{A_e}\lvert\mathcal{E}_z(x,y)\rvert^2 dxdy}{P}\,\dfrac{\Pi}{\lambda}$
Trans. Elec.	$\frac{1}{8\pi}\left(\frac{eE_o}{mc^2}\right)^2\frac{(1+\beta_{oz})^2}{\beta_{oz}}\frac{\gamma_{oz}^4}{\gamma_o^2}\frac{A_e}{A_g}\lambda\quad\left[\frac{1}{2\pi}\frac{eE_o}{mc^2}\frac{\gamma_{oz}^4}{\gamma_o^2}\frac{A_e}{A_g}\lambda\right]$

Mag. Brem.	$\frac{1}{8\pi}\left(\frac{eB_oc}{mc^2}\right)^2\left(1+\beta_{oz}\right)^2\frac{\gamma_{oz}^4}{\gamma_o^2}\frac{A_e}{A_g}\lambda\left[\frac{1}{2\pi}\left(\frac{eB_oc}{mc^2}\right)\frac{\gamma_{oz}^4}{\gamma_o^2}\frac{A_e}{A_g}\lambda\right]$
Compton-Raman	$\frac{1}{2\pi}\left(\frac{eE_o}{mc^2}\right)^2\left(1+\beta_{oz}\right)^2\frac{\gamma_{oz}}{\gamma_o^2}\frac{A_e}{A_g}\lambda\left[\frac{8}{\pi}\left(\frac{e}{mc^2}\right)^2\left(\frac{\mu_o}{\varepsilon_o}\right)^{1/2} S_o\frac{\gamma_{oz}^4}{\gamma_o^2}\frac{A_e}{A_g}\lambda\right]$

The validity of (70,72) for the Cerenkov and Smith-Purcell FELs is evident from Eqs. (49,60) and linear response derivations based on the Vlasov Equation [17,18,20,21]. In these cases α=o and

$$\chi(\omega, s + ik_o) = \chi_{33}(\omega, s + ik_o) = \chi_p(\omega, s + ik_o)$$

(where for the Cerenkov FEL case one substitutes always k_o=0). Since there is no transverse pump (wiggler) in these cases $p_{oz}=p_o$. The coupling coefficient values listed for these cases in rows 1,2 of Table 2 were derived before in Eqs. (50,61). A_e is the electron beam cross section area.

In the longitudinal electrostatic bremsstrahlung FELs a periodic axial electric field is applied on the electron beam as it propagates through a uniform electromagnetic waveguide. The axial modulation of the electron beam current enables interaction with TM electromagnetic modes via the longitudinal component of the electromagnetic mode electric field. A vlasov equation linear response analysis of this problem[11] resulted that the different longitudinal susceptibility terms relate to the plasma susceptibility function (72) in the following way

$$\chi_{33}^{E}(\omega, s, s) = \chi_p(\omega, s) + \alpha^2 \chi_p(\omega, s + ik_o)$$

$$\chi_{33}^{E}(\omega, s, s + ik_o) = \chi_{33}^{E}(\omega, s + ik_o, s) = \alpha\, \chi_p(\omega, s + ik_o) \tag{73}$$

$$\chi_{33}^{E}(\omega, s + ik_o, s + ik_o) = \chi_p(\omega, s + ik_o)$$

where

$$\alpha = \bar{J}_z(s)/\bar{J}_z(s + ik_o)$$

is (near synchronism) the relative amplitude of the -1 order space harmonic of the current wave

$$s \cong ik_{zo} \cong i\,(\omega/v_o - k_o)$$

relative to the amplitude of the zero order harmonic $s \cong i\,(k_{zo}+k_o) \cong i\omega/v_o$. The first order perturbative derivation[11] resulted in

$$\alpha^2 = \left(\frac{1}{2\beta_o^2\gamma_o^3}\,\frac{L}{\lambda}\,\frac{e\phi_o}{mc^2}\right)^2 \tag{74}$$

where $\phi_o = E_o/k_o$ is the amplitude of the periodic electrostatic potential (E_o is the amplitude of the periodic electrostatic fields).

Near synchronism (4) we set in (73) $\chi_p\,(\omega, s \cong ik_{zo}) \cong 0$ and obtain from (66)

$$\varepsilon(\omega, s,\ ik_o) = 1 + (1 + \alpha^2)\ \chi_p\,(\omega, s + ik_o) \tag{75}$$

We also find that out of the many terms in (69), the only term which is not vanishing or negligible near synchronism is the longitudinal-longitudinal tensor component in the first term:

$$K \cong \frac{\omega}{4P} \iint_{-\infty}^{\infty} \chi_{33}^{E}\,(\omega, s, s)\ \left|\mathscr{E}_z\right|^2 dxdy = \kappa\chi_p\,(\omega, s + ik_o) \tag{76}$$

This confirms the validity of (70) for the longitudinal electrostatic FEL with k given in row 3 of Table 2. Since there is no transverse modulating force in this case, again the average axial momentum is equal to the average total momentum of the electron beam as it enters the interacting region P_{oz}=p.

It is instructive to find out that Cerenkov, Smith-Purcell and longitudinal electrostatic bremsstrahlung FELs all have similar expressions for the coupling coefficient . This stems from the fact that they all involve direct longitudinal interaction of an electromagnetic wave component with a synchronous electron plasma wave component. In the Cerenkov scheme the z component of the total electromagnetic mode field $\underline{\mathscr{E}}$ (x,y) can be synchronized and coupled to electron plasma waves. In the Smith-Purcell (or travelling wave tube) type scheme only the electromagnetic mode first order, space harmonic $\_^{(1)}$ (x,y) is synchronous with the electron plasma waves. In the longitudinal electrostatic bremsstrahlung scheme the total electromagnetic mode $\_$ (x,y) is synchronous with only the -1 order space harmonic of the electron plasma wave which has an amplitude α (Eq. 74). In all three cases the coupling coefficient is proportional to the radiation wavenumber $k = 2\pi/\lambda$ and to "relative power" factors of the interacting wave components.

In the case of magnetic bremsstrahlung FEL a periodic magnetic force (Lorentz force) is applied on an electron beam which is propagating through an axially uniform electromagnetic waveguide. The transversely modulated electron beam can then interact with a transverse electromagnetic (TEM) wave propagating along the z direction. Interaction with the beam space charge waves is made possible via the pondermotive force which is the axial force operating on the electrons by the combined effect of the transverse fields of the magnetostatic pump (wiggler) and the electromagnetic wave.

To derive the appropriate expression for the coupling coefficient κ and the susceptibility function of the magnetic bremsstahlung FEL we compare the results of Vlasov equation linear response theories of the magnetic bremsstrahlung FEL[4,10] to the dispersion-gain relation of the present analysis (68-72). In [4,10] the models assumed are of a transversely infinite and transversely uniform electron beam, and a transversely infinite and uniform (helical) magnetostatic field and electromagnetic TEM wave. With these assumptions it turns out that near synchronism the only non-vanishing or non-negligible term of the susceptibility tensor is

$$\chi^E_{33}\ (\omega, s + ik_o,\ s + ik_o) = \chi_p\ (\omega, s + ik_o).$$

Hence

$$\varepsilon(\omega, s, ik_o) = 1 + \chi_p\ (\omega, s + ik_o)/\varepsilon_o \tag{77}$$

Out of the many terms of Eq. (69) the only non-vanishing or non-negligible terms are

$$\chi^E_{ki}\ (\omega, s, s)\ (kj = 1,2)$$

in the first term of the right hand side of (69). These tensor terms are also found to be proportional to $\chi_p\ (\omega, s + ik_o)$. Hence

$$K \cong \frac{\omega}{4P} \sum_{k,j=1}^{2} \iint_{-\infty}^{\infty} \mathscr{E}_k^* \chi_{kj}^E(\omega s, s)\ \mathscr{E}_j\ dxdy = \kappa\ \chi_p\ (\omega, s + ik_o) \tag{78}$$

This confirms the validity of (70) for the magnetic bremsstrahlung FEL with κ given in row 4 of Table 2 (specifically for a helically polarized TEM electromagnetic wave and a helical pump magnetostatic field). The factor A_e/A_g is a "relative power" factor defined by

$$\frac{A_e}{A_g} = \frac{1/2\ \sqrt{\varepsilon_o/\mu_o} \iint_{A_e} \left| \mathscr{E}_t\ (x,y) \right|^2 dxdy}{P} \tag{79}$$

If the electromagnetic field is transversely uniform in the whole waveguide cross-section (TEM wave) then A_g is the waveguide cross-section area. In the more general case A_g is the effective area of the waveguide mode and is defined by (79). It can be then smaller than the waveguide cross section area.

The susceptibility function in (70) for the bremsstrahlung FEL is given by (71,72) with $\alpha=0$. The distribution function $g^{(o)}$ in (72) is centered around $(0,0,p_{oz})$ where p_{oz} is the average axial momentum of the electron beam and $P_{oz} \neq p_o$ because of the transverse modulation of the beam momentum due to the pump (wiggler) field. The transverse momentum components p_x, p_y are the transverse canonical momentum components of the electrons, and are constants of the motion in the wiggler magnetic field[10].

The transverse electrostatic bremsstrahlung FEL scheme is essentially equivalent to the magnetic bremsstrahlung FEL except that a periodic transverse electrostatic force is applied on the electrons by periodic alternating electrodes, replacing the periodic magnetic (lorentz) force which is applied in the magnetic bremsstrahlung FEL by means of static magnets or coils. It was also suggested that high amplitude electrostatic field modulation can be obtained due to the change in the space charge field of an electron beam transversing through a periodically rippled waveguide, and that this can be utilized for a free electron laser scheme.[12,28]

By a simple extension of the linear analysis of magnetic bremsstrahlung FELs[4,10] it is possible to show[43], that the general gain-dispersion relation (70) applies, to a good approximation, to the transverse electrostatic bremsstrahlung FEL as well, with appropriate modification of the coupling coefficient κ (given in Table 2 row 4).

In the Compton-Raman FEL the electron beam is pumped (wiggled) by both the transverse electric and magnetic fields of a time varying electromagnetic wave propagating in counter direction to the electron beam. Since the pump (wiggler) field is time dependent, the analysis of section 4 should be somewhat modified to apply to the Compton-Raman scattering FEL. It is evident that this problem is very similar to the problem of bremsstrahlung FEL but since the pump field varies in time with frequency ω_0 it turns out that the pondromotive force (due to the combined effect of the pump field of ω_0 and the amplified electromagnetic field of frequency ω is varying in time with frequency $\omega-\omega_0$. The longitudinal current and the longitudinal space charge electric field are then characterized with frequency and wavenumber $(\omega-\omega_0, s+ik_0)$, where

$$s \cong ik_{zo} = i\frac{\omega}{c} .$$

It turns out then that the general dispersion-gain relation (70-72) still applies to this case (with α=o) if we substitute in the expression for the plasma susceptibility (72) $\omega\to\omega-\omega_0$. The coupling parameter κ for the Compton-Raman FEL can be derived from a linear response analysis of this problem[16,43]. The expression for κ given in Table 2 row 6 corresponds to the case that both the pump wave and amplified (signal) wave are circularly polarized. The parameter κ is expressed there also in terms of the pointing vector power density of the electromagnetic pump S_0 which is given in this case by

$$S_o = \frac{1}{4}\sqrt{\frac{\epsilon_o}{\mu_o}}\, E_o^{\,2} \tag{80}$$

It should be noted, that in all the transverse modulation FELs (transverse magnetic, electrostatic and Compton-Raman scattering) the electron mechanical momentum distribution f_o (p_x, p_y, p_z) inside the interaction region (neglecting the signal wave effect) has an average transverse momentum (quiver momentum) amplitude $p_o \neq 0$.

This transverse momentum is endowed to the electron beam by the periodic (helical) transverse force of the "wiggler" field, and is vanishing ($p_o = 0$) in the case of the longitudinal FELs. However, the distribution function $g^{(o)}$ (p_x, p_y, p_z) which appears in (72) is generally defined for all FELs in terms of the transverse momentum variables

$$p_{\perp} = \tilde{p}_{\perp} - p_{\perp o}$$

where $\tilde{p}_{\perp}'$ is the electron transverse mechanical momentum at the entrance to the interaction region and $p_{\perp o}'$ is the average quiver momentum. These variables are the canonical momentum components in the case of magnetic bremsstrahlung and Compton-Raman scattering FELs, and are there constant of the motion in the wiggler field (approximately so also for the transverse electrostatic bremsstrahlung FEL). The amplitude of the quiver momentum p_o is given for the different FELs by

$$p_{o\perp}' = \begin{cases} \dfrac{eB_o}{k_o} & \text{for Mag. bremss.} \\[2ex] \dfrac{eE_o}{k_o V_{oz}} & \text{for Elec. bremss.} \\[2ex] \dfrac{dE_o}{\omega_o} & \text{for Comp.-Raman} \end{cases} \tag{81}$$

If the wiggler field is turned on adiabatically then the dependence of $g^{(o)}$ on the transverse "canonical" momentum variables p_x, p_y is the same for all FELs and is equal to the initial transverse momentum distribution of the electron beam far away from the interaction region. The dependence of $g^{(o)}$ on the axial momentum variable p_z is shifted in the case of the transverse modulation FELs and centered in the wiggler field around an average axial momentum.

$$p_{oz} = \sqrt{p_o^2 - p_o^2} \tag{82}$$

where p_o is the average total mechanical momentum. In the longitudinal modulation FELs $p_{oz} = 0$.

The axial average velocity is calculated from (82) using the relation $v_{oz} = p_{oz}/(\gamma_o m)$. The parameter γ_{oz} can be then calculated from (7), or directly from (81) using the formula

$$\gamma_{oz} = \frac{\gamma_o}{\left[1 + (p_o /mc)^2\right]^{1/2}} \tag{83}$$

We have shown that the different FELs - both longitudinal and transverse modulation schemes - satisfy the same gain dispersion relation (70). As mentioned earlier this common relation stems from the fact that the basic interacting waves in all interaction schemes are the same electromagnetic mode and the longitudinal single electron or space charge waves, which are coupled in either case by a different mechanism. We now turn to normalize the common relation (70) and derive the FEL gain expressions.

The common gain-dispersion relation (70) would yield the power gain for any FEL at any wavelength when $\overline{a}(s)$ is inverse Laplace transformed and substituted in

$$\frac{P(z)}{P(o)} = \left|\frac{a(z)}{a(o)}\right|^2 \tag{84}$$

The operating wavelength is determined from the solution of the dispersion equation which is the condition for the vanishing of the denominator in (70).

$$(s-ik_{zo})[1 + \chi\,(\omega,s+ik_o)/\varepsilon] - i\,\kappa\,\chi\,(\omega,s+ik_o)/\varepsilon=o \tag{85}$$

For a weakly coupled system (κ is small), the eigenmodes of the system have wavenumbers close to the eigenmodes of the uncoupled system (κ=o) which are the electromagnetic wave ($s = ik_{zo}$) and the electron beam plasma waves (solutions of the plasma dispersion equation : $1 + \chi/\varepsilon = o$). The approximate wavenumbers of these waves (more precisely - the wavenumbers of the space harmonics which participate in the interaction), are listed in Table 1, giving rise to the radiation conditions listed in the third column of the table. When the wavenumbers of the interacting waves match (synchronism), the dispersion equation (85) vanishes and the appearance of a pole in (70) indicates strong coupling of the electron plasma and electromagnetic waves. We should point out that Eq. 70 and the parameters in Table 2 were all derived with the assumption of operation near synchronism.

The susceptibility function $\chi\ (\omega,s)$ (Eqs. 71,72) can be expressed in terms of familiar functions and parameters. The distribution function $g^{(o)}\ (\underline{p})$ may be substituted in terms of a normalized distribution function of a single variable:

$$g^{(o)}(p_z) \equiv \iint_{-\infty}^{\infty} g^{(o)}(p_x,p_y,p_z)\ dp_x dp_y \equiv \frac{n_o}{p_{zth}}\ \bar{g}\ \left(\frac{p_z - p_{oz}}{p_{zth}}\right) \qquad (86)$$

where n_o is the electron beam density, p_{zth} - the longitudinal momentum spread of the electron beam distribution, and p_{oz} is the average electron beam momentum in the longitudinal (z) direction. In terms of the normalized function $\bar{g}$, the plasma susceptibility (18) can be written as

$$\chi\ (\omega,s)/\varepsilon = \frac{1}{2}\ \frac{k_D'^2}{s^2}\ G'(\zeta) \qquad (87)$$

where

$$G(\zeta) \equiv \int_{-\infty}^{\infty} \frac{\bar{g}\ (x)}{x - \zeta}\ dx \qquad (88)$$

$$\zeta \equiv \frac{i\omega/s - v_{oz}}{v_{zth}'} \qquad (89)$$

$$k_D'^2 \equiv 2\ \frac{\omega_p'^2}{v_{zth}'^2} \qquad (90)$$

$$\omega_p'^2 \equiv (1+\alpha^2)\frac{\omega_p^2}{\gamma_o\gamma_{oz}^2} = \frac{1+\alpha^2}{\gamma_o\gamma_{oz}^2}\frac{e^2 n_o}{m\varepsilon} \tag{91}$$

$$v'_{zth} = \frac{p_{zth}}{\gamma_o\gamma_{oz}^2 m} = \frac{E_{zth}}{\gamma_o\gamma_{oz}^2\beta_{oz} mc} \tag{92}$$

E_{zth} is the longitudinal kinetic energy spread of the electron beam. In stimulated Compton Raman scattering substitute $\omega \to \omega - \omega_o$.

Often the electron distribution is approximated by a shifted Maxwellian distribution. For this case:

$$\bar{g}(x) = \frac{1}{\sqrt{\pi}} e^{-x^2} \tag{93}$$

and

$$G(\zeta) = \frac{1}{\sqrt{\pi}} \int_{-\infty}^{\infty} \frac{e^{-x^2}}{x-\zeta} dx \tag{94}$$

is the so called plasma dispersion function which is tabulated in Ref. [29].

Before we go to the next section where the solution of (70,85) and the laser gain regimes are discussed, let us examine Eq. 85, in the limit of a cold beam ($p_{zth} \to o$). In this limit we get from (89,88) $|\zeta| \to \infty$ and $G'(\zeta) \to 1/\zeta^2$. Substitution in (87) and (70) gives

$$\chi(\omega,s) = -\varepsilon \frac{\omega_p'^2}{(\omega + iv_{oz}s)^2} \tag{95}$$

$$\bar{a}(s) = \frac{\left(s + ik_o - i\frac{\omega}{v_{oz}}\right)^2 + \frac{\omega_p'^2}{v_{oz}^2}}{\left(s - ik_{zo}\right)\left[\left(s + ik_o - i\frac{\omega}{v_{oz}}\right)^2 + \frac{\omega_p'^2}{v_{oz}^2}\right] - i\kappa\frac{\omega_p'^2}{v_{oz}^2}} a(o) \tag{96}$$

and the dispersion relation (85) is

$$\left(s - ik_{zo}\right)\left[\left(s + ik_o - i\frac{\omega}{v_{oz}}\right)^2 + \frac{\omega_p'^2}{v_{oz}^2}\right] - i\kappa\frac{\omega_p'^2}{v_{oz}^2} = 0 \tag{97}$$

This equation is similar to the conventional travelling wave tube dispersion equation[30]. Its physical significance is seen when we take the limit $\kappa = o$ (no interaction). We see that the uncoupled eigenmodes of the system in the cold beam limit are the electromagnetic wave

$$s = ik_{zo} \tag{98}$$

and the slow and fast plasma waves (correspondingly)

$$s + ik_o = i\left(\frac{\omega}{v_{oz}} + \frac{\omega'_p}{v_{oz}}\right) \tag{99}$$

$$s + ik_o = i\left(\frac{\omega}{v_{oz}} - \frac{\omega'_p}{v_{oz}}\right) \tag{100}$$

The dispersion relation (97) can be further simplified into the compact form

$$\delta k\left(\delta k - \theta - \theta_p\right)\left(\delta k - \theta + \theta_p\right) + Q = 0 \qquad 101)$$

where we defined the complex wave number modification due to coupling - δk, by

$$s \equiv ik_{zo} + i\,\delta\, k \tag{102}$$

the synchronism (detuning) parameter:

$$\theta \equiv \frac{\omega}{v_{oz}} - k_o - k_{zo} \tag{103}$$

($\omega \to \omega - \omega_o$ in stimulated Compton-Raman scattering)

the space charge parameter:

$$\theta_p \equiv \frac{\omega_p'}{v_{oz}} \tag{104}$$

and the gain parameter

$$Q \equiv \kappa \frac{\omega_p'^2}{v_{oz}^2} = \kappa\, \theta_p^2 \tag{105}$$

6. FREE ELECTRON LASER GAIN REGIMES

In principle the calculation of the gain of any FEL at arbitrary operating conditions is straightforward, requiring only to perform an inverse Laplace transform of (70), then using (84) and the appropriate coupling parameter from Table 2. In practice the execution of an inverse Laplace transorm may be somewhat difficult in the general case ($p_{th} \neq o$) where the exact plasma dispersion function (88 or 94) must be used in (87). In this general case it may be most useful to evaluate the inverse transform

$$a(z) = \frac{1}{2\pi i} \int_{\gamma - i\infty}^{\gamma + i\infty} \bar{a}(s) e^{sz} ds \tag{106}$$

by numerical integration in the complex field. A computer program for performing this integral was developed.[21] Representative gain curves calculated by this program are shown in Figure 3. As a function of the interation-length-normalized-synchronism-parameter $\bar{\theta}$:

$$\bar{\theta} = \theta \ell = \left(\frac{\omega}{v_{oz}} - k_{zo} - k_o \right) \ell \tag{107}$$

and various values of the normalized thermal spread parameter $\bar{\theta}_{th}$

$$\theta_{th} = \frac{\omega}{v_{oz}} \frac{v'_{zth}}{v_{oz}} = \frac{2\pi}{\beta_{oz}} \frac{v'_{zth}}{v_{oz}} \frac{1}{\lambda} \tag{108}$$

$$\bar{\theta}_{th} = \theta_{th} \ell \tag{109}$$

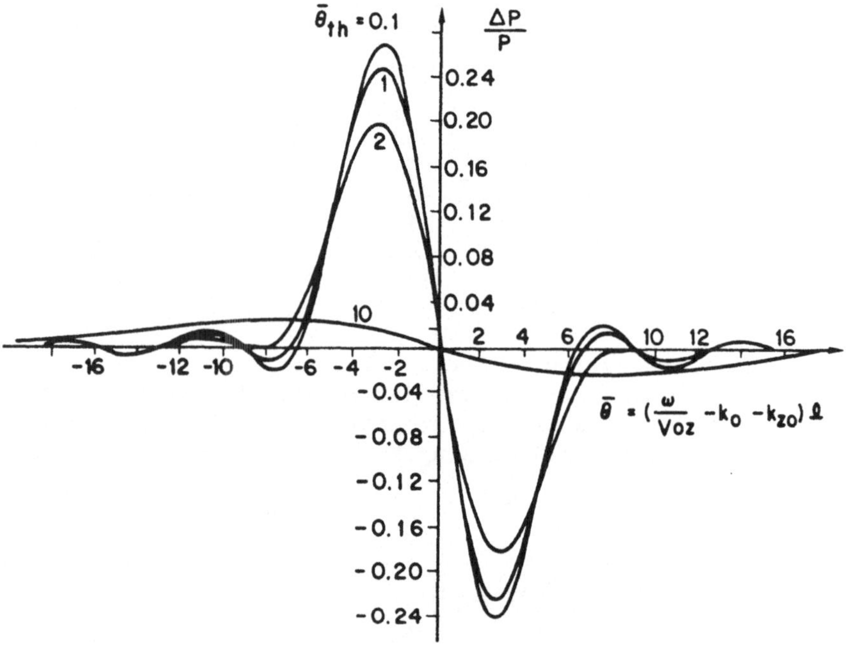

Figure 3. Numerically calculated gain curves of the various values of the normalized thermal spread parameter $\overline{\theta}_{th}$ ($\overline{k}$ = 1, $\overline{\theta}_p$ = 1).

(in Compton-Raman FEL $\omega \to \omega - \omega_0$ and consequently (108) changes into $\theta_{th} = 4\pi/1+\beta_{oz}\ v'_{zth}/v_{oz}\ 1/\lambda$). The termal parameter θ_{th} has the interpretation as the spectral width (evaluated in wave number space) of electron wave numbers $\Delta\ (\omega/v_z)$ in an electron beam with average longitudinal velocity v_{oz} and velocity spread v'_{zth}. As we will see later in this section the beam is considered cold or warm in $\bar{\theta}_{th} \ll 1$ or $\bar{\theta}_{th} \gg 1$ respectively. Figure 3 displays examples in both limits as well as intermediate cases. The normalized coupling and space charge parameter $\bar{\kappa}$ and $\bar{\theta}_p$ which appear in Figure 3 are defined by

$$\bar{\kappa} \equiv \kappa \ell \tag{110}$$

$$\bar{\theta}_p \equiv \theta_p \ell = \frac{\omega_p'}{v_{oz}} \ell \tag{111}$$

In may practical limits exact calculation of the gain curve is not necessary and analytical expressions for the gain may be derived with certain approximations. These different limits (gain regimes) were delineated by a number of authors [4,7,10,17, 18,20] for various kinds of FELs. Indeed they arecommon to all of them, since they result from the same gain-dispersion relation (70). we will briefly describe the gaincharacteristics in these regimes.

Figure 4 displays "phase matching" (mometum conservation) diagrams which help to understand the various gain regimes discussed. It is essential a "blow up" of the phase matching diagram in Figure 2 (b), looking at the details of the phase matching near the electron waves disperation curve at the various gain regimes. In the bremsstrahlung FELs, k_0 in Figure 4 is reversed in direction. Also in the stimulated Compton-Raman scattering $k_0 = \omega_0/c$ is reversed and $\omega/v_{oz} \to (\omega - \omega_0/v_{oz}$. In the cerenkov FEL $k_0 = 0$. The other details remain the same in all FELs <u>Cold beam low gain regime</u>.

As was shown in the previous section, in the cold beam limit ($p_{zth} \to o$) the dispersion equation (85) reduces to (97) or (101) which are simple third degree polinomial equations with a finite number of roots (three). If the roots are found, then the evaluation of the inverse Laplace transform (106) may be straightforwardly calculated by use of the residuum method. This gives

$$\frac{a(z)}{a(o)} = \sum_{j=1}^{3} A_j e^{s_j z} \tag{112}$$

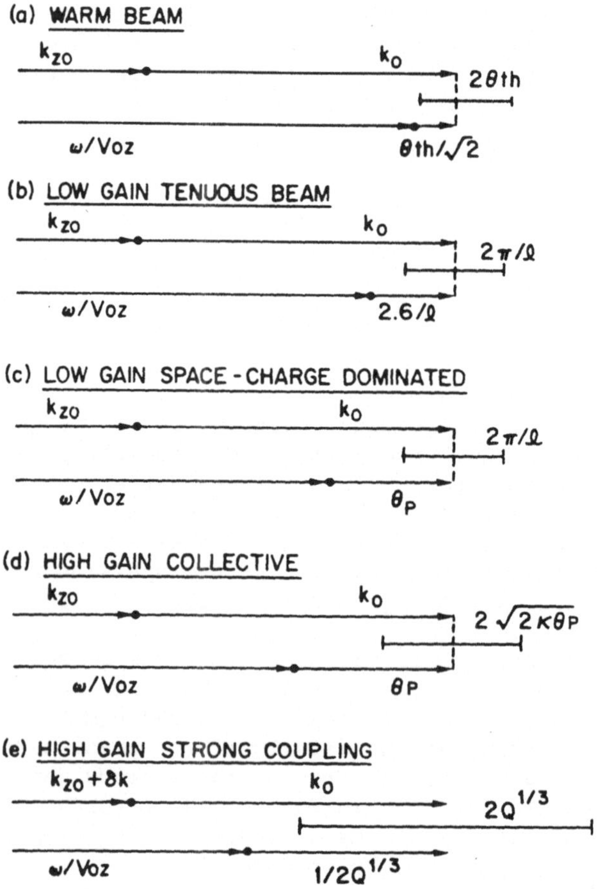

Figure 4. "Phase matching" (momentum conservation) diagrams at the different gain regimes assuming operation a the maximum gain point. The section at the end of each diagram indicates the width of the gain curve in wave-number space. There is gain as long as the mismatch between the arrows is within this section.

where A_j are the residues of (96) at s_j.

The low gain regime occures at the limit when the normalized gain parameter $\bar{Q}$ is very small: $\bar{Q} \ll 1$ (a sufficient condition).

$$\bar{Q} \equiv \bar{\kappa}\, \bar{\theta}_p^{\,2} = Q\ell^3 \tag{113}$$

In this limit the roots of (97) and(101) will be close to the roots of the uncoupled waves (Q = o) - Eqs. 98 -100, and may be expanded to first order in κ or Q around this zero order values. After a lengthy mathematical calculation, [21] substituting (112) in (84) and calculating the power output to first order in κ (or Q), one gets

$$\frac{\Delta P}{P(0)} = \bar{Q}\, F\left(\theta,\bar{\theta}_p\right) \tag{114}$$

where $|\Delta P| \equiv |P(\ell) - P(0)| \ll P(0)$, and

$$F\left(\bar{\theta},\bar{\theta}_p\right) \equiv \frac{1}{2\theta p}\left[\frac{\sin^2\left(\frac{\bar{\theta}+\bar{\theta}_p}{2}\right)}{\left(\frac{\bar{\theta}+\bar{\theta}_p}{2}\right)^2} - \frac{\sin^2\left(\frac{\bar{\theta}-\bar{\theta}_p}{2}\right)}{\left(\frac{\bar{\theta}-\bar{\theta}_p}{2}\right)^2}\right] \tag{115}$$

The function $F(\bar{\theta},\bar{\theta}_p)$ is shown in Figure 5 for various values of the parameter $\bar{\theta}_p$ (111) Since $F(\bar{\theta},\bar{\theta}_p) < 1$ the parameter $\bar{Q}$ indicates an upper limit on the power gain available.

Eqs. 114, 115 give the low gain characteristics of free electron lasers <u>including space charge effect</u>. For a tenuous beam or a short interaction length $\bar{\theta}_p \ll \pi$ and (115) reduces to

$$F(\bar{\theta},0) = \frac{d}{d\bar{\theta}}\left[\frac{\sin(\bar{\theta}/2)}{\bar{\theta}/2}\right]^2 \tag{116}$$

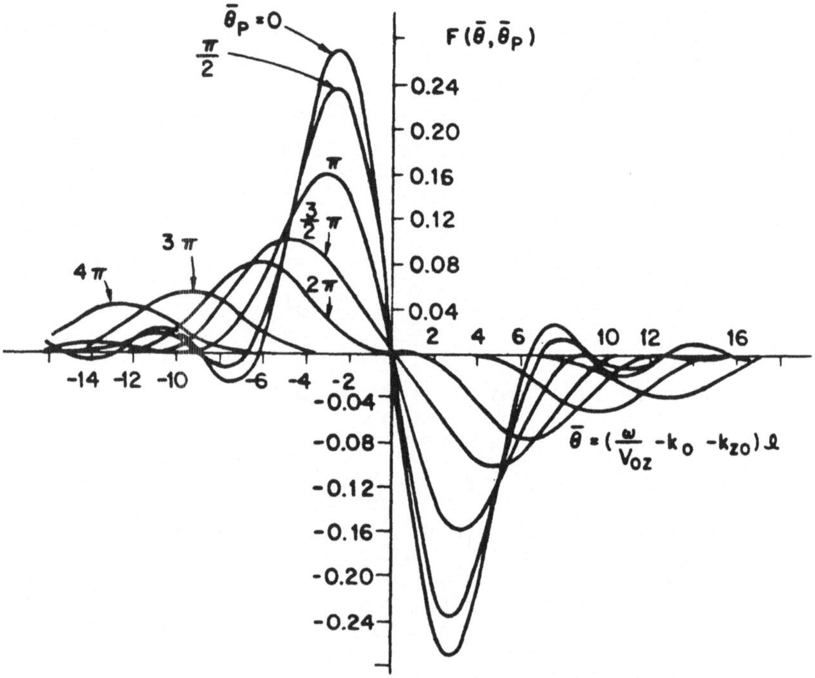

Figure 5. The normalized low gain curves $F(\bar{\theta}, \bar{\theta}_p)$ (Eq. 115) for various values of the normalized space charge parameter $\bar{\theta}_p$.

which is the familiar single electron gain function which appears in the analyses of the various free electron lasers when space charge effect is neglected [2-5,10,19,20]. Notice also that the gain curve for this case (Figure 5, $\bar{\theta}_p = o$) resembles the computer calculated curve of Figure 3 for $\bar{\theta}_{th} = 0.1 \ll 1$ (cold beam). It attains its maximum value at

$$\theta = -\frac{2.6}{\ell} \quad \text{or} \quad \frac{\omega}{v_{oz}} + \frac{2.6}{\ell} = k_{zo} + k_o \tag{117}$$

This situation is illustrated in Figure 4(b).

It is interesting to point out that the expression which was derived by Lousiell et all [31] for the magnetic bremsstrahlung FEL gain in the low gain regime including space charge effect, is actually the Taylor expansion of (115) to first order in $\bar{\theta}_p^{\,2}$:

$$F(\bar{\theta},\bar{\theta}_p) \simeq \frac{d}{d\bar{\theta}} \left[\frac{\sin(\theta/2)}{\bar{\theta}/2}\right]^2 + \frac{\bar{\theta}_p^{\,2}}{6} \frac{d^3}{d\bar{\theta}^3} \left[\frac{\sin(\bar{\theta}/2)}{}\right]^2 \tag{118}$$

This expansion is correct of course only for $\bar{\theta}_p \ll \pi$.

The general space charge limited expression (115) has the interesting physical interpretation as the result of the interference of the electromagnetic wave (98) with the slow space charge wave (99) on one hand and with the fast space charge wave (100) on the other hand. Whenever the electromagnetic wave is phase matched with the slow space charge wave ($\theta \simeq -\theta_p$ or $k_{zo} + k_o \simeq \omega/v_{oz} + \theta_p$), the interference yields maximum net gain. When the electromagnetic wave is pahse matched with the fast space charge wave ($\theta \simeq \theta_p$ or $k_{zo} + k_o \simeq \omega/v_{oz} - \theta_p$) maximum attenuation is obtained. The pahse matching diagram in the space charge dominated low gain regime ($\bar{\theta}_p \gg \pi$) at the maximum gain point:

$$\theta = -\theta_p \quad \text{or} \quad \frac{\omega}{v_{oz}} + \theta_p = k_{zo} + k_o \tag{119}$$

is shown in Figure 4(c).

The gain expression (114) may be of practical importance in situations when low gain operation is expected (as in laser oscillators), and weak coupling parameter (κ) is unavoidable (as in the case of short wavelength operation). In this situation one may try to increase the gain parameter $\bar{Q}$ (103) by increasing the electron density and the interaction length ℓ, thus arriving to the regime $\bar{\theta}_p >> \pi$. For example with beam parameters $n_o = 10^{11}$ cm $^{-3}$, $\gamma_o = \gamma_{oz} = 10$, $v_{oz} \simeq c$, and $\ell = 2$ m one gets $\bar{\theta}_p \simeq 5$, and it is necessary to use (115) to describe the gain.

When we are in the space-charge dominated limit $\bar{\theta}_p >> \pi$ at the maximum gain point (119) the normalized gain function (115) attains a value $F(-\bar{\theta}_p, \bar{\theta}_p) = 1/(2\bar{\theta}_p)$. Thus the maximum gain (114,115) is given by:

$$\left(\frac{\Delta P}{P}\right)_{max} = \frac{\bar{Q}}{\bar{\theta}_p/2} = 2\kappa\theta_p\ell^2 \tag{120}$$

we notice that in this low gain space charge dominated regime the gain grows proportionally to $\sqrt{n_o}$ while in the tenuous beam limit where collective effects were negligible, the gain was proportional to n_o (single electron interaction). Notice also that since $\bar{\theta}_p >> \pi$ was assumed, the parameter $\bar{Q}$ gives again an upper limit on the gain available.

It should be pointed out here that as the electron density is increased, and space charge effects become important, saturation (trapping) effect would start at a lower input power level of the of the electromagnetic radiation. In this case the linear analysis which led to (114) may fail to describe the situation with practical power levels, and a complete nonlinear analysis should be used. [32,33]

Cold beam high gain collective (Raman) regime

For weak enough coupling ($\bar{Q} << 1$) the roots of (101) are all real and small, thus the roots s_j (Eq. 102) are all imaginary and close to the uncoupled wave numbers (98-100). They give then rise to the "interferential" gain expression (114,115). As $\bar{Q}$ is increased, the polynomial equation (101) starts having one real and two complex (conjugate) solutions. Consequently, on of the roots s_j (Eq. 102) must have positive real part (corresponding to gain) and the other two have a negative and a vanishing real part (corresponding to gain) and the other two have a negative and a vanishing real part (corresponding to loss and a constant amplitude respectively). In this limit - "the high gain limit" - we can neglect the interference between the different roots, in (112), (84), and keep only the exponentially growing wave. Then from (84)

$$\ell n \frac{P(\ell)}{P(0)} \approx 2\ell n \; A_j \quad + 2(\mathrm{Re}\, s_j)\ell = - 2 \; (\mathrm{Im}\delta k)\ell \tag{121}$$

where s_j is the exponentially growing root.

In the particular case when the electromagnetic wave is synchronous with the slow space charge wave:

$$\theta = - \theta_p \quad \text{or} \quad \frac{\omega}{v_{oz}} + \theta_p = k_{zo} = k_o \tag{122}$$

and assuming $|\delta k| \ll \theta_p$, Eq. 101 may be approximated by a simple degree equation:

$$(\delta k)^2 = - \frac{Q}{2\theta_p} \tag{123}$$

The root of this equation which corresponds to a growing wave is: $\delta k = - i \sqrt{Q/2\theta_p}$, and substituting in (121) it gives

$$\ell n \; \frac{P(\ell)}{P(0)} = \left(\frac{2\bar{Q}}{\theta_p}\right)^{1/2} = \left(2\kappa\theta_p\right)^{1/2} \ell \tag{124}$$

The gain regime is often termed as the stimulated Raman regime since it involves stimulated scattering of the electromagnetic wave by the slow space charge plasma wave [4, 10, 34, 34, 36]. The phase matching diagram of these waves (eq. 122) is shown in Figure 4(d).

In this gain regime the gain grows with electron density proportionally to $n_o^{1/4}$. Also we point out that the derivation of (124) required the constraints $\theta_p >> |\mathrm{Im}\delta k\ell| >> 1$, thus we see from (124) that the parameter $\sqrt{\bar{Q}}$ gives an upper limit of the gain available in this regime.

Cold beam - high gain - strong coupling regime

In the limit of high gain and strong coupling ("strong pump") [4,7,10,18,20] the FEL parameters satisfy $Q^{1/3} >> \theta_p$ (or equivalently : $\kappa >> \theta_p$). Near synchronizm

$$\theta \simeq o \qquad \text{or} \qquad \frac{\omega}{v_{oz}} \simeq k_{zo} + k_o \tag{125}$$

θ and θ_p are negligible relative to $|\delta k|$ in Eq. 101, which reduces then into a straightforwardly soluble third degree polinomial equation

$$(\delta k)^3 = - Q \tag{126}$$

The root of this equation which corresponds to a growing wave is:

$$\delta k = \frac{1 - i\sqrt{3}}{2} Q^{1/3} \tag{127}$$

and using (121) it gives [4,6,10,18,20]

$$\ell n \frac{P(\ell)}{P(o)} = \sqrt{3}\,\bar{Q}^{1/3} = \sqrt{3}\left(\kappa\theta_p^2\right)^{1/3}\ell \tag{128}$$

Hence in this gain regime the parameter $\bar{Q}$ solely determines the available power gain. The gain depends on the electron beam density in proportion to $n_o^{1/3}$.

Eq. (127) indicates that the real part of the wave number k_{zo} changes appreciably due to the interaction (by $Re\delta k$). Instead of (125) it would be perceptive to draw the phase matching diagram at the maximun gain point (Figure 4e) in terms of the modified wave number $k_{zo} + Re\delta k$

$$Re\delta k = \frac{1}{2} Q^{1/3} \quad \text{or} \quad \frac{\omega}{v_{oz}} + \frac{1}{2} Q^{1/3} = (k_{zo} + Re\delta k) + k_o \tag{129}$$

Warm beam high gain regime

In deriving the cold beam dispersion and gain relations (96,97) we used an asymptotic expansion of (88) which is valid only for $|\zeta| \gg 1$. If this condition is not satisfied, one had to go back and solve (85) with the plasma susceptibility given by (87,88) and not by (95).

Using (102) in (89), the parameter ζ may be written around $s = ik_{z0}$ in terms of the detuning parameter θ (103):

$$\zeta \simeq \frac{\theta - \delta k}{\theta th} \tag{130}$$

where θ_{th} is defined by (108). If $\theta_{th} \gg |Im\delta k|$ (which later yields $\theta_{th}^3 \gg Q$) and $\theta_{th} \gg \theta_p$ (which is equivalent to $k_{zo} + k_o \gg k_D$ - the space charge wave number is much shorter than the Debye length), then it follows from (130) that the requirement $|\zeta| = |(\theta - Re\delta k)^2 + (Im\delta k)^2|^{1/2}/\theta_{th} \gg 1$ cannot be satisfied at any of the synchronizm conditions required for the gain regimes previously discussed. In these conditions we are bound to look for gain in the regime $|\zeta| \lesssim 1$ and solve (85) with $\chi(\omega,s)$ given given by (87,88).

With small enough coupling coefficient κ, it is possible again to solve the dispersion equation (85) by means of a first order expansion of the roots in terms of κ. In the conditions stated above it is possible to show that apart for the electromagnetic-like root (98). all the other zero order solutions of the dispersion equation are complex, corresponding to plasma waves which decay strongly by Landau damping. It is sufficient then to consider only the isolated root $s = ik_{zo}$ (98) which would give an exponentially growing wave.

The first order expansion of (85) around $s = ik_{zo}$ gives

$$\text{Res} = -\delta k_i \simeq \frac{\kappa}{2} \frac{k_D'^2}{(k_{zo}+k_o)^2} \frac{\text{Im}\, G'(\zeta)}{\left|1 - \frac{1}{2}\frac{k_D'^2}{(k_{zo}+k_o)^2} G'(\zeta)\right|^2} \tag{131}$$

where $\delta k_i \equiv \text{Im}\delta k$, and ζ is given by (130):

$$\zeta \equiv \zeta_r + i\zeta_i \tag{132}$$

$$\zeta_r \equiv \frac{\theta - \text{Re}\, k}{\theta_{th}} \simeq \frac{\theta}{\theta_{th}} \tag{133}$$

$$\zeta_i \equiv -\frac{\delta k_i}{\theta_{th}} \tag{134}$$

From the definition (88) we have

$$\text{Im}\, G'(\zeta) = \int_{-\infty}^{\infty} g'(x) \frac{\zeta_i}{(x-\zeta_r)^2 + \zeta_i^2} dx \tag{135}$$

Because of our initial assumptions $|\zeta_i| = |\delta k_i| / \theta_{th} \ll 1$, the Lorentziam inside the intergral (135) is much narrower than the function $g'(x)$. Hence $\text{Im}\, G'(\zeta_r) \simeq \pi g'(\zeta_r)$

$$\text{Im}\, G'(\zeta_r) \simeq \pi g'(\zeta_r) \tag{136}$$

Using the previously assumed inequality $k'_D << k_{zo} + k_o$, (131) can be written in the form:

$$\delta k_i \simeq -\frac{\kappa}{2} \frac{k_D'^2}{\left(k_{zo}+k_o\right)^2} \text{ Im } G'(\zeta_r) = -\frac{\bar{Q}}{\bar{\theta}_{th}^2} \text{ Im } G'(\zeta_r) \tag{137}$$

In terms of the detuning parameter θ the gain is given by [4,10,17,18,20]

$$\ell n \frac{P(\ell)}{P(0)} = -2\delta k_i \ell = 2 \frac{\bar{Q}}{\bar{\theta}_{th}^2} \text{ Im} G'\left(\frac{\theta}{\theta_{th}}\right) = \frac{\bar{Q}}{\bar{\theta}_{th}^2} 2\pi g'\left(\frac{\theta}{\theta_{th}}\right) \tag{138}$$

For a Maxwellian electron distribution (93) the imaginary part of the plasma dispersion function (136) is:

$$\text{Im} G'(\zeta_r) = -2\sqrt{\pi}\,\zeta_r\, e^{-\zeta_r^2} \tag{139}$$

This function is shown in Figure 6. We see that it attains its maximum value when:

$$\zeta_r = -1/\sqrt{2} \quad \text{or} \quad \theta = -\theta_{th}/\sqrt{2} \quad \text{or} \tag{140}$$

$$\frac{\omega}{v_{oz}} + \frac{1}{\sqrt{2}}\theta_{th} = k_{zo} + k_o$$

The diagram of this phase matching condition is shown in Figure 4a.

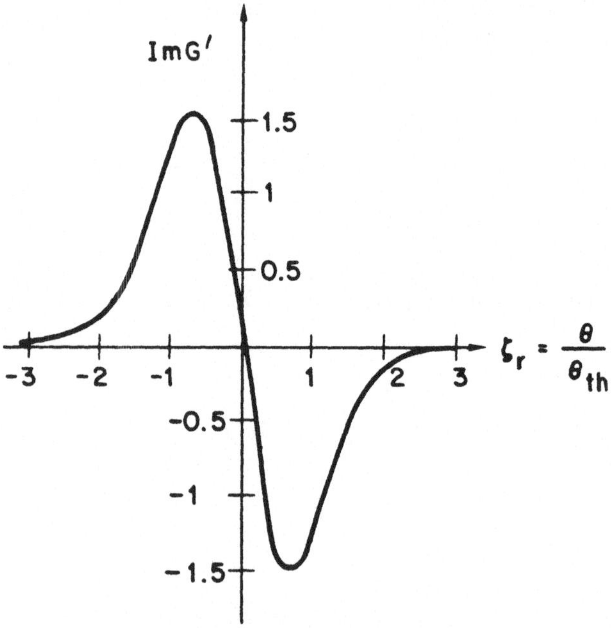

Figure 6. The gain curve function Im $G'(\zeta_r)$ for a maxwellian distribution (Eq. 139)

At the maximum gain point (140) Im $G'(-1/\ 2)\simeq 1.5$, and the maximum gain is

$$\ell n\,\frac{P(\ell)}{P(o)} = 3\,\frac{\bar{Q}}{\bar{\theta}_{th}^{\,2}} = 3\,\frac{\kappa\theta_p^{\,2}}{\theta_{th}^{\,2}}\,\ell \qquad (141)$$

Since we assumed initially $\bar{\theta}_{th} \gg 1$, again the parameter $\bar{Q}$ indicates an upper limit on the power gain available.

The warm beam regime is a regime of single electron interaction[17] and therefore the gain turned out to be proportional to the electron beam density n_o.

Warm beam low gain regime

Assuming in the low gain limit that $\kappa \to o$ and also that space charge effect can be neglected ($\theta th >> \theta_p$ or $k_{zo} + k_o >> k_D^1$), Eq. (70) reduces to:

$$\bar{a}(s) = \left[\frac{1}{s-ik_{zo}} + \lim_{\delta\to o} \frac{i\kappa\chi(\ \omega, s = ik_o)/\varepsilon}{(s-ik_{zo})\ (s-ik_{zo}+i\delta}\right] a(o) \qquad (142)$$

By substituting (87) to (89) in (142) we get the explicit dependence of $\bar{a}$ (s) on s. The inverse Laplace transform (106) can be carried out now through the integral over x (or v_z) in the expression for χ (87,88) and evaluated by means of the residuum method. The resulting after some tedious mathematical expansion (to first order in κ)

$$\frac{\Delta P}{P(o)} = \frac{\bar{Q}}{\bar{\theta}_{th}^{\,2}} \int_{-\infty}^{\infty} \frac{\sin^2(\bar{\theta}'\ /2)}{(\bar{\theta}'\ /2)^2}\ \bar{g}'\left(\frac{\bar{\theta} - \bar{\theta}'}{\bar{\theta}_{th}}\right) d\ \bar{\theta}' \qquad (143)$$

This expression can be interpreted as a convolution between the warm beam gain curve $g'(\theta/\theta_{th})$ (138) and a spectral line shape function sin $^2(\theta/2)^2$ which can be attributed to the wavenumber uncertainty due to the finite interaction lingth. When intergation by parts is applied to (143) the resulting expression will involve convolution between the cold beam gain curve $\frac{d}{d\theta}$ [$\sin^2(\theta/2)^2$] (116) and a line shape funcion $\bar{g}$ ($\theta/\theta_{\ th}$) which corresponds to line broadening due to electron velocity spread. Notice that the linear convolution relation (143) is applicable only in the limit of single electron interaction($k_{zo} + k_o >> k_D$) and low gain ($\theta P/P(0)<<1$). However the derivation of (143) there was no restriction on $\bar{\theta}_{th}$, and it would apply to both cold and warm beams as well as the intermediate regime.

In the limit $\bar{\theta}_{th} >>1$ (warm beam) the line shape function $\sin^2 (\bar{\theta}/2)/[2\pi(\bar{\theta}/2)^2]$ reduces to a "delta" function and (143) reduces to

$$\frac{\Delta P}{P(0)} = \frac{\bar{Q}}{\bar{\theta}_{th}^2} 2 \pi g' \left(\frac{\bar{\theta}}{\bar{\theta}_{th}}\right) \tag{144}$$

This expression is consistent with the high gain expression (138) which gives an identical result when we approximate $\ell n[P(\ell)/P(0) \simeq P/P(0)$. The high gain expression (138) thus applies in the low gain limit as well. The reason is that in the case we had, where the dispersion equation (85) had only a single significant root ($s \simeq ik_{zo}$), the high gain approximation (121) is correct at any gain level.

In the limit of a cold beam ($\bar{\theta}_{th}<<1$) Eq. (143) reproduces the low gain - cold tenous beam limit espression (114,116)

In conclusion of this section we illustrate in Figure 7, the transition among the different gain regimes discussed before. Figure 7 displays the maximum FEL gain $\ell n[P(\ell)/P(o)]_{max}$ as a function of the gain parameter $\bar{Q} = \bar{\kappa}\bar{\theta}_p^2$ (113,105) for various values of beam parameters $\bar{\theta}_p$, $\bar{\theta}_{th}$ (111,109,108). When $\bar{\theta}_{th}$, $\bar{\theta}_p<1$ (curve 1 in Figure 7) the FEL operates in the cold beam tenous beam regime for any value of $\bar{Q}$. For $\bar{Q} <<1$ the FEL operates in the low gain regime $[\Delta P/P(o)]_{max} \approx \ell n[P(\ell)/P(o)]_{max} <<1$ (Eqs.114,116). When the gain parameter increased ($\bar{Q} >>1$) we arrive into the high gain strong coupling regime $\ell n[P(\ell)/P(o)_{max} >>1$ (Eq. 108). For parameters choice $\bar{\theta}_p >1\ \bar{\theta}_{th}$ (curve 2) the FEL operates in cold beam regimes for any value of $\bar{Q}$. For $\bar{Q} <<1$ the operation is in the space charge dominated low gain regime (100). For $\bar{Q} >> 1$ we arrive to the high gain collective regime (104), and if Q is further increased ($\bar{Q} >> \bar{\theta}_p^3$) we finally arrive to the high gain strong coupling regime (108). When $\bar{\theta}_{th} >1$, $\bar{\theta}_p$ (curve 2), the FEL is operating basically in the warm beam regine in both the low and high gain limits (138,144). Only when $\bar{Q}>> \bar{\theta}_{th}^3$ we arrive to the cold beam-high gain-strong coupling regime (108).

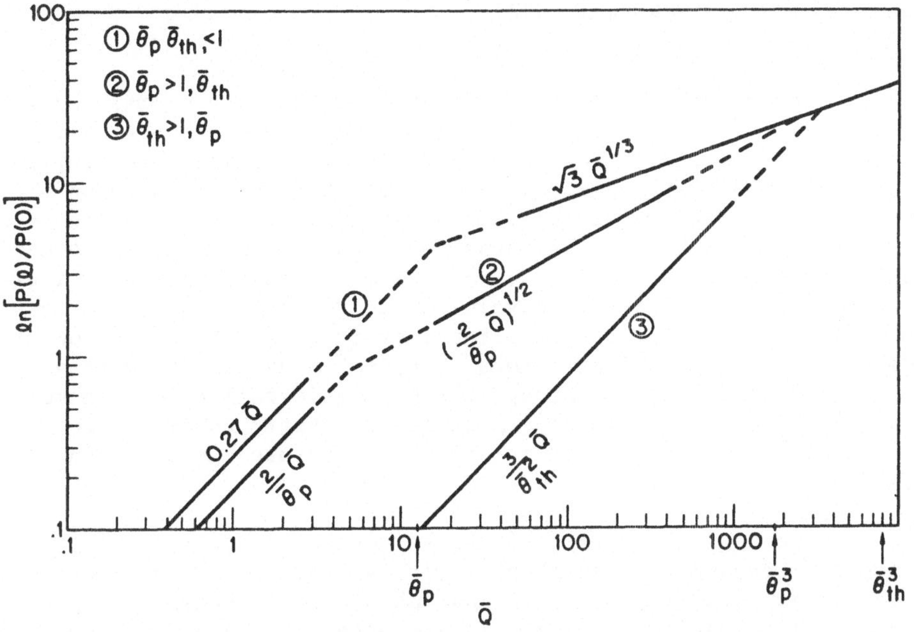

Figure 7. The dependence of FEL gain on the parameter $\bar{Q}$ for different values of beam parameters. Curve 2 is drawn for $\bar{\theta}_p = 12$ and curve 3 for $\bar{\theta}_{th} = 20$.

If we take for instance the familiar example of Magnetic bremsstrahlung FEL, and assume that the dependence on magnetic field B_0 enters predominantly through $\bar{Q}$ (neglecting 81 - 83: $\beta_{oz} \simeq \beta_0$ $\gamma_{oz} \simeq \gamma$) then Figure 7 displays the behoviour of gain as as a function of B_0^2. For low values of B_0 the gain is always proportional to B_0^2. For a large enough magnetic field (strong coupling regime) the gain goes like $B_0^{2/3}$. In the particular case of curve No.2 the gain goes through a (collective) regime where it varies in proportion to B_0.

7. THE GAIN PARAMETER OF FELS

From the discussion above it appears that the normalized gain parameter $\overline{Q}$ (113) is a good figure of merit to characterize the gain of any free electron laser in all gain regimes (114,120,124,128, 138,143). While the maximum gain obtainable depends also on parameters which characterize the beam ($\overline{\theta}_p$, $\overline{\theta}_{th}$) the dependence of the gain on the kind of FEL considered enters in all these equations only through the parameter $\overline{Q}$. Furthermore, the parameter $\overline{Q}$ indicates an upper limit of the gain in all regimes. As we increase $\overline{Q}$ by either increasing the coupling coefficient or the electron beam density, this upper limit grows at a declining rate from roughly $\overline{Q}$, through $\overline{Q}^{1/2}$ (if we go through the high gain collective regime) and finally as $\overline{Q}^{1/3}$ (Figure 7). A simple necessary condition for obtaining appreciable gain in any FEL structure and any gain regime is $\overline{Q} > 1$.

In comparing the gain of the various kinds of free electron lasers discussed in this chapter, it is best to compare their normalized gain parameter $\overline{Q}$. This was calculated by substituting κ from Table 2 in Equation (113) and is presented in Table 3 for the various kinds of FELs.

Since in many cases the current density J_0 is a limiting factor, we expressed in Table 3 the modified plasma frequency in terms of the current density: $\omega_p'^2 = eJ_0/(m\varepsilon_0 v_{oz}\gamma_{oz}^2\gamma_0)$. The parameter r_e which appears in the table is the classical electron radius:

$$r_e \equiv \frac{e^2}{4\pi\varepsilon_o mc^2} = 2.818 \times 10^{-15} \text{ m} \tag{145}$$

The expression given for the longitudinal electrostatic bremstrahlung FEL (row 3) is taken from Ref. [11]. The expression in brackets corresponds to the highly relativistic limit, assuming operation at an angle $\phi_m = (\sqrt{3}\,\gamma_0)^{-1}$ at which the gain parameter is maximal.

When one considers the dependence of the parameter $\overline{Q}$ on the operating parameters of the various FELs, it should be noticed that some of these parameters are dependent on each other. For example, the wavelength λ, period L, velocity v_0 and radiation angle ϕ depend on each other through the radiation relations which are given for the various kinds of FELs in Table 1.

TABLE 3: THE NORMALIZED GAIN PARAMETER Q FOR VARIOUS FELs (The highly relativistic limit expressions are given in brackets).

Cerenkov	$4\pi^2 \frac{J_o}{e}\frac{r_e}{c}\frac{\ell^3}{\lambda}\frac{\frac{1}{2}\sqrt{\varepsilon_o/\mu_o}\iint_{A_e}\left|\mathscr{E}_z(x_e,y_e)\right|^2 dxdy}{P}\frac{1}{\beta_o^3\gamma_o^3}$
Smith-Purcell	$4\pi^2 \frac{J_o}{e}\frac{r_e}{c}\frac{\ell^3}{\lambda}\frac{\frac{1}{2}\sqrt{\varepsilon_o/\mu_o}\iint_{A_e}\left|\mathscr{E}_z^{(1)}(x_e,y_e)\right|^2 dxdy}{P}\frac{1}{\beta_o^3\gamma_o^3}$
Long. Elect.	$\pi\frac{J_o}{e}\frac{r_e^2}{c^2}\lambda\ell^3\frac{\sqrt{\varepsilon_o/\mu_o}}{mc^2}E_o^2\frac{A_e}{A_g}\frac{\sin^2\phi}{\cos\phi}\frac{1}{(1-\beta_o\cos\phi)^4}\frac{1}{\beta_o^5\gamma_o^9}$
	$\left[\frac{27}{16}\pi\frac{J_o}{e}\frac{r_e^2}{c^2}\lambda\ell^3\frac{\sqrt{\varepsilon_o/\mu_o}}{mc^2}E_o^2\frac{A_e}{A_g}\frac{\gamma_{oz}^2}{\gamma_o^3}\right]$
Trans. Mag.	$2\pi\frac{J_o}{e}\frac{r_e^2}{c^2}\lambda\ell^3\frac{\sqrt{\mu_o/\varepsilon_o}}{mc^2}H_o^2\frac{A_e}{A_g}\frac{(1+\beta_{oz})^2}{\beta_{oz}^3}\frac{\gamma_{oz}^2}{\gamma_o^3}$
	$\left[8\pi\frac{J_o}{e}\frac{r_e^2}{c^2}\lambda\ell^3\frac{\sqrt{\mu_o/\varepsilon_o}H_o^2}{mc^2}\frac{A_e}{A_g}\frac{\gamma_{oz}^2}{\gamma_o^3}\right]$
Trans. Elect.	$2\pi\frac{J_o}{e}\frac{r_e^2}{c^2}\lambda\ell^3\frac{\sqrt{\varepsilon_o/\mu_o}}{mc^2}E_o^2\frac{A_e}{A_g}\frac{\gamma_{oz}^2}{\gamma_o^3}$
Compton-Raman	$32\pi\frac{J_o}{e}\frac{r_e^2}{c^2}\lambda\ell^3\frac{S_o}{mc^2}\frac{A_e}{A_g}\frac{\gamma_{oz}^2}{\gamma_o^3}\frac{(1+\beta_{oz})^2}{\beta_{oz}^3}$
	$\left[128\pi\frac{J_o}{e}\frac{r_e^2}{c^2}\lambda\ell^3\frac{S_o}{mc^2}\frac{A_e}{A_g}\frac{\gamma_{oz}^2}{\gamma_o^3}\right]$

In comparing various types of FELs and examining the dependence of their gain on the operating parameters, one should specify which are the independent parameters. In Table 3 the independent parameters are assumed to be λ, v_o and ϕ. The period L (or in the Cerenkov FEL - the index n) should be determined from the relations of Table 1. This choice is most convenient for the purpose of comparison among the different FELs.

The factor $1/2\sqrt{\varepsilon_o/\mu_o}\iint_{A_e} |\mathcal{E}_z(x_e, y_e)|^2 dxdy/P$ which appears in the first three rows of Table 1 can be interpreted as a relative power factor or generalized "filling factor" which indicates which fraction of the electromagnetic wave power participates in the interaction. In Table 3 row 3 the calculated expression for the filling factor of the electrostatic bremsstrahlung FEL [11] was substituted in. Numerical calculation of this factor for Cerenkov FEL was done in [21]. In the case of a Smith Purcell laser structure, the calculation of the filling factor requires the full Floquet mode solution of the electromagnetic wave in the periodic waveguide. While such involved calculation would be avoided in the present work, we may point out that the filling factors of the Cerenkov and Smith-Purcell FELs are only a function of the waveguide and electron beam geometrical dimensions normalized to the radiation wavelength. Therefore if one scales the geometry of the waveguide and the beam in proportion of the operation wavelength, then it appears from the radiation conditions of Table 1, that over a wide range of wavelengths the "filling factor" is only a function of the independent parameters v_o and ϕ, and does not depend directly on the wavelength λ. Consequently Table 3 is an explicit representation of the gain dependence on λ for the various FELs, assuming that the limiting factor is the current density J_o and that the waveguide geometrical dimensions of the laser amplifier are optimized for the operating wavelength. It is instructive to note that the wavelength dependence of the gain parameter $\overline{Q}$ is a proportionality relation for all FELs except the Cerenkov-Smith-Purcell type for which it is an inverse proportion.

The dependence of the gain parameter on the electron beam energy is not explicit in Table 3 for the Cerenkov-Smith-Purcell FELs, since the filling factors are a function of the electron velocity. However, it is probably correct to conclude that the gain parameter Q goes down as the energy is increased in all kinds of FELs, but in the longitudinal interaction FELs (first three rows in Table 3) it drops at a faster rate than in the transverse interaction FELs (last three rows).

We should point out that in particular cases like the magnetic bremsstrahlung FEL operated with a high energy linear accelerator[42], it may be more practical to use different independent parameters to describe the scaling laws of the gain parameter. Since the beam energy can be varied to quite a large extent with such accelerators, it is useful to express γ_{oz} in terms of the wavelength λ and magnet period L. If we use the radiation relation of Table 4 row 4 in the extreme relativistic limit and Eq. (83), the expression for the magnetic bremsstrahlung gain parameter $\overline{Q}_{MB}$ can be writt4en as

$$\overline{Q}_{MB} = 8\sqrt{2}\pi \frac{J_o}{e} \frac{\gamma_e^2}{c^2} \frac{\lambda^{3/2}\ell^3}{L^{1/2}} \frac{\sqrt{\mu_o/\varepsilon_o}\, H_o^2}{mc^2} \left[1 + \frac{p_{oz}^2}{(mc)^2}\right]^{-3/2} \frac{A_e}{A_g}$$

This indicates that for a constant magnet period L, the gain parameter scales with the wavelength like $\lambda^{3/2}$ as argued in [44]. It should also be appreciated that the limits on obtainable current density J_o may be wavelength dependent. As total current is increased and beam area A_e decreased (keeping the filling factor A_e/A_g constant) higher gain is obtainable, but if the electron beam has finite emittance (defined as the product of beam area and angular spread), then the maximum current density may be limited by the maximum allowable beam angular spread which is consistent with proper operation of the FEL in the particular (low gain) regime that it is designed to operate in. This limit is then wavelength dependent and requires more detailed consideration for particular cases.[44]

In comparing magnetic and electrostatic bremsstrahlung FELs it is necessary to point out the basic advantage of the first kind which stems from the fact that the force applied on a relativistic electron by a magnetic field is usually much greater than the force applied by an electrostatic field considering the state of the art of laboratory available periodic magnetic and electric fields (the latter is limited mostly by breakdown effects). However, if one considers free electron lasers also in the non-relativistic regime, it is seen from Table 3 that the gain of electrostatic FELs grows higher at low β_{oz} values.

As a last point in the comparative discussion of free electron lasers gain we should indicate the advantage of transverse interaction FELs compared to the longitudinal interaction FELs

TABLE 4. THE EFFICIENCY COEFFICIENT $\tilde{\eta}$ USED FOR CALCULATING THE EFFICIENCY (148) AT DIFFERENT GAIN REGIMES.

Gain regime	$\tilde{\eta}$
Low gain tenuous beam	$2.6/\ell$
Collective (low or high gain)	θ_p
High gain strong coupling	$\frac{1}{2} Q \quad = \frac{1}{2}\left(\kappa\theta_p^2\right)^{1/3}$
Warm beam high gain	$\frac{\pi^3}{3}\left[g'\left(\zeta_r\right)\right]^4 \frac{Q^3}{\theta_{th}^8} \left(=0.035\frac{Q^3}{\theta_{th}^8}\right)$
Warm beam low gain	$\frac{\pi^2}{3} g'\left(\zeta_r\right) \frac{1}{\bar{\theta}_{th}^2} \frac{\pi}{\ell} \left(=1.5\frac{1}{\theta_{th}^2} \frac{\pi}{\ell}\right)$

in the consideration of waveguide losses of the electromagnetic wave. The effect of waveguide losses were not included in the present model. However, for practical FEL design it may be in some cases an important consideration, which cannot be ignored. To a good approximation it may be argued that the net gain of a free electron laser is equal to the gain of a lossless FEL structure substracted by the waveguide losses of the electromagnetic mode. Since transverse electric or transverse electromagnetic modes are used in transverse interaction FELs suffer usually less waveguide losses than the transverse magnetic modes used in longitudinal interaction FELs, this provides a substantial advantage to the first kind when net gain is being compared.

8. FREE ELECTRON LASER EFFICIENCY

As discussed in a number of articles[32,33,37] high efficiencies of the order of tens of percents are potentially achievable in free electron lasers using various efficiency enhancement techniques like tapering the pump period and intensity,[32,33,45] or using energy retrieval schemes like storage ring[46] or depreseed collector[2,37] (which is currently used in conventional traveling wave tubes). In this section we will consider only the basic efficiency η_o of the various FELs, noting, though, that this efficiency may be increased appreciably when efficiency enhancement techniques are used.

Since the efficiency of radiative energy extraction from the electron beam is limited by saturation effects, a full non linear analysis is necessary to derive the FEL efficiency η_o. Nevertheless it is possible to estimate its value from the linear analysis, by using a simple consideration presented by Sprangle et al[10]. For a magnetic bremsstrahlung FEL with a cold beam, this estimate has been shown to be in excellent agreement with the more rigorous self consistent nonlinear formulation of Ref. [32]. Also it agrees in the low gain regime with the result of Ref. [38] which was obtained with a more elaborate analysis.

Consider any kind of FEL operating at any of the cold beam gain regimes discussed before (the warm beam regime must be discussed separately). The combined excitation of the coupled electromagnetic wave and the electron beam space charge waves has a component which propagates with wavenumber $k_{zo} + k_o + \mathrm{Re}\delta k$ and phase velocity

$$v_{ph} = \frac{\omega}{k_{zo} + k_o + \mathrm{Re}\delta k} \tag{146}$$

The estimate of FEL efficiency is based on the argument that the saturation mechanism is electron trapping in the periodic potential of the excitation, which propagates with phase velocity v_{ph} (146). It can be shown that the maximum deceleration in the electron velocity during the interaction up to full trapping is 2 δv_z, where δv_z is the initial axial velocity difference between the electron beam and the phase velocity of the trapping potential:

$$\delta v_z \equiv v_{oz} - v_{ph} \cong \frac{\mathrm{Re}\delta k - \theta}{k_{zo} + k_o} v_{oz} \cong (\mathrm{Re}\delta k - \theta)\frac{\lambda}{2\pi} \beta_{oz}^2 c \quad (147)$$

Thus the maximum rediative energy extraction efficiency can be claculated as the relative change in the electron beam kinetic energy when its axial velocity decreases by δv_z

$$\eta_o = \frac{\Delta\gamma_o}{\gamma_o^{-1}} \frac{1}{\gamma_o^{-1}} \frac{\partial\gamma_o}{\partial v_{oz}} 2ov_z = \frac{\beta_{oz}^3}{\beta_o^2} \gamma_{oz}^2 (1 + \gamma_o^{-1}) \tilde{\eta} \frac{\lambda}{\pi}$$

$$[\ \eta_o = \gamma_{oz}^2 \ \tilde{\eta} \ \frac{\lambda}{\pi}\ ] \quad (148)$$

where

$$\tilde{\eta} \equiv \mathrm{Re}\delta k - \theta \quad (149)$$

The expression in brackets corresponds to the highly relativistic limit.

The efficiency can now be calculated for any of the cold beam gain regimes described in Section 6 by simply substituting the appropriate θ and $\mathrm{Re}\delta k$ in (148, 149). The values of these parameters will be usually assumed to correspond to the maximum initial gain condition. The efficiency coefficient $\tilde{\eta}$ is in the case of a cold beam exactly the wavenumber mismatch drawn for various gain regimes in diagrams (b) to (e) of Figure 4. It is listed in Table 4 for the different gain regimes. Notice that in the low gain and collective gain regimes $|\mathrm{Re}\delta k| << |\theta|$ and $\tilde{\eta} \simeq -\theta$ and in the high gain strong coupling regime $|\mathrm{Re}\delta k| >> |\theta|$ and $\tilde{\eta} \simeq \mathrm{Re}\delta k$.

The estimate of the laser efficiency in the warm beam high and low gain limits requires a somewhat different consideration. As is indicated from Eqs. (135) and (143) in these gain regimes only a small fraction of electrons in the electron beam distribution participate in the interaction. In the high gain case (Eq. 135) these are the electrons with normalized axial velocities which lie in the regime

$$|x - \zeta_r| = |v_z - v_{ph}|/v'_{zth} < |\zeta_i| = |\delta k_i|/\theta_{th},$$

or that their velocities lie within the spectral width of the wave due to its growth

$$|\omega/v_z - k_r| < \delta k_i .$$

From either of these relations we find that the class of electron velocities, which takes place in the interaction is:

$$|v_z - v_{ph}| \lesssim \delta v_z \equiv v_{ph} \frac{\delta k_1}{k_r} \cong \frac{\omega}{k_{zo} + k_o} \frac{\delta k_i}{k_{zo} + k_o} \cong v_{oz} \frac{\delta k_i}{k_{zo} + k_o} \tag{150}$$

During the interaction the trapped electrons reverse their velocity relative to the phase velocity of the potential wave v_{ph}, until at saturation a local plateau is generated and

$$\bar{g}'\left(\frac{v_{ph} - v_{oz}}{v'_{zth}}\right) = 0$$

(see Figure 8).

In the warm beam low gain case the saturation process is similar, except that the class of electrons which participate in the interaction, is determined by the width of the spectral line shape function $\sin^2(\bar{\theta}'/2)/(\bar{\theta}'/2)^2$ which appears in the convolution (143).

$$|\theta' - \theta| = |\omega/v_z - v_{ph}|\ell < \pi .$$

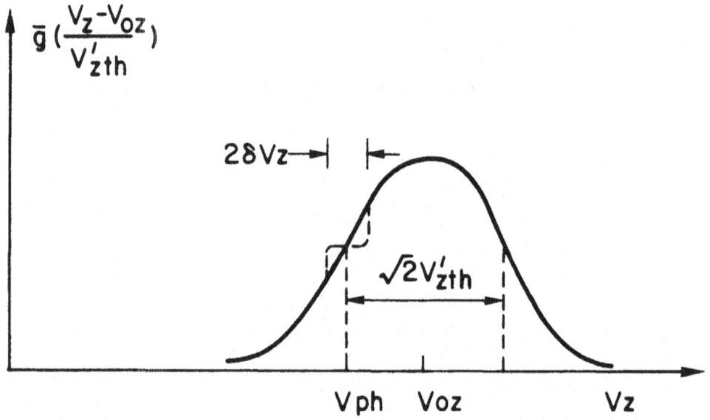

Figure 8. The electron distribution function in the warm beam gain regime drawn in velocity space before (continuous line) and after saturation (broken line).

The wave numbers of these electrons lie within the wavenumber spectral width determined by the finite interaction length:

$$\left|\frac{\omega}{v_z} - v_{ph}\right| < \pi/\ell \quad .$$

These relations can be written as

$$|v_z - v_{ph}| \lesssim \delta v_z = v_{oz} \frac{\pi/\ell}{k_{zo} + k_o} \tag{151}$$

The radiation extraction efficiency of the free electron laser in the warm beam regimes is equal to the relative change in the kinetic energy due to the interaction, where the calculation of the change in beam kinetic energy before the interaction and after saturation is most conveniently done in momentum space:

$$\eta_o = \frac{\Delta E_{KE}}{E_{KE}} = \frac{1}{\gamma_o - 1} \frac{1}{n_o} \int_{p_{zph} - \delta p_z}^{p_{zph} + \delta p_z} (\gamma - 1)\left[g^{(o)}\left(p_{zph}\right)\right] dp_z \tag{152}$$

where

$$p_{zph} = \gamma_{ph} m v_{ph} \text{ and } \delta p_z \cong \gamma_o \gamma_{oz}^2 m \delta v_z \quad .$$

First order expansion of both γ and $g^{(o)}(p_z)$ around $p_z = p_{zph}$ and some mathematical steps give

$$\eta_o \cong \frac{1}{\gamma_o - 1} \frac{d\gamma_o}{dp_{zo}} \left.\frac{dg^{(o)}}{dp_z}\right|_{p_{zph}} \frac{2\left(\delta p_z\right)^3}{3} = \frac{2}{3} \frac{\gamma_o \gamma_{oz}^2 \beta_{oz}}{\gamma_o - 1} g'(\zeta_r) \frac{\left(\delta v_z\right)^3}{c v'^2_{zth}} \tag{153}$$

Using (150,151) and (137) we finally find that in both high and low gain warm beam regimes the efficiency is given again by (148) with the parameter $\tilde{\eta}$ listed in Table 4 rows 4 and 5 for the high and low gain limits respectively. The expressions in parentheses were calculated for an electron beam with a Maxwellian distribution (93) assuming initial tuning to the maximum gain point $\zeta_r=-1/\sqrt{2}$.

The radiation extraction efficiency for all FELs and all gain regimes is estimated by the simple expression (148) with the efficiency coefficient $\tilde{\eta}$ given in Table 4 for the different gain regimes. This expression applies to FEL amplifier structures assuming saturation is reached within the interaction length ℓ. For oscillator structures (in which saturation is always obtained within the interaction length) it applies with the assumption of single mode operation.

The efficiency depends on the interaction length ℓ only in the low gain-tenuous beam regimes (both cold and warm beam). Notice though that the expressions in Table 4 rows 1 and 5 apply with the assumption that saturation is reached right at the interaction length ℓ. In the cold beam low gain and collective regimes (rows 1, 2 in Table 4), the efficiency is proportional to λ and depends only on the beam parameters (velocity and density) and the interaction length. It is thus the same for all FELs considered. Also in the warm beam low gain regime (Table 4 row 5) the efficiency is independent on the kind of FEL considered and depends only on the parameters of the beam and the interaction length. The efficiency in this limit is proportional to λ^3.

In the high gain-strong coupling and high gain-warm beam limits (rows 3, 4 in Table 4) the efficiency depends on the kind of FEL considered through the parameter Q. Its dependence on wavelength and beam velocity varies for different FELs and can be found by substituting in (148) the corresponding $\tilde{\eta}$ parameter, the parameter θ_{th} (108) and the values of the parameter Q as listed in Table 4 for the different FELs. In all cases the efficiency η_o grows with the wavelength λ. This dependence is particularly strong in the high gain warm beam case, where it goes like λ^6 in the Cerenkov-Smith-Purcell FELs and λ^{12} in the bremsstrahlung FELs.

In the case when beam energy retrieval schemes are used the total efficiency can be calculated from the expression for the basic efficiency (148) by multiplying it by an efficiency enhancement factor M. This factor is equal in the case of a pulsed mode storage ring to the number of times that the electron beam can be recycled before its spread becomes intolerable, or in the case of steady state operation, the ratio between the electron energy and the acceleration gap energy. In the case of a depressed collector scheme it is equal to the ratio between the electron beam acceleration and collection voltages. In the case of efficiency enhancement by pump tappering we cannot use at all the estimate (148) and a full nonlinear analysis is necessary [32,33,45].

In conclusion of this section we will briefly discuss the revelations of the different saturation characteristics in the two separately treated limits of warm and cold beams. We saw that while in the warm beam case only a class of electrons (150) or (151) participates in the interaction and gets trapped at saturation, in the cold beam limit all the electrons participate in the interaction and trapping process. As pointed out before in [17, 39], the local plateau formation (Figure 7) is analogous to the saturation and diminishing of population inversion in a class of atoms in Doppler broadened gas lasers. In terms of laser theory, the saturation nature of a warm beam FEL corresponds to inhomogeneous broadening and that of a cold beam to homogeneous broadening[40].

As in gas lasers there is "hole burning" effect in the gain curve of the warm beam FEL. However, there is still an important difference between the two: in contrast to the homogeneously broadened gas laser, the interaction mechanism itself changes the population of velocity classes in the vicinity of the interacting class of electrons. The gain curve of the saturated FEL (the derivative of the saturated distribution curve in (Figure 7) displays from the point of saturation on an effect of "hole burning" at $v_z \cong v_{ph}$. But it also shows a new effect of "hill heaping" at the sides of the "hole" ($v_z \cong v_{ph} \pm \delta v_z$), where the slope of the distribution function is high.

These different saturation behaviours will have explicit expression in FEL oscillation characteristics.[47] In the homogeneous broadening-cold beam regime there will be "mode competition" between the modes of a long cavity laser, all "attempting" to extract power from the same electrons. This will tend to depress multimode operation in the laser. In the inhomogeneous broadening-warm beam regime, there will be at saturation an interesting new effect of "mode cooperation" which will tend to increase the number of oscillating modes in a laser cavity which has a sufficiently dense spectrum of modes. The effect of mode cooperation will tend to wash out any bumps on the electron distribution function and spread it out, and will produce wide band radiation. Notice that this "super-inhomogeneous" saturation behaviour is different from that of inhomogeneously broadened gas lasers, where there is no cooperation between the modes.

9. FREE ELECTRON LASER POWER

One of the most attractive qualities of free electron lasers seems to be the potential capability of high power operation. Since the interaction is done in vacuum and the active medium is not a material, problems of material damage and thermal distortion of the wave front are avoided. Since the radiation extraction efficiency of FELs can be made quite high, the laser power can be quite high too, and limited essentially by the electron beam power which can be drawn into the interaction region.

The radiative power which is generated by the free electron laser is

$$\Delta P = VI\eta_o \tag{154}$$

where η_o is the basic radiation extraction efficiency (148), I is the total beam current and V is the beam acceleration voltage

$$V = \left(\gamma_o - 1\right) mc^2/e \tag{155}$$

For a given efficiency and beam acceleration energy the radiative power generation (154) will depend only on the amount of electron current I which can be made to interact with the electromagnetic wave in the free electron laser. In many circumstances there will be a limit on the current density J_o which can be generated at the FEL input and propagated along the interaction length, hence the limiting amount of interacting current will be given by

$$I = A_e J_o \tag{156}$$

Can the electron beam cross section A_e be increased indefinitely? Appart from various technical limitations we should be aware that the transverse dimensions of the electron beam should be limited by the range over which the periodic statis fields and the electromagnetic mode field have an appreciable value, so that appreciable coupling coefficient (Table 2) exists to carry out the interaction. This limitation stems from the fact that the static periodic fields and the electromagnetic modes should satisfy the Laplace equation

$$\nabla^2 \left\{ \underline{E}_o \,(x,y)\, \sin k_o z;\ \ \underline{B}_o \,(x,y)\, \sin k_o z \right\} = 0 \tag{157}$$

in the case of bremsstrahlung FELs, and the wave equation

$$\left(\nabla^2 + \frac{\omega^2}{c^2}\right) \left\{ \underline{\mathscr{E}}^{(1)} \,(x,y) e \ i(k_{zo} + k_o)z \right\} = 0 \tag{158}$$

for the Cerenkov-Smith-Purcell type (in the Cerenkov case

$$k_o = 0,\ \underline{\mathscr{E}}_1 \rightarrow \underline{\mathscr{E}}) \quad .$$

The solution of Eqs (157,158) in the case of a planar wave-guide structure (Figure 9a) is straightforward, giving for the bremsstrahlung FELs

$$\underline{E}_o \,(x,y);\ \underline{B}_o \,(x,y)\ \alpha e^{-qx} \tag{159}$$

where

$$q = k_o = \frac{2\pi}{L} \tag{160}$$

and for the Cerenkov-Smith-Purcell FELs

$$\underline{\mathscr{E}}^{(1)} \,(x,y)\ \alpha e^{-qx} \tag{161}$$

where

$$q = \left[\left(k_{zo} + k_o\right)^2 - \frac{\omega^2}{c^2}\right]^{1/2} \tag{162}$$

Thus both the static pump field in the case of bremsstrahlung FELs and the slow electromagnetic wave component in the Cerenkov-

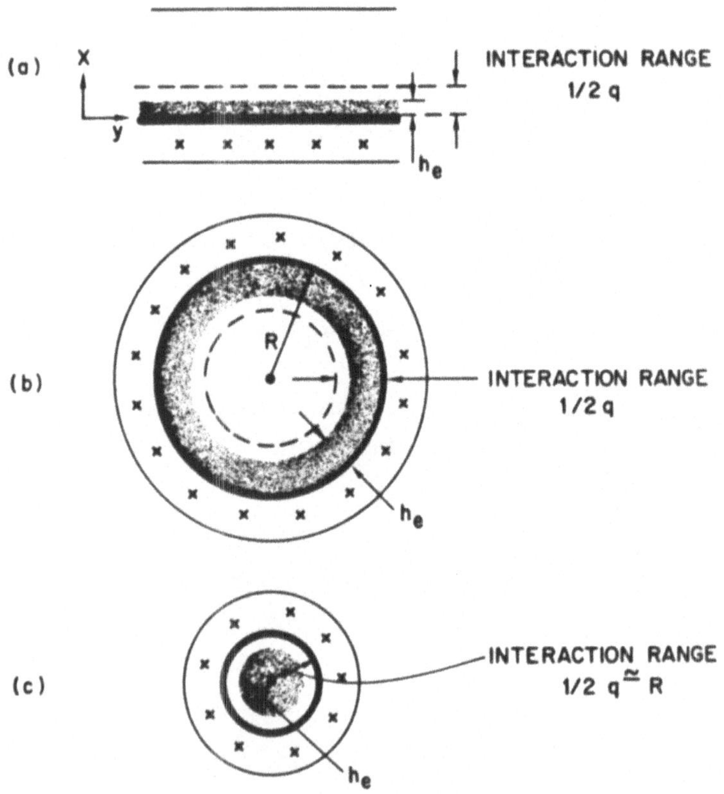

Figure 9. Cross section diagrams of various FEL structures:

(a) Sheet beam in a planar waveguide structure.
(b) Annular beam in a cylindrical waveguide structure.
(c) Solid beam in a cylindrical waveguide structure.

The cross marked regions symbolize the source of periodic static field in bremsstrahlung FELs and the slow wave structure in Cerenkov-Smith-Purcell FELs.

Smith-Purcell FELs decay exponentially away from the source of periodic static field (magnets, coils or electrodes) or from the slow wave structure (periodic waveguide, dielectric waveguide). In order to obtain appreciable coupling coefficient (κ), the electron beam should be passed roughly within the range of one exponential e-folding distance of the squared field intensity: $1/(2q)$. This distance which we will term "the interaction range" can be found in terms of λ, β_{oz} and ϕ by substituting in (160,162) the synchronism condition (4) and the radiation conditions of Table 1. The interaction range for the various FELs are listed in the second column of Table 5. The expressions in brackets correspond to the special case of ($\gamma_o >> 1$) where for the longitudinal bremsstrahlung case we substitute in this limit

$$\phi = \phi_m = \left(\sqrt{3}\gamma_o\right)^{-1} .$$

In all FELs the interaction range $1/2q$ is proportional to the wavelength λ and only when the beam is highly relativistic ($\gamma_o >> 1$) the interaction range can be appreciably larger than a wavelength. This explains why at short wavelengths it was necessary in recent FEL research to go to relativistic electron beams in order to obtain appreciable radiation. We see from Table 5 that in the highly relativistic limit, bremsstrahlung FELs have an advantage of roughly a factor γ_o in the interaction range compared to Cerenkov-Smith-Purcell FELs. There is no difference between them in the nonrelativistic limit ($\beta_{oz} << 1$).

For other waveguide cross section geometries like a cylindrical waveguide (Figure 9b,c) the solution of (157,158) gives other function solutions for the transverse variation of the fields instead of (159,161). For the case of a cylindrical waveguide the fields depend on the transverse coordinates like modified Bessel functions of the argument qr (r - the distance from the cylinder axis). When $r >> 1/q$ the modified Bessel functions behave near the waveguide walls like an exponential function $\sim e^{qr}$. In a crude approximation we may extend the notion of interaction range to waveguides with general cross sections and claim that the interaction region, at which the coupling coefficient κ has an appreciable value, extends along the perifery of the waveguide within one interaction range $1/2q$ away from the walls (see Figure 9a,b). Out of this region the coupling coefficient drops down exponentially.

It is evident that for obtaining strong interaction it is usually preferable to pass the electron beam within the interaction region, where κ has its maximum value. So in the case of a planar geometry (Figure 9a) we will prefer to use a sheet beam of

TABLE 5: INTERACTION RANGE AND UPPER BOUND OF EXTRACTABLE POWER IN FELs

FEL	Interaction range 1/2q		Max. Power ΔP_m	
	$/\left(\frac{\lambda}{4\pi}\right)$		$/\left(\frac{1}{4\pi^2}\frac{mc^2}{e} J_o w_e \tilde{\eta} \lambda^2\right)$	
Trans. Brem.	$\frac{\beta_{oz}}{1-\beta_{oz}} = (1+\beta_{oz})\beta_{oz}\gamma_{oz}^2$	$\left[2\gamma_{oz}^2\right]$	$(1+\beta_{oz})\, \beta_{oz}^4\gamma_{oz}^4\gamma_o$	$\left[2\gamma_o\gamma_{oz}^4\right]$
Long. Brem.	$\frac{\beta_o}{1-\beta_o \cos\phi}$	$\left[\frac{3}{2}\gamma_o^2\right]$	$\frac{\beta_o^4\ \gamma_o^3}{1-\beta_o \cos\phi}$	$\left[\frac{3}{2}\gamma_o^5\right]$
Cerenkov - S.P.	$\beta_o\gamma_o$	$[\gamma_o]$	$\beta_o^4\ \gamma_o^4$	$\left[\gamma_o^4\right]$

dimensions w_e x h_e where w_e is the waveguide width and $h_e \cong 1/(2q)$ and perifery $w_e = 2\pi R$ (Figure 9b). In case that the experimental conditions require the use of a solid cylindrical beam it is preferable (for the sake of obtaining appreciable gain) to keep the waveguide radius not much bigger than $R \cong 1/(2q)$. The beam cross section is then approximately $A_e = \pi/(2q)^2$.

It should be pointed out that at the present time there is little theoretical analysis for operation of various kinds of FELs in regions where either the electromagnetic wave mode or the pump field (in bremsstrahlung FELs) vary transversely[48,49]. Magnetic bremsstrahlung FEL experiments which were carried out in the low gain regime[42] used a thin solid beam which was propagated along the axis of a superconducting magnet (Figure 9c) where the periodic magnetic field can be considered (to first order approximation) transversely uniform. On the other hand experiments which were carried out in the high gain regime with intense relativistic electron beams (for example [36]) usually use an annular electron beam

(Figure 9b) where there is a certain transverse gradient in the periodic magnetic field intensity across the beam cross section. A transverse gradient in the magnetic field intensity may impose limitation on propagating the electron beam in a confined cross section along long enough interaction length. It will also have an effect of increasing the effective longitudinal momentum spread of the beam as is evident from inspection of (81,82,159) (this may be partly overcome by the "gain expansion" mechanism in particular configurations[48]). These effects will impose limits on the width of the electron beam which can be used in particular configurations of bremsstrahlung FELs depending on the electron beam energy and the gain regime of operation. This, as well as technical limitation on propagating the electron beam too close to the waveguide walls, would usually limit the electron beam cross section area A_e to less than the maximum interaction region cross section area w_e x 1(2q). However, it is generally true that in either case the electron beam cross section area cannot be much larger than the interaction region area if appreciable gain is to be attained, hence for the goal of estimating the upper bound of radiative power generation in free electron lasers the estimate that the useful electron beam cross section area is equal to the interaction region area:

$$A_e \cong w_e \times 1/(2q) \tag{163}$$

is an appropriate assumption.

Using (155), (156) and (163) in (154) we get the expression for the maximum available power

$$\Delta P_m = \frac{1}{4\pi^2} \frac{mc^2}{e} J_o w_e \beta_{oz}{}^3 \gamma_{oz}{}^2 \gamma_o \left(\frac{1/2q}{\lambda/4\pi}\right) \tilde{\eta} \lambda^2 \tag{164}$$

This is listed in the third column of Table 5 for the different kinds of FELs. The efficiency coefficient $\tilde{\eta}$ is listed in Table 4 for the different gain regimes. The expressions in brackets in Table 5 correspond to the highly relativistic limit ($\gamma_o >> 1$), where in the case of longitudinal electrostatic bremstrahlung FEL, operation at the maximum radiation angle

$$\phi_m = (\sqrt{3}\gamma_o)^{-1}$$

was assumed.

In practice, the limitations discussed above will not allow realization of the optimal case, where the whole interaction region cross section is filled with electrons. Consequently the practically extractable power falls short of the upper bound for power extraction (164) by a factor $I_0/(J_0\lambda_e \times 1/2q)$, which is the ratio between the actual drawn current and the maximum current which would ideally be passed through the whole interaction region cross section. This factor will be usually closer to 1 for Cerenkov-Smith-Purcell FELs where there are less limitations for filling up the interaction region cross section.

Table 5 and Eq. (164) apply to all FELs except the Compton-Raman scattering FEL. In this case there is no fundamental limit on the interaction region cross sectional area and it is simply determined by the electron beam cross sectional area (assuming it is smaller than the pump electromagnetic beam cross section). The maximum radiative power extractable from the electron beam should be then expressed in terms of the electron beam cross section A_e by substituting (163) into (164):

$$\Delta P_m = \frac{1}{\pi}\frac{mc^2}{e} J_o A_e \beta_{oz}^3 \gamma_{oz}^2 \gamma_o \bar{\eta}\lambda \tag{165}$$

Notice that this is the power extracted from the electron beam only. In the case of a non-relativistic electron beam, a significant amount of extra power may be extracted from the electromagnetic pump as well. The estimation of this power will require a separate discussion.

The lack of interaction width limitation in the case of the Compton-Raman FEL means that the extractable power can be increased in this case by simply increasing the beam cross section area A_e (assuming the current density J_o is a limiting factor). In practice this increase may be limited by limitations on the total beam current and by the amount of realizable pump wave power (recall that for Eq. (165) to be valid the pump power density and the corresponding laser gain should be high enough to lead to saturation within the interaction length ℓ). The lack of interaction width limitation in the Compton-Raman FEL should be especially advantageous for short wavelength FEL operation and finite available beam energy. In this case the interaction width of other FELs diminishes (Table 5 column 2) and the Compton-Raman scattering process has an advantage. A viable option for producing radiation in this limit may be the composition of a two step FEL, in which one kind of FEL produces high power radiation at a relatively long wavelength, and this radiation operates as a pump in an adjacent Compton-Raman laser which produces radiation at a short wavelength. Simultaneous operation of a bremsstrahlung and a Compton-Raman FELs, operating as spontaneous emission amplifiers, was demonstrated in Reference [41].

In the cold beam low gain regime (which may be in particular a useful operation regime in FEL oscillators) and in the collective regime ($\tilde{\eta} = 2.6/\ell$ and $\tilde{\eta} = \theta_p$ respectively) the maximum power is proportional to λ^2 for all FELs, and to λ for Compton-Raman scattering. In both cases it is independent on the FEL gain parameter (or coupling coefficient). Nevertheless it cannot be concluded that laser gain is inmaterial in determining the laser power. It should be recalled that the assumption is in all cases that the laser saturates within the interaction length ℓ. In the case of a very low gain laser this assumption would require an impracticable input power (in the case of amplifier) or impracticable resonater quality factor (in the case of a laser oscillator). Thus, even in the low gain regime the laser gain will indirectly affect the maximum available power through the choice of ℓ.

In the warm beam low gain regime the substitution of $\overline{\theta}_{th}$ (109) and η (Table 4 row 5) results in (164) a λ^5 wavelength dependence of the maximum power available in all FELs except Compton-Raman FEL (165) for which there is a λ^4 dependence.

In the high gain strong coupling and warm beam regimes there is an explicit dependence of the maximum power on the gain parameter Q. From Tables 3, 4, and 5 we find that the maximum power generation of all FELs still grows with λ for all FELs considered. There is a particularly strong wavelength dependence in the high gain warm beam case. This dependence goes like λ^7 in the Cerenkov-Smith-Purcell FELs, like λ^{13} in the bremsstrahlung FELs, and like λ^{12} in the Compton-Raman FEL.

It is instructive to point out that the maximum power generation grows very strongly with beam energy in all FELs and all cold beam regimes. We may say that apart from the possibility of generating short wavelength at high electron beam energies, the high power generation level at this limit is what distincts most markedly modern free electron lasers development from classical microwave tubes.

We see from Table 5, column 3, that in the highly relativistic limit bremsstrahlung FELs have an advantage in maximum power generation by a factor of about γ_0 compared to the Cerenkov-Smith-Purcell FELs, (at least in the low gain and collective regimes). This stems from the fact that the interaction width of the first kind is larger from that of the second kind by the same factor (Table 5 column 2). However, in the nonrelativistic limit ($\beta_0 << 1$) the maximum power generation drops down in these regimes at the same fast rate (like $\beta_0{}^4$) for both kinds of FELs. It should also be recalled that the upper bound of FEL power given by Table 5 can be approached with less difficulties in the case of Cerenkov-Smith-Purcell FELs because of the limitations on electron beam cross section in bremsstrahlung FELs with transverse gradient in the pump.

Like in the discussion of the gain parameter in Section 7 we point out here that the operation condition and the independent parameters should be clearly identified when scaling laws are derived. Table 5 and Eq. (164) give the upper bound on extractable power in terms of the independent parameters λ,ϕ and γ_{oz}. The period L or the index of refraction η (in Cerenkov FEL) are determined from the radiation condition (Table 1). In cases like the magnetic bremsstrahlung FEL operated with a high energy linear accelerator[42] the beam energy may be varied to quite a large extent and it is thus more useful to express γ_{oz} in terms of λ and L using the radiation condition of Table 4 row 4. For the case of a highly relativistic beam with a solid beam cross section operating in the cold beam low gain regime[42] we substitute into (164) $\gamma_{oz} = (L/2\lambda)^{1/2}$, $w_e = \pi \times 1/(2q) = \gamma_{oz}{}^2\lambda/(2\pi)$ and $\tilde{\eta} = 2.6/\ell$. Using (83) we get for this particular case:

$$\Delta P_m = \frac{2.6}{32\sqrt{2\pi}} \frac{mc^2}{e} J_o \left[1 + \left(\frac{p_o}{mc}\right)^2\right]^{1/2} \frac{L^{7/2}}{\lambda^{1/2}\ell} \qquad (166)$$

the upper bound on the power extraction seems to scale like $\lambda^{-1/2}$. It should be noted that the limit on current density J_o and the factor by which the realizable power falls short of the upper bound of power extraction (167) may depend on λ and the other independent parameters. This dependence requires more detailed consideration[44].

We point out that the expression of maximum power generation derived in this section applies without change also for FELs with energy retrieval schemes, but of course fails completely in efficiency enhancement schemes like tapering. However, if the total efficiency η in the latter case is known, the maximum power ΔP_m can be still calculated from Eqs. (154-156) and (163) with η_o substituted by η and $1/(2q)$ given in Table 5.

As an example consider the case of the transverse bremsstrahlung FEL operating in the cold beam low gain regime with parameters values similar to those of the Stanford experiment[42]:

$$\gamma_o = 48,\ L = 3.2\ \text{cm}\ \gamma_{oz} = 39\ (B = 0.24\ \text{Tesla}),\ n_o = 2 \times 10^9 \text{cm}^{-3}$$

$$(J_o = 9.6\ \text{A/cm}^2),\ \lambda = 10.6\mu\ \text{m and}\ \ell = 5.2\text{m} \quad .$$

Eq. (166) gives ΔP_m = 84 KW. In the actual experiment the basic and technical limitations on the electron beam diameter discussed before did not allow the use of an optimal size beam which fills the whole interaction region $J_o \times \pi(1/2q)^2$ = 2A (as assumed in (166) and only a current of 70 mA was passed in the center of the waveguide. Hence the expected power generation falls short from the calculated upper bound of power generation by a reduction factor

$$I_o/|(J_o\pi \times (1/2q)^2| = 0.07\text{A}/2\text{A resulting } \Delta P = 3\ \text{KW}$$

in agreement with the measured value[42].

10. FREE ELECTRON LASER SPECTRAL WIDTH

In all gain regimes considered in Section 4 the gain is a function of the detuning parameter

$$\theta = \frac{\omega}{v_{oz}} - k_{zo} - k_o \quad (103) \quad .$$

In each gain regime there is an optimal value of the detuning parameter θ_m (which is always negative), for which the gain is maximal. The values of θ_m at the different gain regimes are given in Table 6. The phase matching diagrams corresponding to the maximum gain condition are drawn in Figure 4 for the different gain regimes.

When we keep all the laser Operating Parameters constant and only change the radiation wavelength λ, the detuning parameter θ changes from its maximum gain value θ_m. The differentiation of (103) gives

$$\Delta\omega = \frac{\Delta\theta}{v_{oz}^{-1} - v_g^{-1}} \tag{167}$$

where

$$v_g \equiv \frac{d\omega}{dk_z} \tag{168}$$

is the group velocity of the electromagnetic mode in the waveguide.

Equation (167) applies to all FELs. In Cerenkov FEL the evaluation of the group velocity of the mode v_g may require the complete solution of the uncoupled mode dispersion relation in the dielectric waveguide. The group velocity in this case may depend on the index of refraction dispersion of the dielectric $dn/d\omega$ as well as the waveguide dispersion. In all other FEL structures the group velocity can be evaluated straightforwardly by differentiation of Equation 5:

$$v_g = c \cos \phi \tag{169}$$

and substitution in (167) gives

$$\frac{\Delta\omega}{c} = \frac{\Delta\theta}{\beta_{oz}^{-1} - \cos^{-1}\phi} \tag{170}$$

In the case of low order modes in wide cross section waveguides and in the transverse FELs, we have $\phi = o$, and using the radiation condition of Table 1:

$$\frac{\Delta\lambda}{\lambda} \quad \frac{\Delta\theta}{k_o} \tag{171}$$

If $\Delta\theta$ is the range of detuning parameter value around θ_m over which there is still gain, then $\Delta\omega$ or $\Delta\lambda$ are the laser spectral widths in the frequency or wavelength domains.

The parameter $\Delta\theta$ can be evaluated in the different gain regimes by simple inspection of the gain dependence on θ (Eq. 138, Figure 6 in the warm beam limit, and Eq. 115 Figure 5 in the low gain-cold beam limits) and by some algebraic investigation of the dispersion relation (101) in the cold beam-high gain regimes (collective and strong pump). The results are given in Table 6. The wave number mismatch ranges, over which the laser has gain, are also illustrated in Figure 4 for the various gain regimes.

For all FELs the spectral width is given by substituting the appropriate parameter $\Delta\theta$ in Equations (170, 171) or (167). Only in the high gain regimes (rows 3,4, in Table 6) does the spectral width $\Delta\omega$ depend on the particular kind of laser or on the wavelength λ through the coupling parameter κ or the gain parameter Q which are given in Table 2 and 3 respectively.

From the validity conditions of the different gain regimes it comes out that the parameter $\Delta\theta$, and consequently the spectral width, grow up the lower the gain regime is listed in Table 6. Hence the lower limit on the spectral width occurs in the low gain regimes (row 1,2). In Equation (171) it gives

$$\frac{\Delta\lambda}{\lambda} \quad \frac{L}{\ell} \tag{172}$$

TABLE 6. THE DETUNING PARAMETER VALUE OF MAXIMUM GAIN θ_m, AND THE DETUNING PARAMETER RANGE OF GAIN $\Delta\theta$.

Regime	θ_m	$\Delta\theta$
Low gain-tenuous beam	$-2.6/\ell$	$2\pi/\ell$
Low gain-space charge	$-\theta_p$	$2\pi/\ell$
High gain-collective	$-\theta_p$	$2\sqrt{2\kappa\theta_p}$
High gain-strong pump	0	$2Q^{1/3}$
Warm beam	$-\frac{1}{\sqrt{2}}\theta_{th}$	$2\theta_{th}$

This is usually quite a wide spectral width, which grows with wavelength and beam energy (as can be verified by substituting the radiation conditions from Table 1). In a typical example like the Stanford experiment[42]

$$(L = 3.2 \text{ cm} \ \ \ell = 5.2 \text{ m}) \quad , \quad \frac{\Delta\lambda}{\lambda} = 0.6\% \quad .$$

REFERENCES

1. H. Motz, "Applications of the radiation from fast electron beams", J. Appl. Phys. 22, 527-535 (1951).

2. J. M. Madey, "Stimulated Emission of bremsstrahlung in a periodic magnetic field", Appl. Phys. 42, 1906-1913 (1971).

3. F. A. Hopf, P. Meystre, M. O. Scully and W. H. Louisell, "Classical theory of free electron laser", Phys. Rev. Lett. 37, 1342-1345 (1976).

4. N. M. Kroll and W. A. McMullin, "Stimulated emission from relativistic electrons passing through a spatially periodic transverse magnetic field", Phys. Rev. A17, 300-308 (1978).

5. W. B. Colson, "One body analysis of free electron lasers", Physics of Quantum Electronics Vol. 5, p. 157-196 (Ed. S. Jacobs, M. Sargent III and M. Scully) Addison-Wesley Pub. (1978).

6. T. Kwan, J. M. Dawson and A. T. Lin, "Free electron laser", Phys. of Fluids 20, 581-588 (1977).

7. A. Hasegawa, "Free electron laser", Bell System Tech. J. 57, 3069-3089 (1978).

8. A. Bambini, A. Renieri, S. Stenhalm, "Classical theory of a free electron laser in a moving frame", Phys. Rev. A, 19, 2013-2025 (1979).

9. I. Bernstein and J. L. Hirshfield, "Amplification on a relativistic electron beam in a spatially periodic transverse magnetic field", Phys. Rev. A-20, 1661-1670 (1979).

10. P. Sprangle, R. Smith and V. L. Granatstein, Infrared and Millimeter Waves, Vol. 1, p. 279 (Ed. K. J. Button) Academic Press (1979); P. Sprangle, R. Smith "Theory of free electron lasers", Phys. Rev. A21, 293-301 (1980).

11. A. Gover, "A free electron laser based on periodic longitudinal electrostatic bremsstrahlung", Physics of Quantum Electronics Vol. 7, 701-728 (Eds. S. Jacobs, H. Pilloff, M. Sargent III and M. Scully) Addison Wesley Pub. (1980); A. Gover, "An analysis of stimulated longitudinal electrostatic bremsstrahlung in a free electron laser structure", to be published in Appl. Phys. 23 (1980).

12. G. Bekefi and R. E. Shefer, "Stimulated raman scattering by an intense relativistic electron beam subjected to a rippled electric field", 1979 IEEE International Conf. on Plasma Science, Montreal, Canada, Conf. Record, p. 12, 13; G. Bekefi and R. E. Shefer, same title, J. Appl. Phys. 50, 5158-5164 (1979).

13. R. H. Pantell, G. Soncini and H. E. Puthoff, "S-9 stimulated photon-electron scattering", IEEE J. QE-4, 905-907 (1968).

14. V. P. Sukhatine and P. W. Wolff, "Stimulated Compton scattering as a radiation source - theoretical limitations", J. Appl. Phys., 44, pp. 2331-2334 (1973).

15. V. A. Dubrovskii, N. B. Lerner and B. G. Tsikin, "Theory of Compton laser", Sov. J. Quant. Electron., 5, 1248-1253 (1976).

16. P. Sprangle, A. T. Drobot, "Stimulated backscattering from relativistic unmagnetized electron beams", J. Appl. Phys. 50, 2652-2661 (1979).

17. A. Gover and A. Yariv, "Collective and single electron interaction of electron beams with electromagnetic waves, and free electron lasers", Appl. Phys. 16, 121-138 (1978).

18. J. E. Walsh, T. C. Marshall, M. R. Mross and S. P. Schlesinger, "Relativistic electron beam generated coherent sub-millimeter wavelength Cerenkov Radiation", IEEE Transc. M.T.T.-25, 561-563 (1977); J. E. Walsh, "Stimulated Cerenkov Radiation", Physics of Quantum Electronics Vol. 5, 357-380 (Eds. J. Jacobs, M. Sargent III and M. Scully) Addison Wesley Pub. (1978).

19. A. Yariv, C. C. Shih, "Amplification of radiation by relativistic electrons in spatially periodic optical waveguides", Optics Commun. 24, 233-236 (1978).

20. A. Gover, Z. Livni, "Operation regimes of Cerenkov-Smith-Purcell free electron lasers and T. W. Amplifiers", Optics Commun. 26, 375-380 (1978).

21. Z. Livni, A. Gover, Linear Analysis and Implementation Considerations of Feee Electron Lasers Based on Cerenkov and Smith Purcell Effects, Tel Aviv University, School of Engineering, Quantum Electronics Lab. Scientific Report 1979/81 (AFOSR 77-3445).

22. S. J. Smith, E. M. Purcell, "Visible light from localized surface charges moving across a grating", Phys. Rev. 92, 1069 (1953).

23. F. S. Rusin, G. D. Bogomolov, "Orotron - an electronic oscillator with an open resonator and reflecting grating", Proc. of the IEEE, 57, 720-722 (1968).

24. V. K. Korneyenkov, V. P. Shestopalov, "A generation of diffraction radiation with a quasioptical energy output and a fixed distance between the optical cavity mirrors", Radio Eng. Electr. Phys. 22, 148-149 (1977).

25. K. Mizuno, S. Ono, Y. Shibata, "Two different mode interactions in an electron tube with a Fabri-Perot resonator - the Ledatron", IEEE Transac. ED-20, 749-752 (1973).

26. R. M. Phillips, "The Ubitron - a high power millimeter wave TWT", IRE Transac. ED-7, 231 (1960).

27. L. A. Vaynshtain, Electromagnetic Waves, Sovietskoye Radio Moskow (1957) (Russian).

28. M. R. Mross, T. C. Marshall, D. E. Efthimion, S. P. Schlesinger, "Submillimeter wave generation through stimulated scattering with an intense relativistic electron beam and zero frequency pump", 2nd International Conf. on Submillimeter Waves, Puerto Rico, Dec. 1976, Conf. Digest p. 28.

29. B. D. Fried, S. D. Conte, The Plasma Dispersion Function, Acaedemic Press, New York (1971).

30. J. R. Pierce, Travelling Wave Tubes, Van Nostrand, Princeton (1950).

31. W. H. Louisell, J. F. Lam, D. A. Copeland, "Effect of space charge on free electron laser gain", Phys. Rev. A18, 655-658 (1978).

32. P. Sprangle, C. M. Tang, W. M. Manheimer, "The nonlinear theory of free electron laser and efficiency enhancement", Physics of Quantum Electronics Vo. 7, 207-255, (Eds. S. Jacobs, H. Pilloff, M. Sargent III and M. Scully) Addison-Wesley Pub. (1980); P. Sprangle, C. M. Tang, W. M. Manheimer, "Nonlinear theory of free electron lasers and efficiency enhancement", Phys. Rev. A21, 302-318 (1980).

33. N. M. Kroll, P. L. Morton, M. N. Rosenbluth, "Variable parameter free electron laser", Physics of Quantum Electronics Vol. 7, 89-112 (Eds. S. Jacobs, H. Pilloff, M. Sargent III and M. Scully) Addison Wesley Pub. (1980).

34. P. A. Sprangle, V. L. Granatstein, L. Baker, "Stimulated collective scattering from a magnetized relativistic electron beam", Phys. Rev. A-12, 1697-1701 (1975).

35. D. B. McDermott, T. C. Marshall, "The collective free electron laser", Physics of Quantum Electronics Vo. 7, 509-522 (Eds. S. Jacobs, H. Pilloff, M. Sargent III and M. Scully) Addison-Wesley Pub. (1980).

36. D. B. McDermott, T. C. Marshall, S. P. Schlesinger, R. K. Parker, V. L. Granatstein, "High power free electron laser based on stimulated Raman back-scattering", Phys. Rev. Lett. 41, 1368-1371 (1978).

37. L. R. Elias, "High power, cw, tuneable (uv through ir) free electron laser using low energy electron beams", Phys. Rev. Lett. 42, 977-981 (1979).

38. F. A. Hopf, M. Meystre, M. O. Scully, W. H. Louisell, "Strong signal theory of a free electron lasers", Phys. Rev. Lett. 37. pp. 1342-1345 (1976).

39. F. A. Hopf. P. Meystre, G. T. Moore, M. O. Scully, Nonlinear theory of free electron lasers", Physics of Quantum Electronics Vol. 5, pp. 41-114 (Eds. S. Jacobs, M. Sargent III and M. Scully) Addison Wesley Pub. (1978).

40. A. Yariv, Quantum Electronics, John Wiley, New York (1975).

41. V. L. Granatstein, S. P. Schlesinger, M. Herndon, R. K. Parker, J. A. Pasour, "Production of megawatt submillimeter pulses by stimulated magneto-Raman scatterin-", Appl. Phys. Lett. 30, 384-386 (1977).

42. L. R. Elias, W. Fairbank, J. Madey, H. A. Schwettman, T. Smith, "Observation of stimulated emission of radiation by relativistic electrons in a spatially periodic transverse magnetic field", Phys. Lett. 36, pp. 717-720 (1976).

43. A. Gover, unpublished.

44. J. Madey, Lecture in the International School of Quantum Electronics, ERICE, Sicily, (Aug. 1980).

45. D. Pronitz, A. Szoke and V. K. Neil, "One dimensional computer simulation of the variable wiggler free electron laser", Physics of Quantum Electronics Vol. 7, 751-588 (Eds. S. Jacobs, H. Pilloff, M. Sargent III and M. Scully); S. A. Mani, "Free electron laser interaction in a variable pitch wiggler" (ibid pp. 589-622); W. H. Louisell, C. D. Cantrell, W. A. Wegener, "Single-particle approach to free electron lasers with tapered wigglers" (ibid pp. 623-646); C. A. Brau and R. K. Cooper, "Variable wiggler optimization", (ibid pp. 623-646).

46. A. Renieri Report 77.33 CNEN-Centro di Frascati, Edizione Scientifiche C. P. 65, Frascati, Rome, Italy (1977); L. R. Elias, J.M.M. Madey, T. I. Smith, Stanford High Energy Physics Lab Report HEPL-824 (1978), to be published in Appl. Phys.

47. A. Gover, "A predicted effect of mode cooperation and "white light lasing" in warm beam free electron lasers", to be published in Optics Letters Dec. (1980).

48. J. Madey, R. Taber, "Equations of motion for a free electron laser with a transverse gradient", Physics of Quantum Electronics Vol. 7, pp. 741-778 (Eds. J. Jacobs, H. Pilloff, M. Sargent and M. Scully) Addison Wesley Pub. (1980).

49. P. Sprangle and C. M. Tang, "Three dimensional nonlinear theory of the free electron laser", NRL Memorandum Report 4280 (1980); to be published in Phys. Rev. Lett.

50. J. D. Jackson, Classical Electrodynamics, John Wiley & Sons Inc. (1975).

PHYSICS OF THE FREE ELECTRON LASER*

C. Pellegrini

Brookhaven National Laboratory

Upton, NY 11973, U.S.A.

1) INTRODUCTION

The free electron laser is a powerful source of tunable, coherent, electromagnetic radiation which can operate in a wavelength interval extending from about one millimeter to the VUV, and perhaps the soft X-ray, region. The efficiency of energy transfer from the electron beam to the radiation field is of the order of one per cent, but can be as high as twenty to twenty-five per cent in specially designed high power systems. The time structure of the laser beam reflects that of the electron beam. Depending on the type of accelerator used to produce the electron beam, one can build lasers with CW operation, or providing very short pulses in the nanosecond or picosecond region.

Because of all these unique characteristics the interest in the free electron laser has grown rapidly during the last years, after the two experiments at Stanford[1,2] have shown that a free electron laser can be built and operated successfully. The theoretical works preceding and following these experiments have also given us a good understanding of the basic physics of the free electron laser and of the main characteristics of this device.

In this paper we want to discuss the classical theory of the free electron laser and describe its main properties. We will also discuss the relationship between the electron beam parameters, like energy, energy spread and emittance, and the laser parameters. In the last part of this paper we will also describe some of the problems

*Work supported by the U.S. Department of Energy.

encountered in operating a free electron laser in an electron storage ring and the possibility that it offers of operating the laser in the wavelength region below 1000 Å.

We will limit ourselves to the case of low density electron beams, neglecting the possibility of exciting some collective plasma modes. Hence we only consider the free electron laser in the Compton regime, in which a single electron is interacting with the radiation field and the wiggler field. The case in which also the collective electron mode interacts with the other two fields, three wave interaction or Raman scattering, is also very interesting and is discussed in other contributions to this book by H. Boehmer, V. L. Granatstein, and T. C. Marshall.

2) PRINCIPLES OF OPERATION OF A FREE ELECTRON LASER

The basic scheme of a free electron laser is shown in Fig. 1. A plane electromagnetic wave and a bunch of relativistic electrons are traversing a wiggler magnet. At the exit of the wiggler magnet one can obtain a plane wave of higher intensity, while the average electron energy is decreased.

A wiggler magnet produces a field with alternating polarity along its axis. It can be either transverse or helical. For a transverse wiggler, the field can be represented by

$$\underline{B}_w = B_w \hat{x} \cos 2\pi \frac{z}{\lambda_w} \tag{1}$$

where z is the direction along the magnet axis and λ_w is the field period. The two axis x and y are orthogonal to z and $\hat{x}$, $\hat{y}$ are unit vectors in their directions. For a helical wiggler one has

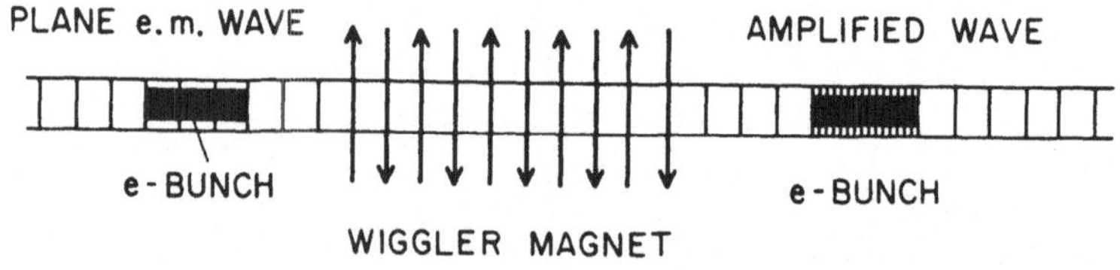

Fig. 1. Schematic representation of a free electron laser.

$$\underline{B}_w = B_w \{\hat{x} \cos 2\pi \frac{z}{\lambda_w} + \hat{y} \sin 2\pi \frac{z}{\lambda_w}\} \tag{2}$$

In the following we will consider the case of a helical wiggler. However, all the results can be applied, with only minor modifications, to the case of a transverse wiggler magnet.

A simplified description of how energy is transferred from the electrons to the radiation field can be given in two ways.[3] In a quantum mechanical description it is easier to consider the interaction between the electron and the radiation field in the frame of reference moving with the electron velocity parallel to the magnet axis. In this frame the wiggler magnetic field can be approximated by a plane electromagnetic wave. Electrons can backscatter the photons of this wave and the scattering can be enhanced (stimulated radiation) by the incoming plane wave.[4,5]

In a classical description the wiggler magnetic field determines the electron trajectory so that work can be done by the plane wave on the electrons. In other terms, the electron trajectory makes a nonzero angle with the direction of propagation of the incoming plane wave and the term $\underline{E} \cdot \underline{v}$ has a nonzero average value.[6]

The initial works[4,5] on the subject used a quantum mechanical picture describing the interaction as stimulated radiation in the wiggler magnetic field or stimulated bremsstrahlung. However, when the radiation wavelength, λ, is larger than the electron Compton wavelength, $\lambda\!\!\!^{-}_c$

$$\lambda > \lambda\!\!\!^{-}_c$$

and the number of photons in a volume $\lambda^2 \lambda_w$, λ_w being the wiggler period, is much larger than one, the classical description is appropriate.[7] The second condition means that the fluctuations in the number of photons emitted or absorbed is small and can be neglected. In the following, we will assume that these conditions are satisfied and use a classical description.

The classical theories of the free electron laser can be divided into two groups. In one we find the calculations of stimulated radiation from a single electron.[6,8] In the second group are the calculations based on the Boltzmann-Maxwell equations or analogous techniques.[9]

In this paper we will use the single electron approach and derive the electron equations of motions in the combined wiggler and radiation field. From these equations we obtain an effective Hamiltonian which can be used to study the behavior of many electrons by means of the Vlasov equation. We then introduce the Maxwell equations in the slow amplitude and phase approximations and obtain a complete set of equations describing the time evolution of the electron beam and the radiation field.

3) SINGLE PARTICLE EQUATIONS OF MOTION

In this section we will derive the equation describing the free electron laser following the approach developed by Colson.[6] We consider one electron moving in the wiggler magnetic field given by (2). We assume the wiggler to have N_w periods so that its total length is $N\lambda_w$. We also assume that the plane electromagnetic wave propagating along the wiggler axis (Fig. 1) is circularly polarized and its field is given by

$$\underline{E} = E_o \{\hat{x} \sin (\frac{2\pi z}{\lambda} - \omega t + \Phi_o) + \hat{y} \cos (\frac{2\pi z}{\lambda} - \omega t + \Phi_o)\} \tag{3}$$

$$\underline{B} = \hat{z} \times \underline{E} \tag{4}$$

The electron is supposed to be relativistic and with a small transverse velocity. If $\underline{\beta}$ is the electron velocity, in units of the velocity of light, we assume

$$\underline{\beta} = \beta_{//} \hat{z} + \underline{\beta_\perp} \tag{5}$$

with

$$\beta_{//} \approx 1, \quad \beta_\perp << 1 \tag{6}$$

We can write the electrons equations of motion as

$$\dot{\underline{P}} = e\underline{E} + e\underline{\beta}\times(\underline{B} + \underline{B_w}) + \text{space charge force}$$
$$+ \text{radiation reaction force} \tag{7}$$

The radiation reaction force can be estimated by comparing the energy loss as synchrotron radiation with the force produced by the wiggler magnetic field, $e\,B_w$[10]. The condition for the radiation reaction to be negligible can then be written as

$$\frac{2}{3}(r_o\ \gamma\ B_w)^2 < e\ B_w \tag{8}$$

where γ is the electron energy in rest mass units, m_oc^2, and r_o is the classical electron radius. Condition (8) can also be written as

$$\gamma^2 < \frac{m_o c^2}{e\, B_w\, r_o} \qquad (9)$$

For B_w = 10 KG, (9) gives $\gamma < 10^6$, which is largely satisfied for the case we want to study, where $\gamma \lesssim 10^3$.

The condition for neglecting the space charge force can be written as[11]

$$E_{sc} < E_o/\gamma \qquad (10)$$

where E_{sc} is the longitudinal space charge electric field. This can also be written in terms of the electron density, n_e, as

$$n_e < \frac{eE_o}{m_o c^2 \gamma} \frac{1}{\lambda\, r_o} \qquad (11)$$

and for $\gamma = 10^3$, $\lambda = 1$ μm, $E_o = 10^6$ V/m, we have $n_e < 10^{13}$ electrons/cm^3. We assume also this condition to be satisfied.

The equation of motion can be further simplified in the case of highly relativistic electrons. In fact, to order $1/\gamma^2$, (7) can be written as

$$\dot{\underline{\beta}} = \frac{e}{m_o c\gamma}\, \underline{\beta} \times \underline{B}_w \qquad (12)$$

$$\dot{\gamma} = \frac{e}{m_o c}\, \underline{E}\cdot\underline{\beta} \qquad (13)$$

so that the wiggler magnetic field determines the electron trajectory and the radiation field the energy exchange.

Solving eq. (12) one finds that to order $\beta_\perp/\gamma$, the electron describes a helical trajectory[12,13] of pitch

$$\delta = \beta_\perp \qquad (14)$$

and radius

$$\rho = \frac{\beta_\perp \lambda_w}{2\pi} \qquad (15)$$

The electron energy remains constant along the trajectory. The perpendicular velocity is

$$\underline{\beta_\perp} = \frac{K}{\gamma} \{\hat{x} \cos(2\pi\ z/\lambda_w) + \hat{y} \sin(2\pi\ z/\lambda_w)\} \tag{16}$$

with the wiggler parameter, K, given by

$$K = \frac{eB_w \lambda_w}{2\pi m_o c^2} \tag{17}$$

Using (16), (3), the equation (13) describing the energy exchange can be written as

$$\dot{\gamma} = - \frac{eE_o \underline{\beta_\perp}}{m_o c} \sin \Phi \tag{18}$$

where the phase Φ is

$$\Phi = (\frac{2\pi}{\lambda_w} + \frac{2\pi}{\lambda}) z - \omega t + \Phi_o \tag{19}$$

Since

$$\dot{z} = \beta_{//} c \tag{20}$$

and

$$\beta_{//} = \left[1 - \frac{1+K^2}{\gamma^2}\right]^{\frac{1}{2}} \approx 1 - \frac{1+K^2}{2\gamma^2} \tag{21}$$

one can obtain from (19) that

$$\dot{\Phi} \approx \frac{2\pi c}{\lambda_w} \{1 - \frac{\lambda_w}{\lambda} \frac{1 + K^2}{2\gamma^2}\} \tag{22}$$

We can define a resonant electron energy, γ_r^2, for which $\dot{\Phi} = 0$. This energy is given by

$$\gamma_r^2 = \frac{\lambda_w}{2\lambda} (1 + K^2) \tag{23}$$

When this condition is satisfied the variation of the optical phase in one wiggler period, i.e. for $\Delta z = \lambda_w$, $\Delta t = \lambda_w/\beta_{//}c$, as seen by the electrons, is equal to 2π.

It is worth noting that electrons of energy γ_r moving in a wiggler of period λ_w and constant K emit spontaneous radiation at the wavelength[12]

$$\lambda = \frac{\lambda_w}{2\gamma_r^2}(1 + K^2) \tag{24}$$

Using (23) we can write the electron equation of motion as

$$\dot{\gamma} = -\frac{e E_o \beta_\perp}{m_o c} \sin \Phi \tag{25}$$

$$\dot{\Phi} = \frac{2\pi c}{\lambda_w}\left\{1 - \frac{\gamma_r^2}{\gamma^2}\right\} \tag{26}$$

These equations can be derived from a Hamiltonian

$$H = \frac{2\pi c}{\lambda_w}\left\{1 + \frac{\gamma_r^2}{\gamma^2}\right\}\gamma - \frac{e E_o \beta_\perp}{m_o c}\cos \Phi \tag{27}$$

4) THE SMALL SIGNAL REGIME

If we assume that the initial electron energy is near to γ_r and that it remains near to γ_r during the crossing of the wiggler, we can simplify the equations (25), (26) and (27). Let us define the quantity

$$\eta = \frac{\gamma - \gamma_r}{\gamma_r} \tag{28}$$

and assume

$$\eta \ll 1. \tag{29}$$

The equations for energy and phase change and the corresponding Hamiltonian can be written now as

$$\dot{\eta} = -\frac{e\,E_o\,\beta_\perp}{m_o c\,\gamma_r}\sin\phi \tag{30}$$

$$\dot{\phi} = \frac{4\pi c}{\lambda_w}\eta \tag{31}$$

$$H = \frac{2\pi c}{\lambda_w}\eta^2 - \frac{e\,E_o\,\beta_\perp}{m_o c\,\gamma_r}\cos\phi \tag{32}$$

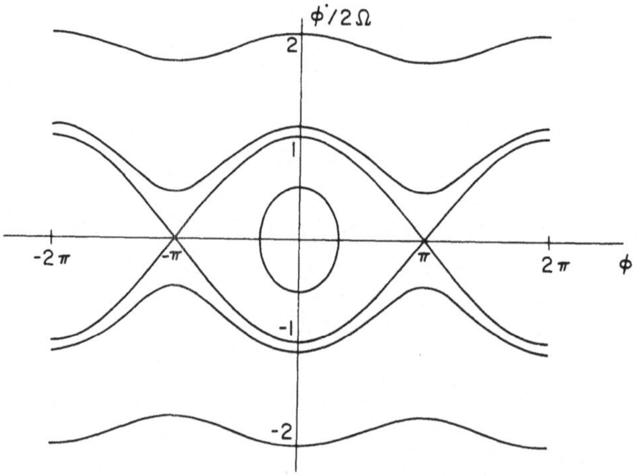

Fig. 2. Phase space electron trajectories.

A phase space represetnation of the Hamiltonian is given in Fig. 2. We can see that the Hamiltonian is the same as that describing the motion of a pendulum, with a region of bounded motion and one of unbounded motion divided by a separatrix. The small amplitude oscillation frequency, Ω, is given by

$$\Omega^2 = \frac{4\pi\ e\ E_o\ \beta_\perp}{m_o\ \gamma_r\ \lambda_w} \tag{33}$$

The maximum energy displacement on the separatrix is defined by

$$\phi_o = 0, \quad \dot{\phi}_o = \pm 2\Omega \tag{34}$$

or

$$\phi_o = 0, \quad \eta_o = \pm \frac{e\ E_o\ \beta_\perp\ \lambda_w}{\pi\ m_o c^2} \tag{35}$$

The electron energy transfer to the radiation field depends on what trajectory the particle is following in phase space, and can be either positive or negative. We are interested in calculating the energy transfer, i.e. the change in electron energy, for an electron bunch. To do this we can either integrate the equations of motion (30), (31) and then average over the distribution of the initial conditions, ϕ_o, η_o, or use an equation describing the time evolution of the electron distribution function, $f(\phi,\eta,t)$, i.e. the Vlasov equation. In both cases we have to add some more information to tell us how the electric field E_o is changing in space and time. For the time being we assume that the change in E_o in a single passage through the wiggler is small and consider E_o to be a constant in equation (30), or in the Hamiltonian (32). This is what we call the "small signal regime". This restriction will be removed in a later section when the Maxwell equations will be added to the particle equations to obtain a complete description of the free electron laser.

5) THE VLASOV EQUATION

We want to study how a given initial electron distribution evolves in time. To simplify our problem we assume that the radiation field intensity, the electron beam density and other relevant quantities are nearly uniform in the direction perpendicular to the wiggler axis. If we call q any one of this quantities we require that they do not change appreciably in a transverse distance, x, of order λ, or

$$\frac{\lambda}{q}\frac{dq}{dx} << 1 \tag{36}$$

This condition is usually well satisfied and this allows us to study our problem as a one dimensional problem and to write the electron distribution function as a function of ϕ, η and t only. The corrections to this approximation produced by the non-zero radius and angular divergence of the electron beam are discussed in another paper, by A. Luccio and C. Pellegrini, in this same book.

To study the evolution of the electron distribution function, $f(\phi,\eta,t)$, we use the Vlasov equation, i.e.

$$\frac{\partial f}{\partial t} + \dot{\phi}\frac{\partial f}{\partial \phi} + \dot{\eta}\frac{\partial f}{\partial \eta} = 0 \tag{37}$$

with $\dot{\phi},\dot{\eta}$ given by (30), (31), which we will rewrite here for convenience as

$$\dot{\phi} = 2\,\omega_o\,\eta \tag{38}$$

$$\dot{\eta} = -\frac{\omega_o}{2}\left(\frac{\Omega}{\omega_o}\right)^2 \sin\phi \tag{39}$$

where

$$\omega_o = 2\pi c/\lambda_w \tag{40}$$

and Ω is given by (33). Using ω_o, Ω, the Hamiltonian can also be rewritten as

$$H = \omega_o\,\eta^2 - \frac{1}{2}\,\omega_o\left(\frac{\Omega}{\omega_o}\right)^2 \cos\phi \tag{41}$$

To discuss the initial value of the distribution function it is important to notice that a change of 2π in the phase ϕ, corresponds to a change in the longitudinal electron position equal to λ. On this scale length the electron density can be considered uniform for all practical experimental conditions and we can assume that the distribution function at the time $t = 0$ is uniform in phase, or

$$f_o(\phi,\eta) = f(\phi,\eta,\ t=0) = \frac{\rho_e\lambda}{2\pi}\,g(\eta) \tag{42}$$

where $g(\eta)$ describes the energy distribution and ρ_e is the longitudinal electron density.

The distribution function (42) is not a steady state solution of (37). In fact a steady state solution is of the form

$$f_{eq} = f_{eq}\,(H) \tag{43}$$

H being the Hamiltonian (41). Hence the electron beam will evolve from the initial nonequilibrium distribution (42) to a final steady state distribution of the form (43). In this process it will exchange energy with the radiation field and it will become nonuniformly distributed in ϕ, i.e. bunched on a scale length λ.

We solve now the Vlasov equation by a perturbation technique, using the fact that the quantity $(\Omega/\omega_o)^2$ is always small compared to one. In fact we have from equation (39) that the maximum change in η in one wiggler period is

$$(\Delta\eta)_M = \pi \quad (\frac{\Omega}{\omega_o})^2 \tag{44}$$

But $\Delta\eta$ is also equal to the relative energy variation, $\Delta\gamma/\gamma$, which we can always assume to be smaller than one.

We assume a distribution function of the form

$$f = \sum_n f_n \, (\phi,\eta,t)(\frac{\Omega}{\omega_o})^{2n} \tag{45}$$

with f_o given by (42). Substituting (45) in (37) and using (38), (39) we obtain

$$\frac{\partial f_n}{\partial t} + 2\,\omega_o\,\eta\,\frac{\partial f_n}{\partial\phi} = \frac{\omega_o}{2}\sin\phi\,\frac{\partial f_{n-1}}{\partial\eta} \tag{46}$$

Integrating (46) we obtain, for n = 1,

$$f_1 = \frac{\rho_e\lambda}{8\pi}\,\frac{1}{\eta}\frac{\partial g}{\partial\eta}\{\cos\phi - \cos(\phi + 2\,\omega_o\,\eta\,t)\} \tag{47}$$

which shows that to first order in $(\Omega/\omega_o)^2$ the distribution is no more uniform in ϕ. The change in energy of the electron beam can be obtained from the average value of η. To first order in $(\Omega/\omega_o)^2$ this is given by

$$\langle\eta\rangle = \langle\eta\rangle_o + (\frac{\Omega}{\omega})^2 \int_o^{2\pi} d\phi \int d\eta\; f_1(\phi,\eta,t)\eta \tag{48}$$

where

$$\langle\eta\rangle_o = \int_o^{2\pi} d\phi \int d\eta\; f_o(\eta,\phi)\eta = \rho_e\lambda \int d\eta\; g(\eta)\eta \tag{49}$$

Using (47) we see that the second term on the r.h.s. of (48) is zero, so that to first order in $(\Omega/\omega_o)^2$ the beam starts to be bunched on the scale of λ but does not yet exchange energy with the radiation field.

The solution of (46) for n = 2 can be written in the form

$$f_2 = \sum_n f_{2n}(\eta,t)e^{in\phi} \tag{50}$$

To evaluate the moments of η, we need to know only the term independent from ϕ, f_{20}. This is given by

$$f_{20} = \frac{\rho_e\lambda}{64\pi} \frac{\partial}{\partial\eta} \left\{\frac{1}{\eta^2} \frac{\partial g}{\partial\eta} [1 - \cos 2\omega_o \eta t]\right\} \tag{51}$$

Using (51) we can evaluate $<\eta>$ to second order in $(\Omega/\omega_o)^2$, obtaining

$$<\eta> = <\eta>_o + \frac{\rho_e\lambda}{16} \left(\frac{\Omega}{\omega_o}\right)^4 (2\omega_o t)^3 \int d\eta\, g(\eta)\, F(2\omega_o\, \eta\, t) \tag{52}$$

where

$$F(x) = \frac{1}{x^3} \{\cos x - 1 + \frac{x}{2} \sin x\} \tag{53}$$

Equation (52) allows us to evaluate the average electron energy change at the wiggler exit, i.e. for $t = \lambda_w N_w/\beta_{//}c$, or, for $\beta_{//} \approx 1$,

$$x = 2\omega_o\, \eta\, t = 4\pi\, N_w\, \eta \tag{54}$$

The gain function F(x) is plotted in Fig. 3. One can see that the electrons lose energy if $x > 0$ and gain energy if $x < 0$. The maximum energy loss is obtained for $x = 2.5$ or $\eta \approx 0.2/N_w$. If the initial electron energy distribution is narrow compared with the width of F(x), or

$$\Delta\eta = \frac{\Delta\gamma}{\gamma} << \frac{1}{5N_w} \tag{55}$$

we can approximate the function $g(\eta)$ in (52) with a δ-function

$$g(\eta) = \delta(\eta - \eta_o) \tag{56}$$

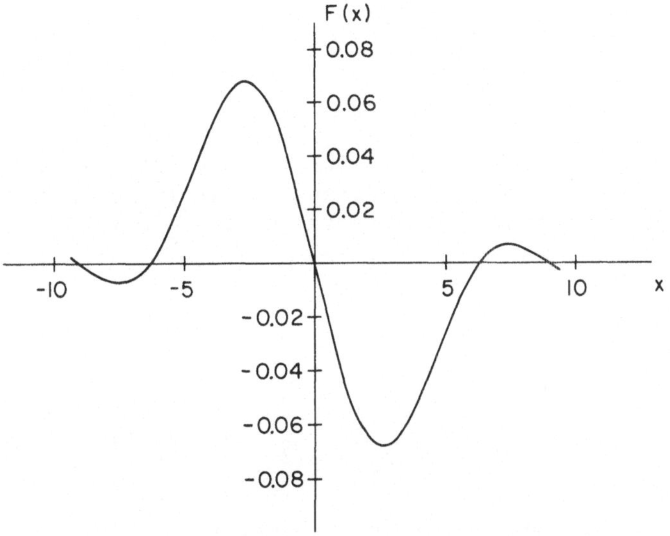

Fig. 3. The gain function F(x).

and obtain for the average energy change

$$\Delta\langle\eta\rangle = \langle\eta\rangle - \langle\eta\rangle_o = 4\pi^3 \rho_e \lambda \left(\frac{\Omega}{\omega_o}\right)^4 N_w^3 F(x_o) \tag{57}$$

with

$$x_o = 4\pi N_w \eta_o \tag{58}$$

Using (47), (51), we can also evaluate other properties of the electron beam during its evolution from the initial state of the form (43). It is interesting to notice that because the Hamiltonian is an even function of η, in the final state we have

$$<\eta>_{\text{final state}} = 0 \tag{59}$$

so that the energy of the electrons is redistributed evenly around the resonant energy. At this point also the small signal gain is zero and there is no energy exchange between the electrons and the radiation field.

6) THE FREE ELECTRON LASER GAIN

From the expression (57) for the average electron loss, we can obtain an expression for the gain, G_0, in the electromagnetic field intensity, I, during one traversal of the wiggler. Defining

$$G_0 = \Delta I/I \tag{60}$$

we obtain

$$G_0 = 32\sqrt{2}\, \pi^2 \lambda^{3/2} \lambda_w^{1/2} \frac{K^2}{(1+K^2)^{3/2}} \frac{i_p}{\Sigma i_A} N_w^3 F(x) \tag{61}$$

where i_p is the peak electron current, $i_A = ec/r_o = 1.7 \times 10^4$A, and Σ is the transverse cross section of the electromagnetic wave. To derive (61) we have assumed that the electron beam is bunched, with a bunch length ℓ_e and transverse cross section Σ_e equal to Σ. It is easy to generalize (61) to the case $\Sigma_e \neq \Sigma$.

From the behavior of the function F(x), with $x = 4\pi N(\gamma-\gamma_r)/\gamma_r$, as illusted in Fig. 3 and from the discussion of Section IV we can obtain the following results:

1. The electrons lose energy if $\gamma > \gamma_r$, gain energy if $\gamma < \gamma_r$; no energy change occurs if $\gamma = \gamma_r$;

2. The maximum gain is obtained for $(\gamma - \gamma_r)/\gamma_r \approx 0.2$;

3. The maximum energy transfer from the electron to the electromagnetic field is

$$(\Delta\gamma)_{MAX} \approx \gamma_r \frac{1}{2N} \tag{62}$$

which gives for the system an efficiency

$$\eta \leq \frac{1}{2N} \tag{63}$$

4. for a nonmonochromatic electron beam the gain is obtained by folding (61) with the electron energy distribution function; if the width of this distribution, σ_e is large compared to 1/2N, the effective gain is strongly reduced and can become nearly zero.

To estimate the order of magnitude of G_0 we assume $F(x) \approx -7 \times 10^{-2}$, $K \approx 1$ and that Σ_p is determined by diffraction limit to $\Sigma \approx \lambda(N_w\lambda_w)$. With these assumptions (61) becomes

$$G_0 \approx \pi^2 \left(\frac{\lambda}{\lambda_w}\right)^{\frac{1}{2}} \frac{i_p}{i_A} N_w^2 \tag{64}$$

For $\lambda = 1\mu m$, $\lambda_w = 5$ cm, $N_w = 100$, we obtain

$G_0 \approx 2.5 \times 10^{-2}\, i_p$, with i_p in Ampere.

This result shows that to obtain a gain of the order of several percent, the electron beam peak current must be of the order of several ampere. We can now summarize the characteristics that the electron beam must have in order to obtain a good amplification of the electromagnetic field:

1. the energy spread must be small, $\sigma_e < \frac{1}{2N}$

2. the transverse cross section must be smaller or equal to that of the electromagnetic wave or $\Sigma_e \leq \Sigma$

3. the angular divergence must be such that[3] $\langle\theta^2\rangle^{\frac{1}{2}} < \frac{1}{\gamma\sqrt{2N}}$

4. the peak current must be of the order of several ampere or larger.

To obtain a free electron laser oscillator we modify the system of Fig. 1 by adding an optical cavity, i.e. two mirrors. One of these mirrors is assumed, for simplicity, to be a perfect reflector, while the other is assumed to transmit a fraction α of the incident light (Fig. 4). We neglect the other possible losses of the light in the cavity.

This system can be a laser oscillator if the gain is larger than the loss, or

$$G \geq \alpha \tag{65}$$

The case $G = \alpha$ describes the steady state operation of the system.

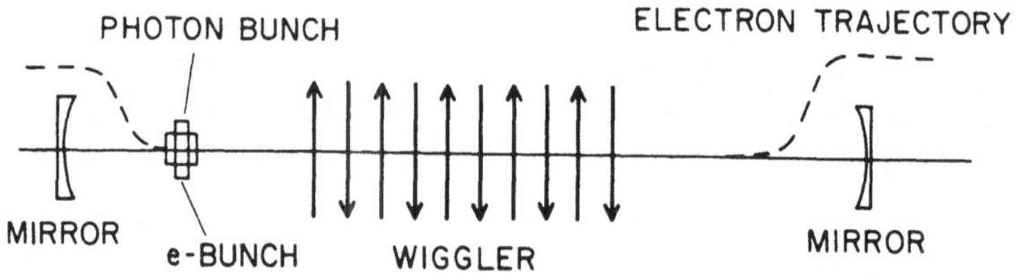

Fig. 4. A free electron laser oscillator.

Condition (65) must be satisfied for the effective gain, corresponding to a beam with its own energy spread. The output laser power in the steady state operation, $G = \alpha$ is given by

$$P_L = \eta\ (E\ i_{av}) \tag{66}$$

where the efficiency, η, has a maximum value of the order of $1/(2N_w)$, $E = m_0 c^2 \gamma$ is the electron energy and i_{av} is the average electron current. Equation (66) shows us that a fraction η of the electron beam power is transformed into laser beam power.

The space-time structure of the laser beam reflects the electron beam structure. We assume that the electron beam is bunched, with a bunch length ℓ_e and a bunch separation L_e. We also assume, for simplicity, that $L_e >> \ell_e$. Since the gain is different from zero only where the electron density is nonzero, one obtains in the laser cavity a radiation pulse of length $\ell_p \approx \ell_e$. The length of the radiation pulse determines the line-width, $\delta\omega$, of the laser radiation

$$\delta\omega = \frac{2\pi c}{\ell_e} \tag{67}$$

or

$$\frac{\delta\omega}{\omega} = \frac{\lambda}{\ell_e} \tag{68}$$

The distance between successive radiation pulses in the output laser beam is determined by the length of the optical cavity, L_{op}, and is equal to $2L_{op}/c$.

To keep the system running, the electron bunch and the radiation pulse must be synchronized and the distance, L_e, between bunches must be a multiple of $2L_{op}$. In the simplest case this synchronization condition can be written as

$$L_e = 2L_{op} \tag{69}$$

7) THE EVOLUTION OF THE RADIATION FIELD

In sections 5 and 6 we have studied the behavior of the free electron laser by studying the electron notion. This is valid if the change in the radiation field intensity during one pass of the wiggler is small. To remove this limitation we must introduce Maxwell equations and couple them with the particle equations.[14-16]

It is convenient to write the Maxwell equations, using the vector potential $\underline{A}$ and in the Coulomb gauge, as

$$\Box \underline{A} = -\frac{4\pi}{c}\underline{J_\perp} \tag{70}$$

where $\underline{J_\perp}$ is the transverse electron current.

We write the vector potential $\underline{A}$ in the form

$$\underline{A} = E_o(z,t)\frac{\lambda}{2\pi}\{-\hat{x}\cos\left[\frac{2\pi z}{\lambda} - \omega t + \phi_o(z,t)\right] + \hat{y}\sin\left[\frac{2\pi z}{\lambda} - \omega t + \phi_o(z,t)\right]\} \tag{71}$$

and assume that the amplitude $E_o(z,t)$ and the phase $\phi_o(z,t)$ are slowly varying functions:

$$\frac{\partial E_o}{\partial t} \ll \omega E_o \qquad \frac{\partial E_o}{\partial z} \ll \frac{2\pi}{\lambda} E_o \tag{72}$$

$$\frac{\partial \phi_o}{\partial t} \ll \omega\phi_o \qquad \frac{\partial \phi_o}{\partial z} \ll \frac{2\pi}{\lambda} \phi_o \tag{73}$$

This means that we can neglect the variation of E_o and ϕ_o in one optical wavelength or period.

Using the slow varying amplitude and phase approximation we can rewrite (70) as two equations for the amplitude and phase

$$\frac{\partial E_o}{\partial z} + \frac{1}{c}\frac{\partial E_o}{\partial t} = -\frac{2\pi}{c} J_1 \quad (74)$$

$$E_o \left\{ \frac{\partial \phi_o}{\partial z} + \frac{1}{c}\frac{\partial \phi_o}{\partial t} \right\} = \frac{2\pi}{c} J_2 \quad (75)$$

where

$$J_1 = J_x \sin\alpha + J_y \cos\alpha \quad (76)$$

$$J_2 = -J_x \cos\alpha + J_y \sin\alpha \quad (77)$$

and

$$\alpha = \frac{2\pi}{\lambda} z - \omega t + \phi_o \quad (78)$$

The transverse electron current can be written in terms of the transverse velocity $\beta_\perp$ and the position $\underline{x}$ of the particles, as

$$\underline{J_\perp} = e \sum_i \underline{\beta_{\perp i}} \, \delta(\underline{x} - \underline{x_i}) \quad (79)$$

where the sum is made over all electrons. Using (16), (76) and (77) we obtain

$$J_1 = e\, c\, K \sum_i \frac{\sin\phi}{\gamma_i} \delta(\underline{x} - \underline{x_i}) \quad (80)$$

$$J_2 = -\, e\, c\, K \sum_i \frac{\cos\phi}{\gamma_i} \delta(\underline{x} - \underline{x_i}) \quad (81)$$

with the relative phase, ϕ, of the electron and the radiation field, defined as in (19).

To make the sum over the particles, we assume that the macroscopic electron density, $\rho(\underline{x})$, describing the bunch shape on a scale large compared with λ, remains unchanged during the wiggler crossing. On a microscopic level, at a scale length of the order of λ, the density changes because of the microbunching produced by the electron-radiation field interaction. We can then evaluate (80), (81) by first summing over the particles contained in a cylinder of radius equal to the beam radius and length, λ, equivalent to a change of 2π in ϕ, and then multiply locally be $\rho(\underline{x})$.

It is convenient to use the variable η instead of γ, and transform the sum into an integral with the help of the distribution function $f(\phi,\eta,t)$. We then have

$$J_1 = ec \frac{K}{\gamma_r} \rho(\underline{x}) \left\langle \frac{\sin \phi}{1+\eta} \right\rangle \tag{82}$$

$$J_2 = - \frac{e\ c\ K}{\gamma_r} \rho(\underline{x}) \left\langle \frac{\cos \phi}{1+\eta} \right\rangle \tag{83}$$

where we define the average value of a function $g(\phi,\eta)$ as

$$\langle g(\phi,\eta)\rangle = \int_0^{2\pi} d\phi \int_{-\infty}^{+\infty} d\eta\ f(\phi,\eta,t)\ g(\phi,\eta) \tag{84}$$

Since $\eta \ll 1$ in the following, we will neglect η compared to one in (82), (83).

We can now write down the full system of equations describing the free electron laser, i.e.

$$\frac{\partial E_o}{\partial z} + \frac{1}{c}\frac{\partial E_o}{\partial t} = - \frac{2\pi}{c} J_1 \tag{85}$$

$$E_o \left\{ \frac{\partial \phi_o}{\partial z} + \frac{1}{c}\frac{\partial \phi_o}{\partial t} \right\} = \frac{2\pi}{c} J_2 \tag{86}$$

$$\frac{\partial f}{\partial t} + \dot{\phi}\frac{\partial f}{\partial \phi} + \dot{\eta}\frac{\partial f}{\partial t} = 0 \tag{87}$$

$$\dot{\eta} = - \frac{\omega_o}{2}\left(\frac{\Omega}{\omega_o}\right)^2 \sin \phi \tag{88}$$

$$\dot{\phi} = 2\omega_o \eta - \dot{\phi}_o \tag{89}$$

With J_1 and J_2 given by (82), (83) and Ω^2 defined by (33).

The full system of equations (85) to (89) can be solved numerically for given initial conditions to determine the time and space evolution of the electron beam and of the radiation field. We will not attempt to do this in this paper, but will consider only a few simple cases where we can obtain an analytical solution.

8) PERTURBATIVE SOLUTION OF THE VLASOV-MAXWELL EQUATIONS

Following the procedure used in Section 5, we assume that the quantity $(\Omega/\omega_o)^2$, the small amplitude pendulum frequency, is small and we look for a perturbative solution of the Vlasov-Maxwell equations (85) - (89). We use the solution obtained in Section 5 to write to first order

$$f = \frac{\rho_e \lambda}{2\pi} g(\eta) + \frac{\rho_e \lambda}{8\pi} \frac{1}{\eta} \frac{\partial g}{\partial \eta} \{\cos \phi - \cos(\phi + 2\omega_o \eta t)\} \tag{90}$$

This solution was obtained assuming that the radiation field amplitude, E_o, is a constant. Since to zero order both J_1 and J_2 are zero, any change in E_o is of first order in $(\Omega/\omega_o)^2$, and would influence f only in the second order term.

To evaluate the change in E_o to first order in $(\Omega/\omega_o)^2$, we first evaluate J_1, J_2 and then integrate (85). Using (90) and (82), (83), neglecting η as compared to one in these last two equations, and assuming a monoenergetic electron beam, i.e. $g(\eta) = \delta(\eta - \eta_o)$, we obtain

$$J_1 = \frac{1}{8}\left(\frac{\Omega}{\omega_o}\right)^2 \frac{ecK}{\gamma_r} \rho(\underline{x}) \{\sin 2\omega_o \eta_o t - 2\omega_o \eta_o t \cos 2\omega_o \eta_o t\} \tag{91}$$

$$J_2 = -\frac{1}{8}\left(\frac{\Omega}{\omega_o}\right)^2 \frac{ecK}{\gamma_r} \rho(\underline{x}) \{1 - \cos 2\omega_o \eta_o t + 2\omega_o \eta_o t \sin 2\omega_o \eta_o t\} \tag{92}$$

We now use (91) and integrate (85). To simplify the integration we assume that the electron density is uniform within a cylinder of radius a and length ℓ

$$\rho(\underline{x}) = \frac{N_e}{\pi a^2 \ell} \tag{93}$$

Integrating (85) we obtain

$$E_o = E_{o,in} + 2\pi \left(\frac{\Omega}{\omega_o}\right)^2 \frac{ecK}{\gamma_r} \frac{N_e}{\pi a^2 \ell} (\omega_o^2 t^3) F(2\omega_o \eta_o t) \tag{94}$$

where the function F(x) is given by (53) and $E_{o,in}$ is the radiation field amplitude at the wiggler entrance. We now write the field at the wiggler exit as

$$E_{o,out} = E_{o,in} + \Delta E_o \tag{95}$$

with ΔE_o given by the second term on the right hand side of (94) evaluated for $\omega_o t = 2\pi N_w$. The small signal gain, relative change of the radiation field intensity in going through the wiggler, can be written as

$$G_o = \frac{2\Delta E_o}{E_{o,in}} \tag{96}$$

and using (94) we obtain again the expression (61).

9) WIGGLER WITH PARAMETERS VARYING ALONG THE AXIS

As dicussed in Section 5 a steady state solution of the Vlasov equation can be obtained in the form

$$f_{eq} = f(H) \tag{97}$$

with

$$H = \eta^2 - \frac{1}{2}\left(\frac{\Omega}{\omega_o}\right)^2 \cos\phi \tag{98}$$

For a solution of this form it follows from (82) that, because H is an even function of ϕ,

$$J_1 = 0 \tag{99}$$

and from (85) that E_o = constant. Hence, in an equilibrium state the gain is zero and there is no energy exchange between the radiation field and the electron beam.

The fact that J_1 is zero in the equilibrium state limits the efficiency of the system to the low value discussed in Section 6. It is, however, possible to avoid this limitation if we modify the particle Hamiltonian (98), introducing a term odd in ϕ so that $\langle\sin\phi\rangle \neq 0$. The way to do this is to use a wiggler in which the period or the magnetic field or both vary along the wiggler length.[15]

For a variable wiggler we can define the resonant energy as a function of the distance along the wiggler axis by using equation (23), as

$$\gamma_r^2(z) = \frac{\lambda_w(z)}{2\lambda}\,[1 + K^2(z)] \tag{100}$$

We will assume that the resonant energy is a slowly varying function of z,

$$\frac{\lambda_w}{2\gamma_r^2}\frac{d\gamma_r^2}{dz} << 1 \qquad (101)$$

The phase, ϕ_r, of a particle having an energy equal to the resonant energy, γ_r, can be defined, with the help of (25), (26) as

$$\sin\phi_r = \frac{d\gamma_r^2}{dt}\left[\frac{2eE_oK}{m_oc}\right]^{-1} \qquad (102)$$

In analogy with what is done in linear accelerators the phase ϕ_r is called the synchronous phase and the particle characterized by ϕ_r, γ_r the synchronous particle.

For a particle with energy γ the equation for the change of energy can now be written, using (16), (18), (102), as

$$\frac{d}{dt}[\gamma^2 - \gamma_r^2] = -\frac{2eE_oK}{m_oc}[\sin\phi - \sin\phi_r] \qquad (103)$$

Assuming that the energy remains always near to the resonant energy, or that

$$\eta = \frac{\gamma - \gamma_r}{\gamma_r} << 1 \qquad (104)$$

we can again write the equations for η and ϕ in a form similar to (30), (31) or (38), (39) as

$$\dot{\eta} = -\frac{1}{2}\omega_o\left(\frac{\Omega}{\omega_o}\right)^2 \{\sin\phi - \sin\phi_r\} \qquad (105)$$

$$\dot{\phi} = 2\,\omega_o\eta \qquad (106)$$

In writing (105) we have neglected a term $2\eta(\dot{\gamma}_r/\gamma_r)$ which is usually negligible if (101) is satisfied.

The Hamiltonian corresponding to (105), (106) is

$$H = \omega_o \eta^2 - \frac{1}{2}\omega_o \left(\frac{\Omega}{\omega_o}\right)^2 \{\cos\phi + \phi \sin\phi_r\} \qquad (107)$$

The last term in (107) is odd in ϕ and allows us to obtain a value of $<\sin\phi> \neq 0$ also in an equilibrium state and so a large energy exchange. Assuming that the energy of an electron remains near to γ_r we can estimate the electron energy change from (102). Assuming $\sin\phi_r \approx 1/2$ we have, using the variable $z = \beta_{//}ct \approx ct$,

$$\gamma_r^2(0) - \gamma_r^2(L_w) \approx \frac{2eE_oKL_w}{m_oc^2} \qquad (108)$$

where L_w is the wiggler length.

If the amplitude E_o of the input radiation field is large enough we can have $\gamma_r(L_w) << \gamma_r(0)$ and most of the electron energy is transferred to the radiation field. For example, for $\gamma_r(0) = 100$, $K = 1$ we need a product E_oL_w os the order of 10^9 volt.

To study in more detail the motion of a particle around the resonant energy let us rewrite the Hamiltonian (107) as

$$H = \eta^2 + V(\phi) \qquad (109)$$

where the potential $V(\phi)$ is given by

$$V(\phi) = -\frac{1}{2}\left(\frac{\Omega}{\omega_o}\right)^2 [\cos\phi + \phi \sin\phi_r] \qquad (110)$$

and is plotted in Fig. 5. In this figure the shaded area represents a potential well where electrons are trapped and oscillate around the synchronous phase. The minimum of $V(\phi)$ corresponds to $\phi = \phi_r$ and the maximum to $\phi = \pi - \phi_r$. The width in ϕ of the potential well is defined by

$$\Delta\phi = (\pi - \phi_r) - \phi^* \qquad (111)$$

where ϕ^* is obtained by solving the equation

$$V(\phi^*) = V(\pi - \phi_r) \qquad (112)$$

The depth of the potential well is defined by

$$\eta^2_{MAX} = V(\pi - \phi_r) - V(\phi_r) \qquad (113)$$

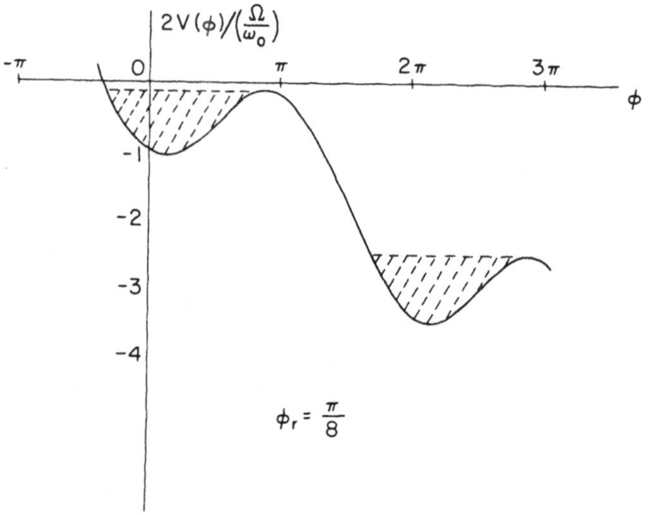

Fig. 5. Potential well in an axially variable wiggler.

or

$$\eta_{MAX} = \left(\frac{\Omega}{\omega_o}\right) \{\cos \phi_r - (\frac{\pi}{2} - \phi_r) \sin \phi_r\}^{\frac{1}{2}} \tag{114}$$

Particles with $\eta < \eta_{MAX}$ and $\phi^* < \phi < \pi - \phi_r$ will oscillate around γ_r and ϕ_r. For these particles the average energy change in traversing the wiggler is equal to the change in the resonant energy, so that the efficiency in the transfer of energy from the electrons to the radiation field can now be written as

$$\eta = \frac{\Delta\phi}{2\pi} \; \frac{\gamma_r(0) - \gamma_r(L_w)}{\gamma_r(0)} \tag{115}$$

For a reasonable choice of parameters one can have $\Delta\phi/2\pi \sim 50\%$ and $\Delta\gamma/\gamma \sim 50\%$ so that the overall efficiency (115) can be of the order of 25%. This is to be compared with an efficiency of the order of 1% for a uniform wiggler.

A detailed description of the properties of axially variable wigglers and of the methods to optimize their design is given in the paper by D. Prosnitz, A. Szoke and V. K. Neil [20]. An alternative way to increase the efficiency by using wigglers in which the magnetic field can change in the direction orthogonal to the wiggler axis is discussed in the paper by J. Eckstein and J. M. J. Madey.

10) INFRARED AND ULTRAVIOLET FREE ELECTRON LASERS

Equations (61), (66), (68) describe the characteristics of the output laser beam. The gain of the system, as well as the linewidth, the output power and the space-time structure of the laser beam are determined by the electron beam characteristics.

We want to discuss now what kind of electron accelerator is needed to produce a beam suitable for a given free electron laser and what are the consequence of this choice on the characteristics of the system. Since the region of the spectrum between 0.2 and 10 μm is already well covered by existing laser, we will limit this discussion to wavelengths $\lambda > 10$ μm and $\lambda \leq 0.2$ μm, although high efficiency, high power free electron lasers might be interesting also in the region between 0.2 and 10 μm.

The two conditions on the electron beam cross section and angular divergence, given in Section 5, can be written as a single condition on the product of the horizontal and vertical electron beam emittance, $E_x E_z$. The emittance is given by the beam area in the phase space plane, $x, \dot{x}$ and is an invariant along a beam transport channel at constant electron energy, so it can be conveniently used to characterize an electron beam size and angular divergence. Assuming that the laser beam cross section, Σ, is determined by the diffraction limit, i.e. $\Sigma \approx \lambda N_w \lambda_w$, this condition can be written as

$$E_x E_z < \pi\lambda^2 \tag{116}$$

Typical values for emittances of electron beams produced today by linear accelerators, microtrons or storage rings are:

i) 30 MeV linear accelerator, $E_x E_z \approx 10^{-12}\, m^2 rad^2$

ii) 30 MeV microtron, $E_x E_z \approx 10^{-11} m^2 rad^2$

iii) 700 MeV storage ring, $E_x E_z \approx 10^{-15} m^2 rad^2$

From these numbers using (116) one obtains that for wavelengths $\lambda \gtrsim 1$ μm it is possible and convenient to use a linear accelerator or a microtron while for $\lambda \lesssim 0.2$ μm it is necessary to use a storage ring.

It is also important to notice that while for $\lambda > 10$ μm the mirrors forming the optical cavity have a reflectivity approaching 100% and the cavity losses can be kept to a value of the order of a few percent, for $\lambda < 0.2$ μm the mirror reflectivity decreases, it can be as low as 20 to 30%, and the cavity loss is correspondingly high. This means that in the IR region a gain per pass of the order of 10% can be enough, while in the UV the gain must be at least one order of magnitude higher. Correspondingly the peak electron current needed in the IR region is of the order of 10A and in UV is of the order of 100A or higher. These values of peak current are consistent with what can be obtained in linear accelerator and microtron for the IR and storage ring for the UV.

To make an order of magnitude estimate for a laser in the IR region, let us assume the wiggler parameters: $\lambda_W = 10$ cm, $N_W = 50$; and the electron beam parameters: $\gamma = 50$, average and peak current 6 mA and 5 A. The radiation wavelength can then vary between $\lambda =$ 30 μm, for $K = 0.7$ and $\lambda = 200$ μm for $K = 3$.

The electron (and laser) beam is assumed to consist of 200 pulses 60 ps long, separated by 50 ns. After these 200 pulses, lasting 10 μs, the beam is off for 2 ms and then starts again. During the 10 μs pulse the electron beam average power is 1.5 MW and the laser beam average power is 15 kW.

Electron storage rings can provide beams with high peak current, $i_p \gtrsim 100$A, high average current, $i_{av} \approx 1$A, and good energy spread $\Delta E/E \lesssim 0.1\%$, and emittance. The wiggler magnet can be inserted in a straight section without substantial alteration to the single particle dynamics.[17] However, in this case some of the equations describing the laser must be modified. In fact, in the linear accelerator case each electron bunch traverses the wiggler only once and the electron distribution function at the wiggler entrance is determined only by the accelerator characteristics. Instead in the storage ring case the same electron bunch is repetitively traversing the wiggler and its characteristics are determined also by the interaction with the radiation field.

As we discussed in Section 5, the electron bunch at the wiggler exist has an energy spread larger than that at the wiggler entrance. Hence we can expect that the electron bunch-radiation field interaction will increase the electron beam energy spread thus reducing the effective gain. This problem has been studied by Renieri[18] who has shown that the interaction is such that the energy spread would continue to grow and the effective gain would become zero. The only mechanism opposing this process is the synchrotron radiation damping which tends to decrease the energy spread. An equilibrium between these two processes can be reached such that the laser output power is determined by the strength of the radiation damping process, which is proportional to the total amount of synchrotron radiation energy, U_o, lost by the electrons in one revolution. Hence for a storage ring case equation (66) must be substituted by

$$P_L = \eta(U_o i_{av}) \tag{117}$$

The effective gain is also reduced so that to obtain a value of η near to $1/2N_w$ one needs a gain calculated for the unperturbed beam larger than the losses by a factor of 5-10.[18]

The synchrotron radiation energy loss per revolution is a strong function of electron energy. For a bending radius, ρ, in the ring one has

$$U_o = \frac{4}{3}\pi \frac{r_o}{\rho} \gamma^4 m_o c^2 \tag{118}$$

so that to increase P_L is very convenient to increase γ.

Table I. Parameter of Storage Ring and Free Electron Laser for Short Wavelength

Electron energy	500 MeV
Ring circumference	62.8 m
Average electron current	1 A
Peak electron current	396 A
Number of bunches	3
R.m.s. electron transverse emittance	1.54×10^{-8} mrad
R.m.s. electron energy spread	10^{-3}
Small signal gain at λ = 627 Å	6
Laser output power	16.7 W
Line width at $\lambda = 10^3$ Å	8×10^{-7}

In Fig. 6 we give, as an example, the wavelength, λ, and small signal gain, G_o, versus the wiggler parameter K, for a free electron laser operating in a storage ring and designed to give high value of G_o at wavelength $\lambda \lesssim 10^3$ Å.[19] Some of the parameters characterizing the storage ring and the laser are given in Table I. The high value

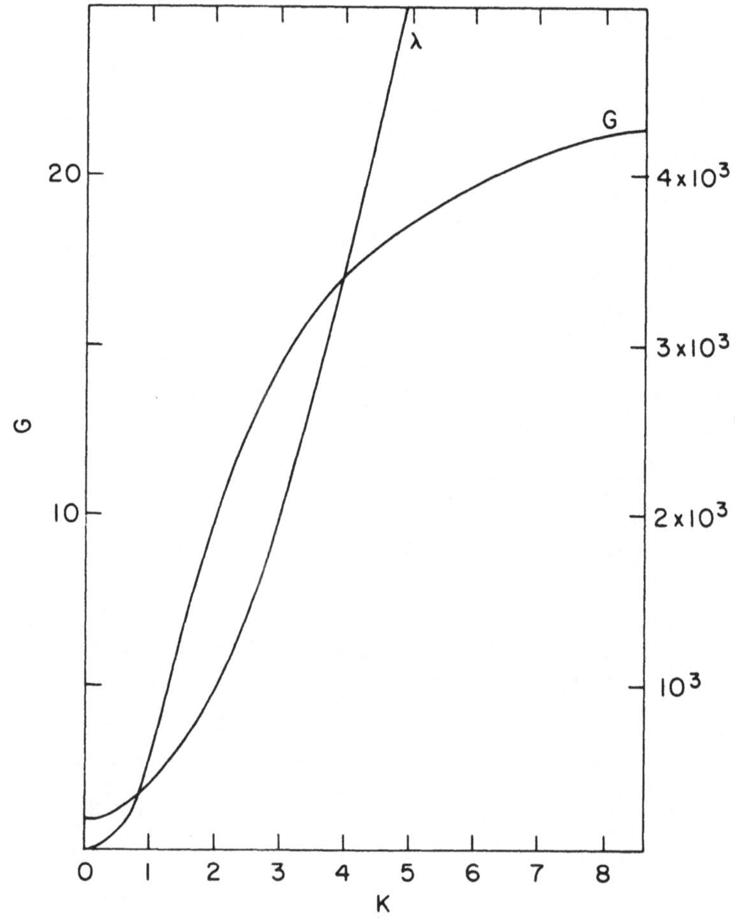

Fig. 6. Wavelength and gain versus the wiggler parameter, K, for a free electron laser in a storage ring.

of the gain near and below 10^3 Å shows the possibility of building a laser in this wavelength region if we can have mirrors with a reflectivity per normal incidence of the order of 50% or larger. The laser output power might be increased by at least one order of magnitude above the value of Table I if we increase the strength of the synchrotron radiation damping by adding an auxiliary wiggler radiator.[17]

REFERENCES

1. L. R. Elias, W. M. Fairbank, J. M. J. Madey, H. A. Schwettman and T.I. Smith, Phys. Rev. Lett 36:717 (1976).
2. D. A. G. Deacon, L. R. Elias, J. M. J. Madey, G. J. Ramian, H. A. Schwettman and T. I. Smith, Phys. Rev. Lett. 38:892 (1977).
3. C. Pellegrini, in Synchrotron Radiation Research, S. Doniach and H. Winick, eds., Plenum Press, New York (1980).
4. J. M. J. Madey, J. Appl. Phys. 42:1906 (1974).
5. J. M. J. Madey, H. A. Schwettman and W. M. Fairbank, IEEE Trans. Nucl. Sci. NS-20:980 (1973).
6. W. B. Colson, Phys. Lett. 59A:187 (1976).
7. J. M. J. Madey and D. A. G. Deacon, in Cooperative Effects in Matter and Radiation, C. M. Bowden, D. W. Howgate and H. R. Robl, eds., Plenum Press, New York (1977).
8. A. Bambini and A. Renieri, Lett. Nuovo Cimento, 21:399 (1978).
9. F. A. Hopf, P. Meystre, M. O. Scully and W. H. Louisell, Phys. Rev. Lett. 37:1215 (1976).
10. J. D. Jackson, Classical Electrodynamics, J. Wiley, New York (1962).
11. V. K. Neil, Lawrence Livermore Laboratory Report UCID-17985 (1978).
12. B. M. Kincaid, J. Appl. Phys. 48:2684 (1977).
13. J. P. Blewett and R. Chasman, J. Appl. Phys. 48:2692 (1977).
14. F. A. Hopf, T. G. Kuper, G. T. Moore and M. O. Scully, in Free-Electron Generators of Coherent Radiation, Physics of Quantum Electronic, Vol. 7, S. F. Jacobs, H. S. Pilloff, M. Sargent III, M. O. Scully, R. Spitzer, eds., Addison-Wesley, Reading (1980).
15. N. M. Kroll, P. L. Morton and M. N. Rosenbluth, in Free-Electron Generators of Coherent Radiation, Physics of Quantum Electronics, Vol. 7, S. F. Jacobs et al., eds., Addison-Wesley, Reading (1980).
16. W. B. Colson and S. K. Ride, in Free-Electron Generators of Coherent Radiation, Physics of Quantum Electronics, Vol. 7, S. F. Jacobs, et al., Addison-Wesley, Reading (1980).
17. J. E. Spencer and H. Winick in Synchrotron Radiation Research, H. Winick and S. Domiach, eds., Plenum Press, New York (1980).
18. A. Renieri, IEEE Trans. Nucl. Sci. NS-26:3827 (1979).
19. C. Pellegrini, Nuclear Inst. and Meth. (in press).
20. D. Prosnitz, A. Szoke and V.K. Neil, UCRL-84799, preprint, Lawrence Livermore Laboratory 1980, prepared for submittal to Phys. Review.

CLASSICAL TRAJECTORIES AND COHERENT STATES OF A FREE ELECTRON LASER

R. Bonifacio, F. Casagrande, and L. A. Lugiato

Istituto di Fisica dell'Universita

Milan, Italy

1. INTRODUCTION

A great deal of theoretical research on the free electron laser (FEL) has occurred since Madey and coworkers first observed FEL operation as an amplifier[1] and as an oscillator[2]. Current trends in theory are reviewed in this paper, together with the experimental efforts.

A many-particle theory of the FEL has been developed by the Arizona group[3] in terms of coupled Maxwell-Boltzmann equations. However, most of the present theoretical understanding is based on the analogy between the motion of the single electron and that of a pendulum[4-7]. In particular, by choosing conditions that lead to a libratory motion of the pendulum, one obtains the well-known expressions for gain and spread[8]. However, the pendulum analogy of the FEL is limited because the frequency of the pendulum is not fixed but depends on the laser field. In section 2, we derive the basic classical equations of motion and briefly discuss the pendulum approximation. In section 3, we formulate conditions that specify the precise limits of validity of the pendulum description and then investigate the behavior of the system when this description fails[9].

A key point for our analysis is the introduction of an "interaction variable" that is the classical counterpart of a suitable harmonic oscillator operator, by which one of us (R.B.) discussed the coherence properties and the photon statistics of the FEL from first principles[10]. In section 4, we review the definition of the coherent states of a FEL as the eigenstates of a properly defined harmonic oscillator operator which acts simultaneously on the electron field Hilbert spaces, and reduces the one-particle FEL

Hamiltonian to that of a single harmonic oscillator driven by a classical field (wiggler) with a nonlinear term quadratic in the photon number. Assuming small fluctuations, this Hamiltonian becomes that of a driven harmonic oscillator with a time dependent frequency. In this approximation the FEL radiates coherently in the Glauber's sense giving a Poisson statistics for the photon number and the electron momentum centered on the classical values.

2. CLASSICAL EQUATIONS OF MOTION AND PENDULUM APPROXIMATION

We start from the Bambini-Renieri-Stenholm one electron Hamiltonian[7]

$$H = p_z^2/2m + \hbar\omega_o(a_L^\dagger a_L + a_W^\dagger a_W) + i\hbar g\,(a_L^\dagger a_W e^{-2ik_o z} - \text{h.c.}) \tag{1}$$

Here, a_L (a_W) represents the laser (wiggler) field and $z(p_z)$ is the coordinate (momentum) of the electron, whose mass m is renormalized by the static magnetic field. The commutation relations are

$$[a_L, a_L^\dagger] = 1, \quad [a_W, a_W^\dagger] = 1, \quad [z, p_z] = i\hbar \tag{2}$$

Hamiltonian (1) describes the FEL in a moving frame which has been properly chosen [5] so that the laser and the wiggler frequencies coincide ($\omega_L = \omega_W = \omega_o = ck_o$). The coupling constant is given by

$$g = (2\pi c/k_o V)\, r_o, \tag{3}$$

r_o being the classical radius of the electron and V the quantization volume.

The free field Hamiltonian can be easily eliminated because it is a constant of the motion. Hence we have

$$H = p_z^2/2m + i\hbar g\,(a_L^\dagger a_W e^{-2ik_o z} - \text{h.c.}) \tag{4}$$

The Hamiltonian (4) has two constants of motion which clarify the basic meaning of the radiation process in the FEL:

$$\begin{aligned} &a_L^\dagger a_L + a_W^\dagger a_W + \text{const} \\ &p_z + \hbar K_o\,(a_L^\dagger a_L - a_W^\dagger a_W) = \text{const} \equiv P \end{aligned} \tag{5}$$

Eliminating $a_W^\dagger a_W$ between eqs. (5) we get

$$p_z/2\hbar k_o + a_L^\dagger a_L = \text{const} \equiv N \tag{5'}$$

Hence we have a scattering process in which photons of momentum $-\hbar k_0$ are transformed (reflected) from the wiggler mode to the laser mode with momentum $\hbar k_0$, so that the total photon number as well as the total momentum are preserved. In particular, for each photon created in the laser mode the electron momentum decreases of $2\hbar k_0$.

The Heisenberg equations of motion are easily obtained from the Hamiltonian, eg. (4).

$$\begin{cases} \dot{z} = p_z/m & (6) \\ \dot{p}_z = -2\hbar k_0 g\,(a_L a_W^{\dagger} e^{2ik_0 z} + \text{h.c.}) & (7) \\ \dot{a}_L = g\, a_W e^{-2ik_0 z}. & (8) \end{cases}$$

From now on we will consider the wiggler field so strong that it can be replaced by a c-number α_W in the Hamiltonian (4). Hence the only meaningful constant of motion in this way remains (5').

The classical equations of the FEL are obtained treating all the operators in eqs. (6)-(8) as c-numbers. In particular, we replace the operator a_L with α_L (t) in (7) and (8). This approximation will be discussed in section 4. Thus our basic equations read:

$$\begin{cases} \dot{z} = p_z/m & (6') \\ \dot{p}_z = -2\,\hbar k_0 g(\alpha_L \alpha_W^{*} e^{2ik_0 z} + \text{c.c.}) = m\ddot{z} & (7') \\ \dot{\alpha}_L = g\alpha_W\, e^{-2ik_0 z} & (8') \end{cases}$$

The relation between the FEL one-particle theory and a pendulum equation is easily seen. In fact, putting

$$\ddot{\theta} = 2\,k_0 z,\quad \alpha_L = |\alpha_L| e^{i\varphi_L},\quad \alpha_W = |\alpha_W| e^{i\varphi_W}\,, \tag{9}$$

eq, (7') reduces to

$$\ddot{\theta} + \Omega^2\,(t)\,\cos\,(\theta + \varphi_L - \varphi_W - \varphi_W) = 0 \tag{10}$$

where

$$\Omega^2\,(t) = 4\,g\omega\,|\alpha_L|\,|\alpha_W| \tag{11}$$

and ω is the Schrödinger frequency $\omega\ 2\hbar\, k_0^2|m$.

Hence we are dealing with a pendulum whose frequency is a function of the field $\alpha_L(t)$, which in turn depends on the coordinate z of the electron as given by eq. (8). To our Knowledge, all analytical treatments of the FEL are limited to the approximation in which both the amplitude and phase of the laser field α_L remain nearly constant in time, so that the analogy between the FEL and a pendulum works. Basically this pendulum regime is obtained when the electron momentum is nearly constant. Consistently with (5) we see that also the photon number cannot change too much, so that we are describing a small gain amplifier. Hence we assume that $\alpha_L(t) \simeq \alpha_L(0)$ in eqs. (7') and (10). This implies that the nonlinear time dependence of the pendulum frequency induced by the field is neglected. More precisely, integrating (8') we have

$$\alpha_L(t) = \alpha_L(0) + g\alpha_W \int_0^t dt' \, e^{-i\theta(t')} \tag{12}$$

If this is inserted into (10), the first term $\alpha_L(0)$ represents the stimulated emission contribution due to the laser field already present at t=0; the second term describes the spontaneous build-up of the field due to the wiggler. In the usual treatments one keeps only $\alpha_L(0)$ in (11) obtaining a pendulum equation with constant frequency. We shall call the approximation of neglecting the spontaneous nonlinear term in (11) the pendulum approximation. We shall now demonstrate from an exact treatment that a well defined threshold value exists for α_W above which such an approximation breaks down and the system behaves as a large gain amplifier evolving from spontaneous emission ($\alpha_L(0)\simeq 0$) to stimulated emission with a large photon number.

Let us define the "interaction variable"

$$\alpha(t) = \alpha_L(t) \, e^{2ik_0 z(t)} \tag{13}$$

which contains both the laser and the electron variables. Noting that $|\alpha|^2 = |\alpha_L|^2$, eq. (5') becomes

$$|\alpha|^2 + p_z/2\hbar\,\kappa_0 = N \tag{14}$$

Once we calculate $\alpha(t)$; its squared modulus gives the photon number and, via (14), the electron momentum p_z. Furthermore, integrating (6') one obtains the electron position, and finally from (8') one gets the field amplitude.

The advantage of the interaction variable (13) is that the three equations (6')-(8') reduce to a unique equation for $\alpha(t)$. In fact, by simple differentiation with respect to time we get (using (8'), (6'), (14))

$$\dot{\alpha} = \dot{\alpha}_L e^{2ik_o z} + 2ik_o \dot{z}\alpha$$
$$= g\alpha_W + 2ik_o \frac{P_z}{m} \alpha = g\alpha_W + 4 i \frac{\hbar k_o^2}{m} (N-|\alpha|^2)\alpha,$$

or finally

$$\dot{\alpha} = 2i \omega (N - |\alpha|^2)\alpha + g\alpha_W \tag{15}$$

For simplicity, we shall take the c-number α_W real and positive in the following.

Eq. (15) has been deduced quantum-mechanically in ref. 10 (see also section 4). It is a first-order differential equation with a cubic nonlinearity and a driving term which represents the wiggler. We note that this equation is formally identical to that of an ordinary laser with an injected signal except that the constant which rules the nonlinearity here is complex, whereas in an ordinary laser is real. As a consequence we have a periodic time behaviour for the FEL whereas the usual laser presents an irreversible approach to a steady state. This deep difference between the two systems is due to the fact that the former is a a Hamiltonian system, whereas the latter is a dissipative open system. It follows that in a FEL the emitted photon number does not have a steady state but must be calculated stopping the time evolution at t=L/c.

3. EXACT TRAJECTORIES

We now describe the exact FEL dynamics in terms of eq. (15) as in ref. 9. For this purpose, it is convenient to introduce some scaled variables which allow reducing the number of parameters in eq. (15) to one, thereby exhibiting the scaling properties of the system:

$$\bar{\alpha} = \frac{\alpha}{\sqrt{N}} , \quad z = \frac{2 g\alpha_W}{\omega N^{3/2}}, \quad T = 2 \omega Nt \tag{16}$$

We have assumed N > 0, which is the most interesting case. Hence eq. (15 becomes

$$d\bar{\alpha} /d\tau = i (1 - |\bar{\alpha}|^2) \bar{\alpha} + z/4 \tag{15'}$$

The parameter z will play a crucial role in the following analysis. Note that it is proportional to the wiggler field α_W, and depends also on the initial values $p_{z,o}$ and $|\alpha_L|_o^2$ of the momentum and of the photon number via $N = (p_{z,o} 2\hbar k_o + |\alpha_L|_o^2)$.

Eq. (15') describes the FEL dynamics in the phase plane of the variables $y_1 = \mathrm{Re}\ \bar{\alpha}$, $y_2 = \mathrm{Im}\ \bar{\alpha}$. We define the variables R, ϕ and p as follows:

$$\bar{\alpha} = Re^{i\phi}, \quad \bar{p} = \frac{p_z}{2\hbar k_o N} \tag{17}$$

We have from (9), (13), (14), (16), (with α_W real),

$$|\alpha_L|^2 = NR^2, \quad \phi = \Theta + \varphi_L = 2\ K_o z + \varphi_L,$$
$$\bar{p} = 1 - R^2. \tag{18}$$

Hence, the distance from the origin in the phase plane directly gives the laser intensity. Furthermore, the points of this plane inside (outside) the circle R=1 correspond to positive (negative) electron momentum.

The trajectories in the phase plane can be found immediately because the system is Hamiltonian and hence the energy is conserved in time. Using the normalized variables, one has

$$(1 - R^2)^2 + z\ R \sin\phi = \bar{E} = \mathrm{const} \tag{19}$$

where $\bar{E}$ is the scaled energy $E/\hbar\omega N^2$.

As already pointed out, the analogy between the FEL and a pendulum works provided both the amplitude and the phase of the laser field α_L remain basically constant in time. From (18), we see that in this case the trajectory is a circle, or part of a circle, centered on the origin in the phase plane. Furthermore, the phase ϕ behaves exactly as the position of the electron. As we are going to show, the necessary and sufficient condition for this pendulum behaviour is that z is suitably small. Let us consider three cases separately.

1) Let us assume that $z \ll 1$ and that also

$$z R / |E| \ll 1 \tag{20}$$

This condition requires that the kinetic energy $(1-R^2)^2$ is much larger than the potential energy $zR \sin\phi$. Hence it amounts to requiring that $zR/(1-R^2)^2 \ll 1$, i.e., using (16), (17) and (18),

$$\frac{2 g \underline{\alpha}_W |\alpha|}{\omega\, p_z/2\hbar k_o)^2} \ll 1 \tag{20'}$$

In the literature, condition (20') defines the unsaturated regime of the FEL [8]. Writing (19) in the form

$$R^2 = 1 \pm \sqrt{\bar{E}}\ (1 - \frac{z R}{\bar{E}} \sin\phi)^{1/2}, \tag{19'}$$

We see that due to (20) the energy curve reduces to the two circles ($y_1 = R\cos\phi$, $y_2 = R\sin\phi$):

$$y_1^2 + \left[y_2 \pm \frac{z/4}{\bar{E}}\right]^2 = 1 \pm \sqrt{\bar{E}} + \frac{(z/4)^2}{\bar{E}} \tag{21}$$

The circle of larger radius lies in the negative momentum region $R > 1$ and the trajectory evolves counter-clockwise. For $(z/4) \ll \sqrt{\bar{E}}$ it is practically centered on the origin. The small circle (which exists for $\sqrt{\bar{E}} < 1$) lies in the positive momentum region $R<1$ and the trajectory proceeds clockwise. It is practically centered on the origin when $1 - \sqrt{\bar{E}} \gg (z/4)^2/\bar{E}$. When these conditions hold, these circles correspond to a pendulum performing a libratory motion. This is the quasi-free particle regime of the FEL[8].
In this regime, in which the FEL works as a small signal amplifier, the spread and the gain can be calculated by a simple perturbative procedure, by performing a proper average over all possible initial positions z_o of the electron [8]. E.G., the expression of the spread in our normalized variables is

$$S(\tau; R_o; z) = \frac{1}{2} \left[\frac{zR_o}{1-R_o^2}\right]^2 \sin^2\left[\frac{1}{2}(1-R_o^2)\,\tau\right] \tag{22}$$

where R_o is the initial normalized photon number. We numerically solved eq. (15') under the conditions $zR_o^2/(1-R_o^2)^2 \ll 1$, $1-\bar{E} \gg (z/4)^2\sqrt{\bar{E}}$. By averaging over the trajectories that start from all the points on the circle of radius R_o (which by (18) amounts to averaging with respect to the initial position z_o) we have calculated the spread and verified expression (18) to an excellent approximation. The same has been done for the gain.

2) Let us assume that $z \ll 1$ but

$$zR/\bar{E} \gg 1 \tag{23}$$

This is the saturated regime, which occurs when R_o is very near to unity. For the sake of definiteness, let us consider the case $R_o=1$ (infinite saturation), in which the electron starts with zero momentum. Hence we have $\bar{E} = zy_2{}^{(0)}$, where $y_2{}^{(0)}$ is the initial value of y_2, so that from (19) we obtain

$$R^2 = 1 \pm \left[z\ (y_2^{(o)} - y_2)\right]^{\frac{1}{2}} \tag{24}$$

Hence for z small enough, the energy curve lies very close to the circle R=1. In this case, it is connected because the argument of the square root can become zero (for $y_2=y_2^{(0)}$). The electron momentum is positive when the point moves to the right and negative when it moves to the left (Fig. 1). This behaviour is similar to that of a pendulum oscillating forth and back. The pendulum behaviour is exactly recovered in the limit z 0, in which the tratory approaches the curve $R=1, y_2 \leq y_2(o)$ (cfr. (24)). Note this limit the frequency Ω of the pendulum tends to vanish, because $\Omega \propto z$ (cfr. (11) and (16)). Also the velocity of the pendulum tends to vanish, because R=1 means $\bar{p}=0$ (see ref. (18)).

3) When z is not small enough, the pendulum picture fails. Let us consider in detail the case $R_o=0$, in which the trajectory starts from the origin of the phase plane. In this case, the initial stage of the radiation emission arises from spontaneous emission (i.e. from the wiggler field α_W), as one sees by setting $\alpha=0$ in the r.h.s. of eq. (15'). Note that this type of spontaneous emission which arises from noise.

For $R_o=0$ one has $\bar{E}=1$ (see (19)) and N coincides with the initial electron momentum $p_{z,o}/2\hbar k_o$. When z is very small, the energy curve for $\bar{E}=1$ is disconnected and consists of the circles (21) (case a of Fig. 2). The outer circle has a radius $R=\sqrt{2}$ and is practically centered on the origin. Here the behaviour is of the type described in case 1): small signal amplifier, in which the stimulated emission dominates the spontaneous one. The inner circle gives the trajectory which starts at $R_o=0$. It has a small radius, equal to $z/4$. Since this circle is strongly asymmetrical with respect to the origin, here the pendulum behaviour is already lost. The spontaneous emission dominates, but the value of the emitted intensity always remains small ($R \leqslant z/2$).

By increasing z, the two circles get somewhat deformed. The center of the outer pseudo-circle moves downward along the y_2 axes, while the center of the inner one moves upward. In correspondence to the threshold value z_T, the north poles of the two pseudo-circles coalesce and the energy curve becomes connected (Fig. 2.b). In order to determine the value of z_T, we consider the intersections of the line $\phi = \pi/2$ with the energy curve for $\bar{E} = 1$. Using (19), we obtain the equation

$$R\,(R^3 - 2\,R + z) = 0 \qquad (25)$$

Eq. (25) has always an unphysical negative solution. The solution $z=z_T$ occurs when the positive solutions of the cubic equation coincide. This condition gives the value

$$z_T = \sqrt{32/27} \simeq 1 \qquad (26)$$

For $z > z_T$, the energy curve remains connected (Fig. 2.c). Starting from $R_o = 0$, we find first a spontaneous emission regime followed by a stimulated emission regime, thereby producing a large gain. Hence, for $z > z_T$ the system behaves as a large gain amplifier. Note that the electron momentum changes sign when the trajectory crosses the circle R=1. Furthermore in this situation one can no longer distinguish between the unsaturated and saturated regimes.

Thus one finds a kind of phase transition for $z=z_T$, in which both the structure of the trajectory and the physical behevaiour change dramatically. This interpretation of z_T as the threshold value of the FEL as a strong signal amplifier is further supported by the fact that approaching the value $z=z_T$ one finds a critical slowing down. In fact, the period of the trajectory becomes longer and longer and diverges for $z=z_T$. This effect is shown in Fig. 3. Actually if one considers the autonomous system for y_1 (T), y_2 (T) which is equivalent to eq. (15'), the point on the trajectory for $z=z_T$ with coordinates ($y_1=0$, y_2 0) turns out to be a critical point.

Some critical remarks are in order before ending this section.

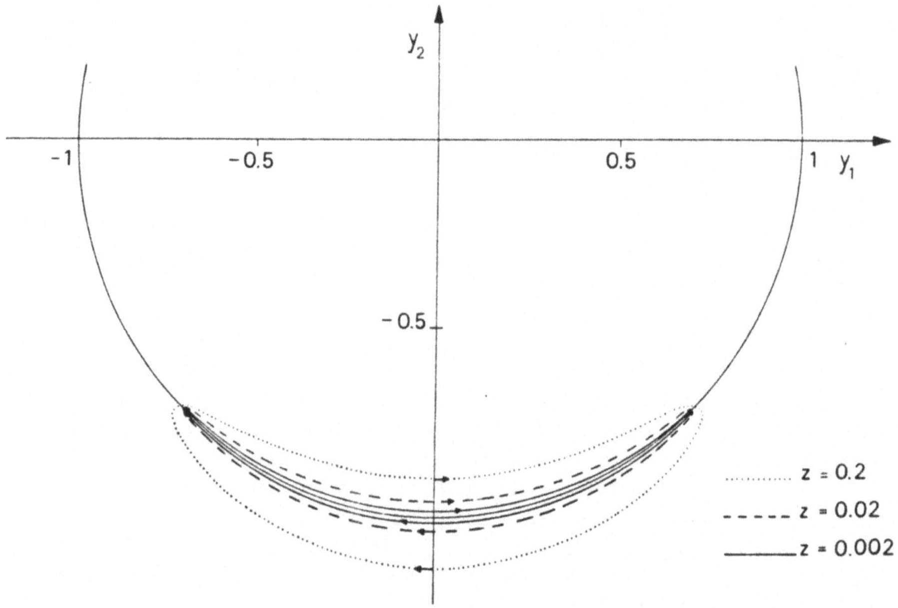

Fig. 1 - Pendulum-like trajectories in the phase plane y_1, y_2 for infinite initial saturation $R_0=1$. By decreasing the value of the parameter z, the trajectories get closer and closer to the circle R=1 and their period T becomes increasingly longer. One finds T=20.3 when z=0.2 (dotted line), T=49.4 and z=0.02 (dashed line) and T=206.6 in the case of z=0.002 (full line).

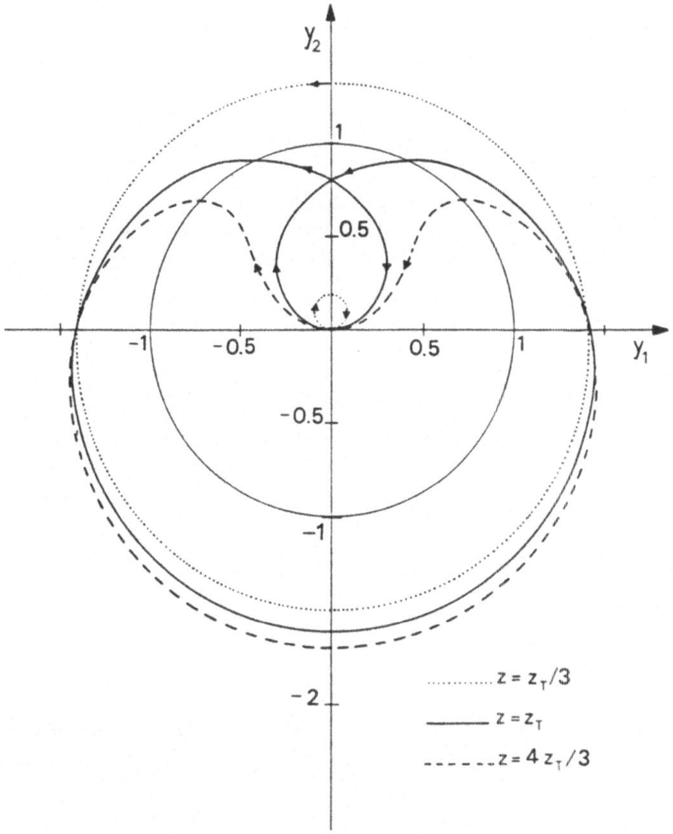

Fig. 2 - Exact trajectories.

a) Below threshold (dotted line). The energy curve is disconnected: the smaller (larger) nearly circular orbit is inside (outside) the circle R=1, where the electron momentum is positive (negative).

b) At threshold (full line): the two orbits coalesce at their north poles.

c) Above threshold: the momentum changes sign crossing the circle R=1. The field builds up from vacuum in all cases except the outer pseudo-circle of a).

a) Ours is a one-particle theory, i.e. we neglect all cooperative effects in the system of electrons, which dominate in the so-called collective or Raman regime of the FEL. In fact, the Stanford FEL [2] uses an electron beam with a low current density, whereas in the Columbia-NRL FEL [11] the beam has a much higher current density exhibiting collective oscillations to be described in the framework of plasma physics [12]. Now in the small gain regime the one-particle approximation is manifestly correct, because the laser field intensity remains practically constant in time, so that the electrons behave as independent pendula with the same fixed frequency. However, in the strong signal regime that we have found for $z>z_T$, the laser intensity changes dramatically and one cannot exclude that the changes in the field induced by one electron can influence the emission of the other electrons. Hence we must take care in comparing our results with experiment. We are presently investigating the role of cooperative effects in our oscillator regime.

b) Since the $R_0=0$ the electron field starts from the vacuum state, one may ask whether the noise fluctuations can dramatically change our picture or not. We have studied this problem by drawing the trajectories that start near the origin of the phase plane. These trajectories turn out to change in a negligible way, especially in correspondence to the south pole of the curves where one has the maximum photon number. Hence the initial noise fluctuations play a minor role in our context, contrary to what happens in the usual laser.

4. COHERENT STATES

Now let us come back to the quantized Hamiltonian (4) in order to define the coherent states and discuss the classical limit of the FEL [10]. It has been noticed [13] that using the constant of motion (5) this Hamiltonian can be written in terms of a single angular momentum operator, defined as

$$J^+ = a_L^\dagger a_W e^{-2ikoz}, \quad J_z = \frac{1}{2} (a_L^\dagger a_L - a_W^\dagger a_W) \tag{27a}$$

The angular momentum commutation relations can be easily verified using (2). In this way H can be written as

$$H = \hbar\omega\,(P - J_z)^2 + i\hbar g\,(J^+ - J^-) \tag{27b}$$

For our purpose it is convenient to define the harmonic oscillator operator

$$A = a_L e^{2ikoz} \tag{28a}$$

Note that the interaction variable (13) is the classical counterpart of (28a). Furthermore,

$$A, A^{\dagger} = 1, \quad A^{\dagger}A = a_L^{\dagger}a_L \tag{28b}$$

Hence, using (5'), (6) and (7), the Hamiltonian (4) can be written as

$$H = \hbar\omega \, (N-A^{\dagger} A)^2 + i\hbar \, g \, (a_W A - h.c.)$$

$$= \hbar\omega p^2 + i\hbar g \, (a_W A^2 - h.c.) \tag{29a}$$

where

$$N - A^{\dagger}A = p = p_z/2\hbar k_o \tag{29b}$$

is the normalized momentum of the electron.

This is the Hamiltonian of two coupled harmonic oscillators with a quadratic nonlinearity for one of them. Hence the FEL dynamics can be discussed in terms of the coherent states $|\alpha_W,\alpha\rangle$ (see refs. 14a,1) defined as the simultaneous eigenstates of a_W and A:

$$a_W|\alpha_W,\alpha\rangle = \alpha_W|\alpha_W,\alpha\rangle, \quad A|\alpha_W,\alpha\rangle = \alpha|\alpha_W,\alpha\rangle. \tag{30}$$

Using the A-operator, the angular momentum (27a) can be rewritten à la Schwinger [15] as $J+ = A^{\dagger}a_W$ and $J_z = \frac{1}{2}(A^{\dagger}A - a_W^{\dagger}a_W)$. Hence the $|\alpha_W,\alpha\rangle$ states are angular momentum coherent states as defined and discussed in ref. 16.

Let us now concentrate on the coherent states $|\alpha\rangle$ of the operator A defined by (28), giving their explicit expression in the electron-field Hilbert space. The field Hilbert space is spanned by the eigenstates $|n\rangle$ of $a_L^{\dagger}a_L$

$$a_L^{\dagger}a_L \, |n\rangle = n|n\rangle, \qquad n=0,1,... \tag{31}$$

Furthermore, thanks to the constant of motion (5') the electron momentum Hilbert space can be restricted to the followint accessible states

$$p|m\rangle = m|m\rangle, \quad \langle m|m'\rangle = \delta_{m,m'} \tag{32}$$

where m+n = N.

In other words, we can assume that the state of the system belongs at all times to the reduced Hilbert space spanned by $|n,m\rangle = |n, N-n\rangle$ provided this is so at t=0. Note that the $|m\rangle$ states correspond to wave functions $|m> = (1/L)e^{2imk_oz}$ which satisfy to periodic boundary conditions on $L=n\ \pi/k_o$. In this way the electron-field states $|n,m>$ can be labeled only by $||n>$ defined as:

$$||n> \equiv |n, N-n> \tag{33}$$

Note that since $e^{\pm 2ik_oz}|m> = |m \pm 1>$ we have

$$A^{\dagger}||n> = n+1\ ||n+1>, \quad A\ ||n> = n\ ||n-1>,$$

$$A^{\dagger}A\ ||n> = n\ ||n> \tag{34}$$

We can now easily define the <u>electron-field coherent states</u> as

$$||\alpha > e^{-|\alpha|^2/2} \sum_{n=0}^{\infty} \frac{\alpha^n}{n!} ||n> \tag{35}$$

Thanks to eqs. (34) it is immediately verified that they satisfy the equivalence relation $A||\alpha> = \alpha||\alpha>$.

In these states both the electron momentum and the laser photon number are well defined since they obey Poisson statistics:

$$P(n) = e^{-|\alpha|^2}|\alpha|^{2n}/n! \tag{36}$$

$$P(m) = e^{-|\alpha|^2}|\alpha|^{2(N-m)}/N-m)!,$$

with

$$<n> = <a_L^{\dagger}\, a_L> = |\alpha|^2, \quad <m> = <p> = N - |\alpha|^2, \tag{37a}$$

and

$$<\delta n^2> = <\delta p^2> = <n> = |\alpha|^2, \tag{37b}$$

where $\delta n = A^{\dagger}A - <n>$, $\delta p = p - <p>$. Note that by definition $<\alpha||a_L||\alpha> = <\alpha||e^{\pm 2ik_oz}||\alpha> = 0$, whereas $<\alpha||A||\alpha> = <\alpha||a_L e^{2ik_oz}||\alpha> = \alpha$. Hence these states are not, strictly speaking, phase coherent states for the field and the electron separately, but represent correlated electron-field states in which only the relative phase is well defined and the photon number and electron momentum obey a Poisson distribution with the same fluctuation.

Furthermore, we note that in general

$$<\alpha||(A^{\dagger})^m A^n||\alpha> = |\alpha|^{2n}\,\delta_{m,n} = <\alpha||(a^{\dagger})^m a^n||\alpha>. \tag{38}$$

Hence these coherent states have the well known factorization property for all the normally ordered products of equal numbers of creation and annihilation operators.

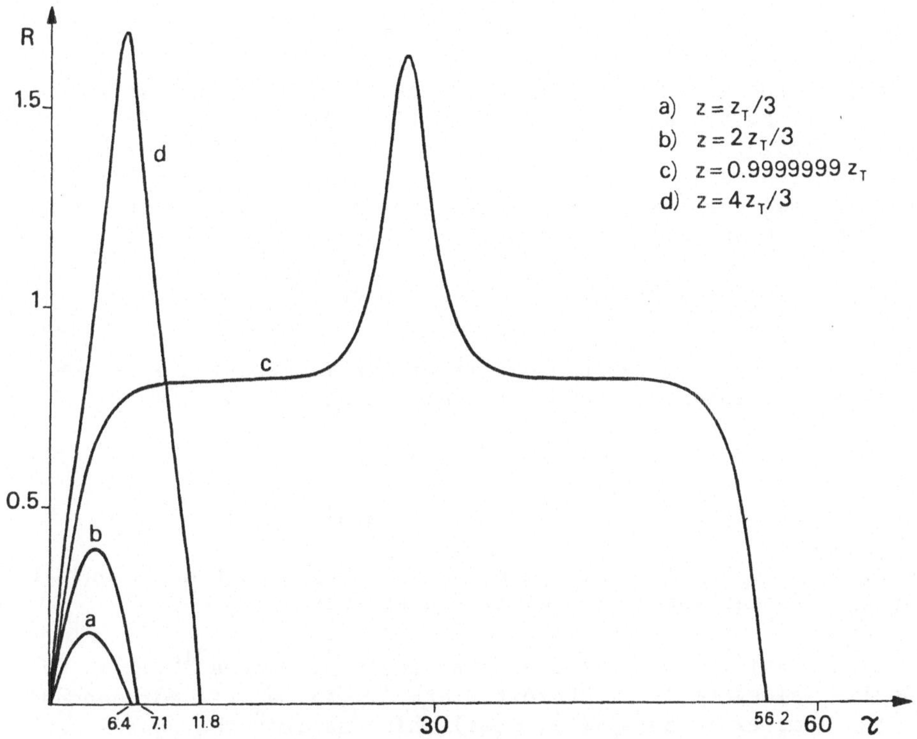

Fig. 3 - Time evolution of the field modulus on one period for different values of z, showing a critical slowing for $z \to z_T$, $R_o=0$, $z_T = 1.088662$.

Hamiltonian (29) does not preserve coherent states since the nonlinear term introduce fluctuations. However, let us define the classical limit of the FEL so that the quantum fluctuations of the photon number nd of the electron momentum are negligible; i.e., we assume.

$$<(\delta n)^2> \quad <n>^2 \;, \quad <(\delta p)^2> \quad <p>^2 \tag{39}$$

Thanks to eqs. (37) we see that inequalities (18) are verified in a coherent state provided that

$$<p>^2 \equiv (N-|\alpha|^2)^2 \quad |\alpha|^2 \quad 1. \tag{40}$$

Note that the condition $<p>^2$ 1 is the classical limit defined e.g. in refs. 5,6 assuming that the field is classical from the very beginning. However, inequality (40a) shows that for $N>>1$ momentum fluctuations are always relevant in the region $|<p>| \lesssim N$. In the classical limit (39) we can approximate the operator p^2 in the Hamiltonian (28) as follows.

$$\begin{aligned} p^2 &= (<p> + \delta p)^2 \simeq <p>^2 + 2 <p>\delta p \\ &= 2 <p> \; p - <p>^2 , \end{aligned} \tag{40b}$$

where we have neglected the square of the fluctuation operator $\delta p = p - <p>$. Inserting (40b) in Hamiltonian (29) we have

$$H = 2\hbar\omega <p> p + i\hbar g \; (a_W A^{\dagger} - \text{h.c.}) \tag{41}$$

$$= 2\hbar\omega \; (<A^{\dagger}A> - N) \; A^{\dagger}A + i\hbar g \; (a_W A^{\dagger} - \text{h.c.})$$

where we have neglected c-numbers since they do not contribute to the equation of motion and we have used (29b).

The Hamiltonian (41) preserves coherent states [14], i.e. if the initial state is a coherent state $|\alpha (t), \alpha_W(t)>$ corresponding to the complex numbers $\alpha(t)$, $\alpha_W(t)$ which satisfy the equations

$$\dot{\alpha} \; (t) = -2i \; \omega(|\alpha|^2 - N)\alpha + g\alpha_W \tag{42a}$$

$$\alpha_W(t) = -g \; \alpha(t) \tag{42b}$$

which must be solved with the initial condition $\alpha(0)=\alpha$ and $\alpha_W(0) = \alpha_W$. If the wiggler field is constant in time, eqs. (42) reduce to eq. (15) which was obtained from the system (6')-(8'), i.e. replacing all operators with c-numbers in the Heisenberg equations (6)-(8).

In conclusion, in the classical limit (39) the FEL quantum state is a coherent state corresponding to a complex time dependent eigenvalue which evolves according to the classical trajectory.

The validity of the semiclassical approximation (400b) and the effect of quantum fluactuations will be discussed elsewhere.

REFERENCES

1. L.R. Elias, W.M. Fairbank, J.M.J. Madey, H.A. Schwettman and T.I. Smith, "Observation of stimulated emission of radiation by relativistic electrons in a spatially periodic transverse magnetic field", Phys. Rev. Lett. 36, 717 (1976);
2. D.A.G. Deacon, L.R. Elias, J.M.J. Madey, G.J. Ramian, H.A. Schwettman and T.I. Smith, "First operation of a free-electron laser", Phys. Rev. Lett. 38, 892 (1977);
3. F.A. Hopf, P. Meystre, M.O. Scully and W.H. Louisell, "Classical theory of a free-electron laser", Opt. Comm. 18, 413 (1976); "Strong-signal theory of a free-electron laser", Phys. Rev. Lett. 37, 1342 (1976); H. Al Abawi, F.A. Hopf and P. Meystre, "Electron-dynamics in a free-electron laser", Phys. Rev. A 16, 666 (1977); H. Al Abawi, F.A. Hopf, G.T. Moore and M.O. Scully, "Coherent transients in the free-electron laser: laser lethargy and coherence brightening", Opt. Comm. 30, 235 (1979); G.T. Moore and M.O. Scully, "Coherent dynamics of a free electron laser with arbitrary magnet geometry. I-General formalism", Phys. Rev. A 21, 2000 (1980; R. Bonifacio, P. Meystre, G.T. Moore and M.O. Scully, "Coherent dynamics of a free electron laser with arbitrary magnet geometry. II. Conservation laws, small-signal theory, and gain=spread relations", Phys. Rev. A 21, 2009 (1980).
4. W.B. Colson, "One-body electron dynamics in a free electron laser", Phys. Lett. A 64, 190 (1977); W.H. Louisell, J.F. Lam, D.A. Copeland and W.B. Colson, "Exact classical electron dynamics approach for a free-electron laser amplifier", Phys. Rev. A 19, 288 (1979); W.B. Colson, contribution to this volume.
5. A. Bambini and A. Renieri, "The free electron laser: a single-particle classical model", Lett. Nuovo Cimento 21, 399 (1978)
6. A. Bambini, A. Renieri and S. Stenholm, "Classical theory of a free-electron laser in a moving frame", Phys. Rev. A 19, 2013 (1979)
7. A. Bambini and S. Stenholm, "Quantum description of free electrons in a laser", Opt. Comm. 30, 391 (1979)
8. R. Bonifacio and M.O. Scully, "Generalized gain-spread expressions for the free electron laser", Opt. Comm. 32, 291 (1980)
9. R. Bonifacio, F. Casagrande and L.A. Lugiato, "Exact one-particle theory of a free-electron laser", Opt. Comm. (in press)
10. R. Bonifacio, "Coherent states of a free electron laser", Opt. Comm. 32, 440 (1980)

11. D.B. McDermott, T.C. Marshall, S.P. Schlesinger, R.K. Parker and V.L. Granatstein, "High-power free-electron laser based on stimulated Raman back scattering", Phys. Rev. Lett. 41, 1368 (1978)
12. P. Spangle and R.A. Smith, "Theory of free-electron lasers", Phys. Rev. A 21, 302 (1980), P. Sprangle, Cha-Mei Tang and W.M. Manheimer, "Nonlinear theory of free-electron lasers and efficiency enhancement" Phys. Rev. A 21, 302 (1980), and referenced quoted therein.
13. G. Dattoli, "A block-like model of the free electron laser", Lett. Nuovo Cimento 27, 247 (1980)
14. a) R.J. Glauber, "The quantum theory of optical coherence", Phys. Rev. 130, 2529 (1963); "Coherent and incoherent states of the radiation field" Phys. Rev. 131, 2766 (1963); b) The relevance of these angular momentum coherent states and of Block states has been discussed by G. Dattoli, A. Renieri, F. Romanelli and R. Bonifacio, "On the coherent states of the free electron laser", Opt. Comm. 34, 240 (1980).
15. J. Schwinger, in "Quantum theory of angular momentum; perspectives in physics", eds. L.C. Biedenharm and H. Van Dam (Academic Press, New York, 1965).
16. R. Bonifacio, Dae M. Kim and M.O. Scully, "Description of a many-atom system in terms of coherent boson states", Phys. Rev. 187, 441 (1969)

ON THE THEORY OF THE FREE ELECTRON LASER

Giuseppe Dattoli, Angelo Marino and Alberto Renieri

Comitato Nazionale Energia Nucleare
Centro di Frascati
C.P. 65
00044 Frascati, Rome, Italy

1. INTRODUCTION AND HISTORICAL REMARKS

The Free Electron Laser (FEL) differs from conventional lasers in having a gain medium consisting of free electrons.

One way to achieve laser action in a FEL is depicted in Fig. 1. A bunch of electrons (e.b.) enters with velocity v a region where a periodic wiggler magnetic field of length L and spatial period λ_q exists. A laser beam (l.b.) injected simultaneously with the e.b. becomes amplified due to the "induced-emission".

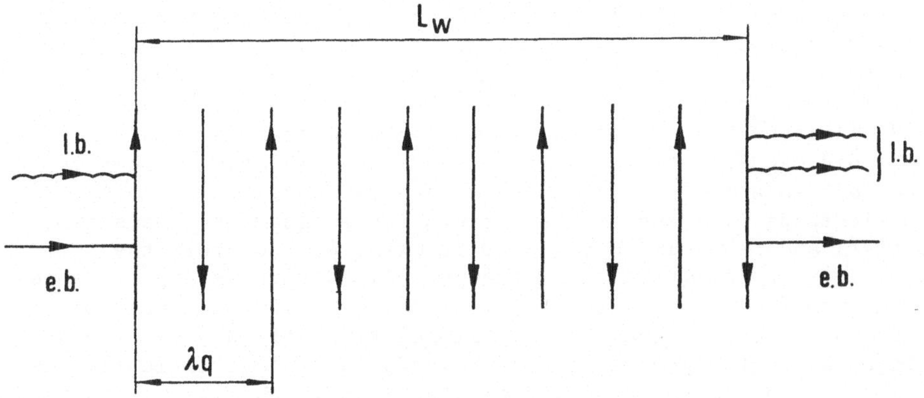

Fig. 1. Schematical FEL layout

The only common expression up to now using conventional laser language is that of "induced emission", but such an expression is not strictly correct since the effect can be understood as a "stimulated-Compton-Scattering" (SCS), i.e. scattering of photons in the presence of radiation.

The idea of an SCS is not a new one, since in 1927 Schrödinger[1] gave a simple explanation of the process and in 1933 Kapitza and Dirac[2] proposed experiments to observe SCS from standing light waves. Later, Motz and Nakamura[3] pointed out that free relativistic electrons, interacting classically with a static periodic electric or magnetic field in a long wave guide, provide a useful source of coherent radiation. Some time after (1971) renewed interest in the argument was aroused by Madey[4], who unified the previous concepts by means of the Weiszäcker-Williams approximation, and obtained a gain formula for the laser process. In the Weiszäcker-Williams approximation, free relativistic electrons interact with virtual quanta of a static periodic magnetic field, and scatter them into real photons of the stimulating field.

After the first sucessful operation of the FEL,[5] much theoretical work was presented to explain the process. In a number of papers, Hopf, Meystre, Scully, Louisell and Al Abawi[6] initialed a multiparticle analysis of the FEL by writing down a classical Vlasov equation by means of which the basic properties of the FEL operation, such as gain and saturation, have been evaluated. Later on, a single particle analysis has been developed; in fact, Baier and Milstein[7] have analyzed the motion of electrons in a static, periodic magnetic field in the presence of radiation to derive the gain equation. Colson[8] has solved the equation of motion of he electrons by means of the Lorentz force equation. Bambini, Renieri and Stenholm[9] have found it expedient to treat the problem in a moving frame where the periodic structure appears as a radiation field, or better a Pseudo Radiation Field (PRF), of frequency equal to that of the "stimulating" field. In such a frame, where the problem can be treated non relativistically, the laser process can be understood as an induced Thomson scattering, resonant by definition, of PRF photons into laser photons.

All of the above quoted papers begin with completely classical hypotheses, i.e., the laser field was assumed to be strong enough to allow a classical treatment. A quantum analysis of the problem has been undertaken, in the framework of a single particle theory in the refs. 10,11. One remarkable result of such an analysis is that it is possible to demonstrate, by means of the density matrix formalism an equivalence between the multiparticle and single particle analysis.[10]

In later work[12] considerations involving the invariance properties of the FEL quantum single mode Hamiltonian have shown that by suitably collecting fields and electron variables, one can express the interaction in terms of angular momentum-like operators; furthermore such an analysis has revealed itself useful in discussing the coherence properties of the emitted radiation.[13,14]

A further development in FEL theory has been recently given by the multimode analysis, motivated by the fact that the laser beam has a multimode structure which follows the electron bunches.[15] Such an analysis has been developed in a number of papers. The single particle approach in a moving frame has been generalized to the multimode case in ref. 16 where both the small and the strong signal regime have been studied with some detail.

Later Bambini, Bonifacio and Stenholm have attacked the problem in ref. 17, always from the viewpoint of the single particle picture, by means of the electromagnetic pulse propagation techniques. Finally, in the framework of a collective picture, Al Abawi, Hopf, Moore and Scully,[18] have studied the propagation of optical pulses.

The above quoted papers, and the references therin, cover the main results of FEL theory up to now. It is the aim of this set of papers to give an account of FEL theory as far as the single particle approach in a moving frame is concerned, in the classical multimode regime. The plan of the lectures is as follows: in section 2 we discuss the main questions connected with FEL multimode analysis; in section 3 we discuss the "interaction" variables, their possible interpretation in terms of the "Bloch-vector" and their time equations of motion; section 4 is devoted to the solution of the "interaction-variables" time equations in the small and strong-signal regimes and section 5 is devoted to concluding remarks. Finally, in Appendix A, we discuss the connection between the moving frame and laboratory reference frame.

2. MULTIMODE FEL THEORY - GENERAL CONSIDERATIONS

It has been already stressed in the introduction that the FEL theory can be formulated in a very simple way in a moving frame, where the problem can be treated nonrelativistically.[9] The main problem connected with the approach of ref 9, as well as with all the first FEL analysis, was the assumption of only one laser mode. It was first observed in ref. 15 that the assumption of only one laser mode is too simplistic since the laser beam has a bunched structure which follows the electron beam bunches.

The extension of the single mode theory in a moving frame to the multimode case was accomplished by two of the present authors in ref 16, with the hypothesis that each laser mode was strong enough to allow a classical description; in other words, the theory was formulated far from the threshold where quantum fluctuations are expected to be important. The laser modes are assumed to be circularly polarized modes of the optical cavity, each with wavelength λ_i $(i = 1, \ldots, n)$.

The wiggler field is provided by a static magnetic field, with spatial period $\lambda_q = 2\pi k_q^{-1}$, and is assumed to be circularly polarized with components†

$$B \equiv \left[B_o \cos(k_q Z),\ B_o \sin(k_q Z),\ 0\right]. \qquad (2.1)$$

The particles of an ultrarelativistic e.b. traverse the wiggler with an axial velocity very close to the light velocity c. The wiggler field as "seen" by the electron possesses an electric as well as a magnetic component, whose ratio is close to 1, and this allows us to treate the wiggler as a radiation field propagating towards the e.b.

As a final point, let us spend a few words on the choice of the moving frame. In reference 9 it was quite natrual to choose the frame where the scattering of wiggler photons into laser ones was "resonant" by definition, i.e., the reference frame where both the laser and wiggler field appeared to the electron as electromagnetic waves moving oppositely along the longitudinal axis with the same frequency. In the multimode case the choice of resonant scattering is meaningless; however, it should be stressed that the choice of the equal frequency frame is only matter of convenience and the analysis can be equivalently carried out in any frame related to the equal frequency one by a Galilean transformation (for further details see ref. 19). For the multimode case, one has to choose a frame where the problem can be treated nonrelativistically.

3. HAMILTONIAN PICTURE

In the reference frame chosen above the problem has been assumed to be nonrelativistic. The longitudinal velocity of the electron in this frame is much smaller than c. The transverse electeon motion, however, owing to the large wiggler field, can become, in principle, relativistic. This effect can be taken into account in the present nonrelativistic formulation by means of a suitable redefinition of the electron mass, which enters the equations of motion (for further comments see the second of refs 9).

The interaction of the e.b. with the laser and wiggler field will be accounted for by the hypothesis that the process is essen-

† It is to be noticed that the assumption of a circularly polarized wiggler is only a matter of convenience; the analysis could instead be carried out for a linearly polarized wiggler.

tially a single particle one. We can easily find an upper limit to the electron density for a single particle analysis of the FEL theory. The basic assumption for a single particle analysis will be that the "plasma-frequency" of the e.b. is much less than the laser frequency we are considering, i.e.,

$$\omega_P = \left[4\pi\overline{N}_e \frac{e^2}{m}\right]^{1/2} \ll \omega; \tag{3.1}$$

where $\overline{N}_e$ is the electron density. From (3.1) one obtains

$$\overline{N}_e \ll \frac{1}{4\pi}\frac{1}{\lambda^2 r_o} \quad (r_o \equiv \text{classical electron radius}) \tag{3.2}$$

which for our purposes is generally valid.

Once (3.2) is satisfied we describe the FEL interaction in the Coulomb gauge, by means of a transverse vector potential $A_\perp$ alone (the scalar potential Φ is neglected since we are in the framework of a single particle theory).

In the moving frame we are considering, we choose to describe the laser field and the wiggler field by their decomposition in terms of harmonic oscillator coordinates$\{\vec{P}, \vec{Q}\}$; thus we write the following nonrelativistic classical single-particle Hamiltonian:

$$H = \frac{1}{2m}\left[\vec{p} - \frac{e}{c}\vec{A}_\perp\right]^2 + \frac{1}{2}\sum_i\left(\vec{P}_i^2 + \omega_i^2\vec{Q}_i^2\right) + \frac{1}{2}\left(\vec{P}_W^2 + \omega_W^2\vec{Q}_W^2\right); \tag{3.3}$$

where $\vec{p}$ is the electron canonical momentum, the summation i is extended to all the n laser modes and W labels the wiggler variables.

The vector potential $\vec{A}_\perp$, expressed in terms of the canonical variables $(\vec{P}, \vec{Q})$, is given by

$$\vec{A}_\perp = \left(\frac{4\pi}{V}\right)^{1/2}\left\{\sum_i \frac{1}{K_i}\left[\omega_i\vec{Q}_i\cos(k_iZ) - \vec{P}_i\sin(k_iZ)\right] + \frac{1}{K_W}\left[\omega_W\vec{Q}_W\cos(k_WZ) + \vec{P}_W\sin(k_WZ)\right]\right\}, \quad V \equiv \text{mode volume} \tag{3.4}$$

where $\left|k_{i(W)}\right|$ are the moduli of the laser (wiggler) wave-numbers (recall that in this frame the laser and wiggler modes appear as oppositely running waves). Assuming circularly polarized fields one has

$$\begin{cases} \vec{P} \equiv \left(\frac{\tilde{P}}{\sqrt{2}}, -\omega \frac{\tilde{Q}}{\sqrt{2}}, 0\right) \\ \vec{Q} \equiv \left(\frac{Q}{\sqrt{2}}, +\frac{1}{\omega} \frac{P}{\sqrt{2}}, 0\right). \end{cases} \tag{3.5}$$

Now we could, in principle, write down the Hamiltonian equations of motion for the electrons and fields in terms of the harmonic canonical coordinates. However, we can achieve a deeper insight in the problem we are treating, by using a simple and physically significant integral of motion in terms of the field action I and phase φ. We can pass from the $(\tilde{P}, \tilde{Q})$ variables to (I, φ) by means of the following contact transformation

$$F = \sum_i \frac{\omega_i}{2} \tilde{Q}_i^2 \operatorname{ctg} \varphi_i + \frac{\omega_w}{2} \tilde{Q}_w^2 \operatorname{ctg} \varphi_w. \tag{3.6}$$

Making use of the standard transformation formulas relating the old to the new coordinates, we write

$$\begin{cases} \tilde{P} = \frac{\partial F}{\partial \tilde{Q}} = \omega \tilde{Q} \operatorname{ctg} \varphi \\ I = -\frac{\partial F}{\partial \varphi} = \frac{\omega}{2} \tilde{Q}^2 \frac{1}{\sin^2 \varphi}. \end{cases} \tag{3.7}$$

Solving for the old coordinates we find

$$\begin{cases} \tilde{P} = (2I\omega)^{1/2} \cos \varphi \\ \tilde{Q} = (2I/\omega)^{1/2} \sin \varphi. \end{cases} \tag{3.8}$$

From the above relations it follows that the Hamiltonian (3.3), written in terms of the new coordinates, can be written as

$$H = \frac{P^2}{2m} + \left(\Omega_w + \omega_w\right) I_w + \sum_i \left(\Omega_i + \omega_i\right) I_i +$$

$$+ 2\sum_i \left(I_w I_i\right)^{1/2} \left(\Omega_w \Omega_i\right)^{1/2} \cos \chi_i^w + \tag{3.9}$$

$$+ 2\sum_{\substack{i,j \\ i<j}} \left(\Omega_i \Omega_j\right)^{1/2} \left(I_i I_j\right)^{1/2} \cos \chi_j^i,$$

where

$$\begin{cases} P = P_Z \equiv \text{longitudinal electron momentum} \\ \Omega_{i(w)} = \dfrac{2\pi c^2 r_o}{V\, \omega_{i(w)}} \\ \chi_i^w = \varphi_w - \varphi_i + \left(k_i + k_w\right) Z \\ \chi_j^i = \chi_j^w - \chi_i^w \end{cases} \tag{3.10}$$

It is to be noted that since the Hamiltonian (3.9) is independent of (x,y), the conjugate momenta P_x, P_y are constants of motion and can be assumed to be zero at all times.

The Hamiltonian (3.9) has a very transparent physical meaning; the first three terms(apart the very small shift factor $\Omega_{i(w)} \ll \omega_{i(w)}$ † which merely redefines the mode frequencies and will be omitted in the following) stand for the electron and free field energy, while the last two terms account for the Laser-Wiggler (L-W) and Laser-Laser (L-L) parts of the interaction.

It is now straightforward to write down the Hamilton equations for the electrons and fields, in terms of the new canonical coordinates. From (3.9) we obtain

---

† It can be shown that the term Ω is due to the electron self energy; futhermore, it is the term which accounts for "spontaneous" scattering in the FEL process. Indeed, it can be shown that the protrons spontaneously emmitted into the laser mode are given by

$$n_L \approx n_w(0)(\Omega\Delta t)^2 \cong n_w(0)/V\ (L_w r_o^2)\ L_w \lambda^2/V.$$

$$\overset{\circ}{z} = \frac{p}{m}$$

$$\overset{\circ}{p} = -\frac{\partial H}{\partial z} = 2\sum_j \left(K_j + K_w\right)\left(I_w\,\Omega_w\right)^{1/2}\left(I_j\,\Omega_j\right)^{1/2} \sin\chi_j^w +$$

$$+ 2 \sum_{\substack{j,i \\ j<i}} \left(K_j - K_i\right)\left(I_i\,\Omega_i\right)^{1/2}\left(I_j\,\Omega_j\right)^{1/2} \sin\chi_j^i$$

$$\overset{\circ}{I}_w = -\frac{\partial H}{\partial \varphi_w} = 2\sum_i \left(\Omega_j\, I_j\right)^{1/2}\left(\Omega_w\, I_w\right)^{1/2} \sin\chi_i^w$$

$$\overset{\circ}{I}_i = -\frac{\partial H}{\partial \varphi_i} = -2\left(\Omega_i\, I_i\right)^{1/2}\left(\Omega_w\, I_w\right)^{1/2} \sin\chi_i^w + \tag{3.11}$$

$$+ 2\sum_{j\neq i} \left(\Omega_i\, I_i\right)^{1/2}\left(\Omega_j\, I_j\right)^{1/2} \sin\chi_j^i$$

$$\overset{\circ}{\varphi}_w = \frac{\partial H}{\partial I_w} = \omega_w + \sum_j \left(\frac{I_j}{I_w}\right)^{1/2}\left(\Omega_w\,\Omega_j\right)^{1/2} \cos\chi_j^w$$

$$\overset{\circ}{\varphi}_i = \frac{\partial H}{\partial I_i} = \omega_i + \left(\frac{I_w}{I_i}\right)^{1/2}\left(\Omega_w\,\Omega_i\right)^{1/2} \cos\chi_i^w +$$

$$+ \sum_{j\neq i} \left(\frac{I_j}{I_i}\right)^{1/2}\left(\Omega_j\,\Omega_i\right)^{1/2} \cos\chi_j^i\ .$$

Furthermore from (3.11) one easily derives

$$p - k_w I_w + \sum_i k_i\, I_i = p_o - k_w I_w(0) + \sum_i k_i I_i(0) \tag{3.12}$$

and

$$I_w + \sum_i I_i = I_w(0) + \sum_i I_i(0) \tag{3.13}$$

where p_o, $I_w(0)$, $I_i(0)$ are the initial electron linear momentum, Wiggler and i-th laser action respectively. The equations (3.12) and (3.13) account for the linear momentum and total action conservation respectively, and are the generalization of the conservation relations derived in refs 9. Relation (3.12) shows that diverting energy from the P.R.F. into laser modes requires that the electron provide the necessary momentum.

Finally let us stress that there is a difference in the single laser mode case[9]. In fact, from total action conservation (eq. (3.13)), we cannot deduce free field energy conservation. Indeed, being the classical action linked to the average number of photons, relation (3.13) tells us that wiggler "photons" are converted into laser ones, but the total number of photons is a conserved quantity.

Let us stress that up to now no approximations have been introduced except those inherent to the model. The choice of the new canonical variables, I and φ, is only matter of convenience, and the eqs (3.11) are the rigorous consequence of (3.9). Then, in principle, their detailed integration would provide the exact solution of our problem.

We can now look for some reasonable physical approximations which will allow us to significantly simplify the integration of the equations (3.11), not easily integrable in their complete form.

We note, however, that, as far as the single laser mode is concerned, the analogues of (3.11) can be exactly solved by means of elliptic functions, and that furthermore the knowledge of the electron motion is equivalent to the knowledge of the evolution of the laser action. In the present case, as follows from a first glance from (3.12) and (3.13), the knowledge of the electron motion would give us only partial information on the evolution of a single laser mode. Thus to follow the i-th laser mode we have to pay attention to the action and phase time-history of a single mode.

Let us first notice that the wiggler field is practically unaffected by the FEL interaction, i.e.,$\delta I_W/I_W(0) \ll 1$. This amounts to saying that the wiggler cannot be signigicantly depleted during the interaction †. This assumption allows us to rewrite the fourth of (3.11) as

† This assumption is appropriate to the wiggler field we are interested in, i.e. those provided by static magnetic field. It could be not strictly correct for "wiggler field" provided by real e.m. fields.

$$\overset{\circ}{I}_i = -2\left(\Omega_w I_w(0)\right)^{1/2}\left[\left(\Omega_i I_i\right)^{1/2}\sin\chi_i^w\right] +$$

$$+ 2\sum_{j\neq i}\left\{\left[\left(\Omega_i I_i\right)^{1/2}\cos\chi_i^w\right]\left[\left(\Omega_j I_j\right)^{1/2}\sin\chi_j^w\right] - \right. \tag{3.14}$$

$$\left. - \left[\left(\Omega_j I_j\right)^{1/2}\cos\chi_j^w\right]\left[\left(\Omega_i I_i\right)^{1/2}\sin\chi_i^w\right]\right\}$$

From the above relation we deduce an obvious but important consequence: <u>the time dependence of the i-th laser mode is fully contained in the terms</u>

$$\left(\Omega_r I_r\right)^{1/2}\begin{Bmatrix}\sin\\ \cos\end{Bmatrix}\chi_r^w .$$

Then the time history of the single i-th laser action is fixed once the evolution of these terms is fixed. From the last of (3.11) it follows that the same conclusion can be derived for the phase variation. Thus, we have to look for a suitable set of new variables. From (3.14) it follows that a convenient choice is†

$$\begin{cases} x_i = \left(\dfrac{I_i}{I_i(0)}\right)^{1/2} \sin\chi_i^w \\[2ex] y_i = \left(\dfrac{I_i}{I_i(0)}\right)^{1/2} \cos\chi_i^w \end{cases} \tag{3.15}$$

† It is to be stressed that for a single mode, a more convenient choice than (3.15) is a set of variables(the FEL Bloch variables) which include the wiggler field; in this connection one can find a set of differential equations which can be easily solved by means of elliptic functions.

which, henceforth, we shall call "interaction-variables", because they contain both the fields and the electron variables. As far as the single mode is concerned the analogues of (3.15) (including the wiggler field action too), have been interpreted as the absorptive and dispersive part of the FEL Bloch vector (see refs 12 for further comments).

It is now straightforward to rewrite (3.11) in terms of the "interaction-variables" (3.15), but their explicit derivation will be omitted for the sake of brevity (see ref. 16). Anyway such equations show that the time evolution of all our canonical variables is determined by the time-history of the "interaction-variables". It is easy to show that by taking the derivate of x_i and y_i with respect to the normalised time $\tau = t/\Delta t$ (Δt being the interaction time) and using the motion invariants (3.12) (3.13), the time-evolution can be derived from the following differential equations

$$\begin{cases} x_i' = -\alpha_i^W + y_i \left\{ \eta_i + D_i \sum_j \tilde{\varepsilon}_j \int_0^\tau x_j d\tau' + R\, D_i \sum_{j,r} \dfrac{\tilde{\varepsilon}_j \tilde{\varepsilon}_r}{D_r} \right. \\ \qquad \left. \int_0^\tau \left(x_j y_r - x_r y_j \right) d\tau' \right\} - \sum_{j\neq i} \alpha_i^j y_j \\ y_i' = -x_i \left\{ \eta_i + D_i \sum_j \tilde{\varepsilon}_j \int_0^\tau x_j d\tau' + R\, D_i \sum_{j,r} \dfrac{\tilde{\varepsilon}_j \tilde{\varepsilon}_r}{D_r} \right. \\ \qquad \left. \int_0^\tau \left(x_j y_r - x_r y_j \right) d\tau' \right\} + \sum_{j\neq i} \alpha_i^j x_j \end{cases} \tag{3.16}$$

where

$$\begin{cases} \alpha_i^{w(j)} = \left[\dfrac{I_{w(j)}(0)}{I_i(0)}\right]^{1/2} \left(\Omega_i \Omega_{w(j)}\right)^{1/2} \Delta t, \\ \eta_i = \left[\left(\omega_w - \omega_i\right) + \left(\omega_w + \omega_i\right) \dfrac{p_o}{mc}\right] \Delta t. \\ D_i = \dfrac{1}{2}\left(1 + \dfrac{\omega_i}{\omega_w}\right), \\ \tilde{\varepsilon}_i = D_i \varepsilon_i \left(\omega_w/\omega_1\right)^{3/2}, \\ \varepsilon_i = \left(\Omega_{Ri}\, \Delta t\right)^2, \\ \Omega_{Ri} = \left[\dfrac{16\pi r_o \omega_i}{mV} \left(I_i(0)\, I_w(0)\right)^{1/2}\right]^{1/2} \\ R_i = \dfrac{1}{8} \dfrac{mc^2}{\Omega_w I_w(0)} \left(\omega_w\, \Delta t\right)^{-2}. \end{cases} \tag{3.17}$$

where Ω_{Ri} has been already defined as the "Pseudo-Rabi-Frequency"[9] of the i-th mode, and ε_i ($\tilde{\varepsilon}_i$) playes the role of a "coupling-constant"; it is to be noticed that the L-L terms give rise to "second order" contributions in ε.

The equations (3.16) have a complicated form; In fact, the only approximation performed has been the neglect of the time variation of I_w; we have retained the L-L contributions.

We can further simplify the eqs (3.16) by taking into account that, for to the wiggler field we are interested in (cfr. footnote preceding eq. (3.14), the wiggler-field action is always much larger than the laser one, i.e.,

$$I_w \ll \sum_i I_i; \tag{3.18}$$

thus we can neglect the strength of the L-L coupling with respect to the L-W one. This allows us to write the eqs (3.16) as

$$\begin{cases} x_i' = -\alpha_i^W + y_i \left\{ \eta_i + D_i \sum_j \tilde{\varepsilon}_j \int_0^\tau x_j \, d\tau' \right\} \\ \\ y_i' = -x_i \left\{ \eta_i + D_i \sum_j \tilde{\varepsilon}_j \int_0^\tau x_j \, d\tau' \right\} \end{cases} \quad . \tag{3.19}$$

The equations (3.19) have now a simpler form than the starting ones (3.16).

The term α_i^W in (3.19) has been understood as the term causing spontaneous scattering from the wiggler filed to the laser one, while the term causing stimulated scattering is the coupling constant $\tilde{\varepsilon}_i$. Note that in (3.16) $\alpha_{i,j}$ and $\tilde{\varepsilon}_j$ $\tilde{\varepsilon}_r$ account, respectively, for the spontaneous and stimulated L-L scattering.

Let us now write the variation of the i-th laser action and phase in terms of the "interaction-variables" (x_i, y_j) as

$$\begin{cases} \delta I_i = -G \dfrac{\tilde{\varepsilon}_i}{D_i} \displaystyle\int_0^1 x_i \, d\tau' \\ \\ \delta\varphi_i = \omega_i \, \Delta t + \dfrac{G \tilde{\varepsilon}_i}{2I_i(0) D_i} \displaystyle\int_0^1 \dfrac{y_i}{x_i^2 + y_i^2} \, d\tau \end{cases} \tag{3.20}$$

where

$$G = \frac{mc^2}{4} \frac{1}{\omega_w^2 \Delta t} \quad . \tag{3.21}$$

4. SOLUTIONS

In deriving equations (3.19) we have performed the following approximations, apart those inherent to the model:

i) Wiggler field practically constant,

ii) Summation over all laser actions much smaller than the wiggler one.

The above two simplifying hypotheses have allowed us to obtain a reasonable form for the differential equations we are interested in.

Equations (3.19) now can be divided into two parts--one describing the linear evolution of the i-th "interaction-variable", and one accounting for the non linear coupling with all the modes via the "coupling-constants" $\tilde{\varepsilon}_i$. For very weak coupling, i.e., for vanishing $\tilde{\varepsilon}_i$, equations (3.19) reduce to those of a forced oscillator with the pulling term α_i^W accounting for the spontaneous scattering in the i-th mode (see ref. 16). However, as already noted, because of the relevance of quantum effects spontaneous scattering cannot be treated consistently in the present classical formalism. We overcome this problem by neglecting the term driving the spontaneous scattering with respect to the stimulated one, i.e., by imposing.

$$\alpha_i^W \ll \varepsilon_i \quad . \tag{4.1}$$

From (3.17 it follows that (4.1) gives a lower limit for a full classical treatment of the i-th laser mode. If we express the action by means of average number of photons, i.e. $I = \langle n \rangle \hbar$, assuming $\omega_i \approx \omega_W = \omega$, (4.1) yields

$$\langle n \rangle \gg \frac{\lambdabar^2}{\lambdabar_e L_W} \tag{4.2}$$

where $\lambdabar_e$, $\lambdabar$, L_W are the reduced Compton wave-length, the laser wave-length and the wiggler length respectively.

We have to stress that condition (4.2) does not imply a classical laser field; indeed, the analysis we performed up to now is completely classical. It is, however, to be noticed that from a full quantum analysis[14] it follows that the electron motion can be thought classical when

$$\frac{\lambdabar^2}{\lambdabar_e L_W} \gg 1, \tag{4.3}$$

which is well satisfied in the optical range we are interested in. Furthermore, once (4.3) holds the structure of the field quantum equations of motion reduce to the classical one.

Neglecting the term α_i^W in (4.9) we can distinguish two cases
(a) Weak coupling $\varepsilon_i \ll 1$
(b) Strong coupling $\varepsilon_i \gtrsim 1$,
which correspond to the small and the strong signal regime respectively. In this section we shall deal with case (a). Condition (a) fixes an upper limit on the i-th mode laser action for the small signal treatment. Connecting (3.17) and (a) we obtain the operating conditions for the small signal classical regime, i.e.

$$\left(\frac{\Omega_w}{\Omega_i}\right)^{1/2} \frac{mc^2}{8} \frac{1}{\omega_i^2\,\Delta t} \ll I_i \ll \left(\frac{mc^2}{8}\right)^2 \frac{1}{\omega_i^4} \frac{1}{I_w} \left(\frac{1}{\Delta t}\right)^4 \cdot \frac{1}{\Omega_i^2} \tag{4.4}$$

from which, by requiring the upper limit to be greater than the lower one, we obtain also the following limit for the wiggler field

$$I_w \ll mc^2/8 \left(\Omega_i/\Omega_w\right)^{1/2} \left(1/\Omega_i\Delta t\right) 1/\Omega_i \left(1/\omega_i\Delta t\right)^2 \tag{4.5}$$

(A) Small Signal Regime

The condition of small coupling (A) allows a perturbative approach to the eqs (3.19). By perturbative approach we mean that we are allowed, owing to the smallness of $\widetilde{\varepsilon}_i$, to perform a series expansion of x_i and y_i up to the first order, in terms of the "coupling-constants", i.e.

$$\begin{cases} x_i = x_i^o + \sum_r \widetilde{\varepsilon}_r x_i^r \\ y_i = y_i^o + \sum_r \widetilde{\varepsilon}_r y_i^r \quad . \end{cases} \tag{4.6}$$

Inserting (4.6) in (3.19), neglecting α_i^w and equating the coefficients of like $\widetilde{\varepsilon}$ power, we obtain

$$\begin{cases} \left(x_i^o\right)' = \eta_i\, y_i^o \\ \left(y_i^o\right)' = -\eta_i\, x_i^o \end{cases} \tag{4.7}$$

and

$$\begin{cases} \left(x_i^r\right)' = \eta_i\, y_i^r + D_i\, y_i^o \int_o^\tau x_r^o\, d\tau' \\ \left(y_i^r\right)' = -\eta_i\, x_i^r - D_i\, x_i^o \int_o^\tau x_r^o\, d\tau' \end{cases} \tag{4.8}$$

which represent, respectively, the unperturbed and the first order perturbed part of the "interaction variables".

The initial conditions are

$$\begin{cases} x_i^o(0) = x_i(0) = \sin \chi_i(0) \\ y_i^o(0) = y_i(0) = \cos \chi_i(0) \\ x_i^{r\neq o}(0) = y_i^{r\neq o}(0) = 0 \end{cases} \tag{4.9}$$

The solution of (4.7) and (4.8) are

$$\begin{cases} x_i^{o,r} = \operatorname{Im} Z_i^{o,r} \\ y_i^{o,r} = \operatorname{Re} Z_i^{o,r} \end{cases} \tag{4.10}$$

where

$$\begin{cases} Z_i^o = \exp\left[j\left(\eta_i\tau + \chi_i(0)\right)\right] \\ Z_i^r = \dfrac{D_i}{2\eta_r^2}\left\{\exp\left(j\chi_-\right)\left[\exp\left(j\eta_-\tau\right) - \left(1 - j\eta_r\tau\right)\exp\left(j\eta_i\tau\right)\right] - \right. \\ \left. \qquad - \exp\left(j\chi_+\right)\left[\exp\left(j\eta_+\tau\right) - \left(1 + j\eta_r\tau\right)\exp\left(j\eta_i\tau\right)\right]\right\} \end{cases} \tag{4.11}$$

$$\eta_\pm = \eta_i \pm \eta_r, \; \chi_i = \chi_i \pm \chi_r$$

It is easy to derive the small signal gain formula, and the phase variation from the above relations. In fact from (3.20) we obtain

$$\left\{\begin{aligned} \delta I_i &= -\frac{G\,\tilde{\varepsilon}_i}{D_i}\int_0^1 \left(x_i^o + \sum_r \tilde{\varepsilon}_r\, x_i^r\right) d\tau \\ \delta\varphi_i &= \alpha_i \int_0^1 \left(y_i^o + \sum_r \tilde{\varepsilon}_r\, y_i^r\right) d\tau + \omega_i \Delta t \end{aligned}\right. \tag{4.12}$$

The equations (4.12) five the action and phase variation of the i-th laser mode as a function (among others χ_+ and χ_i, such "phase-contributions" derived respectively from the first order pertrubed solution x_i^r and from the unpertrubed one x_i^o. To calculate the small signal gain of the i-th laser mode due to the whole e.b. we must take into account the distribution of the e.b. along the z-direction and in energy. Assuming a monoenergetic e.b., with normalized distribution function along z f (z_o), varying very slowly in a wave-length, we can average over the "fast-running" phase contribution χ_+ and χ_i and take the smooth one χ_- as fixed. In this connection one can see that the only contributions to the gain which survive are function of χ_-; indeed, those containing χ_i average to zero, as well as those containing χ_+. Thus the total variation of action and phase due to the whole e.b. will be given by

$$\Delta \begin{Bmatrix} I_i \\ \varphi_i \end{Bmatrix} = N \int_{-\infty}^{+\infty} f(z_o) \left< \delta \begin{Bmatrix} I_i \\ \varphi_i \end{Bmatrix} \right>_{\chi_+,\, \chi_i} dz_o \tag{4.13}$$

where N is the number of electrons in the e.b., and

$$\left< \; \right>_{\chi_+,\, \chi_i}$$

means average over the fast running phases. The slowly varying term is given by

$$\begin{aligned} \chi_- &= \varphi_r\left(t_o\right) - \varphi_i\left(t_o\right) + \left(k_i - k_r\right) z_o = \\ &= 2\left(k_i - k_r\right) z_o + \varphi_r(0) - \varphi_i(0). \end{aligned} \tag{4.14}$$

where t_o is the starting time of the interaction related to z_o by $t_o = - z_o/c$; indeed $\varphi(t_o) = \varphi(0) + \omega t_o$. From (4.13), (4.12) and (4.14) with some algebra we obtain

$$\begin{cases} \Delta I_i = - \sum_r A_{i,r} \left(I_i\ I_r\right)^{1/2} \left[B^c_{i,r} \cos\left(\varphi_i - \varphi_r\right) - \right. \\ \qquad \left. - B^s_{i,r} \sin\left(\varphi_i - \varphi_r\right)\right] \\ \Delta\varphi_i = \omega_i \Delta t + \frac{1}{2} \sum_r A_{i,r} \left(\frac{I_r}{I_i}\right)^{1/2} \left[B^s_{i,r} \cos\left(\varphi_i - \varphi_r\right) + \right. \\ \qquad \left. + B^c_{i,r} \sin\left(\varphi_i - \varphi_r\right)\right] \end{cases} \tag{4.15}$$

where

$$A_{i,r} = 32\pi^2 \frac{\left(r_o c\right)^2}{m} \frac{\omega_w}{\sqrt{\omega_i \omega_r}} D_i D_r \Delta t^3 \frac{NI_w}{V^2} , \tag{4.16}$$

$$\begin{cases} B^{c,s}_{i,r} = \begin{pmatrix} Re \\ -Im \end{pmatrix} B_{i,r} \\ B_{i,r} = E\left(\eta_i, \eta_r\right) \tilde{f}\left[2\left(k_i - k_r\right)\right] \\ \tilde{f}\left[2\left(k_i - k_r\right)\right] = \tilde{f}^c\left[2\left(k_i - k_r\right)\right] + j\,\tilde{f}^s\left[2\left(k_i - k_r\right)\right] \end{cases} \tag{4.17}$$

$$\begin{pmatrix} C\left(\eta_i, \eta_r\right) \\ S\left(\eta_i, \eta_r\right) \end{pmatrix} = \begin{pmatrix} \mathrm{Re} \\ \mathrm{Im} \end{pmatrix} E\left(\eta_i, \eta_r\right)$$

$$E\left(\eta_i, \eta_r\right) = \exp\left(j\eta_i\right)\left\{\frac{1}{\eta_i - \eta_r}\left[\frac{1 - \exp\left(-j\eta_r\right)}{\eta_r^2} - \frac{1 - \exp\left(-j\eta_i\right)}{\eta_i^2}\right] - \right.$$

$$\left. - \frac{j}{\eta_i \eta_r}\right\} . \qquad (4.18)$$

Furthermore $\tilde{f}^{c,s}$ are the cosine and sine Fourier transforms of the distribution function, i.e.

$$\tilde{f}^{c,s}\left(2\left(k_i - k_r\right)\right) = \int_{-\infty}^{+\infty} f\left(z_o\right)\begin{Bmatrix} \cos \\ \sin \end{Bmatrix}\left(2\left(k_i - k_r\right)z_o\right) dz_o \qquad (4.19)$$

As a particularly interesting case let us consider a Gaussian distributed e.b., with r.m.s. σ_z. In such a case we would obtain

$$\tilde{f}^c = \exp\left[-\frac{\left[2\left(k_i - k_r\right)\right]}{2}\sigma_z^2\right], \quad \tilde{f}^s = 0. \qquad (4.20)$$

From the above relations it is easy to understand that once one has assumed a Gaussian distribution for the e.b., the i-th mode mainly interacts with the r-th ones satisfying

$$\left|k_i - k_r\right| \lesssim \frac{1}{\sigma_z} . \tag{4.21}$$

Let us now discuss two particularly interesting cases relevant to the e.b. r.m.s. σ_z :

i) very short e.b..
We assume in this case that σ_z is practically vanishing so that the Gaussian e.b. distribution reduces to a Dirac δ-function; in this connection the cosine Fourier transform reduces to

$$\tilde{f}^c = 1 ; \tag{4.22}$$

this implies that all the modes are coupled.

ii) Uniform distribution along the z-direction.
In this case σ_z is very large (we assume $\sigma_z \to \infty$), so that (4.20) yields

$$\tilde{f}^c = \delta_{r,i} \qquad (\delta \equiv \text{Kronecker symbol}) . \tag{4.23}$$

In such a case in the equations (4.15) only the r = i terms survive, so that each mode evolves independently from the other ones. Specifically, we have

$$\begin{cases} \Delta I_i = - A_{i,i} \operatorname{Re} E(\eta)\, I_i \\ \\ \Delta\varphi_i = \omega_i\, \Delta t + \dfrac{A_{i,i}}{2} \operatorname{Im} E(\eta) , \end{cases} \tag{4.24}$$

where we have put (see eqs (4.18))

$$E(\eta) = -\frac{j}{2}\left[\frac{\eta/2 \cos \eta/2 - \sin \eta/2}{(\eta/2)^3}\right] \cdot \exp(j\, \eta/2) \cdot \tag{4.25}$$

It is to be noticed that Re $E(\eta)$ and $-\operatorname{Im} E(\eta)$ (plotted in Fig. 2), account for the absorptive (action variation) and dispersive (phase variation) behaviour respectively, and are related by the usual dispersion relation.[21]

$$\operatorname{Im} E(\eta) = -\frac{1}{\pi}\int \frac{\operatorname{Re} E\left(\eta_o\right)}{\eta_o - \eta}\, d\eta_o \tag{4.26}$$

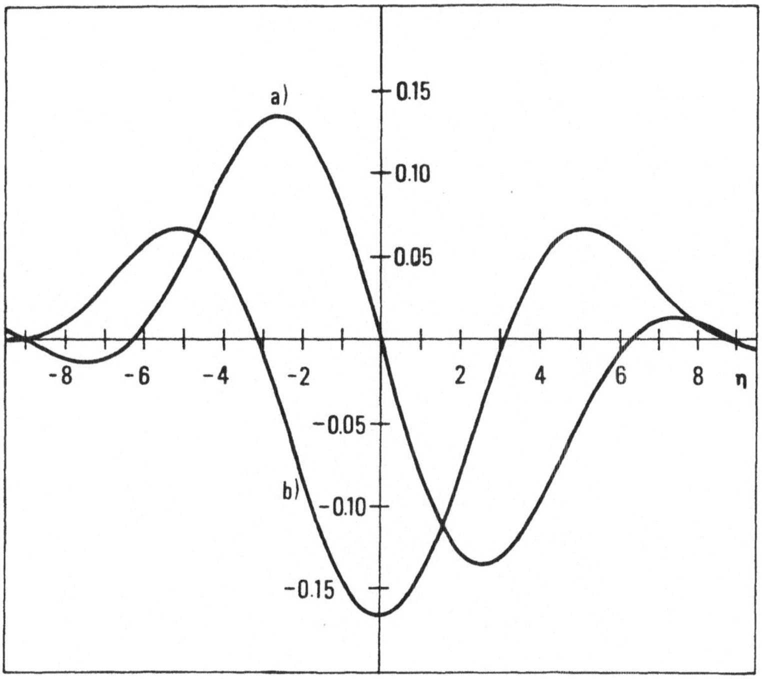

Fig. 2. Absorptive-part ReE(η) curve (a)
Dispersive-part -Im E(η) curve (b)

If the e.b. is not monoenergetic we have to perform a further convolution with respect to the energy distribution, taking into account the dependence of the detuning η_i on the initial momentum ρ_0 (see the second of eqs (3.17)). In this consitions eqs (4.15) hold again, but $B_{i,r}$read

$$B_{i,r} = \int f\left(z_0, P_0\right) dz_0 \, dP_0 \left\{ E\left[\eta_i \left(P_0\right), \eta_r\left(P_0\right)\right] \right.$$

$$\left. \exp\left[+ 2j \left(k_i - k_r\right) z_0\right] \right\} \qquad (4.27)$$

Finally let us evaluate, by means of the procedures used before, the average energy variation and spread induced in the e.b. by the interaction. From equations (3.12) and (3.13) we obtain

$$\delta p = -\sum_i \left(k_w + k_i\right) \delta I_i \tag{4.28}$$

from which we derive the energy dispersion

$$<\delta p^2> = <\sum_{i,j} \left(k_w + k_j\right)\left(k_w + k_i\right) \delta I_i \delta I_i> =$$

$$= \frac{32\,\pi^2 \left(r_o c\right)^2 \omega_w I_w}{V^2} \Delta t^2 \sum_{i,j} \frac{D_i D_j}{\left(\omega_i \omega_j\right)^{1/2}} \left(I_i\, I_j\right)^{1/2} \cdot$$

$$\cdot \left\{ \cos\left[2\left(k_i - k_j\right) z_o - \varphi_i + \varphi_j\right] \cdot \frac{1}{\eta_i \eta_j} \left[1 + \cos\left(\eta_i - \eta_j\right) - \right.\right.$$

$$\left. - \cos\eta_i - \cos\eta_j\right] + \sin\left[2\left(k_i - k_j\right) z_o - \varphi_i + \varphi_j\right] \cdot \frac{1}{\eta_i \eta_j}$$

$$\left.\left[\sin\eta_i - \sin\eta_j - \sin\left(\eta_i - \eta_j\right)\right]\right\} . \tag{4.29}$$

It is easy to verify that Madey's statement [20]

$$<\delta p> = \frac{1}{2} \frac{\partial <\delta p^2>}{\partial p} \tag{4.30}$$

holds, in general, when $k_i \approx k_w$.

Equations (4.28) and (4.30) are useful in dealing with a FEL where the e.b. is recirculated. (Storage Ring FEL[24]).

(B) Strong Signal Regime

We have discussed up to now the essential features of the small signal regime. In this subsection we shall outline with some detail the main feautres of the strong signal one. In dealing with such a regime we assume that the electrons and the fields are strongly pertrubed by the FEL interaction, and this amounts to saying that the "Coupling-constants" $\tilde{\varepsilon}_i$ are large enough to not allow any pertrubative expansion of the kind (4.6). The only allowed approximation is to neglect the spontaneous emission with respect to the stimulated, so that no upper condition like (4.4) and (4.5) can be inferred on laser and wiggler fields.

It could be thaught that because we cannot perturbatively treate eqs (3.19), we are dealing with an insoluble "many-body" problem; fortunately, the situation is different. We have neglected, indeed, the L-L interaction with respect the L-W one, so that for our problem we have performed a kind of "Hartree approximation", which will allow us to give a formal solution to the multimode strong signal regime. Thus, in principle, what we need is only a more complex approach to eqs (3.19).

Let us define the "Collective" variable

$$X(\tau) = \sum_i \tilde{\varepsilon}_i \int_0^{\tau} d\tau' \int_0^{\tau'} d\tau'' \, x_i(\tau'') \, . \tag{4.31}$$

It is easily understood that (4.31) is linked to the longitudinal electron coordinate.

In terms of this quantity, eqs (3.19) read (neglecting the spontaneous emission, i.e., the term α_i^W) :

$$\begin{cases} x_i' = y_i \left[\eta_i + D_i \, X'(\tau)\right] \\ y_i' = -x_i \left[\eta_i + D_i \, X'(\tau)\right] \, . \end{cases} \tag{4.32}$$

The formal solution of the eqs (4.32) is

$$\begin{cases} x_i = \sin\left(\chi_i(0) + \eta_i\tau + D_i\ X(\tau)\right) \\ \\ y_i = \cos\left(\chi_i(0) + \eta_i\tau + D_i\ X(\tau)\right) \end{cases} . \tag{4.33}$$

The problem is now evaluating $\chi(\tau)$. From equations (4.31) and (4.33) we derive the equation of motion for $\chi(\tau)$:

$$X'' = \sum_i \tilde{\varepsilon}_i x_i = \sum_i \tilde{\varepsilon}_i \sin\left(\chi_i(0) + \eta_1\tau + D_i\ X\right) , \tag{4.34}$$

with the initial conditions

$$X(0) = X'(0) = 0. \tag{4.35}$$

Let us redefine the quantities entering the argument of (4.34) by writing the i-th laser mode frequency as

$$\omega_i = \omega_w(1 + i\delta), \quad \left(\delta \ll 1 \text{ being } \frac{\omega_i - \omega_w}{\omega_w} \ll 1\right) \tag{4.36}$$

where δ is a factor depending on the experimental apparatus. We obtain

$$X'' = \sum_i \tilde{\varepsilon}_i \sin\left[\varphi_w - \varphi_i + \left(a_o + b_o\tau\right) + i\left(a_1 + b_1\tau\right) + \right. \\ \left. + \left(1 + i\frac{\delta}{2}\right)X\right], \tag{4.37}$$

where

$$
\begin{cases}
a_o = 4\pi \dfrac{z_o}{\omega_w} & \left(\lambda_w = \dfrac{2\pi c}{\omega_w}\right) \\[2ex]
a_1 = 4\pi \dfrac{z_o}{\left(\lambda_w / \delta\right)} & \\[2ex]
b_o = \dfrac{2p_o}{mc} \left(\omega_w \Delta t\right) & \\[2ex]
b_1 = \delta\left(\omega_w \Delta t\right)\left(\dfrac{p_o}{mc} - 1\right) \approx - \delta\left(\omega_w \Delta t\right), & \left(\text{being } \dfrac{p_o}{mc} \ll 1\right)
\end{cases}
\tag{4.38}
$$

Let us investigate in more detail the parameters (4.38). The quantities a_o and a_1 depend on the longitudinal coordinate z_o of the electron in the beam. We assume that the beam length σ_z is longer than several optical wave lengths λ_w but smaller than λ_w/δ (recall $\delta \ll 1$). In this case a_o varies very rapidly with z_o. (we shall call a_o "fast running phase"), while a_1 is very slowly varing ("slow running phase"). Finally, b_o depends on the electron momentum p_o, and b_1 on the experimental apparatus.

In conclusion, from eqs (4.36) and (4.38), with the initial conditions (4.35), we obtain

$$
X = X\left(\tau \middle| \left\{\tilde{\varepsilon}_j\right\}, \left\{\varphi_w - \varphi_j\right\} \middle| a_o, a_1, p_o\right). \tag{4.39}
$$

Inserting equations (4.35) and (4.29) in eqs (3.18) we derive the action and phase variation of the i-th mode:

$$
\begin{cases}
\delta I_i = - G \dfrac{\tilde{\varepsilon}_i}{D_i} A_x^i\left(\left\{\tilde{\varepsilon}_j\right\}, \left\{\varphi_w - \varphi_j\right\} \middle| a_o, a_1, p_o\right) \\[2ex]
\delta\varphi_i = \omega_i \Delta t + \dfrac{G\,\tilde{\varepsilon}_i}{2I_i(0)} \dfrac{1}{D_i} A_y^i\left(\left\{\tilde{\varepsilon}_j\right\}, \left\{\varphi_w - \varphi_j\right\} \middle| a_o, a_1, p_o\right)
\end{cases}
\tag{4.40}
$$

,

where we have put

$$\left\{\begin{aligned} A_x^i &= \int_0^1 x_i \, d\tau \\ A_y^i &= \int_0^1 \frac{y_i}{x_i^2 + y_i^2} \, d\tau . \end{aligned}\right. \tag{4.41}$$

Finally by averaging first over the "fast running phase" a_0 and, in a second time, over the "slowly varying" one a_1 and the momentum p_0, we obtain the action and phase variation due to the whole electron beam:

$$\left\{\begin{aligned} \Delta I_i &= - G \frac{\tilde{\varepsilon}_i}{D_i} \, B_x^i \left[\left\{\varepsilon_j\right\}, \left\{\varphi_w - \varphi_j\right\}\right] \\ \Delta\varphi_i &= \omega_i \Delta t + \frac{G \, \tilde{\varepsilon}_i}{2I_i(0)} \frac{B_y^i}{D_i} \left[\left\{\varepsilon_j\right\}, \left\{\varphi_w - \varphi_j\right\}\right] , \end{aligned}\right. \tag{4.42}$$

with

$$\left\{\begin{aligned} B_x^i &= \int dp_0 \int dz_0 \, f\left(z_0, p_0\right) \left\{ \frac{1}{2\pi} \int_0^{2\pi} da_0 \, A_x^i \left[\left\{\varepsilon_j\right\}, \left\{\varphi_w - \varphi_j\right\} \middle| a_0, \frac{4\pi z_0}{\lambda_w/\delta}, p_o\right]\right\} \\ B_y^i &= \int dp_0 \int dz_0 \, f\left(z_0, p_0\right) \left\{ \frac{1}{2\pi} \int_0^{2\pi} da_0 \, A_y^i \left[\left\{\varepsilon_j\right\}, \left\{\varphi_w - \varphi_j\right\} \middle| a_0, \frac{4\pi z_0}{\lambda_w/\delta}, p_o\right]\right\} \end{aligned}\right. \tag{4.43}$$

It is worthwhile noticing that the foregoing treatment of our main differential equations has provided a particularly simple tool of analysis, useful for future numerical approaches to the problem. 22

In fact, it has been shown that we need the numerical solution of only one second order differential equation (eq. (4.30)) for the knowledge of action and phase evolution in the strong signal multimode regime.

5. CONCLUSIONS

In this paper we have presented a multimode classical theory of the free electron laser. We have treated it in the most general case by means of a formalism which has provided a particularly simple tool of analysis, allowing us to follow each laser mode evolution without explicit use of the Maxwell equations.

We have worked in a reference frame where the basic amplification laser process can be understood as a scattering of pseudo radiation photons into laser ones. Indeed, the reference frame is chosen so that the electron dynamics can be described nonrelativistically. A particularly useful form has been obtained by introduction of the (x, y) "interaction variables", whose time evolution, contained in eqs (3.19), has allowed us to obtain the action and phase evolution, both in the small and strong signal regime. We want to underline that the "interaction variables" have a well defined physical meaning; in fact, it has been shown that they are linked, for a single laser mode, to the 1 and 2 components of the "FEL Bloch-vector". In references 12 and the classical FEL Bloch equations have been derived and the connection with the interaction variables has been discussed. Furthermore, it can be shown that using the Bloch formalism the single mode FEL equations can be exactly solved in terms of elliptic functions and the solution can be easily derived.

Let us now summarize, with a few comments, our basic approximation.

1) The laser field has been assumed strong enough that no photon fluctuation effects play an appreciable role. In this connection, we have had to look for a condition allowing us to neglect the term driving the spontaneous emission, for which the quantum effects would have been important with respect to the stimulated one. This gives us a lower limit on the i-th laser action. An upper condition on the laser field action is obtained by requiring that the summation extended over each laser action is much smaller that the wiggler term. Thus we have been allowed to write down the "interaction-variable" evoultion neglecting the L-L coupling with respect the L-W one. It is to be noted that this is a largely fulfilled assumption for the wiggler devices we are interested in. For "wiggler-devices" provided by real e.m. fields, the effect due to the L-L coupling could be non-negligible, in principle, with respect to the L-W ones. Therefore, we need a more careful analysis of the problem, In fact, we have to define, besides the already

introduced L-W interaction variables (see (3.15)), a set of L-L ones too and follow their time evolution;[25] furthermore, another parameter relevant to the problem must taken into account, namely the L-L "coupling constant"

$$\left\{ \varepsilon_{ij} = \left(\Omega_{Rij}\Delta t\right)^2 , \; \Omega_{Rij} = \left[\frac{16\pi r_o \omega_i}{mV} \left(I_i(0)\; I_j(0)\right)^{1/2}\right]^{1/2} \right\} .$$

However, the complexities which arise deserve separate treatment.

2) We have worked out our calculations in a moving frame where the longitudinal electron motion has been assumed to be nonrelativistic; such an approximation cannot hold, in principle, for laser frequencies in the blue or UV range.

3) The static wiggler field has been treated in the chosen moving frame as a P.R.F., by means of the Weiszäcker-Williams approximation. However, real e.m. field could provide the "wiggler" field; [23] thus we need a slight modification of the interaction variables to account for the wiggler field variation too.

In conclusion, the approximations above have allowed us to write down the action and phase evolution for each laser mode, in the small and strong signal regimes.

In the case of small signal (i.e. small perturbation) we have been allowed to look for a perturbative solution of eqs (3.19), by expanding the "interaction variables" in terms of the $\tilde{\varepsilon}_i$-parameters, accounting for the coupling of the i-th laser mode to the wiggler one. As far as the strong-signal case is concerned, no perturbative solution is available, thus we have looked for a formal solution of the problem useful for further numerical analysis. In such a framework we have shown that we need the numerical solution of only one second order differential equation.

APPENDIX A

The FEL theory up to now, has been considered developed in a reference frame where the electron motion can be considered non-relativistic; i.e., it has been assumed that in such a frame the longitudinal electron momentum statisfies the following condition:

$$p_o \ll m_o c \cdot \tag{A.1}$$

Let us now, for further development,[26] return to the laboratory frame representation. If the electron energy is E_0, the laboratory frame is moving with respect to the chosen one at a velocity v, close to the light velocity c, namely

$$v_o = c \left[1 - \frac{1}{2} \left(1 + \frac{p_o}{mc} \right)^2 \left(\frac{m_o c^2}{E_o} \right)^2 \right]^{1/2} \tag{A.2}$$

It is now staightforward to express the energy and phase variation, previously derived in the small signal regime in the moving frame, in the laboratory one. To avoid a complicated expression we assume (but it is a largely verified assumption) that the laser bandwidth is sufficiently narrow and we are working near the resonance $\omega_i \sim \omega_w$. In this connection we have

$$\frac{\omega_i - \omega_r}{\omega_i} \ll 1 \qquad \frac{\omega_i - \omega_w}{\omega_w} \ll 1 \quad . \tag{A.3}$$

This assumption ensures that

$$A_{i,r} \sim A = 32\pi^2 \frac{\left(r_o c^2 \right)^2}{m} \Delta t^3 \frac{NI_w}{V^2} ; \tag{A.4}$$

furthermore, η_i as defined in (3.17) becomes

$$\eta_i \sim \left[\left(\omega_w - \omega_i \right) + 2\, \omega_w \frac{p_o}{mc} \right] \Delta t \quad . \tag{A.5}$$

The Lorentz transformation defined through (A.2) allows us to write eqs (4.15)[†]

$$\begin{cases} \Delta W_i = -A \sum_r \left(W_i W_r\right)^{1/2} \left[B^c_{i,r} \cos\left(\varphi_i - \varphi_r\right) - B^s_{i,r} \sin\left(\varphi_i - \varphi_r\right)\right] \\ \Delta\varphi_i = \omega_i \Delta t + \frac{A}{2} \sum_r \left(\frac{W_r}{W_i}\right)^{1/2} \left[B^s_{i,r} \cos\left(\varphi_i - \varphi_r\right) + B^c_{i,r} \sin\left(\varphi_i - \varphi_r\right)\right] \end{cases} \tag{A.6}$$

where we have put (recalling the mass redefinition of the second of Refs 9)

$W_i \equiv \left(\frac{\omega_i I_i}{V}\right) \equiv$ i-th the laser mode energy density

$\varphi_i \equiv$ i-th laser mode phase

$\Delta t \equiv$ interaction time

$$A \equiv 4\pi^2 r_o \sqrt{\frac{2\lambda}{\lambda_q}}\, L\, \lambda\, n_e \frac{k^2}{\left(1 + k^2\right)^{3/2}} \left(\frac{\Delta\omega}{\omega}\right)_o^{-2}$$

$\lambda \equiv$ average laser wavelength

$L \equiv$ wiggler length

$n_e \equiv$ electron density

$\left(\frac{\Delta\omega}{\omega}\right) \equiv \frac{\lambda_q}{2L}$

---

† The equations (A.6) and the following ones are valid for linearly polarized wigglers too, with the only change being a subsitution $B_o/\sqrt{2}$ for B_o in the parameter k.

$$k \equiv \frac{e B_o \lambda_q}{2\pi m_o c^2}$$

$B_o \equiv$ wiggler magnetic field

and finally

$$B^{c,s}_{i,r} = \begin{pmatrix} \mathrm{Re} \\ -\mathrm{Im} \end{pmatrix} B_{i,r}$$

$$B_{i,r} = \int f(z,\varepsilon)\, dz d\varepsilon\, E\left(\eta_i(\varepsilon),\ \eta_r(\varepsilon)\right) \exp\left[+j\left(k_i - k_r\right) z\right] \qquad (A.7)$$

where z is the longitudinal electron coordinate in the laboratory frame, and ε is the particle "off-energy"

$$\varepsilon = \frac{E - E_o}{E_o}, \quad \begin{pmatrix} E = \text{electron energy} \\ E_o = \text{average e.b. energy} \end{pmatrix} \qquad (A.8)$$

With this notation, η_i may be written

$$\eta_i(\varepsilon) = \pi \left(\frac{\Delta\omega}{\omega}\right)_o^{-1} \left\{2\varepsilon - \frac{\omega_i - \omega_o}{\omega_o}\right\} \qquad (A.9)$$

where ω_o is the "resonant frequency", which is defined by the equations

$$\omega_o = \frac{2\pi c}{\lambda_o}, \quad \lambda_o = \frac{\lambda_q}{2}\left(1 + k^2\right)\left(\frac{m_o c^2}{E_o}\right)^2 . \qquad (A.10)$$

From the equations (A.6) it is easily seen that the evolution of laser modes in an optical cavity in the time between the passage of two adjacent electron bunches follows

$$\begin{cases} \Delta W_i = -\left\{ W_i \gamma_T + A(1-\gamma_T) \sum_r \left(W_i W_r\right)^{1/2} \left[B^c_{i,r} \cos\left(\varphi_i - \varphi_r\right) - \right.\right. \\ \qquad \left.\left. - B^s_{i,r} \sin\left(\varphi_i - \varphi_r\right) \right] \right\} \\ \Delta \varphi_i = \omega_i T_e + \frac{1}{2} A \sum_r \left(\frac{W_r}{W_i}\right)^{1/2} \left[B^s_{i,r} \cos\left(\varphi_i - \varphi_r\right) + B^c_{i,r} \sin\left(\varphi_i - \varphi_r\right) \right] \end{cases} \tag{A.11}$$

where we have put

$$\begin{cases} \gamma_T \equiv \text{total optical cavity losses} \\ T_e \equiv \text{time distance between two adjacent electron bunches} \end{cases} \tag{A.12}$$

The second of the equations (A.11) can be written in a simple way, in terms of the optical cavity round trip period T_c.

For synchronism in electron and photon bunches in the cavity it is convenient to choose $T_e \sim T_c$, i.e.

$$T_e = T_c - \delta t \quad \left(\delta t \ll T_c\right) \quad . \tag{A.13}$$

Taking into account the test that the mode frequencies ω_i are correlated to T_c by the equation

$$\omega_i = \frac{2\pi}{T_c} n_i \qquad \left(n_i \equiv \text{integer}\right), \tag{A.14}$$

we have to substitute T_e with $-\delta t$ in the second eqs (A.11).

Finally, the average energy spread (eq. (4.29)) reads

$$\langle\delta\varepsilon^2\rangle = \left(\frac{2\pi r_o}{m_o c^2}\right) \lambda_o^2 \left(\frac{\Delta\omega}{\omega}\right)_o^{-2} \frac{k^2}{\left(1 + k^2\right)^2} \cdot$$

$$\cdot \left| \sum_i \left(W_i\right)^{1/2} \exp\left\{j\left(k_i z - \varphi_i + \frac{\eta_i}{2}\right)\right\} \frac{\sin\left(\eta_i/2\right)}{\left(\eta_i/2\right)} \right|^2$$

while eq. (4.30) becomes

$$\langle\delta\varepsilon\rangle = \frac{1}{2} \frac{\partial}{\partial\varepsilon} \langle\delta\varepsilon^2\rangle .$$

REFERENCES

1. E. Schrödinger: Ann.der Phys. IV Folge 82, 257 (1927)
2. P.L. Kapitza and P.A.M. Dirac: proc. Camb. Phys. 29, 297 (1933)
3. H. Motz: J. Appl. Phys. 22, 527 (1951)
4. J.M.J. Madey: J. Appl. Phys. 42, 1906 (1971)
5. L.R. Elias, W.M. Fairbank, J.M.J. Madey, H.A. Schwettman and T.I. Smith: Phys. Rev. Lett. 36, 717 (1976)
   D.A.G. Deacon, L.R. Elias, J.M.J. Madey, G.J. Ramian, H.A. Schwettman and T.I. Smith: Phys.Rev.Lett. 38, 892 (1977)
6. F.A. Hopf, P. Meystre, M.O. Scully and W.H. Louisell: Opt.Comm. 18, 413 (1976) and Phys. Rev. Lett. 37, 1342 (1976)
   H. Al Abawi, F.A. Hopf and P. Meystre: Phys.Rev.A 16, 666 (1977)
7. V.N. Baier and A.I. Milstein: Phys.Lett. 65A, 319 (1978)
8. W.B. Colson: Phys. Lett. 59A, 187 (1976)
9. A. Bambini and A. Renieri: Lett. Nuovo Cimento 21, 399 (1978)
   A. Bambini, A. Renieri and S. Stenholm: Phys. Rev. A 19, 2013 (1979)
10. A. Bambini and S. Stenholm to appear in J. Phys. and Opt. Comm. 30, 391 (1979)
11. R. Bonifacio and M.O. Scully, lectures given at the winter College on Atomic and Molecular Physics and Quantum Optics Trieste, 23rd January 30th March (1979)
12. G. Dattoli: Report 79.22 C.N.E.N., Centro di Frascati, Frascati, Rome, Italy
    G. Dattoli: Lett. Nuovo Cimento 27, 247 (1980)
13. R. Bonifacio Opt. Comm. 32, 440 (1980)
14. G. Dattoli, A. Renieri, F. Romanelli and R. Bonifacio: Opt. Comm. 34, 240 (1980)
    G. Dattoli, A. Renieri and F. Romanelli: Report 80.22 C.N.E.N. Centro di Frascati, Frascati, Rome, Italy (to appear in Opt. Comm.)
15. A. Renieri: Nuovo Cimento 53B, 160 (1979)
16. G. Dattoli and A. Renieri: Lett. Nuovo Cimento 24, 121 (1979)
    G. Dattoli and A. Renieri: Report 79.37/p C.N.E.N., Centro di Frascati, Frascati, Rome, Italy to be published
17. A. Bambini, R. Bonifacio and S. Stenholm: Opt. Comm. (1980)
18. H. Al Abawi, F.A. Hopf, G.T. Moore and M.O. Scully: Opt. Comm. 30, 235 (1979)
19. S. Stenholm: talk given at LXXIV International School of Physics E. Fermi 2nd Course 1978 "Developments in High Power Lasers and their Applications" Varenna 10th July - 22nd July (1978)
20. J.M.J. Madey: Nuovo Cimento 50B 64 (1979)
21. H.A. Kramers: Atti Congresso Fisica Como, 545 (1927)
    H.A. Kramers: Hand und Jahrbuch der Chem. Physik I, Leipzig (1937)
22. G. Dattoli, A. Marino and A. Renieri: in preparation
23. L.R. Elias: Phys. Rev. Lett. 42, 977 (1979)

24. G. Dattoli and A. Renieri: Report 80.2/p C.N.E.N., Centro di Frascati, Frascati, Rome, Italy to be published in Nuovo Cimento B
25. F. Romanelli: "Teoria del Laser ad Elettroni Liberi" Thesis, Università di Firenze (1980)
26. G. Dattoli, A. Marino and A. Renieri, these proceedings

PULSE PROPAGATION AND LASER LETHARGY IN THE FREE-ELECTRON LASER*

Gerald T. Moore and Marlan O. Scully

University of New Mexico
Institute for Modern Optics
Department of Physics and Astronomy
800 Yale Boulevard NE
Albuquerque, New Mexico 87131, USA

and

Projektgruppe für Laserforschung
Max-Planck-Gesellschaft
zur Förderung der Wissenschaften E.V.
D-8046 Garching/Munich, West Germany

ABSTRACT

We discuss coherent pulse propagation in the free-electron laser (FEL). The effect of laser lethargy is seen to play an important role in the pulsed FEL, as it does in conventional swept-gain amplifiers based on an atomic medium. Numerical calculations of FEL pulse propagation are presented giving good agreement with the Stanford experiment.

In this paper we discuss the physics of a pulsed free-electron laser of the Stanford type.[1,2] We point out that the physics is similar to that of conventional types of swept-gain lasers. In particular, the phenomenon of laser lethargy, which was first discovered[3-5] in calculations involving conventional swept-gain amplifiers, plays a crucial role in FEL pulse propagation.[6-9] A major goal of this paper is to describe this role.

We begin by summarizing, without much mathematics, the basic physics of the pulsed FEL, and then compare the pulsed FEL with

*Research supported by the Office of Naval Research.

conventional swept-gain lasers. Next we describe the role of laser lethargy both in the conventional lasers and in the FEL. Finally, we present numerical calculations of pulse propagation in the regime of the Stanford experiment, and show that good agreement with the experimental data is obtained in such matters as the power-tuning curves and the electron energy distribution.

The Stanford free-electron laser[2] is pumped by picosecond pulses of ultrarelativistic electrons emitted from a linear accelerator. These electron pulses propagate through the bore of a helical (circularly polarized) wiggler magnet of length L and emit radiation in the forward direction whose frequency $\omega_s = ck_s$ is approximately given by the resonance (Doppler up-shift) condition

$$k_s = 2\gamma_s^2 k_q, \tag{1}$$

where $\lambda_q = 2\pi/k_q$ is the wiggler wavelength and γ_s is the Lorentz factor corresponding to the longitudinal motion of the electrons. This radiation, which is emitted into a pulse whose length is of the same order as the electron pulses, is recirculated through the laser by means of an optical resonator and is amplified coherently until saturation leads to a steady state in which the gain per pass equals the loss of the system. This situation is illustrated in Fig. 1. To maintain overlap of the electrons and the light, the resonator length must be tuned so that the transit time T_s of the cavity at the speed of light is close to the repetition rate T_e of the electron pulses coming from the LINAC. The power emitted by the FEL is thus a sensitive function of the delay $\delta t = T_e - T_s = 2\delta z/c$, where δz is the amount by which the resonator length is shortened.

The generation of coherent radiation in the FEL comes about through the mechanism of electron bunching on the optical time scale.[10-15] The electrons feel a force (the ponderomotive force) proportional to the product of the wiggler and laser fields. The phase of this product is a sinusoidal function which propagates at

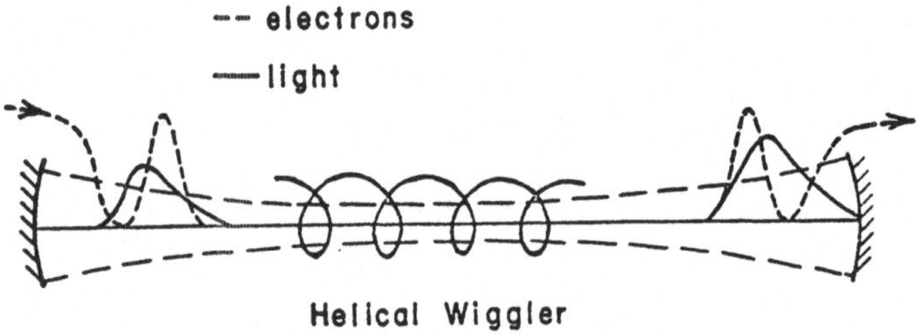

Fig. 1. Schematic diagram of pulse propagation in the FEL.

speed $v_s = \omega_s/(k_s + k_q) = c(1-\gamma_s^{-2})^{1/2}$, which is approximately the speed of the electrons and is slightly less than the speed of light. At the entrance of the wiggler the electrons are unbunched and hence are phased randomly with respect to the ponderomotive potential. They then proceed to fall into the wells of the potential and become bunched, as illustrated in Fig. 2. The motion is analogous to marbles rolling on a level corrugated rooftop or to the motion of simple pendula.[13-15] The electron distribution function is governed by the Boltzmann equation, which must be solved self-consistently in conjunction with the Maxwell equation for the laser field. The first studies of pulse propagation in the FEL[6] utilized an expansion of the Boltzmann distribution function in harmonics of the frequency of the ponderomotive potential. By retaining only the first two terms in this expansion, one obtains a set of "quasi-Bloch" equations which describe the electron motion in close analogy to the density-matrix formalism of conventional laser theory.

An alternative approach[9] which avoids the truncated harmonic expansion and is expected to be more accurate in the highly saturated regime arises[16] through the use of multiple-scaling perturbation theory. In this approach one introduces separate variables to describe the fast optical-frequency oscillations and the slowly varying envelopes of the pulses. The analogy of the electron motion to the motion of pendula is more obvious in this approach. However, the pendulum description must be generalized to take into account the fact that the complex laser field amplitude varies as a function of position and time. The electron motion is then analogous to pendula swinging in a gravitational field which varies, both in magnitude and direction in the plane, as a function of time. Furthermore, electrons in different parts of the electron pulse see different histories of field variations. It is important to stress that these field variations are not known a priori, but must be determined by solving the Maxwell equation for the laser field <u>self-consistently</u> with the equations governing the electron motion. The current driving the Maxwell equation is proportional to the product of the electron bunching amplitude and the wiggler amplitude. The wiggler acts to convert the longitudinal bunching in electron density into a transverse current oscillating at

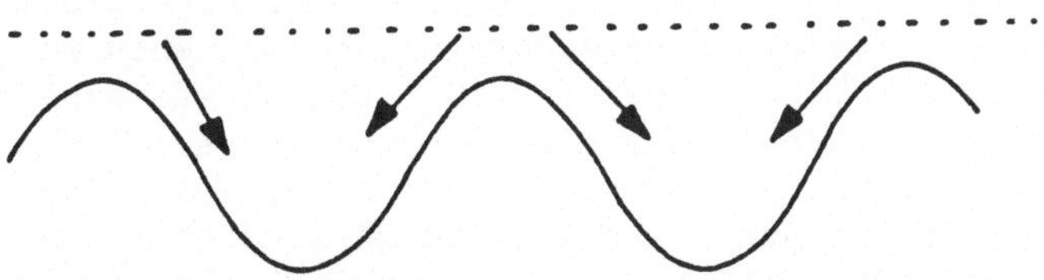

Fig. 2. Formation of electron bunches.

optical frequencies, which is what one needs to generate coherent optical radiation.

On the time scale of the picosecond pulses the electrons remain essentially monoenergetic during their transit through the FEL. Thus, the temporal shape of the electron pulses is essentially constant (to within an optical period). However, the electron pulse speed v_s is significantly less than the speed of light c. In the Stanford experiment the relative time shift $L/2\gamma_s^2 c$ between the electron and optical pulses in traversing the wiggler is of the same order as the pulse widths.

Analogies between conventional lasers and the FEL are quite close, as has been brought out particularly in References 13, 8, and 7. In each case a macroscopic transverse current oscillating at optical frequencies drives the Maxwell wave equation. In conventional lasers this current is generated by oscillating atomic dipoles. In the FEL the oscillating current is produced by bunching the electron density on the optical time scale. An atomic lasing medium is usually best described in terms of the population matrix. The electron distribution in a FEL is analogously described in terms of the Boltzmann distribution function.

Other analogies arise when one considers pulsed operation. Since the Stanford FEL is a swept-gain laser, it is appropriate to compare its operation with the operation of conventional swept-gain lasers. Such devices have attracted considerable attention, particularly for short-wavelength applications where the short natural lifetime of the excited atoms makes it difficult or inefficient to invert the entire lasing medium. The idea is to sweep the population inversion in a pulse moving at or near the speed of light, so as to keep the gain region in the right place at the right time to amplify a co-moving optical pulse. The atoms themselves are not generally localized in the pulse and need not be moving (unlike the case in the FEL, where the electrons move relativistically). Much of our work on swept-gain amplifiers[3-5] has been in connection with possible x-ray laser devices. In such devices the pumping is produced by sweeping an ion beam along a target via a pulsed electric field. Since the sweep speed does not involve signal transmission, it can equal or exceed the speed of light (unlike the case in the FEL).

The effect of laser lethargy was first recognized in our analysis of swept-gain charge-exchange lasers. When we later carried out pulse calculations for the FEL, it was apparent that the same effect was involved. In both cases laser lethargy may be understood as a manifestation of electron inertia. In the case of an atomic medium one pumps atoms to an excited energy level. If these atoms interact with resonant radiation, the atomic density matrix evolves under the influence of this radiation, and eventually an oscillating dipole moment is formed. However, initially the atoms have no dipole moment and cannot emit coherently. Thus, radiation from the atoms is delayed with respect to the time when the atoms begin to interact with external radiation. In pulsed operation the emitted radiation is added to a later part of the pulse than the radiation which stimulated the emission. The net effect is a slowing down of the pulse. In the simplest model of an atomic swept-gain amplifier, the atoms are stationary in the lab frame and the pulse lengths are determined by the spontaneous emission lifetime. Note that the lethargy effect occurs even when the population inversion is swept at the speed of light.

The same lethargy effect occurs in the FEL, except that radiation is produced by electron bunches instead of oscillating atomic dipoles. Electrons entering the wiggler are unbunched and cannot radiate coherently. Bunches start to form when the electrons feel the influence of the ponderomotive force; that is, when they are simultaneously within the wiggler field and the laser field. Because of electron inertia, it takes awhile for the bunches to form and to generate additional radiation. Unlike the case of the atomic swept-gain amplifier, where the atoms may be considered to be at rest, the electrons in the FEL are moving relativistically. Thus, the bunches form at some distance past the entrance of the wiggler, even in cw operation, so that there is no gain at the entrance of the wiggler. In short pulse operation the electron bunches radiate into a later part of the optical pulse than the part which initiates the bunch formation, effectively slowing down the optical pulse. The pulse is slowed down, rather than simply lengthened, because mirror losses damp out the leading edge of the pulse, where there is no gain.

When one calculates pulse propagation in the FEL with the resonator length set for zero delay δt, one finds that after many passes the optical pulse lags behind the electron pulse until the light is just overtaking the electrons near the end of the wiggler. Since the amount of overlap between light and electrons is small, there is little gain, the FEL goes below threshold, and no steady-state operation is possible.

With positive δt one puts the light back on top of the electrons and obtains convergence to a steady state. By a steady-state we mean that the same pulse shape is reproduced at the end of the wiggler after each pass. There is in general some finite range of positive δt over which steady-state operation is possible, and one may calculate power-tuning curves giving the laser output power as a function of δt.

A convenient measure of the lethargic delay between the initiation of bunching and the production of radiation is given by the cavity value of δt which optimizes the FEL output pulse energy. Under conditions of fixed loss (as in the experiment) the optimum δt is an increasing function of the mirror loss. The reason is that for large loss the saturated gain per pass, which must equal the loss in steady-state, is large. Therefore, the added-on (slowed down) light per pass is a substantial fraction of the incident light.

An alternative approach which we sometimes use in numerical calculations is to renormalize the pulse energy after each pass to a prescribed value. In such calculations the loss is not prescribed beforehand, but is inferred to equal the calculated gain per pass in steady state. This approach is useful in isolating the effects of saturation. In particular, if the input pulse energy is sufficiently low, there is no saturation, and one can calculate the threshold characteristics of the FEL. Lethargy is found to be an increasing function of the electron current and a decreasing function of the electron current and a decreasing function of the optical power. The dependence on current is explained by the argument given above: namely, more current implies more gain which implies more slowed-down light added per pass. The dependence on optical power is explained by the fact that high power gives deep wells in the ponderomotive potential and the electrons fall into the wells, get bunched, and begin to radiate quicker. For given current, the dependence of lethargy on the electron pulse length is small. The lethargy effect does not get small for very long pulses except in proportion to the electron pulse length. It is not straightforward to see the relation between pulsed and cw operation by considering the limit of very long electron pulses. For example, the small-signal regime with long electron pulses consists of a short (picosecond) optical pulse located on the peak of the electron pulse. The implication is that one should be cautious in applying cw theory to FEL's with long (nanosecond) electron pulses.

The FEL pulse calculations carried out by our group have been based on two different mathematical approaches. Our first calculations[6] use a self-consistent coupling of the Maxwell wave equation for the laser field with "quasi-Bloch" equations for the electron distribution. The latter are obtained from a truncated expansion of the Boltzmann distribution in harmonics of the bunching frequency. The quasi-Bloch equations give a description of the FEL in close analogy to conventional laser physics, as described by the Bloch equations for two-level atoms. This is summarized (for single-mode operation) in the following chart. The similarity of structure is evident. The principal difference is that the quasi-Bloch equations contain derivatives with respect to the velocity detuning μ. The quantities w_1 and w_2 are the real and imaginary parts of the bunching amplitude, whereas w_3 is the slowly varying part of the distribution.

Quasi-Bloch	Bloch
$\frac{\partial w_1}{\partial z} - \mu w_2 = \kappa \lvert A_i A_s^* \rvert \frac{\partial w_3}{\partial \mu}$	$\frac{\partial u}{\partial t} + \delta v = \kappa \varepsilon w$
$\frac{\partial w_2}{\partial z} + \mu w_1 = 0$	$\frac{\partial v}{\partial t} - \delta u = 0$
$\frac{\partial w_3}{\partial z} = \kappa \lvert A_i A_s^* \rvert \frac{\partial w_1}{\partial \mu}$	$\frac{\partial w}{\partial t} = - \kappa \varepsilon v$

Our more recent pulse calculations[9] have coupled equations describing the electron dynamics (generalized pendulum equations) directly to the Maxwell equation via the use of multiple-scaling perturbation theory. The results calculated using the two approaches are qualitatively the same. Here we present some of our recent numerical calculations, which use a better fit to the parameters of the Stanford experiment.

The basic equations[16] used in our calculations are:

$$d^2\hat{\theta}(z,\tau_0,\theta_0)/dz^2 = -(e^2M^2c^3\gamma_s^2)$$
$$\times \{B_q^* E_s(z,\tau) \exp[i\hat{\theta}(z,\tau_0,\theta_0)] + c.c.\} , \quad (2)$$

$$\sigma^*(z)\partial[\sigma(z)E_s(z,\tau)]/\partial z = (e/2Mc\gamma_s k_q)$$
$$\times B_q I(\tau_0) \frac{1}{2\pi} \int_0^{2\pi} d\theta_0 \exp[-i\hat{\theta}(z,\tau_0,\theta_0)] . \quad (3)$$

In these equations M is the electron mass multiplied by a relativistic correction, B_q is the constant magnetic field amplitude, E_s is the complex amplitude of the laser field, and $I(\tau_0)$ is the electron current. The "slow" variables z,τ, and τ_0 describe the evolution of the system on the scale of the magnet length L and the picosecond pulse envelopes. The variable z is position along the magnet, $\tau = t-z/c$ is retarded time at the speed of light, and $\tau_0 = \tau - z/2\gamma_s^2 c$ is retarded time at the speed of the electrons. The angle $\hat{\theta} = k_q z - \omega_s \tau$ is a "fast" variable describing the position with respect to the ponderomotive potential of electrons which enter the magnet at time τ_0 and initial phase θ_0. The complex function

$$\sigma(z) = (\tfrac{1}{2}\pi\varepsilon_0 c)^{\frac{1}{2}} w_0 [1 + i(2z-L)/w_0^2 k_s] , \quad (4)$$

defined such that $|\sigma E_s|^2$ is the optical power, accounts for diffractive spreading and phase shift, and arises from projecting[16]

Maxwell's equation onto the fundamental Gaussian mode with beam waist w_o appropriate to the Stanford resonator.

The numerical results shown below were obtained using a Gaussian electron pulse and the fixed parameters given in Table 1. The optical power loss per pass is derived from Madey's experiments[17] in which he switched off the electrons and observed the rate of decay of radiation in the cavity. More recently, Madey has reported measurements[18] of the electron pulse shapes. These measurements show that the pulse shapes depend sensitively on the settings of the LINAC and are in general not well represented by Gaussians. This could be a partial explanation of quantitative discrepancies between our calculated results and the experimental data.

TABLE 1

Magnet length	$L = 5.2$ m
Magnet wavelength	$\lambda_q = 3.2$ cm
Magnetic field	$B_q = 2400$ Gauss
Mass shift	$\Delta = 1 + e^2B_q^2/m^2c^2k_q^2 = 1.512$
Electron energy	$\gamma_0 = \Delta^{\frac{1}{2}}\gamma_s = 84.15$
Nominal laser wavelength	$\lambda_s = 2\gamma_s{}^2\lambda_q = 3.417$ μm
Optical beam waist	$w_0 = .1714$ cm
Optical power loss per pass	3.6%
Output mirror transmission	1.5%
Electron pulse length	FWHM = 2.44 psec
Peak current	$I_0 = .315$ amp

In Fig. 3 we show the calculated and experimental power-tuning curves. The qualitative behavior is very similar. Quantitative agreement is to within factors of two or three. In particular, the highly asymmetric character of the experimental curve is reproduced, displaying a sudden shutdown of lasing as the cavity is lengthened toward the zero-delay point. The width of the theoretical curve is wide compared to experiment.[17] We have tried choosing parameters so as to match the experimental tuning range, but the electron distributions thus obtained are too narrow.

In Fig. 4 we show the steady-state optical pulse shape $|E_s(L,\tau)|^2$ obtained at the optimum delay point of Fig. 3. We find a narrowing (coherence brightening) of the optical pulses as the power level increases, associated with a broadening of the optical spectrum, which is given by the square of the Fourier transform of

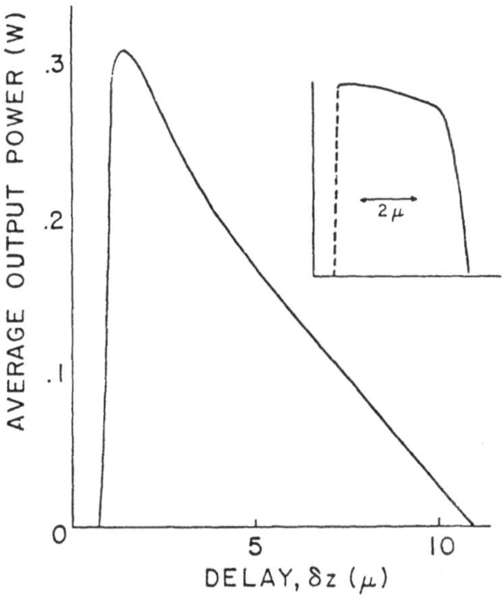

Fig. 3. Calculated power output as a function of shortening δz in the cavity length. Insert shows the experimental power-tuning curve.

E_s. In our earlier calculations[6-8] using the quasi-Bloch equations, we commonly found multiple-pulse structure (pulse ringing) in the saturated regime. In certain circumstances[7] we also observed limit-cycle oscillations in which the optical pulse shape and energy fluctuated periodically over several passes through the resonator. Our more recent calculations[9] do not show pulse ringing or limit cycles, but these effects have now been found in the calculations of several other workers[19] who have analyzed time-dependent FEL operation. Whether such effects really exist still remains an open question.

Certain qualitative features of the optical pulse shapes in the steady state bear mentioning: Light is never present behind the electrons at the end of the magnet. However, light may extend ahead of the electrons in an exponential tail. This generally happens when δt is larger than optimal. Low-energy pulses have a moderate amount of chirp, with the leading edge at a lower frequency than the trailing edge. There is also chirp in the frequency structure of highly saturated pulses, but this structure is generally more complex.

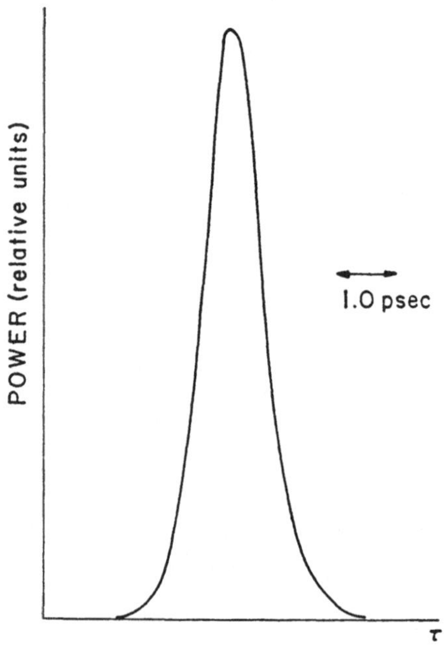

Fig. 4. Optical pulse intensity for operation at the optimal point in Fig. 3.

The optical spectrum of the pulse in Fig. 4 is shown in Fig. 5. It is convenient to measure deviations of the (angular) frequency ν from the resonance value ω_s in terms of the quantity $\delta\nu_0 = 2c\gamma_s^2/L$, which is a measure of transit-time broadening in the FEL. The small-signal gain as a function of frequency for a plane-wave mode is a well-known[20,10,13] antisymmetric expression centered at $\nu = \omega_s$ and peaked at $\nu = \omega_s - 2.6\ \delta\nu_0$. However, when diffractive effects are taken into account, the small-signal gain for a Gaussian mode[16,21,9] is given by

$$G(\nu) \propto \frac{d}{d\nu} \left| \int_0^L dz \exp\left(-i \frac{\nu-\omega_s}{\delta\nu_0} \frac{z}{L}\right) / \sigma(z) \right|^2 . \tag{5}$$

For the Stanford resonator this function is peaked at $\nu = \omega_s - 4.4\ \delta\nu_0$. When we perform pulse calculations in the small-signal regime, we find that, as one would expect, the peak of the optical spectrum is located at this same frequency. As one raises the optical power, the calculated spectrum is observed to broaden, and the peak shifts to lower frequencies. This makes sense, since we

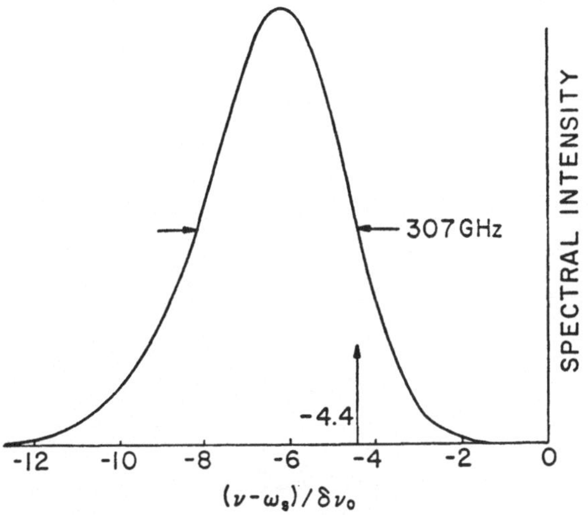

Fig. 5. Optical spectrum of the pulse in Fig. 4. Arrow indicates position of spectral peak in the small-signal regime.

are slowing down the electrons and broadening their energy distribution; the Doppler condition tells us that slower electrons radiate at longer wavelengths. It is puzzling that the experimental spectrum, although it shows some power broadening, does not show any shift in the position of its peak as the power is varied by tuning the cavity length. The experimental peak has been inferred to be at $\nu \simeq \omega_s - .9\ \delta\nu_0$ by assuming that the measured spontaneous emission spectrum peaks at $\nu = \omega_s$. However, if one takes into account the strong directional dependence of the spontaneous emission and the finite aperture and alignment uncertainty of the detector, it is likely that the measured peak is at a frequency significantly less than ω_s.

The electron energy distribution corresponding to the pulse in Fig. 4 is shown in Fig. 6, together with the Stanford experimental distribution. The calculated spectrum is formed by making a histogram of values of $Ld\hat{\theta}/dz$ for electrons arriving at $z = L$. The results shown here assume that the electrons are initially monoenergetic with $d\hat{\theta}/dz = 0$ at $z = 0$. We have also done calculations with an initial energy spread chosen to match the experiment, but find that this has little effect on the results obtained. To obtain good statistics, electron-energy data for many trips are combined in Fig. 6. Grid orientations used in performing the

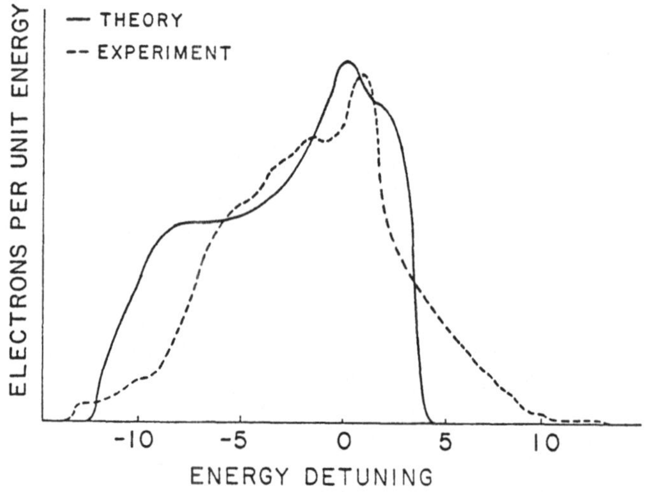

Fig. 6. Theoretical and experimental electron energy distributions. The theoretical distribution corresponds to the pulse in Fig. 4. Abscissa is in units of $Ld\hat{\theta}/dz$.

integration over θ_0 in Eq. (3) are randomized to get a good statistical sample of electrons.

The agreement between theory and experiment is seen to be very good. This shape of electron distribution is typical of what we obtain at moderate levels of saturation. At higher optical powers a prominent peak is formed by electrons whose energy has been reduced from the initial value. This peak is consistent with the single-mode phase-space picture[14] of electron motion, in which trapped electrons start with an energy above resonance and rotate in the phase plane to an energy below resonance.

In conclusion, we have seen that the phenomenon of laser lethargy plays an important role in the physics of the pulsed FEL, as it does also in swept-gain amplifiers using an atomic medium. The operation of the pulsed FEL has been analyzed numerically by solving fully self-consistent equations for the laser field and the electrons. Agreement with the Stanford experimental data is quite good, although there are still unresolved questions concerning the width of the power-tuning curve and shifts in the peak of the optical spectrum as a function of power. The shape of the electron energy distribution (which was not predicted correctly by earlier non-self-consistent calculations)[14] has been obtained in excellent agreement with experiment.

REFERENCES

1. L. R. Elias, W. M. Fairbank, J. M. J. Madey, H. A. Schwettman, and T. I. Smith, Phys. Rev. Lett. 36:717 (1976).
2. D. A. G. Deacon, L. R. Elias, J. M. J. Madey, G. J. Rasmian, H. A. Schwettman, and T. I. Smith, Phys. Rev. Lett. 38:892 (1977).
3. F. A. Hopf, P. Meystre, M. O. Scully, and John F. Seely, Phys. Rev. Lett. 35:511 (1975).
4. F. A. Hopf and P. Meyśtre, Phys. Rev. 12:2534 (1975).
5. F. A. Hopf, P. Meystre, and D. W. McLaughlin, Phys. Rev. 13:777 (1976).
6. H. Al-Abawi, F. A. Hopf, G. T. Moore, and M. O. Scully, Opt. Comm. 30:235 (1979). Early accounts of this work were given at: Tenth International Quantum Electronics Conference, Atlanta (1978); Optical Society of America Meeting, San Francisco (November, 1978); Winter Conference on the Physics of Quantum Electronics, Snowbird, Utah (January, 1979); Stanford FEL Workshop (see ref. 9).
7. Enrico Fermi School on "Developments in High Power Lasers and Their Applications", Varenna, Italy, 1978 (proceedings to be published), article by G. T. Moore, M. O. Scully, F. A. Hopf, and P. Meystre.
8. Free-Electron Generators of Coherent Radiation, Proc. Conf. on the Physics of Quantum Electronics, Telluride, Colorado, 1979, eds. S. F. Jacobs, H. S. Pilloff, M. Sargent, M. O. Scully and R. Spitzer, Vol. 7 (Addison-Wesley, 1980), article by F. A. Hopf, T. G. Kuper, G. T. Moore, and M. O. Scully, p. 31.
9. T. G. Kuper, G. T. Moore, and M. O. Scully, Opt. Comm. 34:117 (1980); G. T. Moore and M. O. Scully, talk delivered at the Stanford FEL Workshop, Stanford University (March, 1979).
10. F. A. Hopf, P. Meystre, M. O. Scully, and W. H. Louisell, Opt. Comm. 18:413 (1976).
11. F. A.Hopf, P. Meystre, M. O. Scully and W. H. Louisell, Phys. Rev. Lett. 37:1342 (1976).
12. H. Al-Abawi, F. A. Hopf, and P. Meystre, Phys. Rev. A16:666 (1977).
13. Novel Sources of Coherent Radiation, Proc. Conf. on the Physics of Quantum Electronics, Telluride, Colorado, 1977, eds. S. F. Jacobs, M. Sargent, and M. O. Scully, Vol. 5 (Addison-Wesley, 1978), article by F. A. Hopf, P. Meystre, G. T. Moore, and M. O. Scully, p. 41.

14. W. B. Colson, Phys. Lett. A64:190 (1977); W. H. Louisell, J. Lam, D. A. Copeland, and W. B. Colson, Phys. Rev. A19:288 (1979). See also articles in Ref. 12 by W. B. Colson, p. 157, and N. Kroll, p. 115.
15. A. Bambini and A. Renieri, Lett. Nuovo Cimento 31:339 (1978); A. Bambini and A. Renieri, Opt. Comm. 29:244 (1978); and A. Bambini, A. Renieri, and S. Stenholm, Phys. Rev. A19:2013 (1979).
16. G. T. Moore and M. O. Scully, Phys. Rev. A21:2000 (1980).
17. Lectures by J. M. J. Madey in Ref. 7.
18. Lectures by J. M. J. Madey at this school.
19. W. B. Colson and S. K. Ride, Phys. Lett. 76A:379 (1980); W. B. Colson and S. K. Ride in Ref. 8, p. 377; A. Bambini, R. Bonifacio and S. Stenholm, Opt. Comm. 32:306 (1980); G. Dattoli and A. Renieri, Lett. Nuovo Cimento 24:121 (1979); lectures given at this school by W. B. Colson, J. Ekstein, and A. Renieri.
20. R. H. Pantell, G. Soncini, and H. E. Puthoff, IEEE J. Quan. Elect. 4:905 (1971); V. P. Sukhatme and P. W. Wolff, J. Appl. Phys. 44:233 (1973).
21. R. Bonifacio, P. Meystre, G. T. Moore, and M. O. Scully, Phys. Rev. A21, 2009 (1980).

FREE ELECTRON LASER WAVE AND PARTICLE DYNAMICS

William B. Colson

Quantum Institute
University of California Santa Barbara
Santa Barbara, CA 93106

INTRODUCTION

In a free electron laser, a beam of relativistic electrons passes through a static periodic magnetic field to amplify a superimposed coherent optical wave (Figure 1). Here, the lasing process has been reduced to its most fundamental form and is manifestly classical in nature. This point is at the root of many of the free electron lasers potential advantages over conventional atomic lasers; many properties of atomic lasers such as efficiency, are limited by quantum mechanics. This new laser is free from the bonds constraining atomic lasers to a particular wavelength and therefore is continuously tunable. The optical cavity contains only light, radiating electrons and the magnetic field so that intense optical fields may propagate without the degrading non-linear effects (self-focussing, etc.) of denser media. The advanced technology of high-energy electron accelerators and storage rings promises efficient recirculation of the beam energy.

The earliest coherent radiation sources, radar and microwave electron tubes, used classical non-relativistic electron beams to amplify long wavelength radiation (10 cm to 0.1 cm). These devices satisfied a wide range of applications with hundreds of varied designs, but it was not possible to generate shorter wavelengths until the early sixties when atomic and molecular lasers were developed. A necessary technical advance at the time was the replacement of "closed" microwave cavities with "open" optical resonators. J.M.J. Madey's conception[1] of the free electron laser in 1971 showed how relativistic electrons and "open" resonators could extend the advantages of electron tubes to the optical regime.

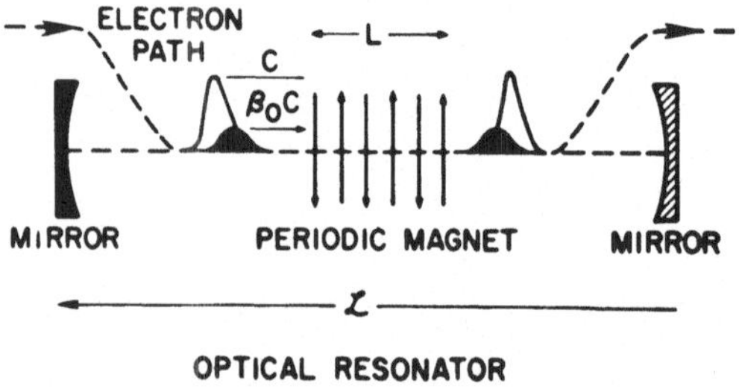

Fig. 1. Successive electron pulses travel through the periodic magnet with z-velocity $\beta_0 C$; the optical pulse is amplified as it slowly passes over oscillating electrons.

Madey and his collaborators at Stanford University demonstrated free electron laser amplification[2] in 1976 and laser oscillation in 1977. In the oscillator experiments, a nearly monoenergetic 43 MeV electron beam from a superconducting linear accelerator was passed through a 5.2 m long helical magnet with a field strength of $B = 2.4$ kGauss and wavelength $\lambda_0 = 3.2$ cm. Short 4 picosecond electron pulses of 1 amp peak current produced 2×10^7 W/cm^2 peak optical power at $\lambda = 3.4\ \mu$ wavelength with circular polarization.

The fundamental physics of free electron lasers is now well unde stood; several theoretical viewpoints adequately describe its behavio Semi-classical quantum theory, or quantum electrodynamics,[1,4-6] explains the laser action as stimulated Compton back-scattering of the virtual photons in the periodic magnet, or equivalently, as stimulated magnetic Bremsstrahlung. In this view, the finite length magnet and the resulting electron kinematics allow stimulated emissio to exceed absorption. Viewed classically,[7] the electron beam is a cold relativistic plasma;[8-10] dispersion relations from the Boltzmann equation can properly characterize the evolution of the electron distribution in the optical wave. The most fruitful and widely used theory calculates the dynamics of individual electrons[11-13] as they are affected by the fields in the laser cavity; the total transverse current then drives Maxwell's non-linear wave equation.[14] Reference 15 and this volume review most of the current theoretical and experimental work on the free electron lasers.

In the next section we develop the equations governing wave and electron dynamics. The electron phase-space, and the optical wave evolution are each examined separately. Finally, the short pulse problem of Stanford's laser is reviewed.

Formalism

In the FEL oscillator, mirrors are placed at each end of the interaction region to store radiation; fresh electrons are either supplied continuously or injected to overlap the rebounding optical pulse. As electrons enter the laser cavity, they are acted on by the static magnetic field, and the oscillating electric and magnetic components of the nearly free optical plane-wave; interparticle Coulomb forces are small for the high energy, low density beam of the Stanford experiment. The magnet guides an electron through N periodic oscillations as it travels the length of the magnet ($L=N\lambda_0$) with z-velocity $\beta_z c(\beta_z \approx 1)$; the small transverse accelerations produce a small amount of spontaneous radiation carrying the polarization of the magnet geometry: circular polarization for a helical magnet, linear polarization for alternating poles. The emission is confined to within an angle $\approx \frac{1}{2}\gamma$ (γmc^2 is the electron energy) about the forward motion, and within a narrow ($\approx \frac{1}{2}N$) spectral line-width about the fundamental $\lambda=\lambda_0(1-\beta_z)\approx\lambda_0(1+\kappa^2)/2\gamma^2$ for $\gamma>>1$ and $\kappa=eB\lambda_0/2\pi mc^2$ where $e=|e|$ and m are the electron charge magnitude and mass, B is magnetic field strength, c is the speed of light. If $\kappa \lesssim 1$, as is usually the case, there will be a small amount of emission into a few well-separated higher harmonics. The Stanford experiment gives typical values for these parameters, and has demonstrated the tunable characteristic of the laser frequency by varying the accelerator energy. In future machines the tunable wavelength range is estimated to be about a decade; this is primarily determined by the dynamic range of the electron source.

The radiation from multiple passes of the electron beam is stored in the resonant cavity. Maxwell's wave equation governs the evolution of a light wave in the presence of an electron current:

$$\left(\vec{\nabla}^2 - \frac{1}{c^2}\frac{\partial^2}{\partial t^2}\right) \vec{A}(\vec{x},t) = -\frac{4\pi}{c}\vec{J}_\perp(\vec{x},t) \qquad (1)$$

where $\vec{A}$ is the radiation vector potential, and $\vec{J}_\perp$ is the transverse current density (cgs units). When the laser is "turned on", the optical wave grows from spontaneous emission to a large amplitude wave with a well-defined phase. After the coherent wave is established, its amplitude and phase can still evolve in time. The following waveform was chosen to represent the laser optical wave during these stages of evolution:

$$\vec{A}(\vec{x},t) = \frac{E(z,t)}{k} (\sin(kz-\omega t+\phi(z,t)),\ \cos(kz-\omega t+\phi(z,t)),0) \quad (2)$$

where $E(z,t)$ is the wave amplitude; the carrier frequency is $\omega=kc$, and the phase is $\phi(z,t)$. When the amplitude and phase of this wave are held fixed, (2) describes a plane wave traveling in the z-direction.

The waveform (2) contains no dependence on x and y; a proper description would give it some finite transverse dimension. In order to address the essential physics of the problem, we choose to avoid this complication by describing dynamics well within the optical wave (an appropriate "filling factor" is included in the definition of the electron density to handle the overlap between the optical mode and the electron beam[1]).

The dynamics of electrons in the combined static and radiation fields are governed by the Lorentz force equations. A helical magnetic field of the form

$$\vec{B}_{mag} = B(\cos k_o z,\ \sin k_o z, 0) \quad (3)$$

produces the optical polarization in (2) and $\lambda_o=2\pi/k_o$ is the magnet wavelength. The radiation electric and magnetic fields are obtained from the vector potential using the slowly varying amplitude and phase approximation explained below. When both of these fields are inserted into the transverse components of the relativistic Lorentz force equations, their contributions nearly cancel in comparison to the magnet (2): $\beta_z B >> (1-\beta_z)E$ when $\beta_z \approx 1$. If injecting perfectly, the large scale, or macroscopic, helical motion is then $\vec{\beta}=\beta_o\hat{z}+\vec{\beta}_\perp$ where $\vec{\beta}_\perp=-|e|\vec{B}_{mag}/\gamma mc^2k_o$. This motion alone appears uninteresting, but it allows efficient energy exchange with the purely transverse radiation field (2) if near "resonance": $\beta_o k_o \approx k(1-\beta_o)$.

Substituting $\vec{\beta}_\perp$ into the fourth component of the Lorentz force we have

$$\frac{d\gamma}{dt} = \frac{e\kappa E}{\gamma mc} \cos(\zeta+\phi) \quad (4)$$

and

$$\frac{d\zeta}{dt} \equiv \frac{\nu c}{L} = c(\beta_z k_o - k(1-\beta_z)) \quad (5)$$

where $\gamma^2=(1+\kappa^2)/(1-\beta_z^2)$, and $\zeta\equiv(k+k_o)z-\omega t$ is the electrons phase within an optical wavelength. If an electron has a velocity such that $\nu=0$, then exactly one wavelength of light is passing over the electron as it travels through one magnet wavelength.

$\nu_0 \equiv \nu(t=0)$ is determined from initial conditions and is called the resonance parameter. E and ϕ are to be interpreted as the local radiation field and phase in the superimposed macroscopic (covering several optical wavelengths) part of the optical beam; the phase-space coordinates (ζ,ν) describe the evolution of electrons on a microscopic scale $(\lesssim\lambda)$. The coordinates (ζ,γ) may also be used since $\nu \approx 4\pi N(\gamma-\gamma_R)/\gamma_R$ near resonance where $\gamma_R^2 = k(1+\kappa^2)/2k_0$ $(\gamma \gg 1)$. The number of periods N is usually large, a few hundred, so that small changes in γ give large changes in ν. This point means that, in general, fractional changes in γ are small during a single pass through the laser, and to a good approximation (4) and (5) become

$$\ddot{\zeta} = \left(\frac{2e^2 B E}{(\gamma_0 mc)^2}\right)\cos(\zeta+\phi) = \left(\frac{\Omega c}{L}\right)^2 \cos(\zeta+\phi) \tag{6}$$

where $\gamma_0 mc^2$ is the initial electron energy, and $k \approx (1+\kappa^2)k_0/2\gamma_0^2$ has been used. 4Ω is the height of the closed orbit separatrix in the dimensionless pendulum phase-space.

Electron dynamics (6) have now been put into a form where we see that the fundamental phase-space is that of the simple pendulum. While exact for low gain, where E and ϕ are nearly constant, the pendulum phase-space is only slightly modified when more complicated effects are self-consistently included. It therefore has been and remains a valuable tool for experiments and theorists.

The optical wave evolves on a slower time scale than do individual electrons. The changes in E and ϕ then act back to slightly alter the phase-space paths guiding electrons. The amplitude and phase of the wave evolve slowly over an optical wavelength $(\dot{E} \ll \omega_r E$, etc.); a faster evolution would diminish the coherence and monochromicity of the radiation. The left-hand side of (1) can therefore be rewritten by inserting (2), and neglecting terms containing two derivatives, either spatial, temporal, or both. The remaining terms are "fast" rotating vectors with "slow" coefficients. Equations which are truly slowly-varying can be constructed by projecting the wave equation onto two unit vectors, $\hat{\varepsilon}_1 = (\cos\psi, -\sin\psi, 0)$ and $\hat{\varepsilon}_2 = (\sin\psi, \cos\psi, 0)$ to get

$$\left(\frac{\partial E}{\partial z} + \frac{1}{c}\frac{\partial E}{\partial t}\right) = -\frac{2\pi}{c}\vec{J}_\perp \cdot \hat{\varepsilon}_1$$

$$E\left(\frac{\partial \phi}{\partial z} + \frac{1}{c}\frac{\partial \phi}{\partial t}\right) = \frac{2\pi}{c}\vec{J}_\perp \cdot \hat{\varepsilon}_2 . \tag{7}$$

The second-order partial differential equation (1) has now been reduced to two first-order differential equations (7); one describing the evolution of the amplitude of the wave, the other describing the evolution of its phase. When there is no source current ($\vec{J}_\perp=0$), E and ϕ satisfy the free-space wave equation.

For relativistic electrons, the transverse radiation force is very small, so the electron's transverse velocity (and therefore the transverse current) is determined almost entirely by the static magnetic field. We project the single-particle currents, $e\vec{\beta}_\perp c$, onto the two unit vectors $\hat{\varepsilon}_1$ and $\hat{\varepsilon}_2$.

The total beam current is the sum of all single-particle currents. The electrons can be labelled by their initial positions and velocities (or, equivalently, resonance parameters); this definition is unique, and rigorously defines the electron beam current (Jean's theorem). In experimental situations, the electron pulse is large compared to an optical wavelength, so on a microscopic scale the electrons are initially spread uniformly over each wavelength of light. Although particle redistribution (bunching) does occur within an optical wavelength, it does not affect the average density in any macroscopic section of the beam several wavelengths long. Similarly, although the energy spread of the injected electron beam would generally not be large enough to result in distortion of the pulse as it travels down the magnet, it may be large enough to result in a significant spread in resonance parameters. On a macroscopic scale neither the bunching mechanism nor an initial velocity spread alter the macroscopic electron pulse shape, and it travels undistorted through the interaction region. Microscopically, however, an electron's resonance parameter, ν_0 and initial position within a wavelength of light ζ_0 (i.e., its coordinates in the pendulum phase-space) are crucial in determining the result of its interaction with the wave. The beam current density in a volume dV (which is large compared to an optical wavelength, but small compared to the pulse size) is found by averaging over ν_0 and ζ_0, then weighting this result by the macroscopic particle density $\rho(z)$ within that volume element. Indicating the appropriate microscopic average by $\langle\,\rangle$ the equation becomes

$$\left(\frac{\partial E}{\partial z} + \frac{1}{c}\frac{\partial E}{\partial t}\right) = -2\pi e\kappa\; \rho(z-\beta_0 ct) \left\langle \frac{\cos(\zeta+\phi)}{\gamma} \right\rangle \tag{8}$$

$$E\left(\frac{\partial \phi}{\partial z} + \frac{1}{c}\frac{\partial \phi}{\partial t}\right) = 2\pi e\kappa\; \rho(z-\beta_0 ct) \left\langle \frac{\sin(\zeta+\phi)}{\gamma} \right\rangle \tag{9}$$

where $\rho(z-\beta_0 ct)$ is the density of the traveling electron pulse. Within the slowly varying amplitude and phase approximation, macroscopic sections of the electron beam (those covering several

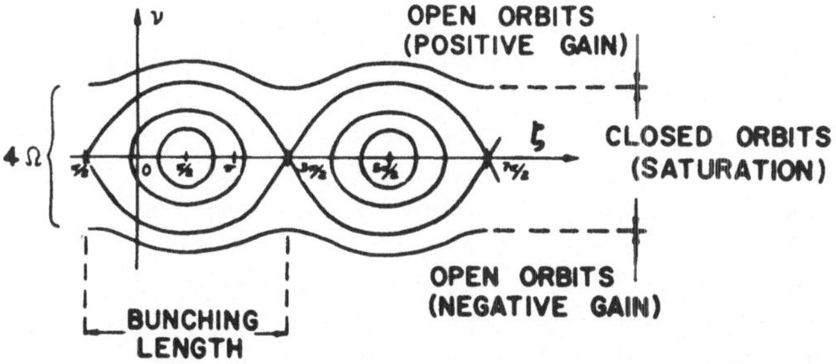

Fig. 2. The pendulum phase-space $(\zeta(t),\nu(t))$ is periodic in the optical wavelength which defines the bunching length. Electrons evolve along their paths in either the open or closed orbit regions. The optical field strength E determines the height of the closed orbit region 4Ω.

optical wavelengths) can be accurately represented in the periodic pendulum phase-space by a single section of phase-space λ long. Equations (8) and (9) are coupled through the pendulum equation (6), or the more general equations (4) and (5). In more complicated magnet structures, the electron equations (4) and (5) may be altered, but the wavelength equations (8) and (9) retain a similar form.

In their general form, the non-linear equations (8) and (9) are valid for low-gain and high-gain systems, in weak or strong optical fields. They describe the evolution for an arbitrary electron pulse shape, and the resulting amplitude and phase (and therefore the structure and spectrum) of the optical pulse. The remainder of this work explores the content of these equations.

ELECTRON PHASE SPACE EVOLUTION

The electron physics can be understood by appealing to the electron phase-space diagrams. Consider the microscopic current within a small volume of the beam. If the pendulum equation coefficient $\Omega=(2B_0E)^{\frac{1}{2}}eL/\gamma mc^2$ were truly constant, the electron phase-space would be exactly that of a single pendulum, as shown in Figure 2. Two sample electrons are included in the figure;

each electron's initial conditions determine the evolution of its "velocity" $\nu(t)=\dot{\zeta}(t)L/c$ and "position" $\zeta(t)$, and therefore constrain it to follow a particular path in phase-space. The height of the "closed-orbit region", 4Ω, is determined by the optical field strength, and is important in determining the character of electron evolution and hence the laser gain process. A large Ω "traps" a large area of the phase-space in closed orbit paths.

The fully coupled equations indicate that an electron's evolution is not governed by the exact pendulum equation, but by a "self-consistent" pendulum equation; at any instant in time, however, an electron's motion can be determined from the pendulum phase-space defined by the values of Ω and ϕ at that instant. The phase-space picture therefore remains a valuable tool in understanding beam evolution. Figures 3-5[16] show the self-consistent evolution of a monoenergetic beam with the approximate parameters of Stanford's laser (except that the optical pulse is assumed to be long and the optical cavity mode structure has not been included). The electrons phase-space paths are almost indistinguishable from pendulum paths; the self-consistent separatrix is included for reference. Since the electrons are spread uniformly over an optical wavelength, and the pendulum phase-space is periodic in the optical wavelength; it is only necessary, then, to consider a sample of electrons distributed uniformly over one optical wavelength.

In Figure 3, all electrons are injected with $\nu_o=2.6$ for maximum gain in weak fields.[11] With the optical power only 10^5 W/cm^2, all electrons fall in the open orbit region. The beam acquires a small energy spread, and some bunching about $\zeta=\pi$ can be detected. The gain equation and electron distributions have previously been derived in this regime by expanding the pendulum equation in powers of E.[11]

In Figure 4, the initial optical field is stronger (10^6 W/cm^2); the closed-orbit region has expanded, and now contains some of the electrons. The energy spread is larger, and bunching is more evident. Note that in both Figures 3 and 4 the $\zeta=\pi$ is overpopulated at the end of the laser to amplify E in (8).

In Figure 5, the field is large enough (10^7 W/cm^2) that saturation begins to occur: electrons gain and lose energy in a nearly symmetric way, and the gain (originally $\sim$15%) has dropped to $\sim$5%. When the laser oscillator reaches the point where gain per pass equals the loss per pass, it runs in a steady state. Note here that the $\zeta=\pi$ phase is overpopulated before the end of the laser! The electrons become spread again at the end of the laser and do not efficiently drive the wave. In the Stanford experiment, only a small fraction ($\approx\nu_o/4\pi N\approx 0.1\%$) of the electron beam energy

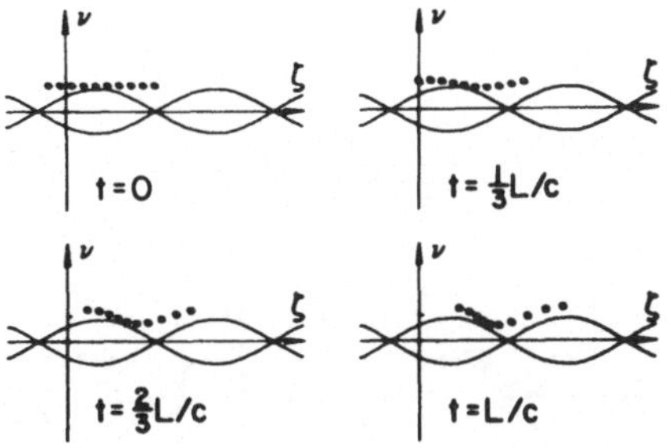

Fig. 3. In weak optical fields (power=10^5 W/cm^2), electrons evolve in the open-orbit region and acquire a small spread in energy.

is extracted at saturation.

The character of the gain process is analogous to the energy exchanged between two weakly coupled pendula (the optical wave and electron beam). For very short times, little energy can be transfrred, and for very long times, the exchange averages to zero. But for the appropriate finite time, defined by ν_0 in our case, energy flows in one direction only, giving a net transfer to the optical wave. Note that the energy density in a relativistic electron beam can be quite large; any reasonable fraction that can be transferred to an optical wave produces a sizable laser field.

Electron dynamics may also be derived from a self-consistent pendulum potential; $V(\zeta)=-(\Omega c/L)^2\sin(\zeta+\phi)$ and $\ddot{\zeta}=-V'(\zeta)$ give (6). The potential changes slowly and self-consistently with Ω and ϕ coupled to the wave equations. From this viewpoint, when electrons enter at the resonant velocity $c\lambda_0/(\lambda_0+\lambda)$, they are initially stationary on the $V(\zeta)$-surface ($\nu_0=0$); an equal number of particles "roll" ahead and back exchanging equal amounts of energy with the optical wave. There is no gain in this case. If electrons enter at a slightly higher velocity, then all electrons are initially "rolling" along the ζ-axis of the corregated V-surface. For optimum

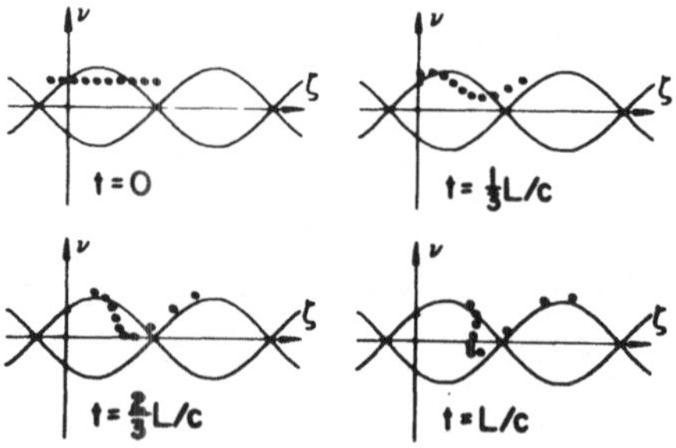

Fig. 4. In stronger fields (power=10^6 W/cm^2) Ω is larger; bunching becomes evident at the end of the laser.

gain conditions the "rolling" is slow; none will "roll" past more than one crest during the interaction time, and the electron beam will then lose energy to the optical wave. This is the gain mechanism. During amplification the initially monoenergetic, uniform beam becomes bunched at the optical wavelength and spread in energy; the fractional energy spread is $\delta\gamma/\gamma \simeq \Omega/\pi N$ for weak fields and $\approx 1/4N$ at saturation. Maximum gain for weak fields occurs when the "rolling" velocity is $\nu_0=2.6$. Absorption is predicted and observed for $\nu_0 \lesssim 0$. For typical parameters, each ampere of beam current within the optical mode cross section gives a few percent gain; one to one-hundred amps of peak current can be provided by accelerators or storage rings. The useful energy range is roughly ten to several hundred MeV; this spans a range of wavelengths from submillimeter to x-rays. Higher energies (with the best feasible magnets) result in very low gain.

After many passes of the electron beam, the intracavity optical amplitude becomes large, $V(\zeta)$ becomes large, and saturation occurs. When $\Omega \gtrsim 2.6$, there is no value of ν_0 which can prevent the nearly symmetric falling of particles into the potential troughs, the electrons become "trapped," and gain decreases. In future experiments the deep troughs may be put to an advantage, increasing the energy

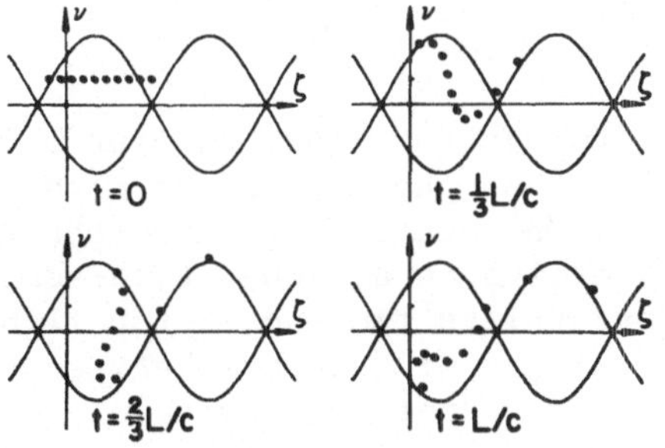

Fig. 5. Saturation occurs when the fields become so strong (power= 10^7 W/cm^2) that nearly all electrons are "trapped" in the closed-orbit region.

extracted from the electron beam and extending the laser performance. At large field strengths, electrons are trapped in the beginning stages of the magnet; the magnet (called a "tapered wiggler") is designed with a slowly decreasing wavelength so the guiding phase-space paths move down. Computer simulations show that about half the electrons remain trapped in the deep decelerating "buckets" with ≈10% (possibly 50%) energy extraction. This is the same mechanism (in reverse) used in linear accelerators; in fact, a periodic magnet with a slowly increasing wavelength and a powerful laser pulse may be used as a particle accelerator. The possibility of modified magnet geometries is an important flexibility in free electron laser design; in an atomic laser, this would correspond to altering the atomic structure, seen by an excited electron, during the emission process.

Optical Wave Evolution

After concentrating on electron dynamics, we now consider the result wave evolution. In a small section within long pulses, we can take ρ, E, and ϕ to be spatially uniform in z. Furthermore, assume small electron energy extraction ($\gamma \approx \gamma_0$) and low gain so that the pendulum equation is valid. The phase averages are difficult to perform, since the solutions to the full pendulum equation ζ are elliptic integrals. The averages start at zero since the phases are uniformly populated, but the electrons response to the wave leads to non-zero values. In weak optical fields, the pendulum equation and averages may be expanded to first-order in E and the integrals performed; this gives a maximum gain $G_{max} = 0.27 e^4 B^2 \rho \lambda_0 (L/\gamma_0 mc^2)^3$ when ν_0=2.6.

In reference 15 (Colson and Ride, Chapter 13) an approximate solution was found for these phase averages with higher order E-dependence included. On each pass through the magnet, E and ϕ are nearly constant; they evolve on a slower time-scale than the electrons. A more appropriate time-scale for light is the evolution over many passes. The light will bounce between the mirrors many times, and if electrons are continuously supplied (or injected in pulses to overlap the optical pulse), it will grow during each pass until saturation. Its growth, dE, over a number of round-trips, dn (which is $\gtrsim$1, but small compared to the characteristic evolution time), is δEdn, where δE is the growth per pass. δE will have contributions from two sources: the electron beam interaction, and the inherent losses of the optical cavity. The net growth over a single pass is found by integrating (8) and (9) from t=0 to t=L/c. E and ϕ change very little over this time-scale and the resulting integrals produce terms proportional to E and E^3 with constant coefficients. The long term behavior of E and ϕ are described by

$$\frac{dE}{dn} = \alpha E - \beta E^3$$
$$\frac{d\phi}{dn} = \alpha' - \beta' E^2 \qquad (10)$$

where n is the number of round-trips of the light in the resonator (pulses must be long enough so that every part of the pulse evolves in the same way) and the lowest order coefficients are

$$\alpha = \frac{2e^4 B^2 \lambda_0 L^3 \rho}{(\gamma_0 mc^2 \nu_0)^3} \left(1 - \cos \nu_0 - \tfrac{1}{2}\nu_0 \sin \nu_0\right) - \frac{1}{2Q}$$
$$\alpha' = \frac{2e^4 B^2 \lambda_0 L^3 \rho}{(\gamma_0 mc^2 \nu_0)^3} \left(\sin \nu_0 - \tfrac{1}{2}\nu_0 (1 + \cos \nu_o)\right) \qquad (11)$$

Note that 2α is the gain per pass and is identical with the gain coefficient derived using energy conservation;[11] Q^{-1} is the fractional power loss per pass. α and α' are exact in the weak-field, low-gain limit and are therefore fundamental results for the free electron laser. The coefficients β and β' are lengthy expressions not presented here; they are written out in reference 15. Furthermore, they are less fundamental since they are dependent on the specific higher order approximation scheme. Both α and β are antisymmetric functions of ν_o centered about resonance ($\nu_o=0$), but they are not exactly the same shape. Both α' and β' are symmetric functions of ν_o and also differ in detailed shape. β and β' are proportional to $[(2e^8B^4\lambda_oL^7\rho)/(\gamma_o mc^2\nu_o)^7]$ times a function of ν_o.

The differential equations (10) can be solved for the amplitude and phase of the wave after any pass n.

$$E^2(n) = E_o^2 e^{2\alpha n}\left(1+\frac{\beta E_o^2}{\alpha}e^{2\alpha n}\right)^{-1}$$

$$\phi(n) = \phi_o + \alpha' n - \frac{\beta'}{2\beta}\ln\left(1+\frac{\beta E_o^2}{\alpha}e^{2\alpha n}\right) \qquad (12)$$

where E_o and ϕ_o are the initial amplitude and phase and it is assumed that the laser starts far from saturation, $E_o^2 \ll \alpha/\beta$. The phase initially accumulates as $\alpha' n$; then after saturation (when $e^{2\alpha n} \gg 1$) $\phi(n) \rightarrow (\alpha' - \alpha\beta'/\beta)n$. In the early stages of evolution, the power grows exponentially then asymptotically approaches the constant value $\alpha c/4\pi\beta$.

In the laser E and λ are not externally prepared, but evolve as determined by the system parameters: ρ, B, λ_o, γ_o, N and Q. We chose $\rho=1.9\times10^9$ cm^{-3}, B=2.4 kGauss, $\lambda_o=3.2$ cm, $\gamma_o=85$, N=160, and $Q^{-1}=0.35$ to describe the Stanford laser.[2] The changing phase $\phi(n)$ is to be interpreted as an evolving laser frequency $\omega=2\pi c/\lambda$ which slowly changes the resonance parameter ν. From the definition of ν and the form of the optical wave (2), we identify $\nu(n)=\nu_o+\partial\nu(n)/\partial n|_{\nu=\nu_o}$. The shift in ν is small (only 2% of its initial value of 2.6 for maximum gain). We previously neglected this shift in the low gain limit; here we see that this was justified. During growth the shift is away from resonance (and the maximum gain point at 2.6); after saturation ν moves nearly back to the maximum gain point.

The formulation above enables us to describe the onset of free electron laser operation as a second-order phase transition. Equation (10) may be rewritten in the form $\dot{E}=-\partial\Phi(E)/\partial E$, where $\Phi(E)=-\alpha E^2/2 + \beta E^4/4$. The dynamic equation for the amplitude of the laser field is then described by the overdamped motion of the coordinate E

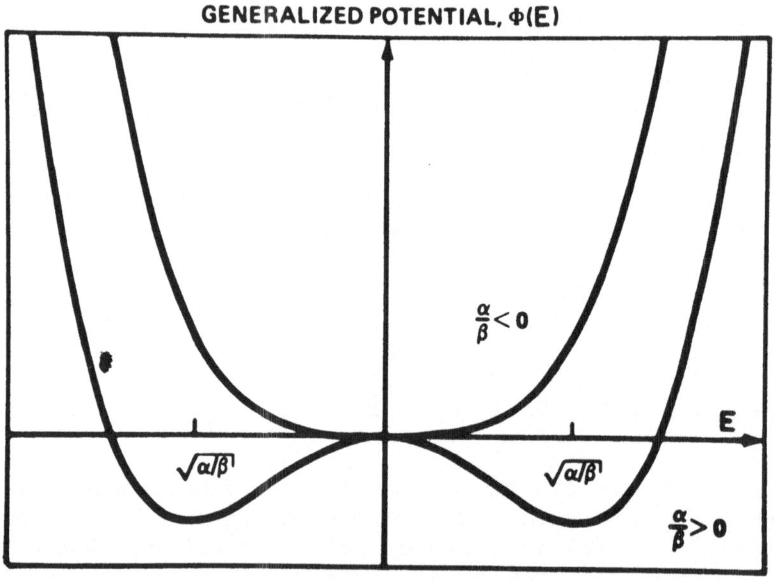

Fig. 6. If the electron current is raised above threshold, the generalized potential $\Phi(E)$ changes shape; fluctuations drive the field from zero to the new steady-state minima $\pm(\alpha/\beta)^{1/2}$.

in the generalized potential $\Phi(E)$. It is evident in Figure 6 that the behavior of E depends critically on the sign of α. If $\alpha<0$, losses in the system exceed the gain, and the field amplitude fluctuates near zero. At the critical current density $\rho\beta_0 c$ (for which $\alpha=0$) the laser reaches threshold; for greater currents the potential takes on a different form and fluctuations cause evolution to a new steady-state configuration at $E^2=\alpha/\beta$. It is important we found that $\nu(n)\approx\nu_0$ for all n so that only the amplitude evolution needs to be followed in developing $\Phi(E)$; therefore, β and β' need only be accurate near $\nu\approx\nu_0=2.6$ (as they are), the maximum gain point in weak fields.

The potential $\Phi(E)$ has the same form as the thermodynamic potentials which describe ferroelectricity, ferromagnetism (in the Ginzburg-Landau formulation), superconductivity, and laser action in atomic lasers.[17] Each of these phenomena can be described by a mean-field theory, with the result that the system changes from a disordered state to an ordered state when an external parameter attains some critical value. This low-gain analysis of the free electron laser is also a self-consistent mean-field theory. The "order parameter" is E (analogous, for example, to the magnetization in a ferromagnet). If E is small, the photon density ($E^2/4\pi\hbar k_r c$) is small, and the laser phase ϕ has no long range order. If the system has "lased", E^2 goes to its large value of α/β; the laser phase becomes ordered over many optical wavelengths (the coherence length) producing a classical electromagnetic wave. The role of spontaneous emission has been included in a quantum mechanical description of this same evolution.[5]

Short Pulse Evolution

Now that the electron and wave evolution has been explored for long uniform beams, we consider the short pulse dynamics. The system of equations (6), (8) and (9) can be solved to take into account the spatial structure of both the optical and electron pulses. The behavior of the free electron laser is, in fact, modified by short-pulse effects.[18] The shape of the optical pulse, its Fourier transform (which shows a laser line shift), and the optical pulse "slippage" over the slightly slower electrons, are all sensitive to the pulse length. This is not particularly surprising, since each of these depends on the overlap between the optical pulse and the electron beam--which for short pulses is continually changing. In Stanford's system, for example, as the short ($\sim$1 mm) pulses travel together down the 5.2 m magnet (Figure 1), the optical pulse gradually passes part of the way (.5 mm) over the electron pulse. Each section of the optical pulse sees a varying electron density; similarly, each section of the electron pulse sees a varying optical field. The evolution is therefore quite complex.

In the Stanford laser, radiation is stored in an over-moded resonant cavity, $\mathcal{L}$ =12 m long. The mode geometry causes a small change in the field's amplitude along the laser, and a more significant change in the phase of the wave. The self-consistent separatrix shifts with the phase ($\Delta\phi$/pass$\approx\pi/4$), but the qualitative behavior of the electrons remains the same. The undriven wave's amplitude and phase vary as they propagate along the laser magnet axis. E varies in proportion to $(1+(2z-L)^2/\mathcal{L}^2)^{-\frac{1}{2}}$ and the optical phase changes by $-\tan^{-1}((2z-L)/\mathcal{L})$. These undriven changes in the wave are included at each step in the systems evolution. This mode also changes the coupling in (8) and (9) and is included by introducing a "filling factor" $F(z)=F_o/(1+(2z-L)^2/\mathcal{L}^2)$ which multiplies ρ everywhere. F is the ratio of the electron beam area to the optical mode area and equals $F_o=.082$ at $z=L/2$.

In the working laser oscillator, the optical pulse remains in the resonator, bouncing between mirrors at either end. On each round-trip 3.5% of the pulse's power is lost at the mirrors. To maintain the pulse, a fresh electron beam with fixed gaussian shape is injected every cycle, and timed to overlap the rebounding optical pulse. The evolution of a low amplitude, coherent wave, can be followed through many hundreds to a thousand cycles in the resonator.

It would appear that to "synchronize" each electron pulse with the rebounding optical pulse (to have it overlap the optical pulse in the same way on each pass), the electrons should be injected every $2\mathcal{L}/c$ seconds. But while $2\mathcal{L}/c$ is the bounce time of a photon, it is not the bounce time of the centroid of the optical pulse. Since there is more gain at the end of a free electron laser than at the beginning,[11] the trailing edge of an ultra-short optical pulse experiences more amplification than its leading edge. The net effect is that the centroid of the optical pulse passes over the electrons at a speed less than $c(1-\beta_o)$, and would therefore intercept the next electron pulse later than the expected $2\mathcal{L}/c$. If the experimenter does not compensate for this effect, the optical pulse centroid will continually move back, and after many passes will no longer adequately overlap the electron pulse; when this occurs, the absorption per pass exceeds the gain, and the equilibrium oscillator power is zero.

In the Stanford experiment, the resonator length was varied until maximum steady-state power was achieved. The experimenters found that the power was sensitive to changes on the order of microns, but did not know the absolute length of the cavity within microns. We can now infer that the resonator must have been slightly shortened by $\Delta\mathcal{L}$, to decrease the $2\mathcal{L}/c$ bounce-time, and compensate for the slower speed of the optical pulse. Figure 7 shows the steady-state power as a function of $\Delta\mathcal{L}$. Our result is in fair agreement with the curve found experimentally, but is wider for the precise parameters reported.[19] One of the less well-known experimental quantities is the electron pulse length (a factor of two uncertainty), which could clearly have a serious effect on the slippage curve width. The average current is actually well-known, so that uncertainties in the electron pulse length translate into uncertainties in ρ and therefore, gain. We have seen numerically that larger gain widens the curve in Figure 7 (50% gain/pass and 10% loss/pass make the curve ≈30 microns wide) so that less gain from a slightly longer electron pulse could conceivably give closer agreement to experiment. At this point, however, there are too many uncertainties to meaningfully

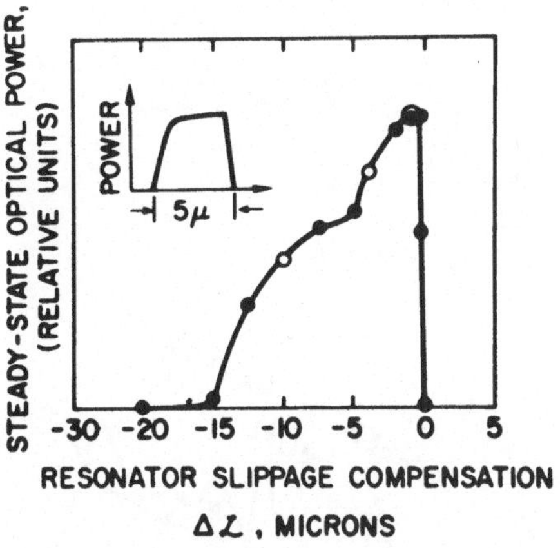

Fig. 7. The length of the resonator must be adjusted to compensate for reshaping of the optical pulse. The steady-state power is a sensitive function of the length, $\Delta\mathcal{L}$. The inset shows Stanford's experimental curve, which is similar to the theoretical curve, but not as wide.

pursue better agreement; future experiments will improve this situation.

Once the cavity has been properly adjusted, the free electron laser can evolve to, and operate in, a steady state. In Figure 8 are shown two optical pulses evolving in the Stanford laser; after several hundred passes the pulse evolution slows considerably to a steady state. The multiple peaked structure is typical of short pulses with small slippage compensation $\Delta\mathcal{L}$ ($\lesssim$2 microns); in this case the optical pulse "rides" near the middle of the electron pulse. If $\Delta\mathcal{L}$ is larger ($\gtrsim$2 microns), the optical pulse is longer with no structure and "rides" near the front of the electron pulse. The Stanford group has not yet had the opportunity to measure the structure of the optical pulse; when the measurement is performed,

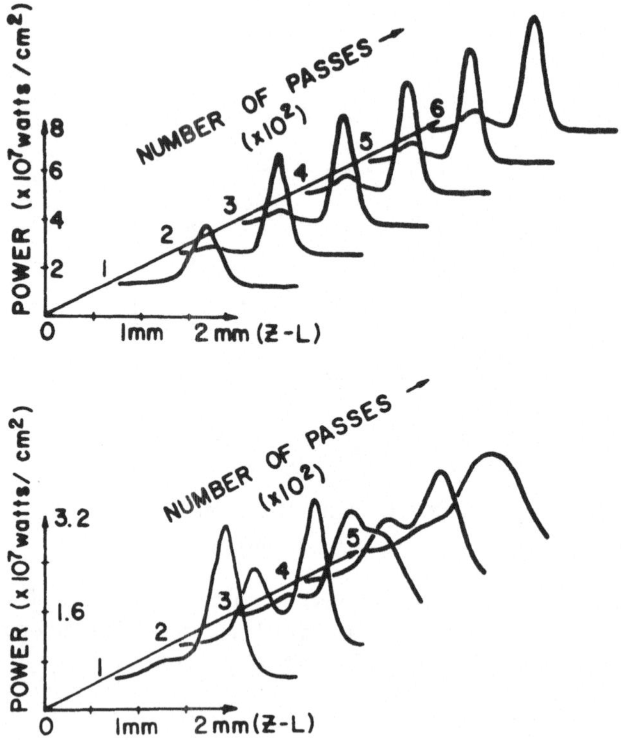

Fig. 8. In (a), the laser runs with small $\Delta\mathcal{L}$ (=½ micron) to steady-state, and in (b) the laser runs with larger $\Delta\mathcal{L}$ (= 2 micron).

it will be a good test of the predictive powers of this analytic technique.

The multiple peak structure in Figure 8 does, however, explain an observed feature of the power spectrum. The peaks (∿0.8 mm apart) would correspond to approximately a 60 GHz modulation in the laser line; the Stanford experiments do report a clear 60 GHz modulation which is indirect evidence for this multiple peak structure.

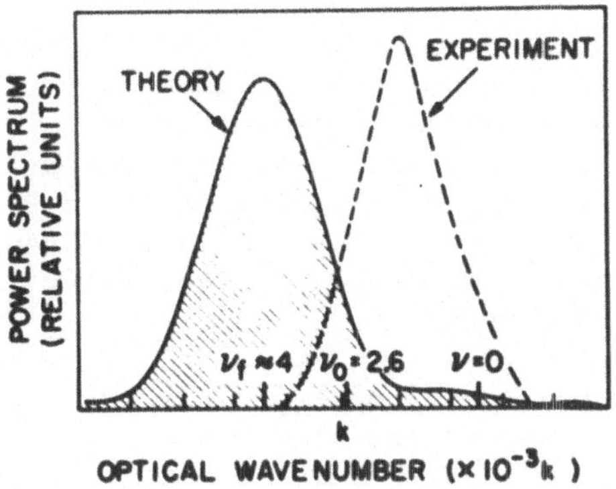

Fig. 9. The power spectrum dP/dk (laser lines) is obtained by taking the spatial Fourier transform of the pulse. The steady-state line-center is calculated to move from $k_r(\nu_o=2.6)$ to $\nu_f\approx 4$ (largely independent of $\Delta\mathcal{L}$), the resonance parameter for maximum strong-field gain. We suggest that the experimental line-center is placed too close to resonance ($\nu=0$) due to detector misalignment (by $\Theta\approx 0.0007$ radians) during the spontaneous emission measurement.

Furthermore, as $\Delta\mathcal{L}$ is increased, the modulation is observed to disappear and the power spectrum narrows; each feature is predicted by the theory here.

The spatial Fourier transform of a steady-state optical pulse yields the laser power spectrum dP(k)/dk as shown in Figure 9. A low amplitude pulse should start at $\nu_o=2.6$, the resonance parameter for maximum weak-field gain; this determines the carrier wave-number k. After many passes the pulse amplitude becomes large and the power spectrum shifts to $\nu_f\approx 4$, the resonance parameter for maximum strong-field gain.[15] The structural cause for the shift is a linear phase change along most of the pulse profile so that $\varphi\approx\delta kz$ where $\delta k/k\approx -0.0015$; the resonance parameter as observed in a detector outside the laser cavity is then shifted by $\delta\nu=-2\pi N\delta k/k=1.5$.

The theoretical width and shape of the power spectrum are in excellent agreement with experiment,[20] but the experimental laser line appears to be shifted towards resonance as determined by comparison with the spontaneous emission line-center. A possible cause for this discrepancy is a slight misalignment of the detector from the magnet axis during the spontaneous emission measurement.[21] This alignment is so delicate that $\Theta \approx 0.0007$ radians (well within the relativistic emission cone of angular width $\gamma^{-1} \approx 10^{-2}$ radians) would shift the spontaneous emission line-center up by $\delta\nu \approx 2.5$ (since $\delta\nu = 2\pi N\gamma^2\theta^2/(1+\kappa^2)$) and make the laser line appear to be shifted towards resonance by the amount shown in Figure 9. Note that there is no other determination of resonance in the Stanford experiment and such a misalignment can only cause the laser to appear shifted towards resonance as found.

No matter how the laser pulse grows, for electrons $\beta_z(\approx 1)$ remains nearly constant and the electron pulse retains its shape. On each pass all electrons are injected with the same energy, but as they respond to the local optical field a small microscopic energy change alters their resonance parameter. In reference 15 the resultant energy distribution is shown; experimental agreement is consistent with the resolution of the spectrometers.[19] For small $\Delta\mathcal{L}$, the fraction energy spread is $\Omega/\pi N$. We found that large $\Delta\mathcal{L}$ produces anamolously small electron distributions by about a factor of two. The electron moves out of the back of the optical pulse prematurely on each pass since the pulse is "riding" on the front of the electron pulse. This early decoupling fails to spread electron energies the expected amount.

The author wishes to acknowledge support by NASA Grant NSG-7490 and many helpful discussions with S.K. Ride, J.M.J. Madey, and J. Eckstein and the numerical assistance of K. Lind and R. Zarnowski.

REFERENCES

1. J.M.J. Madey, J. Appl. Phys. 42, 1906 (1971).
2. L.R. Elias, W.M. Fairbank, J.M.J. Madey, H.A. Schwettman, T.I. Smith, Phys. Rev. Lett. 36, 717 (1976); reported in Physics Today, p. 17, February (1976).
3. D.A.G. Deacon, L.R. Elias, J.M.J. Madey, G.J. Ramian, H.A. Schwettman, T.I. Smith, Phys. Rev. Lett. 38, 892 (1977); reported in Scientific American, p. 63, June (1977).
4. R.H. Pantell, G. Soncini, and H.E. Puthoff, IEEE J. Quantum Electronics 4, 905 (1968).
5. W.B. Colson, Phys. Lett. 59A, 187 (1976); W.B. Colson, Ph.D. Thesis, Stanford University (1977).
6. W. Becker and H. Mitter, Z. Physik B35, 399 (1979).
7. F.A. Hopf, P. Meystre, M.O. Scully and W.H. Louisell, Phys. Rev. Lett. 37, 1342 (1976).

8. N.M. Kroll and W.A. McMullin, Phys. Rev. 17A, 300 (1978).
9. T. Kwan, J.M. Dawson and T. Lin, Phys. of Fluids 20, 581 (1977).
10. P. Sprangle, Cha-Mei Tang, W.M. Manheimer and R.A. Smith, Naval Research Lab memorandum Report 4033 and 4034.
11. W.B. Colson, Phys. Lett. 64A, 190 (1977); W.B. Colson, "Physics of Quantum Electronics Vol. 5, Chapter 4," ed. S. Jacobs, M. Sargent, and M. Scully, Addison-Wesley Publishing Co., (1978).
12. A. Bambini and A. Renieri, Nuovo Cimento 21 , 399 (1978).
13. V.N. Baier and A.I. Milstein, Phys. Lett. 65A, 319 (1978).
14. W. Colson and S.K. Ride, Phys. Lett. 79A, 379 (1980).
15. "Physics of Quantum Electronics, Vol. 7", Ed. S. Jacobs, H. Pilloff, M. Sargent III, M. Scully and R. Spitzer, Addison-Wesley Publishing Co., (1980).
16. Similar figures in reference 15 (Colson and Ride, Chapter 13) do not have the separatrix properly positioned along the ζ-axis.
17. H. Haken, Synergetics, Springer-Verlag, (1977); V. de Giorgio and M.O. Scully, Phys. Rev. 2A, 1170 (1979).
18. F.A. Hopf, T.G. Kuper, G.T. Moore, and M.O. Scully, Physics of Quantum Electronics Vol. 7, Chapter 3", ed. S. Jacobs, H. Pilloff, M. Sargent III, M. Scully and R. Spitzer, Addison-Wesley Publishing Co., (1980).
19. J.M.J. Madey, Final Technical Report to ERDA, Contract EY 76-S-03-0326 PA 48 and PA 49 (1977).
20. In reference 15 (Colson and Ride, Chapter 13), the theoretical laser line was improperly presented as overlapping the experimental line; this was due to a numerical mistake in evaluating the Fourier transform of $\phi(z)$.
21. J.M.J. Madey, Private Communication.

ANALYSIS OF THE SINGLE PASS FREE ELECTRON LASER: THE MULTIMODE SMALL SIGNAL REGIME

Giuseppe Dattoli, Angelo Marino and Alberto Renieri

Comitato Nazionale Energia Nucleare
Centro di Frascati
C.P. 65
00044 Frascati, Rome, Italy

1. INTRODUCTION

The aim of the present paper is to review the essential features of the multimode small signal analysis of the single pass FEL by studying the main aspects of the dynamics and of the laser parameters such as the operating conditions suitable for having laser action, gain, spectrum, lethargy and so on.

The discussion of the above topics will benefit from the notations and results of Ref. 1. The formalism we developed in the above quoted reference deals with an expansion in longitudinal modes of the optical cavity described in terms of the energy densities W_i and phases φ_i (see Eq. (A.11) of Ref. 1).

We emphasize, however, that although such an expansion is in itself useful and mathematically tractable, as far as the small signal regime is concerned, the number of modes, and thus of equations, to be dealt with is very high (typically $10^3 \div 10^4$)★.

The modal expansion, therefore, becomes rather unpleasant for further physical insight and more and more cumbersome (even if feasible) for further mathematical development.

To overcome these difficulties we looked for a new expansion containing both the features of generality of the model one and new physical insight into the dynamical behaviour. To this aim we

★We recall that the number of active modes is given by $N = L_c/2\sigma_z$; thus for $L_c \sim 20$ m, $\sigma_z \sim 1$ cm we have $N \sim 10^3$ (L_c = cavity length, σ_z = electron beam length).

have introduced[2] a suitably defined expansion for FEL operation in terms of "Super Modes" (S.M.). In this lecture we review with some comments the main results of the above quoted paper.

A Super Mode can be defined as the configuration of spatial modes which reproduce themselves after one wiggler passage.

Thus each spatial mode belonging to one S.M. must obey the following equations

$$\begin{cases} \Delta W_i = \alpha W_i \\ \Delta \varphi_i = \Psi \end{cases} \tag{1.1}$$

where α and Ψ are identical for all the modes and can be understood as the gain and the advance in phase per pass respectively. The small signal variations of W_i and φ_i have been derived[1] (see Eqs (A.11) and A.13)) and read

$$\begin{cases} \Delta W_i = -W_i\gamma_T - A(1-\gamma_T)\sum_r (W_i W_r)^{1/2} \\ \qquad \left[B^c_{i,r}\cos(\varphi_i-\varphi_i) - B^s_{i,r}\sin(\varphi_i-\varphi_r)\right] \\ \Delta\varphi_i = -\omega_i\delta T + \frac{A}{2}\sum_r \left(\frac{W_r}{W_i}\right)^{1/2} \\ \qquad \left[B^s_{i,r}\cos(\varphi_i-\varphi_r) + B^c_{i,r}\sin(\varphi_i-\varphi_r)\right] \end{cases} \tag{1.2}$$

where

$$\begin{cases} \delta T = T_c - T_e \;, \; \omega_i = \text{i-th mode frequency} \\ A = 4\pi^2\left(\frac{2\lambda_o}{\lambda_q}\right)^{1/2}\left(\frac{L_w\lambda_o}{\Sigma_L}\right)\left(\frac{I}{I_o}\right)\frac{k^2}{(1+k^2)^{3/2}}\left(\frac{\Delta\omega}{\omega}\right)_o^{-2} \;; \end{cases} \tag{1.3}$$

furthermore γ_T is the cavity loss and (T_c, T_e) are the cavity round trip period and the electron bunch-bunch time distance, λ_q and $\lambda_o = (\lambda_q/2)(1 + k^2)(m_o c^2/E)^2$ are the wiggler wave-length and the resonant one (respectively), L_w is the wiggler length, Σ_L is the laser beam (l.b.) cross section, I, $I_o = ec/r_o$ are the average electron current and the Alfvén one, $k = (e/2\pi)B_o\lambda_q/m_o c^2$ is the wiggler parameter, and $(\Delta\omega/\omega)_o = \lambda_q/2L_w$ is the homogeneous band-width. A particularly important role is covered by the terms $B^{c,s}_{i,r}$ which account for the coupling between the longitudinal modes:

$$\begin{cases} B^{c}_{i,r} = \mathrm{IRe}\, B_{i,r} \\ B^{s}_{i,r} = -\,\mathrm{IIm}\, B_{i,r} \end{cases} \tag{1.4}$$

$$\begin{cases} B_{i,r} = \displaystyle\int f(z)\, E(\nu_i, \nu_r) \exp\left[j(\nu_i - \nu_r)\frac{1}{\mu_c}\frac{z}{\sigma_z}\right] dz \\ E(\nu, \nu_o) = \exp(j\nu) \left\{ \dfrac{1}{\nu - \nu_o}\left[\dfrac{1 - \exp(-j\nu_o)}{\nu^2} - \dfrac{1 - \exp(-j\nu)}{\nu_o^2}\right] \right. \\ \qquad \left. - \, j/\nu\nu_o \right\}, \quad (j = \sqrt{-1}) \end{cases} \tag{1.5}$$

where

$$\begin{cases} \nu_i = \pi \left(\dfrac{\Delta\omega}{\omega}\right)_o^{-1} \dfrac{\omega_o - \omega_i}{\omega_o}, \quad \omega_o \quad \dfrac{2\pi c}{\lambda_o} \\ \mu_c = \dfrac{(\Delta\omega/\omega)_c}{(\Delta\omega/\omega)_o} = \dfrac{\lambda_o/2\sigma_z}{(\Delta\omega/\omega)_o} \,. \end{cases} \tag{1.6}$$

Finally, the function f(z) accounts for the electron beam (e.b.) distribution along the longitudinal direction.★

★ We assume that the energy spread and the e.b. emittance are small enough that the line is homogeneously broadened (for further comments see Ref. 4, Sect. 3.1).

We could now proceed as in Refs. 3,4 by introducing a complex quantity

$$x_i = (W_i)^{1/2} \exp (j\varphi_i) \; ; \tag{1.7}$$

This is very useful because it allows us to combine Eqs (1.1) into a single one, which can be written in vector form. The structure of the equation we obtain is identical to that derived in Refs. 3, 4, with the obvious difference that the first is concerned with the electron configurations "undistorted" by the FEL interaction,[2] and the latter with the steady-state regime with recirculated e.b.[3,4].

Such an equation can be integrated numerically to obtain the main physical features. Let us stress, however, that such a procedure is rather formal and no dynamical behaviour clearly emerges before we perform the numerical integration. Let us therefore look for a reformulation of our main equations in terms of a quantity which has a well defined physical meaning. To this aim we introduce a quantity linked to the electric laser field, rather than to the longitudinal cavity modes, defined by ($k_i = \omega_i/c$)

$$\begin{aligned} \zeta(s) &= \sum_i (W_i)^{1/2} \exp\left[j\left(\varphi_i - (k_i - k_o)\, s\right)\right] \\ &= \sum_i x_i \exp\left(\frac{j\nu_i}{\mu_c \sigma_z} \frac{s}{L_c}\right) . \qquad (L_c = cT_c) \end{aligned} \tag{1.8}$$

Because the laser field is assumed to be circularly polarized it is easily realized that $\zeta(s)$ is proportional to $E_x + jE_y$ ($E_{x,y}$ being the electric field transverse components). Furthermore, the laser power density profile because of (1.8) may be written

$$\frac{dP_L}{ds} = c\,|\zeta(s)|^2 \; . \tag{1.9}$$

In this connection Eqs (1.1) can be rewritten as (for further details see Ref. 2)

$$q_\gamma \, \zeta(s) + (\Delta\Theta) \frac{d\zeta(s)}{ds} = - j \frac{(2\pi)^{3/2}}{\mu_c \, \Delta^2} \int_0^{\Delta} ydy \, \zeta(s+y) \int_{s+y}^{s+\Delta} f(z)dz. \tag{1.10}$$

where

$$q_\gamma = \frac{1}{g_0} \left\{ \frac{\alpha + \gamma_T}{1 - \gamma_T} + 2j \left[\Psi - \frac{\pi}{2} \left(\frac{\Delta\omega}{\omega}\right)_0^{-1} g_0 \Theta \right. \right\} \tag{1.11}$$

and

$$\left\{ \begin{array}{l} \Theta = - \frac{2}{\pi} \left(\frac{\Delta\omega}{\omega}\right)_0 \frac{\omega_0 \delta t}{g_o} \\ \\ \Delta = \mu_c \sigma_z \equiv \text{e.b. - l.b. slippage (see next section for further physical insight on this parameter).} \end{array} \right. \tag{1.12}$$

In deducing the equation above we have imposed the following periodicity condition

$$\zeta \left(- \frac{L_c}{2}\right) = \zeta \left(\frac{L_c}{2}\right) = 0 \quad . \tag{1.13}$$

Let us now consider the physical meaning of Eq. (1.10). We have to stress that it is an integro-differential equation fully equivalent to Eqs (1.1), and thus its solutions give those field amplitudes which reproduce themselves after a passage throughout the wiggler. Furthermore, recalling that $\zeta(s)$ is nothing but the electric field, Eq. (1.10) states that its value at a given point s depends on the value which the field assumed at a point s + y, where y ranges from 0 to the e.b. - l.b. slippage Δ due to the different velocities between electrons and photons.

Some further remarks on the mathematical structure of Eq. (1.10) are now in order. We note that we can rewrite it as an integral equation of the type★

★Note that the Super Modes will be eigenfunctions of Eq. (1.14)

$$q_\gamma \zeta(s) = \int K\,(s,\, s_o)\ \zeta\,(s_o)\ ds_o \tag{1.14}$$

where the kernel K (s, s_o) can be rewritten as follows

$$K(s,\, s_o) = \Delta\ \Theta\ \delta'\,(s - s_o) - j\ \frac{(2\pi)^{3/2}}{\mu_c\ \Delta^2} \int_{s_o}^{s+\Delta} f(z)dz\ . \tag{1.15}$$

$$\theta(s_o - s)\ \vartheta\,(s - s_o + \Delta)\ (s_o - s)\cdot\ ,$$

$\delta'(s - s_o)$ and $\vartheta(s - s_0)$ being the derivative of the Dirac function and the step function, respectively.

We emphasize that the operator $\hat{K}$ defining the kernel is not an Hermitian one, but fulfills the following condition

$$\left[\hat{K},\ \hat{K}^+\right] = 0\quad , \tag{1.16}$$

i.e., it commutes with its self-adjoint; in other words, it is a normal operator.[5] The spectrum of eigenvalues of a normal operator is not complete but overcomplete, but we can overcome this difficulty by considering a restriction of $\hat{K}$ to a suitable subspace where we can define a complete basis.

2. NUMERICAL RESULTS

Equation (1.10) is very general and we must specialize it to the particular conditions relevant to the single pass machines.

Let us therefore recall that in such machines the radiofrequency accelerating system bunches the electrons in short packets, so that if σ_z is the r.m.s. e.b. length we have

$$\sigma_z << L_c\quad . \tag{2.1}$$

Finally, since the number of modes is very large, the discrete spectrum we are analyzing can be approximated with a continuous one. Such an approximation amounts to requiring

$$\Delta << L_c \quad . \tag{2.2}$$

In the framework of the above approximations we can now discuss the numerical analysis of Eq. (1.10).

Let us notice that our main physical quantities will be functions of the parameters Θ and μ_c.

We now will discuss the behaviour of $\mathbb{R}e\, q_\gamma^r$ and $\mathbb{I}m\, q_\gamma^r$ linked, respectively, to the gain and advance in phase per pass of the S.M. by the relations (see Eq. (1.11))

$$\begin{cases} \alpha^r = g_o\,(1-\gamma_T)\ \mathbb{R}e\, q_\gamma^r - \gamma_T \\ \psi^r = \dfrac{g_o}{2}\left[\mathbb{I}m\, q_\gamma^r + \pi\left(\dfrac{\Delta\omega}{\omega}\right)_o^{-1}\Theta\right] \end{cases} , \tag{2.3}$$

where the superscript r labels the order of the S.M.. In Fig. 1 we plot $\mathbb{R}e\, q_\gamma^r$ ($r = 1\ldots 4$, $\mathbb{R}e\, q_\gamma^{r+1} < \mathbb{R}e\, q_\gamma^r$) vs Θ at fixed $\mu_c = 1$. We notice that positive values of these quantities correspond to

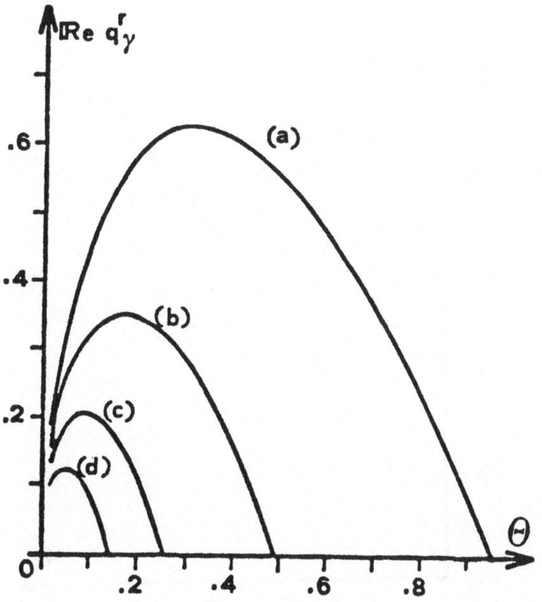

Fig. 1. $\mathbb{R}e\, q_\gamma^r$ vs Θ for the first four eigenstates ($\mu_c = 1.0$), (a) r = 1, (b) r = 2, (c) r = 3, (d) r = 4

$0 < \Theta < \Theta_M$, thus Θ being linked to $-\delta T$ we must require that the optical cavity round trip must be shorter than the time distance between two adjacent electron bunches. This situation is not new for us. We encountered an analogous behavior in the study of the Storage-Ring (S.R.) operation of FEL[3,4], and is due to the so-called "lethargy" phenomenon (already investigated in Refs 6,7). Roughly speaking we have that the interaction between the e.b. and l.b. is such that the l.b. is pushed back towards the trailing edge of the e.b..

For the sake of completeness we plot vs Θ in Fig. 2 (for $\mu_c = 1,2$) the part accounting for the dispersive behaviour of the first S.M., i.e., $\mathrm{Im}\, q_\gamma^1$. Note that to the various order of the S.M. (r = 1... 4) correspond (practically) identical values.

Up to now we have exploited the "lethargy"-behaviour of FEL which, as already stressed,[3] is due to the "group-slippage" between electrons and photons. We have now to exploit the consequences of the different velocities between the l.b. and the e.b..

It is readily understood that if the electrons and the photons are initially shifted by a certain amount $\bar{s}$ we will have more or less gain depending on the fact that the l.b. exploits a more or less conspicuous part of the e.b..

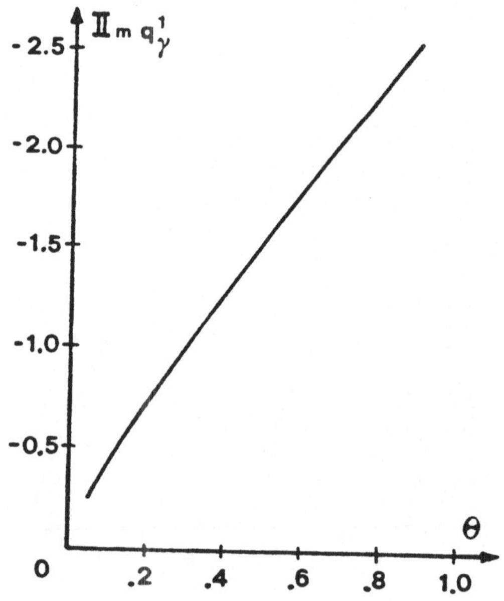

Fig. 2. $-\mathrm{Im}\, q_\gamma^1$ vs Θ ($\mu_c = 1.0$)

This situation is clearly depicted in Fig. 3 where we have plotted vs Θ (for $\mu_c = 1,2$) $\mathrm{IRe}\, q_\gamma^1$ and the quantity $\bar{\tau}$ defined as

$$\bar{\tau} = \frac{\bar{s}}{\sigma_z} \tag{2.4}$$

$\bar{s}$ being the shift between the l.b. and the e.b. at the interaction starting time.

From the figure it appears that the maximum of $\mathrm{IRe}\, q_\gamma^1$ corresponds to a value of $\bar{\tau}$ which is $-\mu_c/2$, that is $\bar{s} = -\mu_c \sigma_z/2$. Why this corresponds to the optimum shift is readily understood if one recalls that the e.b. - l.b. slippage (due to the different velocities of the two beams) after a wiggler passage, is just given by $\Delta = \mu_c\sigma_z$ (see Refs 3,4 for further details)*

*One might ask whether analogous conditions hold for the other orders of eigenvalues; the answer is that the situation is complicated by the fact that the definition of the "optimum" $\bar{\tau}$ depends on the appearance in $|\zeta(s)|^2$ of various peaks (see below for further comments).

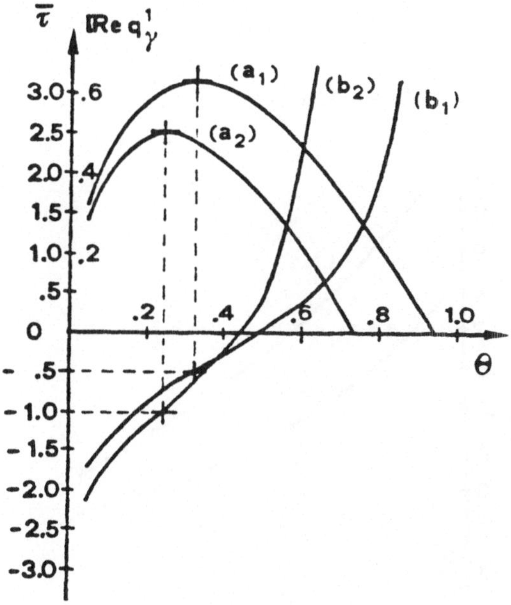

Fig. 3. Curves (a_1), (a_2): $\mathrm{IRe}\, q_\gamma^1$ vs Θ
Curves (b_1), (b_2): $\bar{\tau}$ vs Θ
(a_1) (b_1) ($\mu_c = 1.0$)
(b_2) (b_2) ($\mu_c = 2.0$)

We can now stress the different behaviour with respect to the case of the S.R.. Indeed as to the S.R. we assumed $\mu_c \ll 1$, i.e., $\sigma_z \gg L_w/\ \lambda_0/\lambda_q$, which corresponds to an optimum $\bar{\tau} \sim 0$ (see Refs. 3,4).

Further improvements in the above considerations come from the Fig. 4, where we have plotted as a function of μ_c the maximum of $\mathrm{I\!Re}\, q_\gamma^1$ and the maximum Θ (Θ_M) for which we have positive gain. It can be noticed that to higher values of μ_c correspond smaller values of σ_z compared to the vacuum phase slippage Δ. At the same time, the working region for Θ becomes smaller, that is the FEL operation is more critical (mirror positioning, e.b. timing etc.).

After having discussed the behaviour of the "absorptive" and "dispersive" part as functions of our main parameters Θ and μ_c, let us now discuss the main features concerned with the laser spectrum, bandwidth and so on.

To this aim let us introduce the Fourier transform of $\zeta(s)$, which in the continuous spectrum approximation reads

$$x(\nu) = \frac{1}{2\pi\Delta} \int_{-\infty}^{+\infty} \zeta(s) \exp\left[- j\nu \frac{s}{\Delta}\right] ds \quad . \tag{2.5}$$

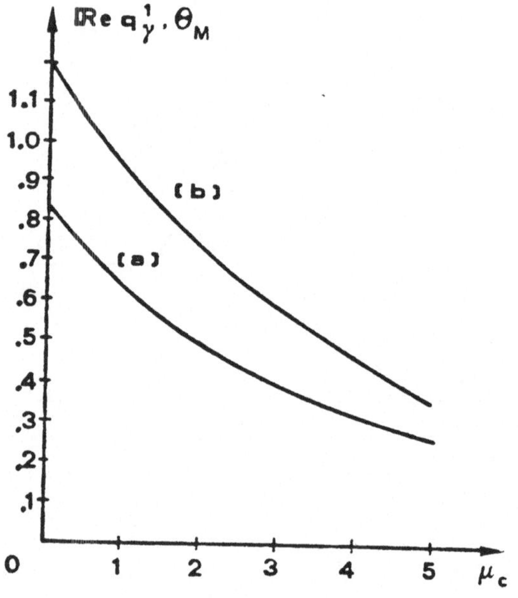

Fig. 4. Curve (a) $\mathrm{I\!Re}\, q_\gamma^1$ (MAX) vs μ_c
Curve (b) Θ_M vs μ_c

In Figures 5,6 we show $|x(\nu)|^2$ and $|\zeta(s)|^2$, respectively, for the eigenvalue r = 1 at $\mu_c = 1$ for different values of Θ (= .1, .4, .7). The dotted curve in Fig. 5 refers to the spontaneous emission spectrum,[8] while that in Fig. 6 to the e.b. longitudinal distribution. From Figure 5 it appears that for increasing Θ we have decreasing bandwidth, while from Fig. 6 it appears that as the l.b. is pushed ahead of the e.b. its length decreases.

Further improvement in the above behaviour is depicted in Fig. 7, where we have plotted vs Θ at $\mu_c = 1$ the quantities σ_ν and σ_τ, defined as the standard deviation of ν and τ; these are related to the laser bandwidth and length by

$$\begin{cases} \left(\frac{\Delta\omega}{\omega}\right)_L = \frac{1}{\pi}\left(\frac{\Delta\omega}{\omega}\right)_o \sigma_\nu \\ \sigma_s = \sigma_z \sigma_\tau \quad . \end{cases} \tag{2.6}$$

From Figure 7 we also find the features qualitatively exhibited by Figs 5,6, i.e., with increasing Θ the laser bandwidth decreases and the l.b. length increase.

Finally, in Figures 8,9 we illustrate $|x^r(\nu)|^2$ and $|\zeta^r(s)|^2$ (r = 1, 2, 3) at fixed Θ (=.2) and μ_c(=1). A remarkable feature is that when the order of the S.M. increases, $|\zeta^r(s)|^2$ shows an increasing number of peaks, so that more and more sidebands appear in the laser spectrum $|x^r(\nu)|^2$.

Some comments are now in order.

Recalling the relation which links the $\mathrm{Re}\, q_\gamma^r$ to the gain of the supermode, we have to remark that once one fixes a particular value of Θ and the level of the cavity losses, one can ignore those supermodes which play no role. It is, furthermore, very likely that the eigenstates above threshold interfere and thus the the resulting spectrum exhibits a number of peaks. This argument is, obviously, only qualitative. A correct analysis of this interference requires knowledge of the amount of the contribution of each eigenvalue to the spectrum. We remark that the quantitative knowledge of each contribution requires a strong signal analysis.

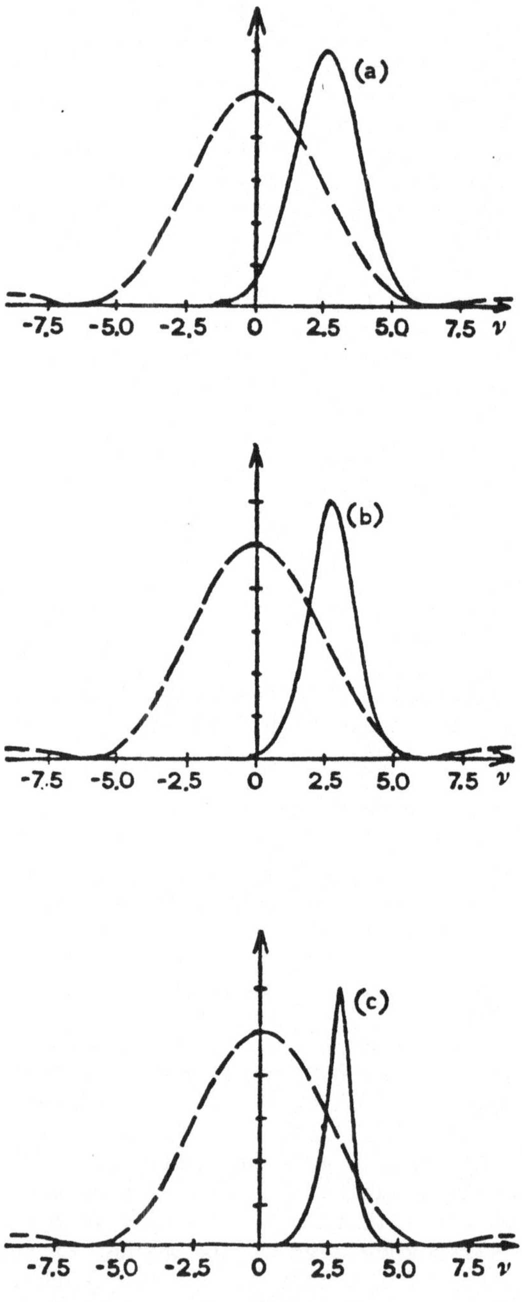

Fig. 5. Spectrum relative to the first eigenstate for $\mu_c = 1.0$ dotted curve $\equiv$ spontaneous emission spectrum (a), $\Theta = .1$, (b) $\Theta = .4$, (c) $\Theta = .7$

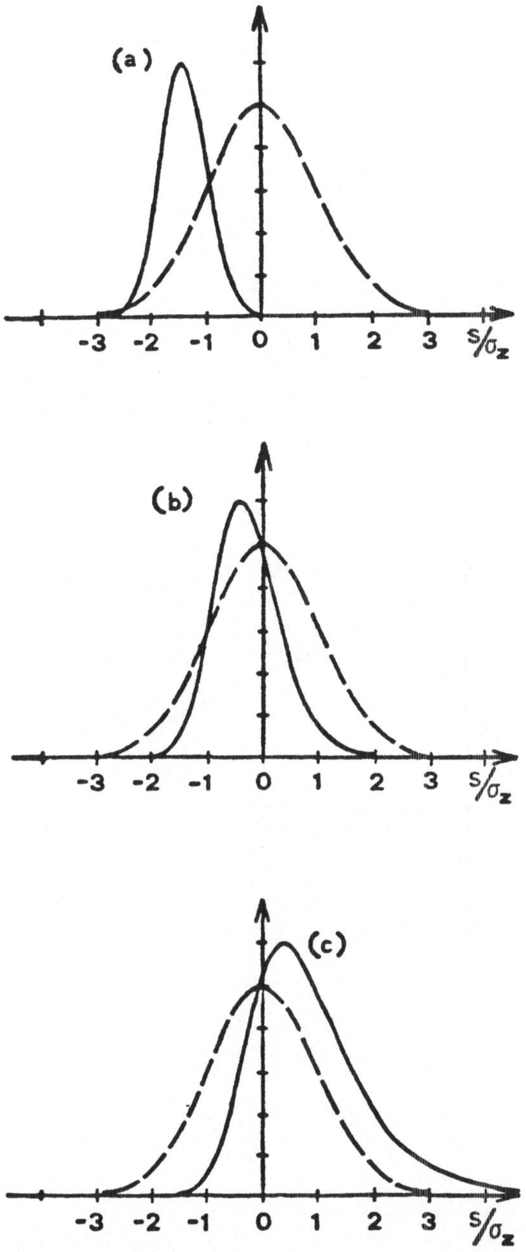

Fig. 6. Power density profile relative to the first eigenstate for μ_c = 1.0 dotted curve ≡ e.b. density profile
(a) Θ = .1, (b) Θ = .4, (c) Θ = .7

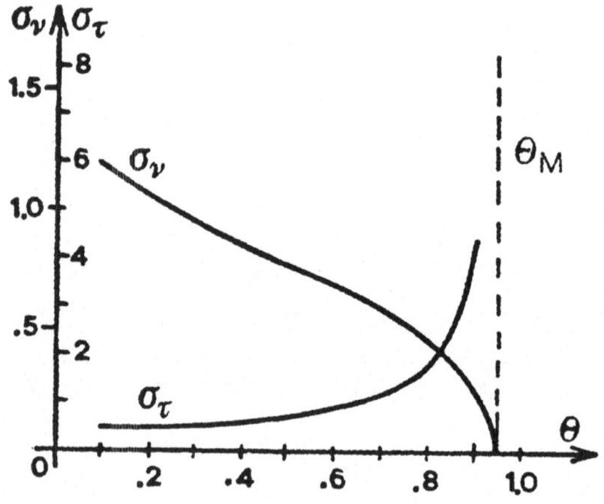

Fig. 7. Curve (a): σ_ν vs Θ (μ_c = 1.0, first eigenstate)
Curve (b): σ_τ vs Θ (μ_c = 1.0, first eigenstate)

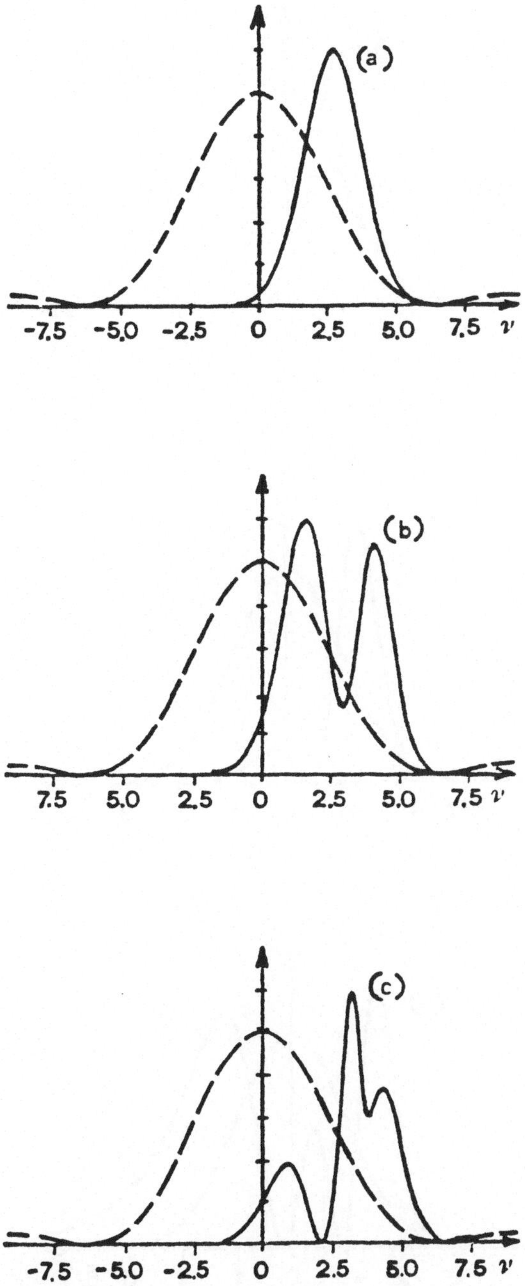

Fig. 8. Spectrum relative to the first three eigenstates for $\mu_c = 1.0$, dotted curve ≡ spontaneous emission spectrum (a) r = 1, (b) r = 2, (c) r = 3

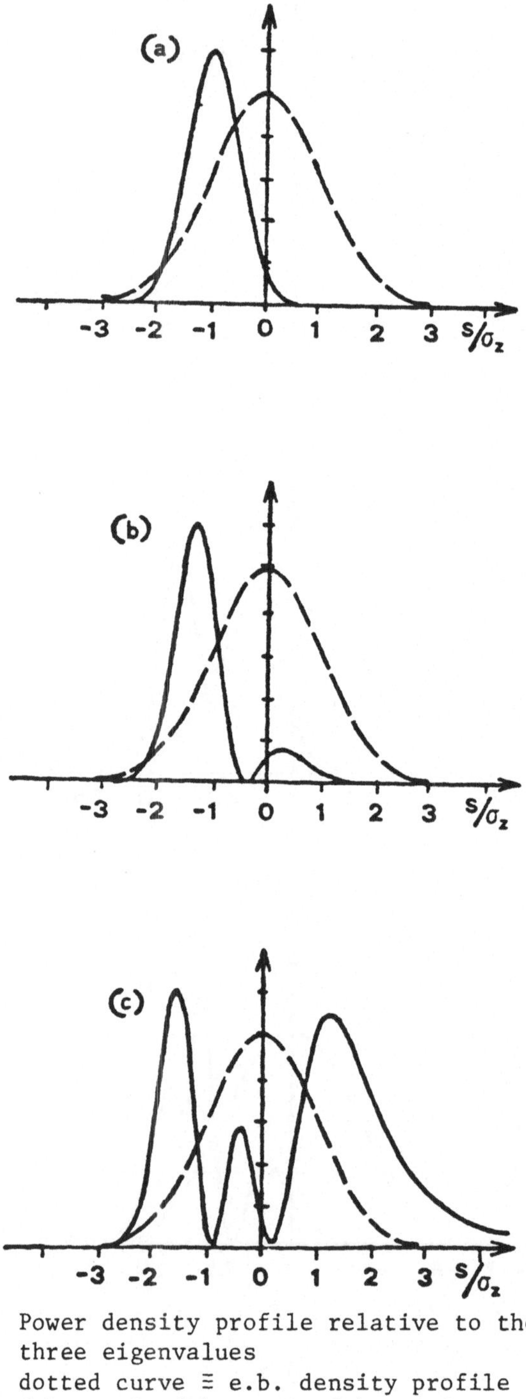

Fig. 9. Power density profile relative to the first three eigenvalues
dotted curve ≡ e.b. density profile
(a) r = 1, (b) r = 2, (c) r = 3

REFERENCES

1. G. Dattoli, A. Marino, A. Renieri, Lectures given at the Course "Physics and Technology of Free Electron Lasers", Erice (1980) to appear in the proceedings of the school, C.N.E.N. Report 80.23/cc, Centro di Frascati, Frascati, Rome, Italy
2. G. Dattoli, A. Marino, A. Renieri, C.N.E.N. Report 80.29/p, Centro di Frascati, Frascati, Rome, Italy, to appear in Opt. Commun.
3. G. Dattoli, A. Renieri, C.N.E.N. Report 80.2/p, Centro di Frascati, Frascati, Rome, Italy, to appear in Nuovo Cimento B
4. G. Dattoli, A. Marino, A. Renieri, Lectures given at the Course "Physics and Technology of Free Electron Lasers", Erice (1980) to appear in the proceedings of the school, C.N.E.N. Report 80.22/cc, Centro di Frascati, Frascati, Rome, Italy
5. M. Reed, B. Simon, "Method of Modern Mathematical Physics", Ed. by Academic Press (1978) Vol. IV, p. 203
6. H. Al-Abawi, F.A. Hopf, G.T. Moore, M.O. Scully, Opt. Commun. 30, 235 (1979)
7. W.B. Colson "Free Electron Laser Wave and Particle Dynamics" Lecture given at the Course "Physics and Technology of Free Electron Lasers", Erice (1980), to appear in the proceedings of the school
8. G. Dattoli, A. Renieri, Nuovo Cimento Lett. 24, 121 (1979)
G. Dattoli, A. Renieri, C.N.E.N. Report 79.37/p, Centro di Frascati, Frascati, Rome, Italy, submitted for publication

THE NON-LINEAR THEORY OF THE FREE ELECTRON LASER WITH TRANSVERSE DIMENSIONAL EFFECTS

P. Sprangle
Naval Research Laboratory
Washington, D.C. 20375

Cha-Mei Tang
JAYCOR
Alexandria, VA 22304

ABSTRACT

This paper treats finite transverse dimensional effects associated with i) the wiggler field, ii) electron beam and iii) radiation beam in a steady state free electron laser (FEL) in an amplifying configuration. Our formulation incorporates various efficiency enhancement schemes such as spatially contouring the wiggler field as well as accelerating the electron beam. Three dimensional effects associated with the wiggler field introduce an axial velocity shear which is equivalent to a beam temperature. For a given amplitude input signal, the transverse gradients in the wiggler places a restriction on the electron beam radius. The restriction on the radius insures that a substantial fraction of beam electrons will be trapped in the ponderomotive brackets. We also find that the radiation field experiences diffraction and refraction effects in such a way that the superposition of the input radiation field and excited radiation field do not destructively interfere. Finally, a 3-D numerical illustration of a 10.6 μm FEL with enhanced efficiency is given.

INTRODUCTION

Numerous publications have treated the 1-D free electron laser (FEL) mechanism.[1-16] As of the writing of this paper, 3-D effects in the FEL have received little attention.[17] It is the purpose of this paper to present a general non-linear 3-D formulation of the FEL in the steady state amplifying configuration including the various efficiency improvement schemes.[7,9,10]

Three dimensional effects associated with the wiggler field introduce important modification in the electron dynamics. When a cold, axially propagating electron beam enters a physically realizable wiggler field (one having transverse spatial variations and satisfying $\nabla x B_W = \nabla \cdot B_W = 0$), a radial variation, i.e., shear, in the axial electron velocity results and is equivalent to a beam temperature. If the effective beam temperature is large, it will significantly reduce the fraction of electrons trapped in the ponderomotive potential buckets.

The effects of finite transverse dimensions of the radiation and electron beam are interrelated. The radiation beam, which experiences both diffraction as well as refraction, can be represented as the superposition of the input field and excited field. In the absence of a detailed 3-D analysis and considering only diffraction effects, it has been argued that these two fields could destructively interfere on axis. This would result in a decrease in the depth of the ponderomotive wave and cause detrapping of the electrons. Our analysis of the problem shows that the combination of refraction and diffraction effects do not result in destructive interference.

A number of efficiency enhancement schemes for the FEL have been identified. Improved efficiency can be achieved by any or all of the following methods: i) contouring, spatially in the longitudinal direction, the amplitude and/or wavelength of the magnetic wiggler field, ii) accelerating the electron beam by applying an external D.C. accelerating electric field. By applying one or more of these efficiency enhancement schemes, the phase of the electrons trapped in the ponderomotive wave can be adjusted so that electron kinetic energy is converted into radiation. Our formulation will include and show the equivalence of the above enhancement schemes.

PARTICLE DYNAMICS - DERIVATION OF PHASE EQUATION

Our model of the FEL configuration is shown in Fig. 1. The generalized linearly polarized wiggler and radiation field are represented by the following vector potentials.

$$\underset{\sim}{A}_w(y,z) = A_w(z) \cosh (k_w(z)y) \cos(\int_0^z k_w(z')dz')\hat{e}_x \qquad (1a)$$

$$\underset{\sim}{A}_R(x,y,z,t) = A(x,y,z) \sin (\frac{\omega}{c}z - \omega t+\varphi (x,y,z))\hat{e}_x \qquad (1b)$$

where $A_w(z)$ and $k_w(z) = 2\pi/\ell_w(z)$ are the slowly varying amplitude and wavenumber of the wiggler field, ℓ_w is the wiggler wavelength and A and φ are the slowly varying amplitude and phase of the total radiation field. We also include an external accelerating D.C. electric field $\underset{\sim}{E}_{ac}(z) = - \partial\phi_{ac}(z)/\partial z\ \hat{e}_z$. In all cases of interest $|\underset{\sim}{A}_w| >> |\underset{\sim}{A}_R|$ by many

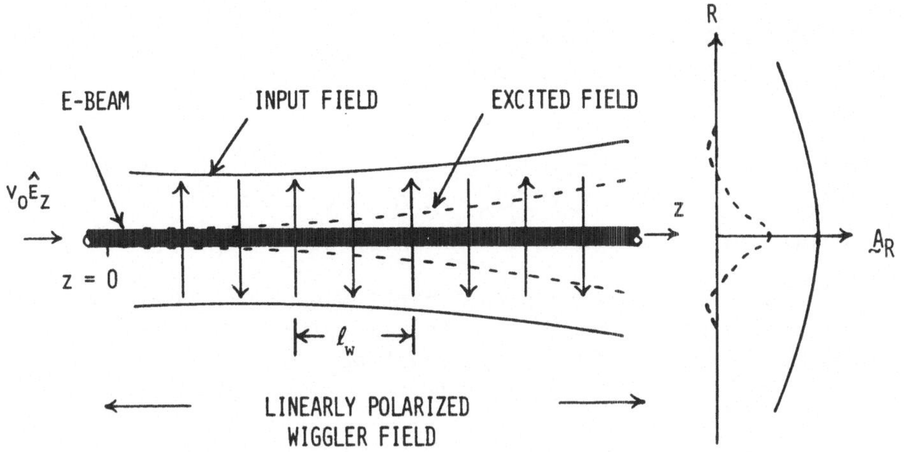

Fig. 1. Schematic of electron and radiation beam in 3-D FEL amplifier configuration.

orders of magnitude, furthermore, it is necessary to have $k_w r_b << 1$ where r_b is the electron beam radius.

To describe the electron dynamics we start with the equation of motion for the axial electron velocity which can be written as

$$\frac{dV_z}{dt} = \frac{|e|}{\gamma_z^2 \gamma m_o} \frac{\partial \phi_{ac}}{\partial z}$$

$$- \frac{|e|^2}{2\gamma^2 m_o^2 c^2} \left[\frac{\partial}{\partial z} \left(A_w \cosh(k_w y) \cos\left(\int_0^z k_w dz'\right)\right)^2 \right. \tag{2}$$

$$\left. + 2k_w A_w \cosh(k_w y)\, A \cos\left(\int_0^z \left(\frac{\omega}{c} + k_w\right) dz' - \omega t + \varphi\right) \right]$$

where $V_z = V_z(x,y,z,t)$ is the axial electron velocity, $\gamma_z = (1-V_z^2/c^2)^{-1/2}$, $\gamma = \gamma_z \gamma_{o\perp}$, $\gamma_{o\perp} = (1 + (|e|\, \underset{\sim}{A}_w(z)/(m_o c^2))^2)^{1/2}$.

In obtaining Eq. (2) we have taken $\omega \approx \gamma_z^2(1 + V_z/c)\, ck_w$ and the x component of electron momentum to be $(|e|/c)\,(\underset{\sim}{A}_w(y,z) + \underset{\sim}{A}_R(x,y,z,t))\cdot\hat{e}_x$. At this point we perform a transformation from the Eularian independent variables t,x,y,z to Lagrangian independent variables t_o, x_o, y_o, z where t_o, x_o, y_o are a particle's time and transverse coordinates at the entrance to the interaction region, i.e. $z = 0$[7].

The equation governing the relative phase between the electrons and the ponderomotive wave (see Ref. (7b) for 1-D derivation) is given by the generalized pendulum like equation

$$\frac{d^2\tilde{\Psi}}{dz^2} = \frac{d^2\varphi}{dz^2} + \frac{\partial k_w}{\partial z} + \frac{|e|\omega/c}{\tilde{\gamma}_z^2\tilde{\gamma} m_o c^2}\frac{\partial\phi_{ac}}{\partial z}$$

$$- \frac{|e|^2\omega/c}{2\tilde{\gamma}^2 m_o^2 c^4}\left[\frac{\partial}{\partial z}\left(A_w \cosh(k_w\tilde{y})\cos\int_0^z k_w dz'\right)^2 \right. \tag{3}$$

$$\left. + 2k_w A_w \cosh(k_w\tilde{y})\, A(\tilde{x},\tilde{y},z)\cos\tilde{\Psi}\right]$$

where $\tilde{\Psi} = \tilde{\Psi}(z,t_o,x_o,y_o) = \int_0^z (\omega/c + k_w(z'))dz' - \omega\tilde{t} + \varphi(x,y,z)$ is the phase, $\tilde{t} = t_o + \int_0^z dz'/\tilde{V}_z$, $\tilde{\gamma} = \tilde{\gamma}_z\gamma_{o\perp}$, $\tilde{\gamma}_z = (1 - \tilde{V}_z^2/c^2)^{-1/2}$, $\tilde{V}_z = \omega/(\omega/c + k_w(z) - d\tilde{\Psi}/dz + d\varphi/dz)$ is the electron axial velocity, $\tilde{x} \approx x_o + \beta_{o\perp}k_w^{-1}\sin\int_0^z k_w dz'$ and $\tilde{y} \approx y_o \cos K_o z$ are the zeroth order transverse electron coordinates, $\beta_{o\perp} = v_{o\perp}/c$, $v_{o\perp} = |e|A_w/(\gamma_o m_o c)$ is the wiggle velocity, $\gamma_o = (1 - v_o^2/c^2)^{-1/2}$, v_o is the magnitude of the total particle velocity and $K_o = \beta_{o\perp}k_w/\sqrt{2}$. Equation (3) completely describes the non-linear particle dynamics in terms of the fields $\underset{\sim}{A}_w$ and $\underset{\sim}{A}_R$. Noting the structure of Eq. (3), it becomes clear that contouring the wiggler wavelength and/or amplitude, i.e., $\partial k_w/\partial z$ or $\partial A_w^2/\partial z$, is directly equivalent to applying an accelerating field, i.e.

$\partial\phi_{ac}/\partial z$.

The axial velocity shear, due to the wiggler gradient is given by $\Delta V_{shear} = c(\beta_{o\perp}k_w y_o/2)^2$ (see Fig. 2), while the corresponding longitudinal energy spread is $\Delta\mathcal{E}_{shear} = \gamma_o^3(\Delta V_{shear}/c)m_oc^2$. The initial depth of the trapping potential[10] is $|e|\phi_{trap}/(\gamma_o m_o c^2) = 2\sqrt{2}\ \gamma_{oz}\beta_{o\perp}\ (A/A_w)^{1/2}$. In order to trap a substantial fraction of

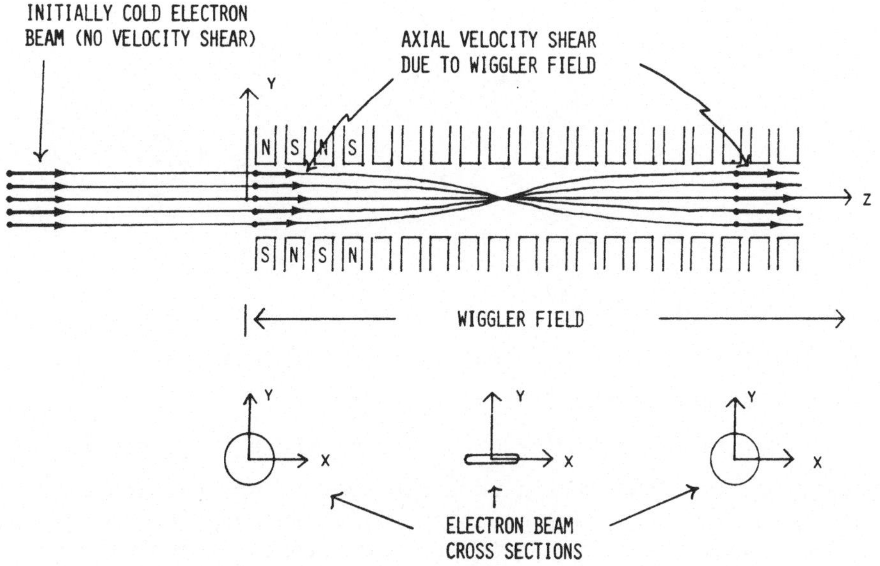

Fig. 2. Periodic focusing of electron beam and axial velocity shear due to the transverse spatial gradients in the magnetic wiggler field.

the electrons we require $\Delta \mathcal{E}_{shear} < |e|\phi_{trap}$, this limits the electron beam radius, maximum y_o, to

$$r_b < (\gamma_o k_w)^{-1} \left(\frac{8\sqrt{2}\,\gamma_{oz}}{\beta_{o\perp}}\right)^{1/2} \left(\frac{A}{A_w}\right)^{1/4} . \quad (4)$$

An initially axially symmetric electron beam injected into the wiggler field, Eq. (1a), will experience a periodic focusing in the y-direction (see Fig. 2). The focusing wavelength along z is $2\pi K_o^{-1} \gg \ell_w$.

Equation (3) neglects space charge effects. This is appropriate if the beam density satisfies

$$n_o \ll (k_w^2 \gamma_{zo}^4 A_w A)(2\pi\gamma_o m_o c^2)^{-1} \quad (5)$$

Equation (5) can be derived by comparing the ponderomotive term to the space charge term in the pendulum equation of ref. (7b).

EVOLUTION OF TOTAL RADIATION FIELD

The radiation field satisfies the wave equation $(\nabla^2 - c^{-2}\partial^2/\partial t^2)\underset{\sim}{A}_R = -4\pi c^{-1} J_x \hat{e}_x$ where the current density is given by

$$J_x(x,y,z,t) = -\frac{\omega_b^2}{4\pi c}\int_{-\infty}^{\infty} dt_o \int_{-\infty}^{\infty} dx_o \int_{-\infty}^{\infty} dy_o$$

$$\theta(x_o,y_o)\delta(x-\tilde{x})\delta(y-\tilde{y})\delta(t-\tilde{t})\,\frac{A_w(\tilde{y},z)}{\tilde{\gamma}} \quad (6)$$

where $\omega_b = (4\pi|e|^2 n_o/m_o)^{1/2}$, and $\theta(x_o,y_o)$ is a function which describes the initial electron beam profile. The radiation field in (1b) can be represented in the form $\underset{\sim}{A}_R(x,y,z,t) = (2i)^{-1}a(x,y,z)\exp i\,(\omega z/c-\omega t)\hat{e}_x$ + c.c., where $a = A\exp(i\varphi)$ is the complex field amplitude which is a slowly varying function of z. Substituting (6) into the wave equation yields an equation for a(x,y,z), which can readily be solved using Fourier transform techniques. The homogeneous solution, a_1, yields the well known Gaussian radiation field, while the particular solution (excited field) is found to be

$$a_2(x,y,z) = -\frac{i}{4\pi}\frac{\omega_b^2}{c^2}\int_0^z dz'\int_0^{2\pi/\omega}\frac{dt_o}{2\pi/\omega}$$

$$\int_{-\infty}^{\infty} dx_o \int_{-\infty}^{\infty} dy_o\, \theta(x_o,y_o)\,\frac{A_w'}{\tilde{\gamma}'}\cosh(k_w'\tilde{y}') \tag{7}$$

$$\frac{e^{i((x-\tilde{x}')^2+(y-\tilde{y}')^2)\frac{\omega/c}{2(z-z')}}}{z-z'}\; e^{-i(\tilde{\psi}'-\varphi')}$$

where $a = a_1 + a_2$ and the primes on quantities denote functions of z'. Equations (3) and (7) describe self-consistently the nonlinear 3-D steady state FEL amplifier.

ILLUSTRATION OF 3-D EFFECTS

For purposes of illustration we can neglect gradients in the wiggler, i.e., $k_w y_o \ll 1$ and replace $\tilde{x}$, $\tilde{y}$ in (7) by x_o, y_o. For a low gain FEL, i.e. $|a_1| \gg |a_2|$, and a plane wave input field, a_1, the phase $\tilde{\psi}$ is very nearly a function of z and t_o only. Choosing a Gaussian electron beam profile, i.e. $\theta(x_o,y_o) = \exp(-(x_o^2 + y_o^2)/r_b^2)$, (7) becomes

$$a_2(r,z) = -\frac{i}{4}\frac{\omega_b^2/c^2}{\gamma_o}\, r_b^2 \int_0^{2\pi/\omega}\frac{dt_o}{2\pi/\omega}\int_0^z dz'$$

$$A_w(z')\, e^{i\varphi(r,z')}\left(\frac{z-z'+iz_o}{(z-z')^2+z_o^2}\right) \tag{8}$$

$$\exp -i\left(\tilde{\psi}(z',t_o) - z_o\left(\frac{z-z'+iz_o}{(z-z')^2+z_o^2}\right)\frac{r^2}{r_b^2}\right)$$

where $z_o = r_b^2\,\omega/2c$ is the effective Rayleigh length associated with the excited radiation. A square or Lorentzian electron beam profile can also be readily integrated in (7). The 1-D limit of (8) is obtained by letting z_o or r_b approach ∞. We will limit ourselves at this point to a constant parameter wiggler and consider only an external accelerating potential. Furthermore, we will make the constant phase, resonant particle approximation. In this approximation all particles are assumed to have the same constant phase, $\tilde{\Psi}_R$. The electron beam in this approximation consist of a pulse train of macro particles separated in distance by $2\pi V_{oz}/\omega$. The rate of change, on axis, of the resonant particle energy is

$$\partial(\gamma_R m_o c^2)/\partial z = |e| \frac{\partial \phi_{ac}(z)}{\partial z} - \frac{|e|^2 \omega/c}{2\gamma_R m_o c^2} A_w A(r=0,z) \cos\tilde{\Psi}_R \tag{9}$$

To obtain the total radiation field we first evaluate $a_2(r,z)$ under the assumption that $|\varphi| << 1$ (this will be shown later to be valid). The amplitude and phase of the <u>total</u> field, on axis, are

$$A(r=0,z) = A_{in} + \alpha_o^2 A_w \left[\tan^{-1}\left(\frac{z}{z_o}\right)\cos\tilde{\Psi}_R - \ln\left(\frac{z^2+z_o^2}{z_o^2}\right)^{1/2}\sin\tilde{\Psi}_R\right] \tag{10a}$$

$$\varphi(r=0,z) = -\alpha_o^2 (A_w/A) \left[\tan^{-1}\left(\frac{z}{z_o}\right)\sin\tilde{\Psi}_R + \ln\left(\frac{z^2+z_o^2}{z_o^2}\right)^{1/2}\cos\tilde{\Psi}_R\right] \tag{10b}$$

where $\alpha_o = \omega_b r_b/2c\sqrt{\gamma_o}$ and $A_{in} = |a_1|$. The stationary phase $\tilde{\Psi}_R$ is obtained from (3).

NUMERICAL EXAMPLE

As an example of a 10.6 μm FEL utilizing a CO_2 laser as an input field, we choose an electron beam of energy 25 MeV ($\gamma_o = 50$), current of I = 5 A and radius (Gaussian profile) of $r_b = 0.5$ mm. Such a beam has a peak density on axis of $n_o = 1.3 \times 10^{11}$ cm^{-3} ($\omega_b = 2.0 \times 10^{10}$ sec^{-1}). The constant parameter wiggler has a magnitude of $B_w = 5.0$ kG and wavelength of $\ell_w = 2.8$ cm which gives $A_w = 2.2 \times 10^3$ statvolt/cm. The wiggle velocity is $v_{o\perp} = 2.6 \times 10^{-2}c$ and the input CO_2 power density is taken to be $P_{in} = 4 \times 10^8$ W/cm^2 which gives $A_{in} = 0.30$ statvolt/cm. Note that the inequalities in (4) and (5) are well satisfied.

Our first illustration is one in which the accelerating potential is zero, hence, $\tilde{\Psi}_R = -\pi/2$ and the particle energy remains constant. The solid curves in Figs. 3a, b and 4a, b are the numerical results of Eq. (8). Figure 3a shows the radiation amplitude, A/A_{in}, as a function of axial position at various radial positions, i.e., r = 0, $r = r_b$ and $r = 4r_b$; and Fig. 3b shows the radial variation of the radiation amplitude A/A_{in} at z = 1 m and 4 m. The gain in the radiation amplitude at z = 4 m is maximum on axis and is 0.17. The net radiation energy flux along the z axis (integrated from r = 0 to r = ∞) is constant since for large r the radiation amplitude is less than the input amplitude. The index of refraction, in this case, is greater than unity, $n = 1 + (c/\omega)\partial\varphi/\partial z > 1$, hence the input field tends to focus while the electron beam will defocus. Figure 4a shows the total radiation phase φ as a function of axial position at various radial positions; and Fig. 4b shows the radial cross section of total radiation phase φ at z = 1 m and 4 m. The maximum value of φ is along the z axis and is approximately 0.067 rad which certainly satisfies our approximation used in (8) to obtain (10).

Our next illustration includes an accelerating potential $\phi_{ac}(z)$ such that $\cos\tilde{\Psi}_R = 0.3$. The dashed curves in Figs. 3a,b and 4a,b are the numerical results of Eq. (8) for this example. The gain in radiation amplitude on axis at z = 4 m is 0.185, see Fig. 3. Since the energy gained in propagating the electron beam through the potential ϕ_{ac} is converted into radiation, the efficiency can be defined as

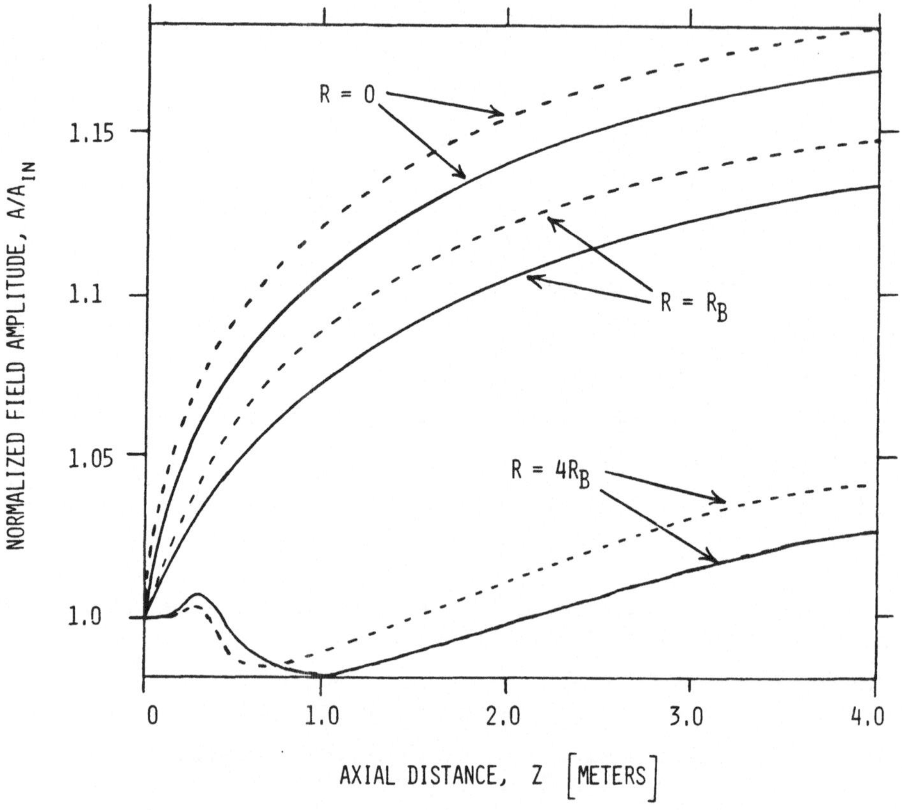

Fig. 3a. Normalized radiation amplitude, A/A_{in}, as a function of z at various radial positions (solid curves for $\tilde{\Psi}_R = -\pi/2$ and dashed curves for $\tilde{\Psi}_R = -1.27$). A_{in} is the amplitude of the input field, which is a plane wave.

$$\eta = |e| \, (\phi_{ac}(z) - \phi_{ac}(0))/\gamma_o m_o c^2$$

$$= -\left(\frac{|e|}{m_o c^2}\right)^2 \frac{\omega/c}{2\gamma_o^2} \int_o^z A_w A(r = 0, z') \cos \tilde{\Psi}_R \, dz'. \tag{11}$$

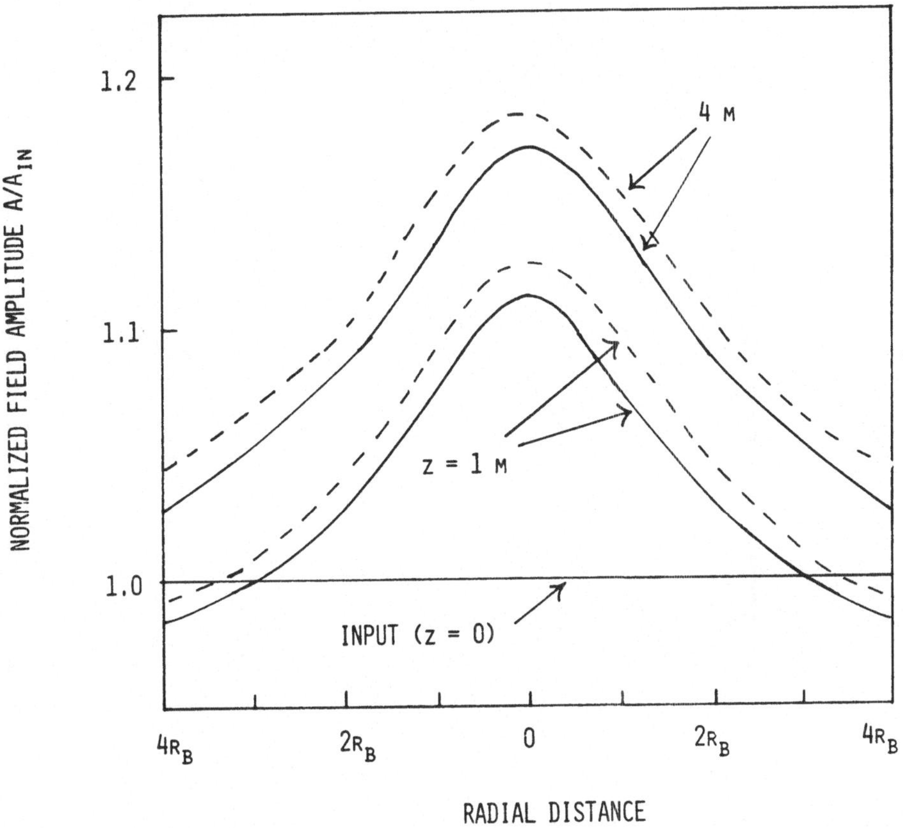

Fig. 3b. Normalized radiation amplitude, A/A_{in}, as a function of radius at z = 1 meter and 4 meters (solid curves for $\tilde{\Psi}_R = -\pi/2$ and dashed curves for $\tilde{\Psi}_R = -1.27$).

The efficiency at the end of z = 4 m is ~ 3.6%. The dashed curves of Fig. 4a show the phase φ as a function of z, and the dashed curves of Fig. 4b show the phase φ as a function of radius. For large z, n is less than unity on axis (defocusing) and becomes greater than unity for large r (focusing). Equations (10a, b) are in excellent agreement with the above numerical illustrations for r = 0.

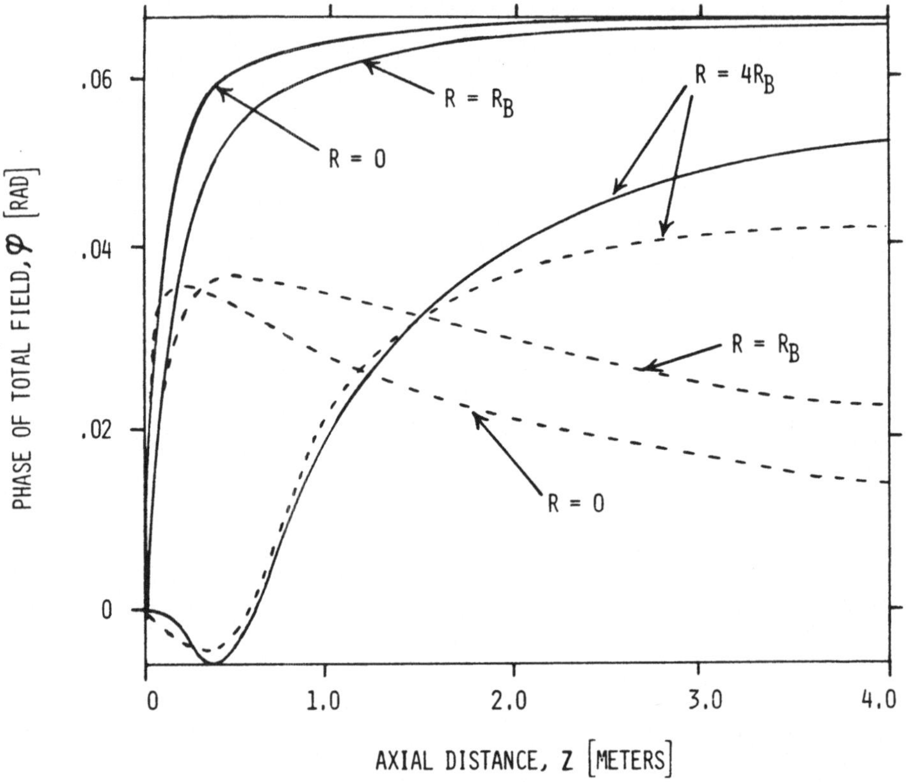

Fig. 4a. Total radiation phase, φ, as a function of z at various radial positions (solid curves for $\tilde{\psi}_R = -\pi/2$ and dashed curves for $\tilde{\psi}_R = -1.27$).

We have seen that the radiation beam can exhibit rather complicated behavior in both the longitudinal and transverse directions. In general the growth of the radiation field along the z axis is due to a combination of refraction effects as well as gain effects (conversion of beam energy into radiation). Destructive interference between the input and excited wave is not a critical issue at least for the parameters considered.

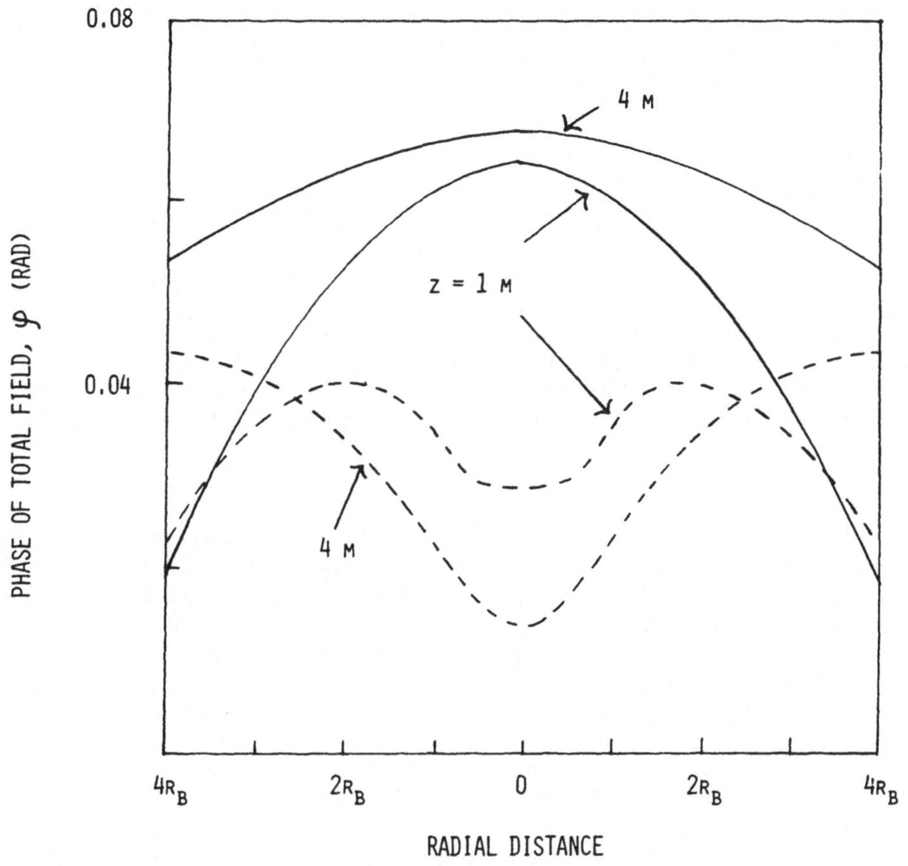

Fig. 4b. Total radiation phase, φ, as a function of radius at $z = 1$ meter and 4 meters (solid curves for $\tilde{\Psi}_R = -\pi/2$ and dashed curves for $\tilde{\Psi}_R = -1.27$).

ACKNOWLEDGMENTS

The authors appreciate useful discussions with I. B. Bernstein and W. M. Manheimer. The authors would also like to acknowledge support for this work by DARPA under Contract No. 3817.

REFERENCES

1. J.M.J. Madey, J. Appl. Phys. 42, 1906 (1971).
2. P. Sprangle and V.L. Granatstein, Appl. Phys. Lett. 25, 377 (1974).
3. W.B. Colson, Phys. Lett. A, 59, 187 (1976).
4. P. Sprangle, R. A. Smith and V.L. Granatstein, NRL Report No. 3911 (1978) and Infrared and Millimeter Waves, Edited by K. Button (Academic Press, New York, 1979), Vol. 1.
5. N. M. Kroll and W.A. McMullin, Phys. Rev. A 17, 300 (1978).
6. P. Sprangle and A. T. Drobot, J. Appl. Phys. 50, 2652 (1979).
7a. P. Sprangle, Cha-Mei Tang and W. M. Manheimer, Phys. Rev. Lett. 43, 1932 (1979), and Phys. Rev. A 21, 302 (1980).
7b. P. Sprangle and Cha-Mei Tang, Proc. 4th Int'l Conf. on Infrared and Near Millimeter Waves, S. Perkowitz, Ed. Miami Brach, 10-15 Dec 1979, p. 98.
8. P. Sprangle and R.A. Smith, NRL Report No. 4033 (1979) and Phys. Rev. A 21, 293 (1980).
9. Free-Electron Generators of Coherent Radiation, Physics of Quantum Electronics, Vol. 7, (1980), Edited by S. F. Jacobs, H. S. Pilloff, M. Sargent III, M. O. Scully and R. Spitzer, Addison--Wesley Publishing Co.
10. N. M. Kroll, P. Morton and M. N. Rosenbluth, JASON Tech. Report JSR-79-15, 1980.
11. G. Bekefi and R. E. Shefer, J. Appl. Phys. 50, 5158 (1979).
12. T. Kwan and J. M. Dawson, Phys. Fluids 22, 1089 (1979).
13. A. T. Lin and J. M. Dawson, Phys. Rev. Lett. 42, 1670 (1979).
14. R. C. Davidson, H.S. Uhm and R. E. Aamodt, Plasma Fusion Center, MIT Report PFC/JA-80-4 (1980).
15. A. A. Kolomenskii and A.N. Lebedev, Soc. J. Quantum Electron 8, 879 (1978).
16. L. Friedland and J. L. Hirshfield, Phys. Rev. Lett. 44, 1456 (1980).
17. 3-D effects in the FEL have been discussed by i) J. Slater, DARPA FEL Review, Arlington, Va., Dec 3-4, 1979, ii) P. Sprangle and Cha-Mei Tang, DARPA/ONR/AFOSR FEL Program Review, LASL, April 24-25, 1980, and iii) Y. P. Ho, Y. C. Lee and M. N. Rosenbluth, Sherwood Meeting, Tucson, Arizona, April 23-25, 1980.

A THREE DIMENSIONAL CALCULATION OF ELECTRON ENERGY LOSS IN A VARIABLE PARAMETER FREE-ELECTRON LASER*

Alfredo Luccio and Claudio Pellegrini

Brookhaven National Laboratory
Upton, N.Y. 11973, U.S.A.

ABSTRACT

The motion of an electron beam through a longitudinally varying period undulator for a single-pass free-electron laser is studied. It is shown that, under certain conditions, the electrons are trapped in an optical "bucket", and deliver on the average more energy to the laser field than in constant period undulators of comparable parameters. The limits set by a finite beam emittance and energy spread are studied in detail by three-dimensional computer simulation.

I. INTRODUCTION

A single-pass free-electron laser (FEL) using a wiggler magnet with either the period, λ_w, and/or the magnetic field, B_w, varying along the magnet axis has been proposed recently.[1] The main advantage of this system over a "conventional" free-electron laser, having a constant period and magnetic field wiggler, is in the higher efficiency of the energy transfer from the electron beam to the laser radiation field. This efficiency, which is of the order of 1% in a conventional FEL, can be of the order of 30% in a variable wiggler FEL.

The theory of the variable wiggler FEL is based on a one dimensional model, in which the electron motion transverse to the laser axis is assumed to be given, and only the motion parallel to the axis is studied. In this paper, we want to study the effect on the laser efficiency of the electron transverse motion and evaluate the electron energy loss for a beam having a spread in angle and in the transverse position at the wiggler entrance.

* Work supported by the U.S. Department of Energy.

The complete three dimensional equations of motion for an electron interacting with the laser field and the wiggler field are integrated numerically. We only consider the case of a small gain regime, assuming that the laser field intensity remains constant. We also limit ourselves to the case of a helical wiggler. The results are compared with the one dimensional model. The effect of the initial position and angular spread can, to a good approximation, be considered equivalent to an increase in the energy spread. the limits for this increased energy spread that must not be exceeded in order to avoid a loss in efficiency are nearly the same as in the one dimensional model.

II. THE EQUATIONS OF MOTION

To describe the energy exchange between the electrons and the radiation field, we use the single particle model developed by Colson.[2] We write the electron equations of motion in the wiggler field, B_W, and laser field, E_L, B_L, as

$$\frac{1}{c}\frac{d(\underline{\beta}\gamma)}{dt} = \underline{E}_L^* + \underline{\beta}\times(\underline{B}_L^* + \underline{B}_W^*) \tag{1}$$

$$\frac{1}{c}\frac{d\gamma}{dt} = \underline{E}_L^* \cdot \underline{\beta} \tag{2}$$

where β is the electron velocity in units of the speed of light, c; γ is the electron energy in units of $m_o c^2$, the rest energy; the fields $\underline{E}^*$, $\underline{B}^*$ are normalized as

$$\underline{E}^* = e\,\underline{E}/(m_o c^2) \tag{3}$$

$$\underline{B}^* = e\,\underline{B}/(m_o c^2) \tag{4}$$

We decompose the velocity in a component, $\beta_{\shortparallel}$, parallel to the wiggler axis, z, and in the components β_x, β_y along the directions x, y orthogonal to the wiggler axis. We assume

$$\beta_{\shortparallel} \simeq 1;\ \beta_x, \beta_y << 1 \tag{5}$$

We use for the laser and the helical wiggler fields the following expressions

$$\underline{E}_L^* = E_L^* \left\{ \hat{x} \sin\left[\frac{2\pi}{\lambda}(z-ct) + \phi_o\right] + \hat{y} \cos\left[\frac{2\pi}{\lambda}(z-ct) + \phi_o\right] \right\} \tag{6}$$

$$\underline{B}_L^* = \hat{z} \times \underline{E}_L^* \tag{7}$$

$$\underline{B}_w^* = B_w^*(z) \left\{ \hat{x} \cos\phi_w + \hat{y} \sin\phi_w - \hat{z} \frac{2\pi}{\lambda_w} \left[x \sin\phi_w - y \cos\phi_w \right] \right\} \tag{8}$$

$$\phi_w = \int_0^z \{2\pi/\lambda_w(z')\}\, dz' \tag{9}$$

To write the wiggler field in the form (8) we have made the following assumptions:

i) the field intensity B_w^* and the period are slowly varying functions of z, so that

$$\frac{d\lambda_w}{dz} \ll 1; \quad \frac{\lambda_w}{B_w} \frac{dB_w}{dz} \ll 1;$$

we have neglected all terms containing these derivatives;

ii) we considered an expansion of the field in powers of x and y near to the wiggler axis, neglecting all terms quadratic or of higher order in $2\pi x/\lambda_w$, $2\pi y/\lambda_w$.

To determine the electron motion and energy change, we will use the equation (2) for γ and the equation (1) for β_x, β_y. The parallel velocity is then determined by the relationship

$$\beta_\parallel = \left[1 - \frac{1}{\gamma^2} - \beta_x^2 - \beta_y^2 \right]^{\frac{1}{2}} \tag{10}$$

For convenience we will use the variable

$$z = \int_0^t c\, \beta_\parallel(t')\, dt' \tag{11}$$

instead of t, the "polar" quantities

$$\beta_\perp e^{i(\alpha+\phi_w)} = \beta_x + i\,\beta_y \tag{12}$$

$$r\, e^{i\theta} = x + i\,y \tag{13}$$

and the relative phase of the electron and the radiation field

$$\phi = \frac{2\pi}{\lambda}(z-ct) + \phi_w + \phi_o \tag{14}$$

The complete set of equations can now be written as $(f' = \frac{df}{dz})$

$$\gamma' = \frac{\beta_\perp E_L^*}{\beta_\parallel} \sin\phi \tag{15}$$

$$\phi' = \frac{2\pi}{\lambda_w}\left[1 - \frac{\lambda_w}{\lambda}\frac{1-\beta_\parallel}{\beta_\parallel}\right] + \alpha' \tag{16}$$

$$\beta_\perp' = \frac{2\pi K}{\lambda_w}\sin\alpha + \frac{1-\beta_\parallel-\beta_\perp}{\gamma\beta} E_L^* \sin\phi \tag{17}$$

$$\alpha' = \frac{2\pi}{\lambda_w}\left[\frac{K}{\gamma\beta_\perp}\cos\alpha - 1\right] + \frac{K}{\gamma\beta_\parallel}\left(\frac{2\pi}{\lambda_w}\right)^2 r \sin(\phi_w - \theta) + \frac{1-\beta_\parallel}{\gamma\beta_\perp\beta_\parallel} E_L^* \cos\phi \tag{18}$$

$$r' = \frac{\beta_\perp}{\beta_\parallel}\cos(\phi_w + \alpha - \theta) \tag{19}$$

$$r\theta' = \frac{\beta_\perp}{\beta_\parallel}\sin(\phi_w + \alpha - \theta) \tag{20}$$

with

$$K = B_w^* \lambda_w / 2\pi \tag{21}$$

III. REDUCTION TO THE ONE DIMENSIONAL MODEL

We have solved the system of equations (15) to (21) by numerical integration. It is, however, useful to reduce the system to the form used in the one dimensional model and to describe briefly some of the main results obtained from it.

The equations of the one dimensional model are obtained if one assumes

$$\beta_\perp' = 0, \qquad \beta_\perp = \frac{K}{\gamma} = \text{constant} \tag{22}$$

$$\alpha' = 0, \qquad \alpha = 0 \tag{23}$$

In this case, the particle trajectory is a helix of pitch angle $\beta_\perp$ and radius $\rho = \beta_\perp \lambda_w / 2\pi$.[3]

Using (10) we have

$$\beta_{\parallel} = \left[1 - \frac{1+K^2}{\gamma^2} \right]^{\frac{1}{2}} \simeq 1 - \frac{1+K^2}{2\gamma^2} \tag{24}$$

and from (15), (16)

$$\gamma' = \frac{KE_L^*}{\gamma} \sin\phi \tag{25}$$

$$\phi' = \frac{2\pi}{\lambda_w} \left[1 - \frac{\lambda_w}{\lambda} \frac{1+K^2}{2\gamma^2} \right] \tag{26}$$

Following reference (1) we define a resonant energy

$$\gamma_r^2 = \lambda_w \, (1+K^2)/2\lambda \tag{27}$$

and a synchronous phase, ϕ_r

$$\sin\phi_r = \frac{d\gamma_r^2}{dz} / 2\,K\,E_L^* \tag{28}$$

The particles inside the "optical bucket" of width $(\Delta\phi, \Delta\gamma)$ will oscillate around ϕ_r, γ_r, and their average energy variation will be equal to the change in γ_r. In the following, we assume that the change in γ_r is due to a change in λ_w while K remains a constant. For a wiggler of length L_w, one obtains

$$\gamma_r^2(L_w) - \gamma_r^2(0) = 2\,K\,E_L^*\,L_w \tag{29}$$

The efficiency is then defined as

$$\eta = \frac{\gamma_r(0) - \gamma_r(L_w)}{\gamma_r(0)} \frac{\Delta\phi}{2\pi} \tag{30}$$

having assumed that the initial distribution of the phase ϕ is uniform in the interval 0-2π. Assuming ϕ_r to be in the interval

π, $\frac{3}{2}\pi$, we have

$$\Delta\phi = 3\pi - \phi_r - \phi^* \tag{31}$$

where ϕ^* is a solution of the equation

$$V(\phi^*) = V(3\pi - \phi_r) \tag{32}$$

with

$$V(\phi) = \cos\phi + \phi \sin\phi_r \tag{33}$$

The bucket "height"

$$\Delta\gamma_B = \gamma - \gamma_r$$

is given by

$$\left(\frac{\Delta\gamma}{\gamma}\right)_B = \left[\frac{E_L^* K \lambda_w}{2\pi\gamma_r^2}\right]^{\frac{1}{2}} \{V(3\pi - \phi_r) - V(\phi_r)\}^{\frac{1}{2}} \tag{34}$$

Notice that in the case K= constant, the quantity $(\Delta\gamma/\gamma_r)$ is independent of z.

All these results are valid for particles injected with initial velocity $\beta_{xo} = K/\gamma$, $\beta_{yo} = 0$, and on axis. For particles injected off axis, the motion will be a superposition of the helical motion and an oscillation of reduced wavelength[3]

$$\lambda_f = \sqrt{2}\gamma\lambda_w/2\pi K$$

The amplitude of this oscillation is equal to the initial displacement r, for a particle injected off axis, or to $\sqrt{2}\gamma\lambda_w\Delta\beta/2\pi K$ for a particle injected at an angle. A particle executing this oscillation will experience a different magnetic field from a particle on axis

$$\frac{\Delta B_w}{B_w} = \frac{1}{2}\left(\frac{2\pi}{\lambda_w}\right)^2\left[r_o^2 + \left(\frac{\gamma\lambda_w}{\sqrt{2}\pi K}\Delta\beta_\perp\right)^2\right] \tag{35}$$

The change in B_w is equivalent to a change in K and hence to a change in γ_r

$$\left(\frac{\Delta\gamma}{\gamma}\right)_{eff} = \frac{K^2}{1+K^2}\frac{\Delta K}{K} \tag{36}$$

and, using (35)

$$\left(\frac{\Delta\gamma}{\gamma}\right)_{eff} = \frac{K^2}{1+K^2}\,\frac{1}{2}\left(\frac{2\pi}{\lambda_w}\right)^2\left[r_o^2 + \left(\frac{\lambda_w}{\sqrt{2}\pi K}\,\Delta\beta_\perp\right)^2\right] \tag{37}$$

If the electron beam has an energy spread $\Delta\gamma/\gamma$, we can define a total energy spread

$$\left(\frac{\Delta\gamma}{\gamma}\right)_T = \left[\left(\frac{\Delta\gamma}{\gamma}\right)^2 + \left(\frac{\Delta\gamma}{\gamma}\right)^2_{eff}\right]^{\frac{1}{2}} \tag{38}$$

and assume that when the condition

$$\left(\frac{\Delta\gamma}{\gamma}\right)_T < \left(\frac{\Delta\gamma}{\gamma}\right)_B \tag{39}$$

is satisfied, the efficiency is still given to a good approximation by (34).

The validity of this assumption can be tested comparing (34) with the results obtained from the numerical integration of (15) to (21). This will be done in the next section.

Before closing this section, we notice that introducing the transverse electron beam emittance

$$\varepsilon = \pi\, r\, \Delta\beta \tag{40}$$

and assuming that the electron beam has cylindrical symmetry, we can rewrite (37) as

$$\left(\frac{\Delta\gamma}{\gamma}\right)_{eff} = \frac{1}{2}\left(\frac{2\pi}{\lambda_w}\right)^2\frac{K^2}{1+K^2}\left[r^2 + \left(\frac{\gamma\lambda_w}{\sqrt{2}\pi^2 K}\right)^2\frac{\varepsilon^2}{r^2}\right] \tag{41}$$

IV. NUMERICAL RESULTS

The Eqs. (15) through (21) have been integrated numerically, for various initial values of the quantities γ, ϕ, $\beta_\perp$, α, and r. The wiggler wavelength λ_w was assumed to vary linearly along the (helical) wiggler according to

$$\lambda_w = \lambda_{wo}\,(1 - z/\sigma) \tag{42}$$

With this, and from the definition of Eq. (28), the resonant phase angle becomes

$$\phi_r = \text{arsin}\left[\frac{-\lambda_{wo}}{2\sigma\lambda\varepsilon_o}\right] \tag{43}$$

A set of the free-electron laser parameters chosen for the calculations is shown in Table 1.

Table 1. Free-electron Laser Parameters

Wiggler wavelength (initial)	$\lambda_{wo} = 2$ cm
Wiggler overall length	$L_w = 2.5$ m
Wiggler taper length {Eq. (42)}	$\sigma = 5$ m
Wiggler parameter {Eq. (21)}	$K = 1$
Radiation wavelength	$\lambda = 10$ μm
Resonant "energy" {Eq. (27)}	$\gamma_{ro} = 44.733$
Normalized rad. field {Eq. (3)}	$E_L^* = 341\ m^{-1}$

As already mentioned in Sect. III, if the electrons are injected into the wiggler at a radius given by[2]

$$r_o = x_o = \lambda_{wo}^2/4\pi^2\rho_o \tag{44}$$

where $\rho_o = m_o\gamma_o c/eB_o$ is the cyclotron radius in the field B_o, and with a transverse velocity given by (22), (23), they move along a helical path close to the wiggler axis.

Figures 1 and 2 show a set of calculated curves of the energy γ vs. distance z along the wiggler in the two cases of $\sigma = 5$ m and 15 m, for a monochromatic bundle of 16 electrons with initial energy equal to the resonant energy; $\gamma_0 = \gamma_{ro}$, injected exactly on the helical path defined by Eqs. (22), (23), (44), and with 16 equidistant starting phases ϕ_o between 0 and 2π.

Here, the resonant energy γ_r decreases with z according to

$$\gamma_r = \gamma_{ro}\sqrt{\lambda_w/\lambda_{wo}} \tag{45}$$

The capture of some of the electrons in an energy-phase bucket around γ_r shows clearly.

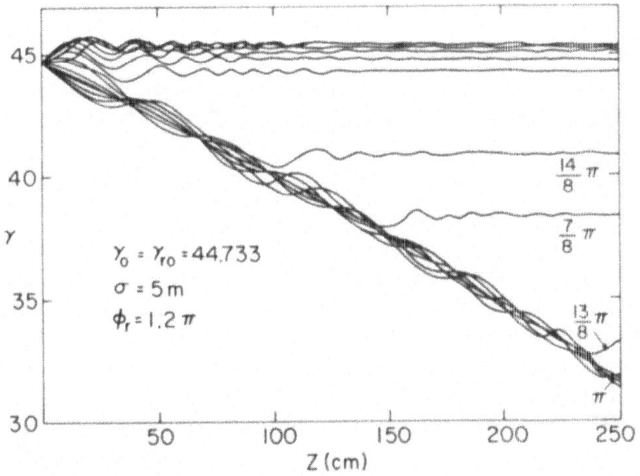

Fig. 1. Normalized energy γ vs. distance z anong the wiggler. Monochromatic electron bundle with $\gamma_o = \gamma_{ro} = 44.733$ injected on helix. Case $\sigma = 5m$.

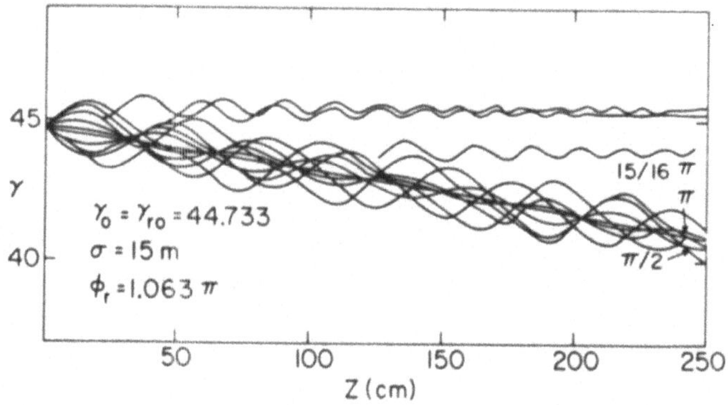

Fig. 2. Normalized energy γ vs. distance z along the wiggler. Monochromatic electron bundle with $\gamma_o = \gamma_{ro} = 44.733$ injected on helix. Case σ = 15 m.

Figures 3 and 4 show the trajectories of the representative points of the electrons in phase-space for the same cases. Here, the values of bucket parameters, from Eq. (43) and from Eqs. (31), (34), given in Table 2, appear to agree well with the numerical results. However, both the bucket height and width shown in Figs. 3, 4 are of the order of 10% smaller than the theoretical values.

Table 2. Phase Space Bucket Parameters

		σ = 5 m	15 m
Resonant "energy"	γ_{ro} =	44.733	
Resonant phase	ϕ_r =	1.200 π	1.063 π
Energy height of bucket	$\left(\frac{\Delta\gamma}{\gamma_r}\right)_B$ =	1.63%	2.78%
Limit angles for bucket	ϕ^* =	0.867 π	0.475 π
	$3\pi-\phi_r$ =	1.800 π	1.937 π
Phase width of bucket	$\Delta\phi$ =	0.933 π	1.462 π
Efficiency Eq. (30)	η =	13.7%	6.4%

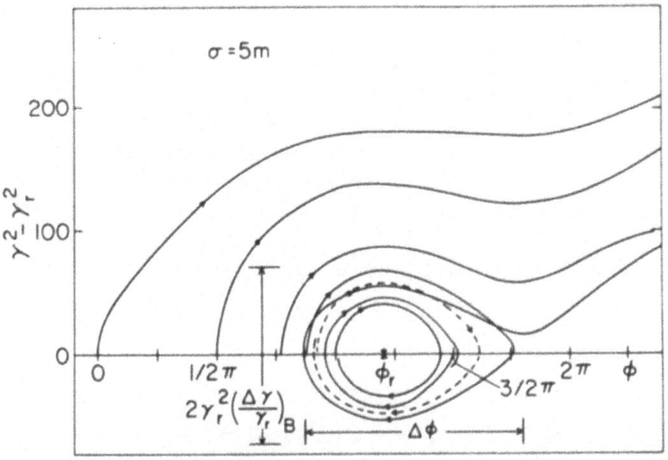

Fig. 3. Phase space trajectories of electrons injected on a helix. σ = 5 m. The capture width appears somewhat smaller than Δϕ shown (see Table 2).

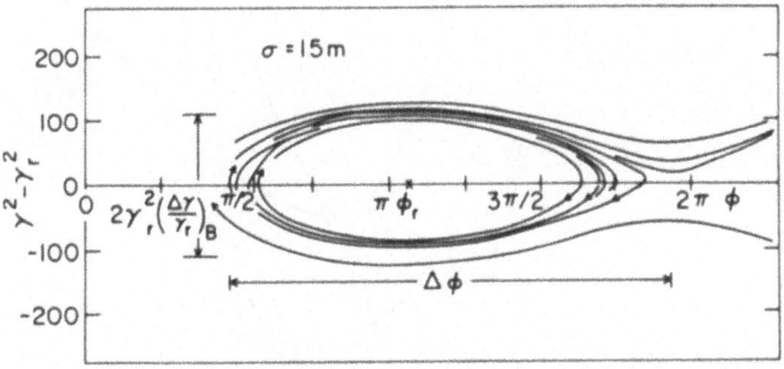

Fig. 4. Phase-space trajectories of electrons injected on a helix. Case σ = 15 m.

Once established the conditions for capture and energy loss of paraxial electrons with resonant energy, we moved on to find how a real beam, with finite emittance, behaves. As it has been shown in Sect. III, a finite radius of the beam and a transverse velocity spread around the values given by Eqs. (44), (22), (23), should act as an additional energy spread.

In our calculations, we assigned a parabolic shape to the density distribution of the beam in the γ, r, $\beta_\perp$ and α spaces, while the density in phase ϕ was always taken as uniform. To reduce the number of cases, we limited ourselves to consider separately the γ, r, $\beta_\perp$ and α sections of the four-dimensional phase space density volume.

Results for σ = 5 m are shown in Figs. 5 through 8.

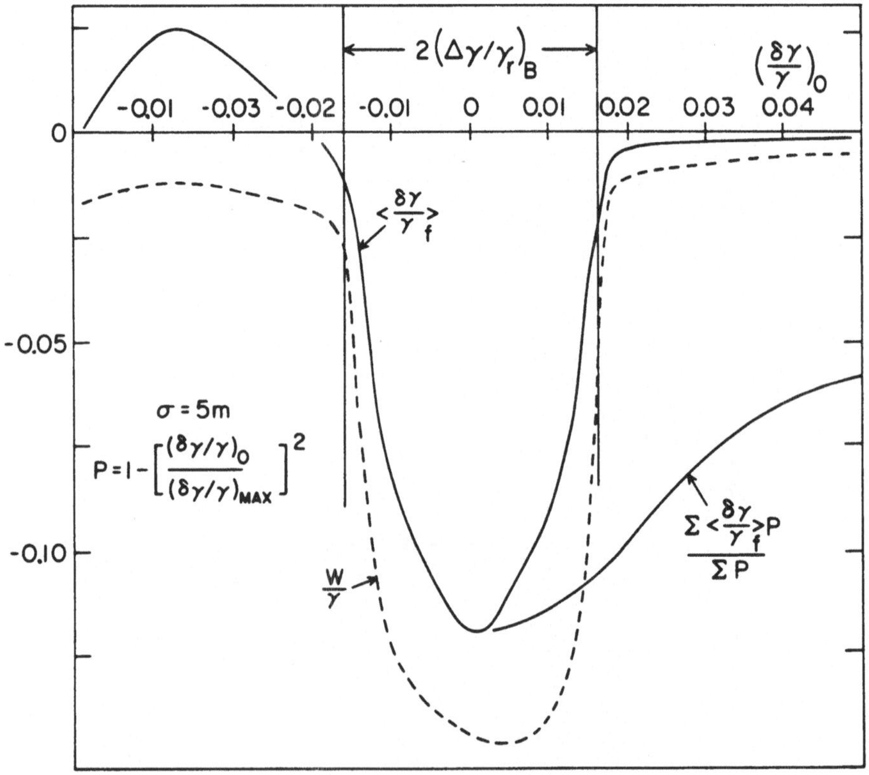

Fig. 5. Average energy loss $<\delta\gamma/\gamma>$, relative statistical standard deviation W/γ and weighted average energy loss as a function of the initial energy. Electrons injected on helix. σ = 5 m. For comparison, the "height" of the bucket from Table 2 is also shown.

Figure 5 shows the average (respect to the initial phase, ϕ) energy loss $\langle\delta\gamma/\gamma\rangle$ and its statistical standard deviation W, as a function of the initial energy.

The bucket effective height, defined at the half maximum point, obtained from Fig. 5 is $(\Delta\gamma/\gamma_0) \simeq 1.31\%$. This value is almost 20% smaller than the value $(\Delta\gamma/\gamma)_B$ given in Table 2. Correspondingly, the maximum energy transfer, averaged over the phase, is 12% instead of the value of 13.4% given in the Table.

The values of W show also a pronounced peak within the bucket, where, because of the capture mechanism, the statistical distribution of final energies exhibits two well separated maxima.

In the same Fig. 5, we give also the average energy loss, weighted on a parabola. This shows that in the present case, a paraxial

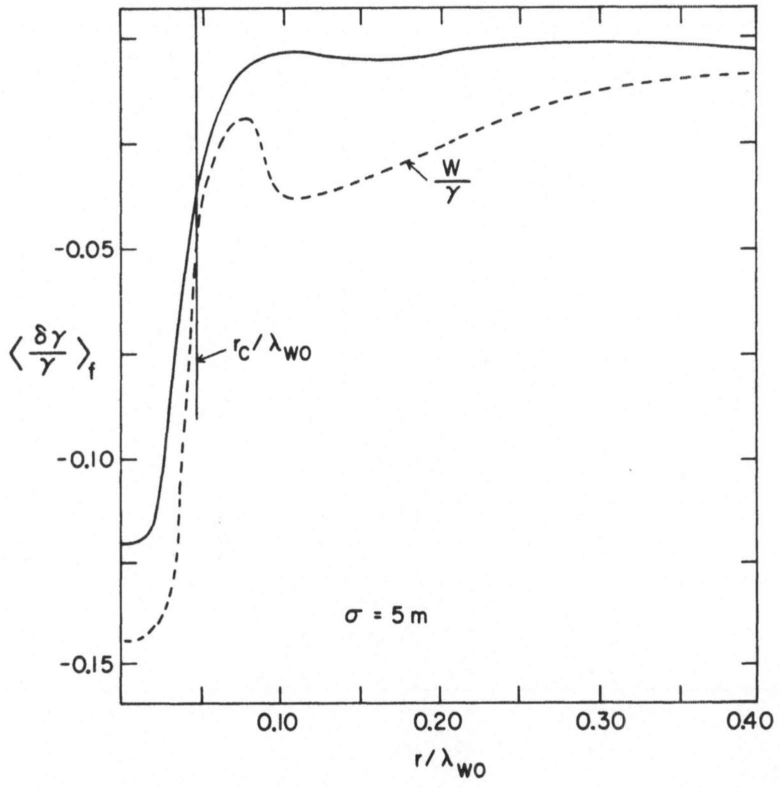

Fig. 6. Average energy loss $\langle\delta\gamma/\gamma\rangle$ and relative statistical standard deviation W/γ as a function of initial radius, for $\beta_{\perp 0} = K/\gamma$ and $\gamma_0 = \gamma_{r0}$. $\sigma = 5$ m. The "capture" radius is also shown for comparison.

beam with 5% initial energy spread loses still on the average 6% of its energy to the radiation field. If the initial energy spread is equal to $(\Delta\gamma/\gamma_r)_B$, the average energy loss is only reduced from 12% to 11%.

Figure 6 shows analogous results for a beam with finite radial extension. According to Eq. (37), a spread in radius is equivalent to a spread in energy

$$\left.\frac{\Delta\gamma}{\gamma}\right|_{rad} = \frac{1}{2}\,\frac{K^2}{1+K^2}\left(\frac{2\pi r}{\lambda_w}\right)^2 \tag{46}$$

which gives, by assuming $\Delta\gamma/\gamma = 1.63\%$, a limiting value

$$r_{capt}/\lambda_w = 0.0406$$

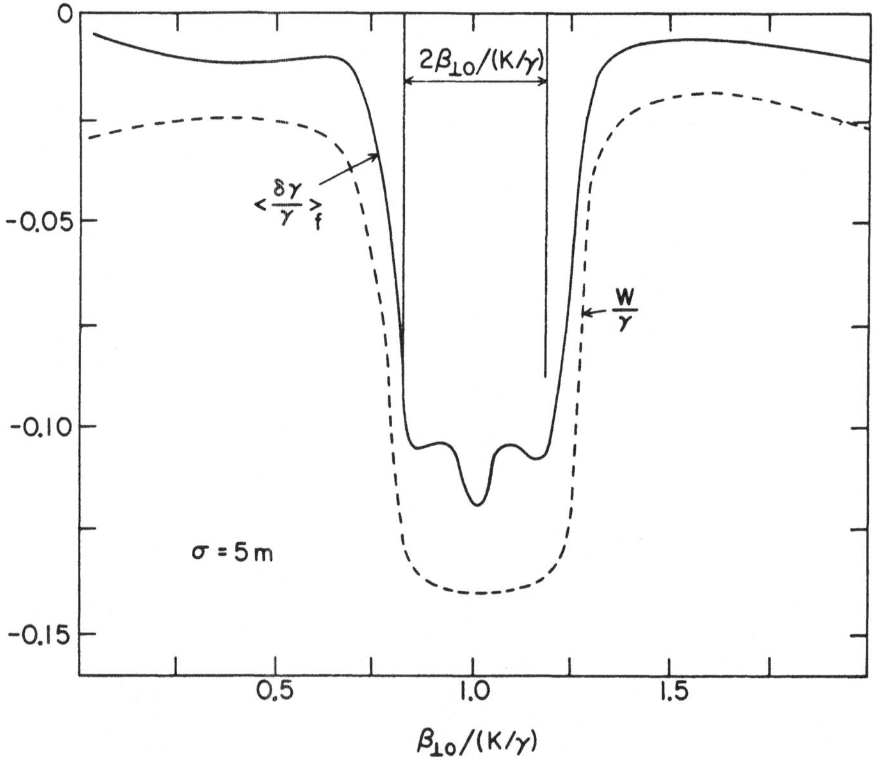

Fig. 7. Average energy loss $<\delta\gamma/\gamma>$ and relative statistical standard deviation W/γ as a function of the initial transverse velocity, for $x_o = r_o$ and $\gamma_o = \gamma_{ro}$. σ = 5 m. The "capture transverse angle" is also shown.

in good agreement with the results.

Figures 7 and 8 show the effect of a transverse velocity dispersion around the values (22), (23). From Eq. (37) it appears that the limiting value for capture of $\beta_\perp$ equals the ratio of r_{capt} to the reduced wavelength of the long wavelength oscillations of the electron trajectories in the wiggler.

Hence

$$\beta_{\perp_{capt}} = \frac{K}{\gamma} \pm \frac{r_{capt}}{\lambda_f} = \left(1 \pm \frac{2\pi}{\sqrt{2}} \frac{r_{capt}}{\lambda_w}\right) \cdot \frac{K}{\gamma} \tag{47}$$

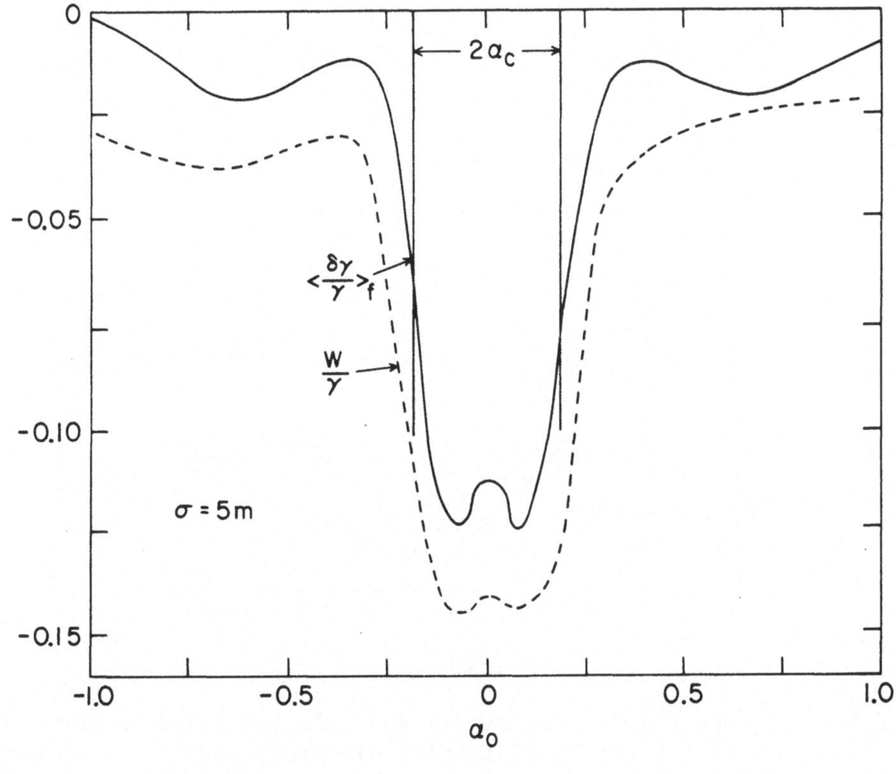

Fig. 8. Average energy loss $<\delta\gamma/\gamma>$ and relative statistical standard deviation W/γ as a function of initial slant angle α_0, for $x_0 = r_0$ and $\beta_{\perp 0} = K/\gamma . \sigma = 5$ m. α_{capt} is also shown.

and, with the present values,

$$\beta_{\perp,\text{capt}} / (K/\gamma) = 1 \pm 0.180$$

again in good agreement with the results.

The same holds for the angle α that the vector $\beta_\perp$ makes with the tangent to the ideal paraxial helix. Limits for α, when $\beta_{\perp 0} = K/\gamma_0$, are

$$\alpha_{\text{capt}} = \pm \frac{2\pi}{\sqrt{2}} \frac{r_{\text{capt}}}{\lambda_w} = \pm 0.180$$

and they appear to bound well the calculated capture peak.

Both Figs. 7 and 8 show a fine structure in the behaviour of $\Delta\gamma/\gamma$ vs. $\beta_\perp$ and vs. α. They are not surprising since we did not

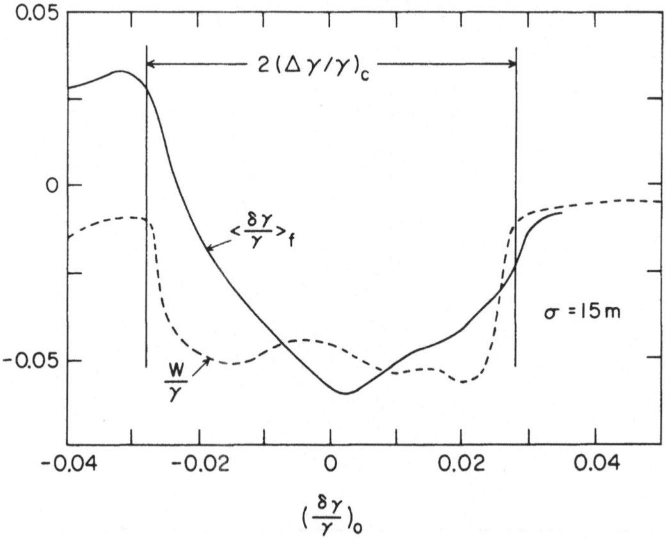

Fig. 9. Average energy loss $<\delta\gamma/\gamma>$ and relative standard deviation W/γ as a function of initial electron energy. σ = 15 m.

average the results over the complete phase-space emittance of the electron beam, but limited ourselves to representative points on the axes. No systematic study of this fine structure was made.

Results of the numerical calculations for the case σ = 15 m are shown in Figures 9 through 12.

Figure 9 shows $<\delta\gamma/\gamma>$ and W/γ as a function of the initial electron energy, for electrons injected exactly on a helix. In the figure, the value of the capture energy spread ("height of the bucket") from Table 2

$$\left(\frac{\Delta\gamma}{\gamma_r}\right)_B = \pm\ 2.78\%$$

is also shown.

Figures 10 and 11 show $<\delta\gamma/\gamma>$ and W/γ as a function of r_o and

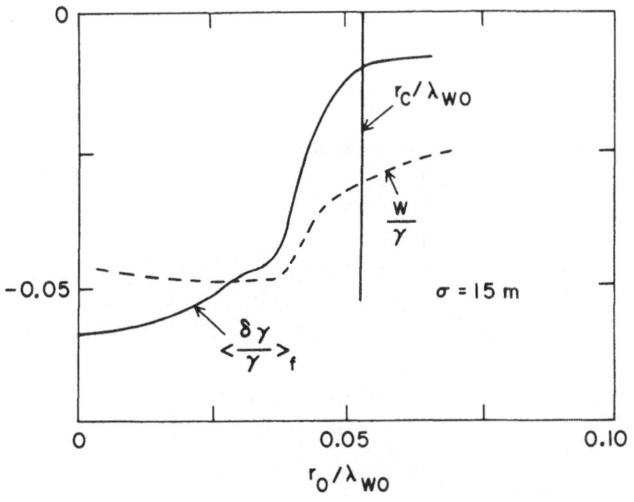

Fig. 10. Average energy loss $<\delta\gamma/\gamma>$ and W/γ as a function of injection radius. σ = 15 m.

$\beta_{\perp o}$ respectively, with their capture ranges

$$r_c/\lambda_{wo} = 0.053$$

$$\left(\frac{\beta_{\perp o}}{K/\gamma}\right)_c - 1 = \pm\ 0.234$$

Outside these ranges, the calculations gave as result oscillations in $\langle\delta\gamma/\gamma\rangle$ as large as 3%. However, by averaging over starting points on the phase-space ellipses

$$\left(\frac{2\pi r}{\lambda_{wo}}\right)^2 + \frac{\gamma^2}{K^2}(\Delta\beta_{\perp o})^2 = \text{const} \tag{49}$$

the $\langle\delta\gamma/\gamma\rangle$ was reduced to 0.8%.

Figure 12 shows $\langle\delta\gamma/\gamma\rangle$ and W/γ as a function of the slant angle α_o. Capture range for α is also shown.

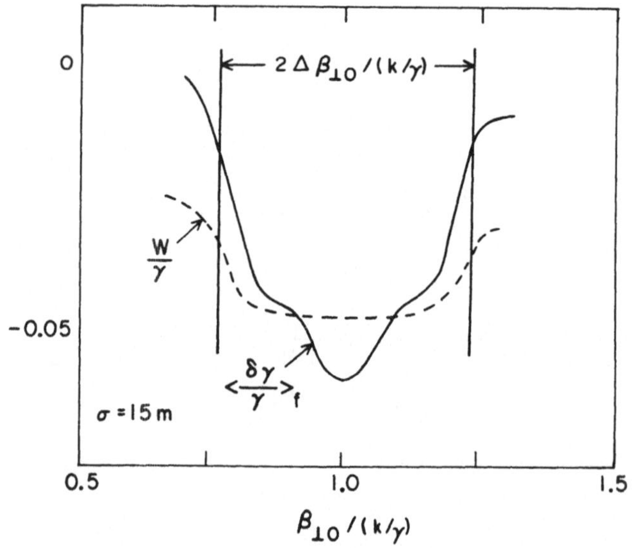

Fig. 11. Average energy loss $\langle\delta\gamma/\gamma\rangle$ and standard deviation W/γ as function of initial transverse velocity $\beta_{\perp o}$. σ = 15 m.

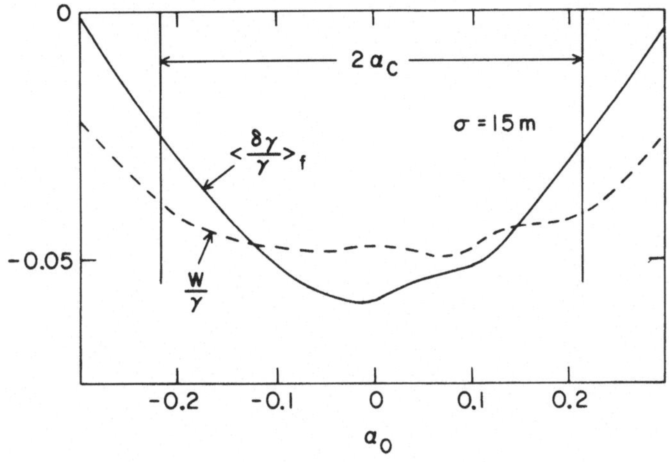

Fig. 12. Average energy loss $<\delta\gamma/\gamma>$ and standard deviation W/γ as a function of initial slant angle α_o. σ = 15 m.

REFERENCES

1. N.M. Kroll, P.L. Morton, and M.N. Rosenbluth,"Physics of Quantum Electronics", Vol. 7, Chap. 4, Addison-Wesley Publ. Co., (1980).
2. W.B. Colson, Phys. Letters, A59:187 (1976).
3. J.P. Blewett and R. Chasman, J. Appl. Phys., 48:2692 (1977).

AN EXPERIMENT OF FREE ELECTRON LASER EFFICIENCY ENHANCEMENT WITH A VARIABLE WIGGLER

H. Boehmer, M.Z. Caponi, J. Munch, G. Neil and N. Schoen

TRW DSSG
One Space Park
Redondo Beach, CA 90278

1. INTRODUCTION

The demonstration of a tunable, visible wavelength, high power high efficiency laser beam is the ultimate goal of a free electron laser program. However, the current theoretical and experimental data base is so limited that it cannot support the demonstration of the long range program goals in one step. Rather, it requires the evolutionary demonstration of one or more intermediate levels of performance. In particular, it is necessary to verify experimentally the theoretical predictions of high efficiencies for free electron lasers operating in the trapped electron regime with adiabatically tapered magnetic wiggler fields and to determine the influence of operating parameters. An experiment designed to achieve this goal is outlined in this paper. A fraction of the beam electrons is trapped in the ponderomotive well formed by the wiggler and an external EM signal that is to be amplified. The wiggler amplitude and/or wavelength is then modified to extract energy from the trapped electrons. The theory of operation is sufficiently well developed at this time to permit the design of the experimental hardware required. In addition, the experiment has been designed for maximum flexibility to provide for experimental optimization and design variations which may be predicted by more elaborate analytical and numerical models currently under development.

In Section 2, we provide the theoretical background for this design, followed by descriptions of the linac, electron beam handling, optical system and wiggler in Sections 3.1, 3.2, and 3.3, respectively.

2. FREE ELECTRON LASER CONCEPTUAL DESIGN

A Free Electron Laser (FEL) device generates stimulated radiation by the interaction of a relativistic electron beam and an external electromagnetic wave or pump.[1] The intensity of the emission increases with the pump amplitude[2] and it is advantageous to use an external ripple magnetic field ("wiggler") as the external pump. The stimulated radiation wavelength λ_s is proportional to the wiggler wavelength λ_w downshifted by a factor proportional to the relativistic electron beam energy squared ($\varepsilon = \gamma mc^2$). For example, for a sinusoidal wiggler of amplitude B_w,

$$\lambda_s \simeq \frac{\lambda_w}{2\ ^2} \quad (1 + 1/2\, a_w^2), \tag{1}$$

where $a_w = \dfrac{qB_w}{mc} \quad \dfrac{\lambda_w}{2\pi}$.

Recently[3], it has been suggested that the FEL efficiency can be dramatically enhanced by adiabatically tapering the wiggler field amplitude and/or wavelength. In this manner, the energy and shape of the ponderomotive potential well ("bucket") formed by the interaction of the wiggler field with the electrons and the amplified laser signal can be spatially varied (cf. Figure 1). The electrons that are intially trapped in the bucket tend to remain trapped if the motion is sufficiently adiabatic. As the bucket energy decreases, the trapped electrons mean energy is reduced. The extracted electron energy goes into radiation providing the amplification of the input laser signal. The relevant details of this concept and the corresponding theoretical approach have been discussed in References 3 and 4, to which we refer the reader. For convenience, we include in Appendix 1 the one dimensional single particle equations that serve to describe the variable wiggler FEL and form the core of the conceptual design. The total efficiency of the process depends on the number of electrons trapped in the bucket (trapping efficiency $= \eta_t \simeq J_t/J$) and the deceleration efficiency of the trapped electrons which is approximately equal to the deceleration efficiency of the "resonant" or synchronous particle (cf, Figure 1) η_D,

where $$\eta_D = \frac{\gamma_r(z) - \gamma_r(0)}{\gamma_r(0)} \tag{2}$$

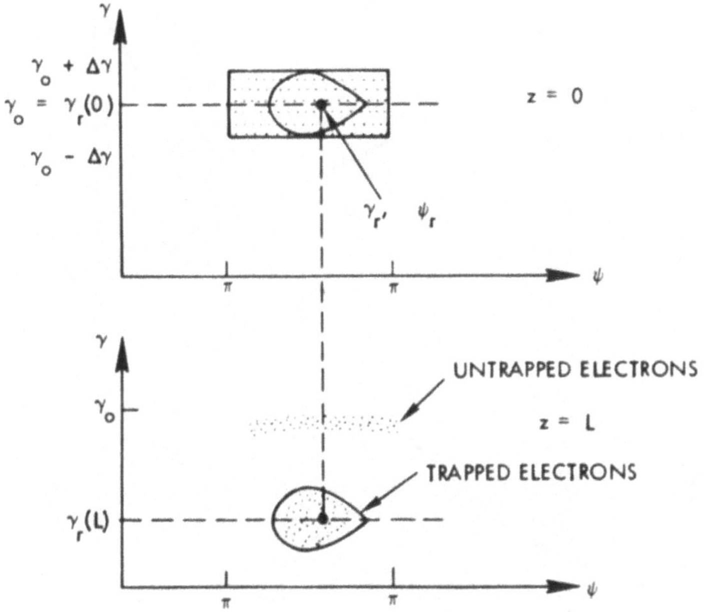

Figure 1. A diabatic decrease in energy of bucket and trapped electrons.

The total energy spread is mainly determined by the "longitudinal" energy spread due to the variation of the rf cycle and the 'effective' transverse energy spread. This transverse energy spread is due to the random transverse motion of the electrons[5] (a consequence of the finite electron beam emittance) and the radial nonuniformity of the wiggler field that is seen by a finite radius electrom beam. For this experiment ($\varepsilon \leq 8$ mm mrad) it can be shown that the effective transverse energy spread due to the wiggler radial nonuniformity is dominant. An optimum design requires $\Delta\gamma/\gamma_{long} \simeq \Delta\gamma/\gamma_{transverse}$. We have found $\Delta\gamma/\gamma_{transverse} \simeq .01$ to be appropriate for the proposed experiment characteristics. This choice restricts the magnetic field amplitude to $B_w \leq 3.1$ kG (cf Figure 2).

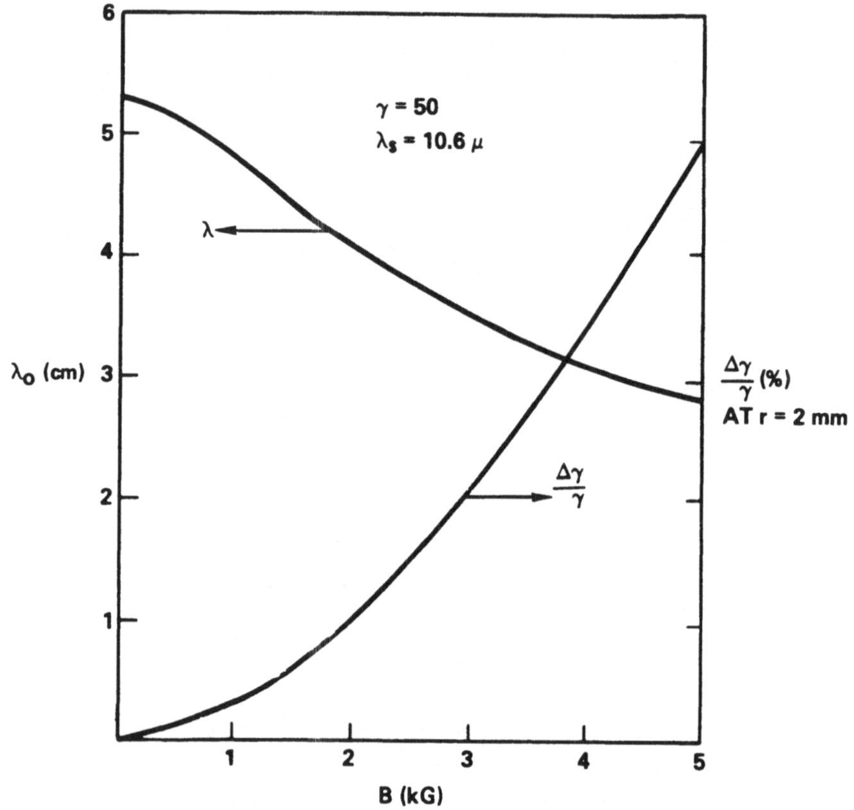

Figure 2. Wiggler Wavelength and Effective Perpendicular Energy Spread due to Field Non-Uniformity vs. B for r_b = 2 mm.

2.2 SINUSOIDAL WIGGLER

The choice of a sinusoidal wiggler introduces the question of particle untrapping due to longitudinal oscillations. In order to avoid such a negative effect the wiggler parameters should be chosen so that the excursion of an electron due to the longitudinal oscillation is small than a radiation wavelength. This condition can be expressed in terms of the wiggler parameters as

$$k_s \Delta z = \frac{a_w^2}{2(1 + a_w^2)} < 1. \qquad (3)$$

This expression shows that for $a_w >> 1$, $k_s\Delta z \rightarrow 1/2 < 1$. In order to have a less marginal condition however, a_w can be chosen on the order of 1 or smaller. This restricts $B_w < 3.1$ kG for $\gamma = 50$ and λ_s radiation wavelength = 10.6 μm. Note that this restriction is the same as that required to produce $\Delta\gamma/\gamma_{transv.} < .01$ at $r_b \sim 2$ mm.

Another design consideration regarding the wiggler parameters is given by the design choice of a_w = constant, or B_w = constant, or λ_w = constant, etc.

These choices have been discussed in detail in Ref. 3. Table 1 shows the results obtained using three possible design characteristics for the same initial conditions consistent with our proposed experiment parameters. Essentially, it is found that the final efficiencies are not very sensitive to the design choice (a_w = constant or B_w = constant, etc). However, the choice λ_w = constant where the magnetic field magnetic magnitude is adiabatically decreased, is experimentally simpler. This choice, λ_w = constant, has been utilized for the design. It should be noted that by keeping λ_w = constant and varying B_w, the longitudinal phase velocity v_{ph} of the bucket remains constant $|(\omega_s/k_w + k_s) = (k_s c/k_c + k_s) \simeq$ constant . The decrease in total energy is due to the reduction in "effective electron mass" or "perpendicular energy" denoted by the term a_w^2 in equation (1).

2.3 LASER SIGNAL

The laser signal (driver) power density is chosen as the maximum experimentally available that can be produced by high repetition rate. In addition it must satisfy the following conditions

1) Its magnitude should be sufficiently high to produce an energy extraction efficiency from the trapped electrons larger than 5%.

2) When it interacts with the wiggler field it should produce a "bucket" such that the bucket height is larger or equal than the total experimental electron energy spread.

3. EXPERIMENT SYSTEMS

As shown in Figure 3, the system consists of three major subsystems:

- The electron beam system with RF linac, beam line and focusing and deflection magnets.
- The magnetic wiggler.
- The CO_2 laser driver and 10.6μ detection system.

3.1 ELECTRON BEAM SYSTEM

A RF linac located at EG&G in Santa Barbara, CA will be used. It has an energy range of 1 - 30 MeV with 25 MeV nominal operating point. The current is 40A in a 50 psec micropulse. The emittance was measured to be 4.3 mm mrad. Quadrupole, driver and steering magnets have been designed to focuse the beam into the magnetic wiggler with a 2x8 mm elliptical waist, the smaller dimension being in the direction of the wiggler field.

One item that does require careful consideration is the optics of the wiggler itself. Only by careful tapering of the entrance and exit magnetic fields it is possible to prevent the addition of transverse momentum to the electrons. An exponential taper of approximately 10 wavelengths was found to be optimum. The role of space charge in the wiggler region was found to be secondary but another effect-the role of alternating gradient focusing by the wiggler-plays an important role in confining the electrons along the B field direction. The problem has been solved analytically and numerically. The effect of the wiggler field is neutral in the plane perpendicular to the B field but yields harmonic oscillator solutions parallel to B. An equilibrium radius may be derived which is equivalent to the radius for a B_z guide field of strength $\sqrt{2}B_w$. This provides a strong confining force capable of maintaining a beam thickness of less than 4 mm.

3.2 OPTICAL SYSTEM

The optical system consists of a laser driver, beam propagation optics and optical diagnostics as shown in Figure 4. This system is of fundamental importance to our FEL experiment, because the laser driver and the magnetic wiggler together permit the extraction of coherent energy from trapped electrons in the electron beam, and the optical diagnostics will enable us to measure accurately the efficiency achieved.

TABLE 1

DESIGN CONSIDERATION FOR EXPERIMENT AND SCALING COMPARISON BETWEEN

CHOICES OF a_w = CONST. or B_w = CONST or λ_w = CONST.

LASER PARAMETERS	ELECTRON BEAM PARAMETERS
$\lambda_s = 10.6\ \mu$	$\gamma = 50$, $E_b = 25$ MeV
$P_{in} = 252\ \mathrm{MW/cm^2}$	$r_b = 3\mathrm{mm}$, $\varepsilon = 8$ mm mrad
	$I = 40$ A, $\Delta\gamma/\gamma = .01$

WIGGLER PARAMETERS

Initial ($z = 0$)	Final ($z = 6$ m)		
$a_w = 1$	a_w = const		
$B_w = 3.04$ KG	$B_w = 3.274$	B_w = const	$B_w = 2.69$kG
$\lambda_w = 3.53$ cm	$\lambda_w = 3.273$ cm	$\lambda_w = 3.38$ const	λ_w = const
	$\eta_D = 3.83\%$	$\eta_D = 3.74\%$	$\eta_w = 3.61\%$

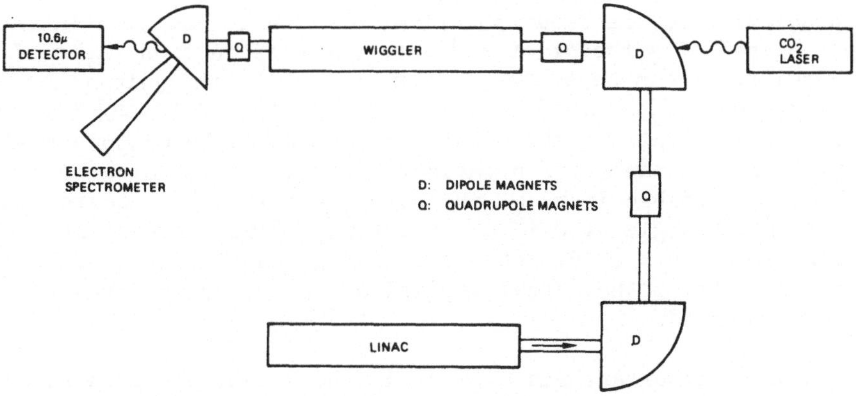

Figure 3: Schematic of the FEL System

The laser driver chosen for the initial experiments consists of an injection mode-locked Gen Tec model DD300 TEA CO_2 laser. This choice was made based on a number of characteristics, including pulse repetition rate, maximum power density and availability with details given below. The theoretical predictions require a power density of 400 MW/cm^2 in the FEL interaction regime to produce a 5% decleration efficiency. The Gen Tec laser device can delives 200 MW/cm^2 average power, but our plans include provisions for extending the system to achieve 400 MW/cm^2 in the final experiment.

The power density of 200 MW/cm^2 is achieved by injection mode locking the laser. The Gen Tec operates at atmospheric pressure, and the shortest pulse, limited by the homogeneously broadened bandwidth, is 1.3 ns.

The laser is injection mode locked by a 1 ns long pulse chopped from a CW CO_2 laser using a CdTe Pockels cell and polarizer. The gain medium in the Gen Tec and the Pockels cell have to be synchronized with each other and the electron beam. This will be accomplished by driving the system from the delayed injection pulses of the electron beam gun.

To measure the CO_2 laser beam energy increase due to the FEL interaction one must accurately measure the input laser energy and the output energy. This is not trivial since the laser pulses vary in energy from pulse to pulse. A severe problem is that the amplification occurs only during the 50 psec beam pulse time, i.e., during a small fraction of the laser pulse of 1.3μ sec length.

The most promising method to measure the ratio of the amplified to input energy is by the use of crossed polarizers as shown in Figure 5. If the linear polarization of the input laser is rotated at 45° to the gain axis of the FEL device, and if the output is analyzed through a linear polarizer with its transmission axis perpendicular to the input polarization then a reduction in the input signal of a factor of 10^3 due to the extinction ratio of a good polarizer can be achieved. The effect of the gain is to increase one component of the input polarization thereby rotating the net polarization. This rotation leads to a measurable component along the axis of the analyzing polarizer. Using the above numbers, the smallest detectable increase in energy is $\sim 1 \times 10^{-3}$, corresponding to a rotation angle of 0.1° and a corresponding increase in the amplitude of 1.4×10^{-3}. Thus for a 30 ps amplification duration in a 1.3 ns beam, the actual amplification can be measured to a resolution of 6%.

Another important measurement is the actual power gain, since it is the power density that affects the degree of trapping achieved. One would therefore like to measure accurately the duration of the

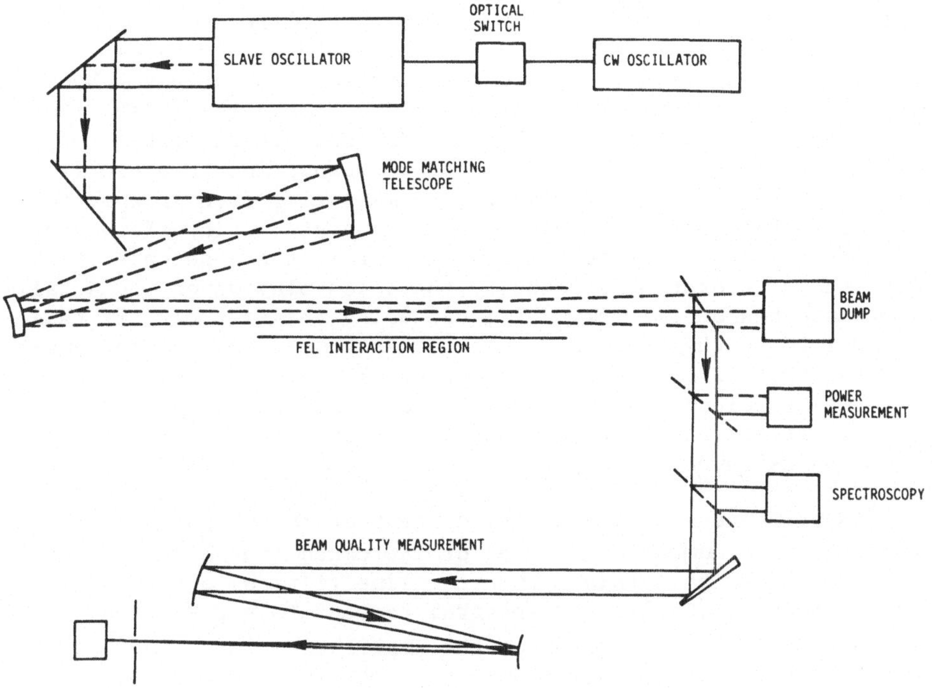

Figure 4: Schematic of the Optical System

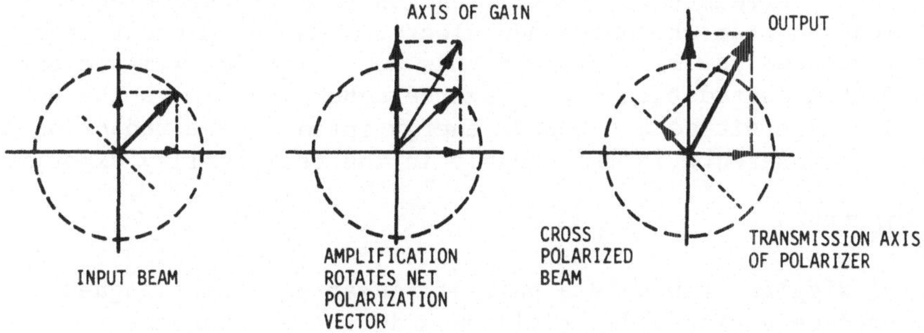

Figure 5: Separation of Input Beam and Output Emission by Crossed Polarizers.

gain, and preferably also the distribution of gain within this pulse, since the instantaneous electron beam current is not known. A Michelson's interferometer can be used to measure the pulse length as shown in Figure 6. By measuring the path difference in the two interferometer legs (L_2-L_1) over which interference fringes from the photon bunch are observed, one can measure not only the bunch length but also the variation of amplitude within the pulse. For a perfectly monochromatic bunch, the time resolution within the photon bunch is $1/\nu$ or 3×10^{-14} sec, corresponding to a precision of 10^{-3} for 30 ps. This would require measuring each fringe, and therefore a high data acquisition rate in a stable interferometer. If the frequency of the output changes, or if the coherence of the photon bunch is otherwise reduced, the deconvolution of the detailed pulse shape can become very difficult using this method. If this turns out to be the case, second harmonic generation in, for example, a $CdGeAs_2$ crystal can be used to detect the interference signal, since the efficiency of second harmonic generation is proportional to the square of the instantaneous intensity, and hence higher when the two pulses overlap.

Measurements of the spectral content of the output are also planned using a number of devices of increased precision. Rough spectroscopy can be obtained by using line filters, and a grating spectrometer can resolve the spectral content to a precision of 10^{-3}, limited by the 30 ps duration of the amplified pulse. For heterodyne spectroscopy to work, a detector bandwidth greater than 100 GHz is needed, ruling out all but the tungsten whisker diodes which are rather impractical devices, but which we shall use if deemed necessary in the course of the experiment.

Using the standard power in the bucket technique or an IR vidicon (Pyrocon) the beam quality of the amplified output will be investigated, again in the crossed polarized mode. The purpose of this experiment is to verify the near to diffraction limited beam quality predicted for amplification from trapped electrons. Amplification from the untrapped electrons may also occur at a greatly reduced power level, and this amplification will be expected to exhibit a degradation in transverse coherence due to the pump field inhomogeneities and e-beam energy spread. The reduction in transverse coherence is measureable in the beam quality experiment.

3.3 WIGGLER

The wiggler chosen is a pair of linear permanent magnet arrays which produce a sinusoidal field on axis. The arrangement of the magnetic field vectors is shown in Figure 7. The arrows in each magnet indicates the magnet polarization. The field at the symmetry axis is given by:

$$B = 2B_s \sum_m \cos(nkz) \exp(-\frac{nkd}{2}) \quad A$$

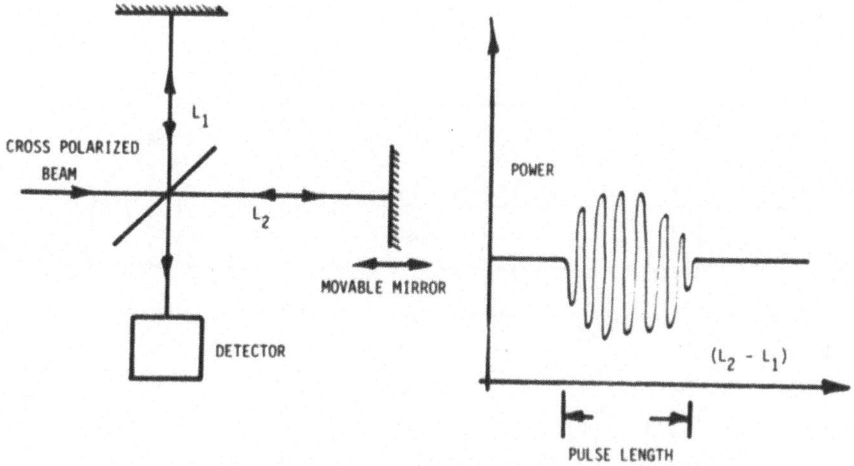

Figure 6: Photon Correlation Measurement of Pulse Length Using Michelson Interferometer

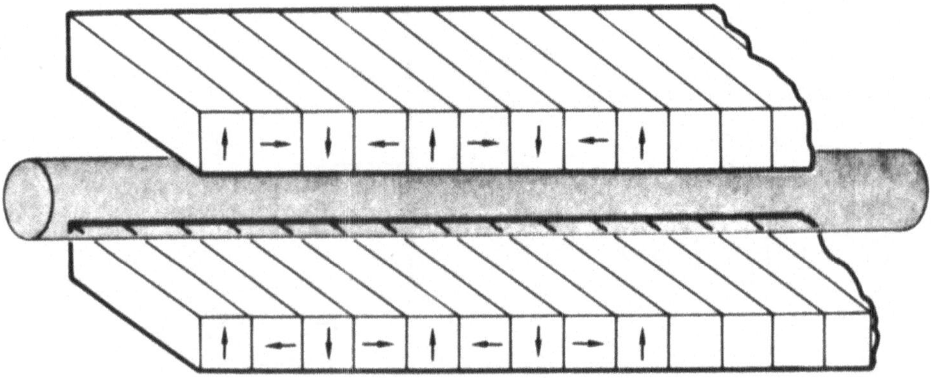

Figure 7: Permanent Magnet Wiggler

where B_s is the magnet surface field of 8.5 kG, n = 1 + 4 m, d the separation between the arrays, and A a factor that depends on the numbers of magnets per wavelength, the filling factor and the length of the magnets. The field strength (B) that can be produced by this arrangement depends on the wavelength of the field desired and the spacing between the planes plus a factor depending on the spacing between the magnets. The magnetic field strength and the wavelength must also self-consistently satisfy the resonant condition for 10.6μm. The solution of the self-consistent B and λ is shown in Figure 2. Away from the midplane the magnetic field rises according to a hyperbolic cosine; the effect is shown in the same figure. Nonuniformities off axis in the central plane have been determined by computer calculations to be less than .5% over the inside diameter of the beam line (1.5 cm).

REFERENCES

1. H. Motz. Journal of Applied Physics 22, (1951) 527.

2. N. M. Kroll and W. A. McMullin, Phys. Rev. A17, 300 (1978); P. Sprangle and R. Smith, NRL Memorandum Report 4033, Jan 1979, M. Z. Caponi, J. Munch and H. Boehmer to be published at J. Quantum Electronics VII - Addison-Eley, 1980.

3. N. Kroll, P. Morton, M. Rozenbluth - Free Electron Lasers with Variable Parameter Wigglers - Submitted to Phys. Rev. A. - Sec. 4 (1979). Also cf.1 papers by same authors at the Telluride Workshop to be published in J. Quantum Electronics VII - Addison-Weley 1980.

4. P. Sprangle, C. M. Tang and W. M. Manheimer - The Nonlinear Theory of Free Electron Lasers and Efficiency Enhancement, NRL Memorandum Report 4034. Also: Telluride Workshop, J. Q. Electronics VII - Addison Weley 1980.

5. V. V. Neil, Emittance and Transport of Electron Beams in a Free Electron Lasers, Jason Technical Report JSR-79-10, SRI Int. 79.

APPENDIX I

DESIGN CODE EQUATIONS

ELECTRON MOTION:

ENERGY $$\frac{d\gamma_i}{dz} = \frac{e_s a_w}{2\gamma} \sin\psi$$

PHASE $$\frac{d\gamma_i}{dz} = k_w - \frac{k_s}{2\gamma_i^2} \left(1 + \frac{a_w^2}{2} + \frac{e_s a_w}{k_s} \cos\psi + \frac{e_s^2}{2k_s^2}\right) + \frac{d\phi}{dz}$$

RADIATION FIELD:

AMPLITUDE $$\frac{de_s}{dz} = \frac{z_0 q a_w}{2\,mc^2} J \left< \frac{\sin\psi}{\gamma} \right|_i \rightarrow \alpha e_s$$

PHASE $$\frac{d\phi}{dz} = \frac{z_0\, q\, a_w}{2mc^2 e_s} J \left< \frac{\cos\psi}{\gamma} \right>_i$$

$$e_s = \frac{qE_s}{mc^2},\quad E = E_s \cos\left[\int k_z dz - w_s t + \phi(z)\right]$$

$$\psi = (k_w + k_s)\, dz - w_s t + \phi$$

FEL DEVELOPMENT PROGRAM AT LOS ALAMOS SCIENTIFIC LABORATORY*

R.W. Warren†

Los Alamos Scientific Laboratory
Los Alamos, New Mexico 87545

INTRODUCTION

The Los Alamos Scientific Laboratory (LASL) has embarked upon an ambitious FEL research program that will last several years, and includes at least three separate experiments. In this paper we shall briefly discuss the overall program, but shall emphasize the first "gain" experiment in which we are now involved in design and construction. We will cover mostly the practical features of the experiments and will point out the problem areas we have found.

LASL is especially well suited for FEL experiments becuase of its experience and expertise in both accelerator and laser technology. The accelerator expertise matured during the design and construction of LAMPF, the half-mile-long, 800-MeV proton accelerator used to produce mesons. The laser expertise developed to serve the needs of the large inertial-fusion program at LASL. This program uses many kinds of laser systems, but concentrates on the topics of most immediate importance to us; that is, 10.6 μm wavelength, good beam quality, high power, and short pulses. Many of the personnel involved in the FEL experiments were selected from the accelerator and laser programs and thus we have had good information exchange with them.

LASL's goal in the three-phase FEL program is to build an efficient, high-average power, 10.6 μm laser. Because it is difficult to achieve high efficiency with a single pass of an electron beam

*Work performed under the auspices of the U.S. Department of Energy with the support of the Air Force Office of Scientific Research, the Defense Advanced Research Projects Agency and the Naval Air Systems Command.
†AT-1, Industrial Staff Member from Westinghouse Research Laboratory.

through a wiggler, either the same beam must be passed through the wiggler several times or the energy remaining after a single pass must be recovered in some way. Several passes from a single beam is difficult because of the accumulating degradation in the beam's quality. On the other hand, great ingenuity must be exercised to find a recovery technique with acceptable efficiency and cost.

RACETRACK EXPERIMENT

The technique that LASL will use to achieve high efficiency is based upon an electron accelerator that is divided into two sequential parts. The first section is conventional but the second serves two functions: it accelerates electrons on a first pass and decelerates them on a second pass to recover a part of their energy. Figure 1 shows this process in some detail. Starting at the injector, electrons are accelerated to 90 keV in the gun, then, after bunching, they are accelerated to 7 MeV in the first linac section. At this poin they are inserted into the main loop, enter the second linac section, and gain an additional 13 MeV. After undergoing a 180° bend, they pass through the wiggler and transfer several percent of their energy to 10.6 μm light. Following another 180° bend, they reenter the second linac, but this time the electron bunches enter with different phase relative to the accelerating fields. The phase is right to decelerate the electrons and transfer their kinetic energy back into rf energy in the cavities of the accelerator. In this way, 13 MeV will be recovered from the beam. The remaining energy is lost in the beam dump. We expect this technique to increase the efficiency of operation by a factor of four. Even higher efficiencies are possible and may be attempted if time permits.

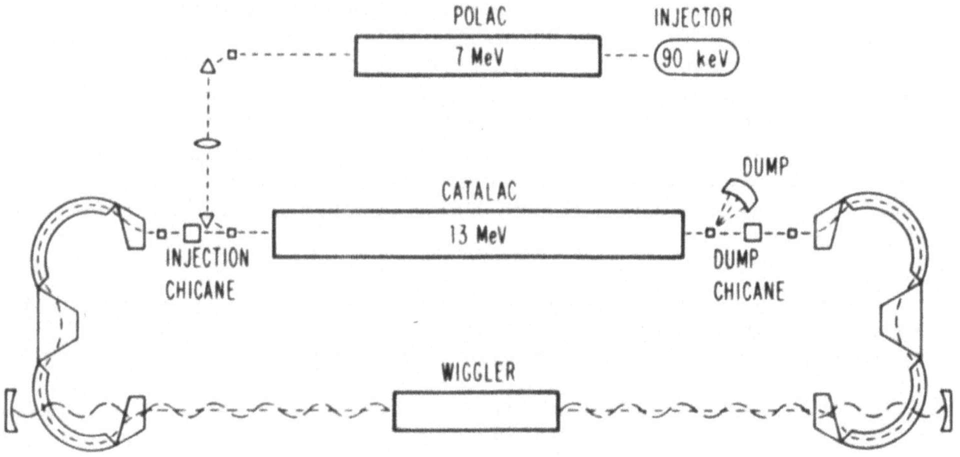

Fig. 1. Racetrack energy recovery system.

We know that the main features of this recovery technique work, but, clearly, many problems must be solved concerning the bending and focusing of the electron beam and the overall stability of the system during the simultaneous acceleration and deceleration of the beams. For this part of the program, we expect to rely heavily upon the expertise of accelerator specialists.

OSCILLATOR EXPERIMENT

Moving back a step to the second "oscillator" experiment, we also will depend very heavily upon the expertise of the laser specialists. In this experiment we will investigate the build-up of laser oscillations. The wiggler must perform adequately at all power levels from noise up to the gigawatt level. The accelerator must generate a train of micropulses whose length exceeds the build-up time of the laser oscillations. The optical beam quality must be well characterized and its dependence upon the various parameters of the electron beam and wiggler must be understood. The intricacies of optical pulse propagation in the electron beam-wiggler medium must be unraveled.

GAIN EXPERIMENT

Moving back another step, we come to the "gain" experiment on which we are now working. During this phase we will design and construct a system that can produce a significant optical gain and energy extraction efficiency when operating at very high optical power levels. The peak power in our optical beam is 1 GW; in our electron beam, 0.4 GW. These high power levels and the large ratio of optical to electron power put us into a different operating region from previous workers. Our main task is to design and to test the components, in particular the tapered wiggler, which make such operation possible. In the following pages, we will discuss these major components, emphasizing their design problems and the choices made to avoid or to deal with them. Figure 2 shows the general layout of the experiment.

Accelerator

A 5 MeV electron linac recently has been used in LASL experimental programs. This linac is being upgraded now to 20 MeV by adding cavity sections similar to those already in use. Table I shows specifications of the present and upgraded accelerator. The operating cycle of the linac will be as follows: the injector will be gated "on" every second or so to inject into the accelerator a pulse of electrons about 5 ns long. The accelerator, by its nature, will bunch this macropulse into a series of five or so 20-ps micropulses separated by about 1 ns and will accelerate them to 20 MeV. Gain and energy extraction can occur in the wiggler only when these micropulses overlap the laser beam both in space and time.

Of particular importance to the experiment are the current and the quality of the electron beam; for example, its spread in energy, space, and momentum. The energy spread can be reduced by better bunching and by reducing the loading of the accelerator cavities by the beam, but high beam currents and long macropulses worsen these effects. The spatial and momentum spread are associated mostly with the injector's gun. These values are adequate for the experiments at 10.6 μm, but operation at shorter wavelengths probably will require improvements.

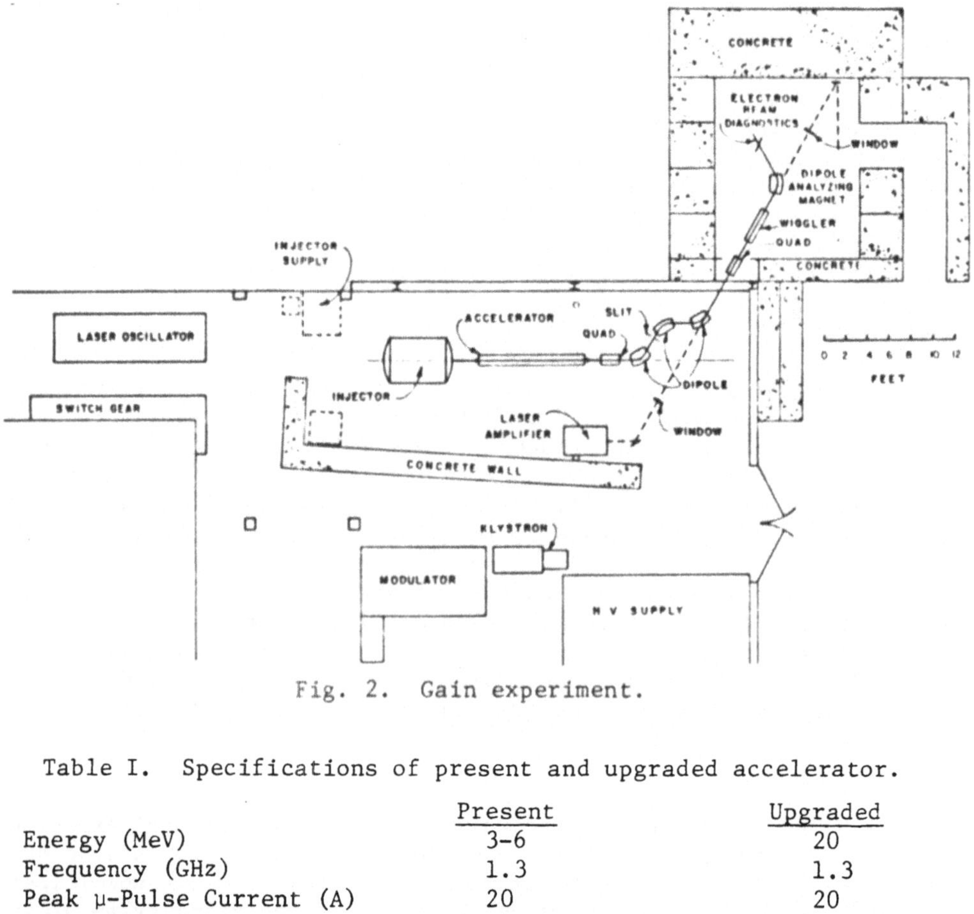

Fig. 2. Gain experiment.

Table I. Specifications of present and upgraded accelerator.

	Present	Upgraded
Energy (MeV)	3-6	20
Frequency (GHz)	1.3	1.3
Peak μ-Pulse Current (A)	20	20
Operation	Steady State	Stored Energy
Energy Spread from Single-Bunch Beam Loading (%)	± 1	± 1
Energy Droop per μ-Pulse (%)	∿ 0	0.1
Emittance at 20 A (mrad-mm)	∿ 10π	∿ 2π
Accelerator Length (m)	0.5	2.5
Pulse Length (ns)	300-1000	5

Figure 2 shows two major jogs in the electron beam line; their major purpose is to allow the joining of the electron and laser beams at the entrance to the wiggler and their subsequent separation at its exit. A secondary purpose of the second jog is to disperse the electron beam so that its energy-loss spectrum can be analyzed. The first complex jog includes three bends from three dipoles. The overall bend is doubly achromatic to maintain beam quality, but at the intermediate point in Fig. 2, where a slit is shown, the beam will be dispersed strongly. A system of slits and beam stops will be used there, to block the expected low-energy tail and to pass only a narrow distribution on to the wiggler.

Laser

In the gain experiment we wish to employ parameters that also are characteristic of the oscillator experiment. We therefore need a laser whose power output is as high as possible. Additionally, we need an optical beam of high quality, but only require a pulse length as long as the electron beam's macropulse, that is, 5 ns. Following these three major constraints, we have designed the CO_2 laser oscillator-amplifier combination, shown in Fig. 3. It will deliver a single longitudinal mode, 1-GW, 1-5 ns pulse of the P(20) line of CO_2 with good spatial and temporal quality.

In Fig. 3 five major elements can be noted: 1) the 2-m cavity oscillator containing both a high-pressure high-gain medium and a low-pressure low-gain smoothing tube, used to select a single mode for oscillation; 2) the electro-optical switch used to pass only a short part (1-5 ns) of the oscillator's output; 3) the four-pass intermediate amplifier, which uses the same gain medium as the oscillator; 4) the combination of spatial filter, isolator, and beam expander used to improve the beam's quality, protect the intermediate amplifier from reflected light, and prepare the beam for the final amplifier; and 5) the final amplifier, a commercially available three-pass Lumonics 600 amplifier.

From the final amplifier, the polarized light will be led to the wiggler by copper mirrors and a salt window. The sizes of the windows and mirrors and their spacings from the wiggler are determined by their damage thresholds.

The focal length of the final mirror is determined by the energy density desired at the focal spot and its depth of focus within the wiggler. These are chosen in a complex way to maximize the gain of the wiggler.

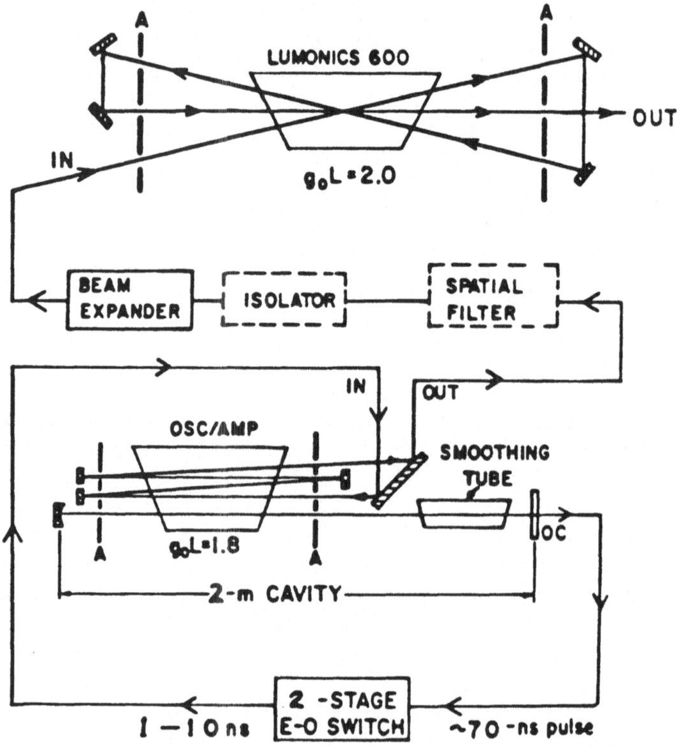

Fig. 3. The 10.6 μm laser system.

Wiggler

In designing the wiggler, we have deliberately sacrificed gain to enhance our diagnostic abilities and to achieve flexibility. In particular, we have provided room to monitor the position and shape of the electron beam inside the wiggler and anticipate altering the taper in magnetic field during the experiment. The use of hundreds of $SmCo_5$ permanent magnets of a fixed size but variable spacing facilitates such alterations. However, we are concerned that variations in magnet strengths will affect the field and bend the electron beam or deflect it away from the laser beam. Accordingly, we have designed a complex test program to evaluate each of the magnets used in the wiggler. From these tests we will select magnets so that adjacent ones cancel to a high degree the aberrations of their neighbors. Finally, we will test the completed wiggler by measuring the deflection of a fine current-carrying wire that will simulate the electron beam. Residual field aberrations will be corrected by a series of 20 "trim" coils mounted outside the wiggler.

To achieve as high a field as possible, the magnets are placed in the vacuum chamber very near the electron and laser beams. Vacuum properties of sintered $SmCo_5$ have been questioned, but preliminary tests show no need for concern. Because of the brittle nature of the magnets and because of their presence in the vacuum system, a mounting technique was needed that avoided complex machining operations or complicated hardware. The technique chosen depends upon two perpendicular grooves cut in each end of the bar magnets. It includes long metal bars machined with regularly spaced teeth that mesh with one groove at each end of each magnet. The spacing between magnets can be changed by inserting a new bar with a different tooth spacing. There are two grooves at each end to allow the square cross-section magnets to be held with any orientation differing by 90°. Special entrance and exit regions are provided at each end of the wiggler to allow the electron beam to enter and to leave the wiggler with no unbalanced deflections at the transitions.

Diagnostics

There are three measurements that are especially important to the experiment's success: one to assure that the electron and laser beams overlap in the wiggler, one to determine the average energy extracted from the electron beam and the spectrum of the depleted electrons, and one to measure the optical gain and the quality of the generated light.

Overlap of the beams will be assured with a system of apertures, fluorescent screens, and an auxiliary He-Ne laser. First, two small apertures are placed at the entrance and exit of the wiggler, and the He-Ne laser is adjusted until a maximum of its light penetrates the wiggler and both apertures. Next, two larger apertures are inserted and the same maximizing adjustment is made with the CO_2 laser. Finally, the electron beam is turned on and a fluorescent screen is inserted into its path while the He-Ne laser is operating. Two spots will be seen on the screen: the fluorescent spot from the electron beam and the scattered-light spot from the He-Ne laser. If the electron beam is moved to superimpose the spots, it will be correctly aligned with the CO_2 laser. In this way, the electron beam can be aligned at the three screens provided; that is, at the entrance, exit, and midpoint of the wiggler. We will adjust the beam position with the 20 trim coils wound around the wiggler and by other steering coils located further up the beam line.

The average energy and the spectrum of the depleted electrons will be determined after they have been bent and dispersed by the final magnet. The beam will strike a fluorescent screen and will be viewed by a vidicon. The calculated energy spread is large enough to be resolved easily.

We expect the measurement of optical gain and beam quality to be difficult because the gain is small (2%) and it occurs only during the electron beam's micropulse, that is, only about 2% of the time that the laser is operating. The gain measurement would be easier if it could be performed only during a micropulse, but it is difficult to resolve the gain's temporal variations with existing instruments. If no attempt is made to resolve them, but an average gain is measured, it will only be about 10^{-3}. It at first appears attractive to use some technique to subtract the input laser pulse from the output light so as to detect only the gain contribution. Various schemes have been suggested using polarizers or narrow-band absorption cells, but they all suffer from two serious problems: low signal level, and an indirect connection between the detected signal and the desired gain. We intend to install a very flexible optical measurement system that will allow us to test several different gain-measuring techniques and to perform other measurements on the output beam quality. The details of this system are still being worked out.

ACKNOWLEDGEMENT

The work discussed above was performed by a team composed of B. Newnam, W. Stein, L. Young, and the author under the direction of C. A. Brau. The author is grateful for their help in preparing this paper.

FELIX - A PROPOSED EXPERIMENTAL FACILITY FOR FREE ELECTRON LASER INVESTIGATIONS

M.W. Poole

Science Research Council
Daresbury Laboratory
Warrington
England

INTRODUCTION

Since the announcement of successful operation of a free electron laser at Stanford there has been world-wide interest in the further development of such FEL sources. In the United Kingdom a study group was set up at Daresbury Laboratory to advise the SRC on the status of such developments and how it might best make a contribution to them.

Although there has been good progress on the theoretical understanding of the basic physics, experimental development of the FEL lags well behind these accomplishments, despite the recent funding of several projects. No comprehensive set of operating data yet exists, and questions raised by the Stanford experiments remain unanswered. Modifications to the standard FEL operating regime have been proposed, but none of these has so far been tested.

One reason for the slow rate of experimental progress is that suitable electron sources are scarce and extremely expensive. The Daresbury study group has carried out a design study of an experiment based on a powerful linear accelerator that is at present in store but which could be recommissioned at a relatively low cost. This project is now under consideration for funding.

THE PURPOSE OF FELIX

The primary purpose is to contribute to the basic understanding of the free electron laser by provision of a test-bed experimental facility. It is hoped to explore a wide range of FEL behaviour and relate this to theoretical predictions. FELIX must therefore be an flexible enough to operate with much variation of basic parameters, and it also requires a diagnostic system that can provide reliable and comprehensive data. No such facility presently exists or is planned elsewhere and it is believed that it could have a long useful lifetime.

The facility would be complementary to other approved projects, such as the insertion of FELs into existing electron storage rings, and could provide information that greatly assists the difficult interpreta tion of recirculation experiments. FELIX would also supply data that could be confidently extrapolated to single pass FELs at quite different operating wavelengths.

A secondary, but important, benefit of FELIX would be to provide the extensive experience of FEL operation that will be essential if they are to be pursued as future sources. The combination of electron accelerator technology and optical techniques is very unusual and can be found in few laboratories at present.

The Daresbury SRS is a synchrotron radiation source that will provide intense VUV and x-ray beams, but it is also planned to include an infra-red beam line. If FELIX is approved it would be a unique source of infra-red radiation and has already attracted the attention of research scientists; it is believed that a strong scientific case can be assembled for infra-red sources of the FEL type, because of their high power and tunability. Although this is not the major purpose of FELIX it would almost certainly be used for exploratory work on research applications; some of the potential users have also offered to assist with FEL diagnostic experiments, for example by the provision of fast detectors.

DESIGN PHILOSOPHY

A major criterion in the design of FELIX has been the adoption of conventional technology wherever possible, together with the incorporation of maximum flexibility into the system. This approach is essentia for the rapid construction and commissioning of a reliable test bed facility that can be easily modified.

Superconducting magnet technology has been deliberately avoided as a result of the above criterion. Furthermore helical magnet geometries although having some theoretically attractive features, have also been rejected; it is more difficult to generate even moderately high fields, the geometry is not easily modified and access to the interior of the magnet is much more restricted. A permanent magnet is under considera-tion, but would only be adopted if it can demonstrate equally convenien control to an electromagnet for output tuning experiments. Whatever the final choice, it is intended to construct a magnet with a modular design that will allow later changes.

Good diagnostic equipment is essential for the comprehensive experimental programme, and this has been a major factor in the choice of FELIX operating wavelengths. The availability of probe laser lines with strong output has led to a nominal central wavelength of 115 μm, and a tuning range from 75-150 μm should adequately demonstrate behaviour as a function of wavelength. The chosen wavelengths are also well suited to the characteristics of an electron source available at Daresbury.

Finally any such test facility should be of a dedicated nature, avoiding interaction with alternative experimental demands on the equipment, and should be as economical as possible. No major items of FELIX equipment will be shared, it will be situated in its own enclosed area and there will be substantial cost savings from the use of existing equipment, particularly the electron source.

MAIN PARAMETERS

Gain Optimisation

The main parameters must be chosen to produce a gain large enough to result in the power building up to a high level during the electron beam pulse. The functional dependence of the gain can be written as follows

$$G \quad \alpha \quad \lambda^{1.5} \quad \lambda_0^{-2.5} \quad \frac{K^2}{(1+K^2/2)^{1.5}} \quad I \quad L^3 \quad \Sigma^{-1}$$

where
λ = laser output wavelength
λ_0 = magnet period
K = deflection parameter (= 0.9337 $\lambda_0 B$ cm.T)
I = peak electron beam current
L = magnet length (= $N\lambda_0$, N periods)
Σ = optical mode area

The smallest optical mode area is achieved by use of a confocal arrangement of mirrors, and for such a radiation mode

$$G \quad \alpha \quad \lambda^{0.5} \quad \lambda_0^{-2.5} \quad \frac{K^2}{(1+K^2/2)^{1.5}} \quad I \quad L^2$$

The length of the FEL should clearly be maximised, but is limited by cost, available space and technological difficulty. For FELIX a value L = 5m has been chosen as a suitable compromise. A high value of peak current should also be chosen but this will be discussed in a later section.

It is easy to show that there is an optimum value of K = 2 arising from the above functional dependence. The magnet period should be made as small as possible.

Periodic Magnet

For the specified value of K, any reduction in λo causes a corresponding increase in the peak magnetic field B. A further limitation on λo is set by the rapidly increasing field inhomogeneity as it is decreased, related to the aperture between the magnet poles.[1] This aperture is set by the acceptable diffraction losses from the confocal cavity, which are greatest at the longest operating wavelength. A tota aperture of 50 mm between the magnet poles, including some allowance fo a vacuum chamber, results in losses ~1% at each mirror at 150 μm, which is considered to be acceptable.

The acceptable field inhomogeneity is set by the permitted resultant line broadening, arising from the finite electron beam cross-section. With the known emittance of the electron source a broadening of the natural (homogeneous) line width by less than 10% results from a value of λo = 200 mm. The associated peak magnetic field is then ~0.1T, a moderate level that can be readily achieved.

Electron Source

The required electron beam energy is determined by the choice of λo. It varies from 22-32 MeV and is 25 MeV at the nominal FELIX wavelength.

The available electron source is the 43 MeV linear accelerator (linac) that was previously in use as the injector to the 5 GeV electrc synchrotron NINA. It provided a current of almost 1A within a 1 μs lon pulse, and the peak current within the radio-frequency bunches was up t 250 A. However the energy spread within the beam was in excess of 1%, and it is therefore proposed to construct an energy compression systen that will also allow the electron bunch length, and accordingly the pea current, to be varied as required.

Any FEL suffers pulse slip due to the different axial velocities of the electron and radiation pulses through the system. On FELIX the slip is ~ 10 ps, comparable to the electron bunch length in the absence of the energy compression system. Reduction of the beam energy spread to ~ 0.1% results in an electron bunch length of about 60 ps, which should overcome any pulse slippage problems. However it will be possible to operate with bunches short enough to investigate slippage problems of the sort encountered at Stanford. The basic FELIX parameters assume this long bunch length and a corresponding peak current of 20A.

Optical pulses reflect between the mirrors of the cavity with a transit time chosen to coincide with the arrival of electron bunches. With a space between mirrors in excess of 8m only 18 passes through the interaction region would be possible during the existing linac pulse duration of 1 μs, insufficient for the build-up of desirable power levels. It is therefore intended to modify the linac to produce a pulse of at least 4 μs, bringing the saturation peak power level of 10 MW within reach.

Output Power and Tuning

With the parameters already discussed the gain per pass through FELIX at the nominal wavelength can be calculated to be 22χ %. The factor χ arises from uncertainty over the value of average radiation mode area to be used, with $\Sigma = \lambda L/\chi$. For a TEMoo gaussian mode it can be shown[2] that the average value of Σ is minimised with $\chi = \sqrt{3}$, and such a value also appears in reasonable agreement with the quoted gain of the Stanford 3.4 μm oscillator experiment[3].

The resultant value for FELIX is a gain per pass ~ 37%, equivalent to a build-up factor ~ 10^9 over the initial 3.5 μs of the electron pulse.

The most important FELIX parameters are summarised in Table 1.

The output wavelength from FELIX can be readily tuned by variation of either electron energy or magnetic field. In the former case the gain scales as $\lambda^{\frac{1}{2}}$, giving a variation of ±15% about the central value over the range 75-150 μm. For magnetic field variation at constant energy, the gain varies by a similar amount.

Table 1. Principal FELIX Parameters

Output wavelength	75-150	μm
Gain per pass *	0.37	
Optical pulse length	60	ps
Homogeneous broadening	2%	
Oscillator linewidth *	0.5%	
Peak power at saturation *	10	MW
Average power at saturation *	10	W
Magnet length	5	m
Magnet period	200	mm
Peak field on axis (K=2)	0.11	T
Peak electron current	20	A
Electron energy	32-22	MeV
Macro electron pulse	4	μs
Repetition rate	25	Hz
Optical cavity length	8.1	m

* At nominal central wavelength (115 μm) and energy (25 MeV)

COMPONENTS OF FELIX

Electron Linac

A number of modifications to the linac have already been discussed. The addition of an r.f. cavity to the gun to allow linac operation at 408 MHz, with a corresponding peak current increase, was employed routinely when the linac operated as the NINA injector.

An achromatic dispersion system at the output of the accelerator will debunch the electron beam, and the resultant phase spread can be used in a short section of waveguide to reduce the energy spread by an order of magnitude. A final spread of ±0.1% should be possible with the bunch length increased to 60 ps. Once again, this system has been previously operated on the linac.

A pulse length of 4-5 μs will be achieved by modifications to the pulse forming networks on the klystron modulators. No further increase will be possible due to the rating on the klystrons themselves, and the linac repetition rate must be reduced from 50 Hz to 25 Hz if existing power supplies are used.

The anticipated beam emittance at 25 MeV is 5 mm.mrad, and adiabatic damping causes this emittance to vary inversely with energy. The present value gives acceptable beam sizes throughout the FELIX system, but any reduction would be of some value. Furthermore any reduction of emittance (ε) permits a corresponding decrease in λ_0 ($\alpha\ \varepsilon^{1/3}$) for the same broadening of linewidth, with a consequent increase in the gain ($\alpha\ \varepsilon^{-1}$) and reduction in the attainable wavelength of the system. Use of a high emissivity gun cathode may therefore be considered as a future development of FELIX.

Beam Transport System

The dispersion section comprises four small 45° bending magnets with moderate field capability. A collimator slit in the centre of this system allows selection of any momentum bite from the dispersed beam, a useful additional facility.

The proposed FELIX layout requires a 90° bend in the beam line before entering the laser magnet. An achromatic bend requires two bending magnets and a central symmetry quadrupole. Two quadrupole doublets control the beam size through the system, and the presence of the symmetry quadrupole requires one of these to be immediately adjacent to the laser, within the optical cavity. There is very flexible control of both horizontal and vertical beam size, and a maximum radius of 7 mm can be achieved.

A final bending magnet after the laser directs the electrons into a faraday cup, and will also be used as an electron spectrometer. The beam line also incorporates all the necessary vacuum components and electron beam diagnostics.

Laser System

A preliminary design of the periodic magnet at the heart of the system assumes the use of an electromagnet. A variable magnetic field up to 0.15T will be provided to allow tuning experiments with FELIX, necessitating up to ~ 3000 A-turns for each of the one hundred pole coils. The cheapest magnet design would employ air-cooled coils dissipating a total of about 4 kW, but a more compact water-cooled version allows greater geometric flexibility in the magnet and also has the potential for further increases in field value if required. The modular magnet design would have detachable poles to allow experiments with modified pole width, pole profile or even periodicity and aperture. A spare high current power supply from the SRS project would be available.

The confocal resonator will be defined by polished copper mirrors with very high reflectivity, and the mirror separation will be controlled by stepping motor to high accuracy ($\lesssim$ 10 μm). Windows to terminate the vacuum system in the periodic magnet will be manufactured from silicon or some other suitable material.

Diagnostics

A CO_2 - pumped methanol laser is commercially available with a CW output ~ 100 mW at 119 μm, and has other useful lines in the wavelength range of interest. In addition to the measurement of FEL gain this laser can be used to seed the cavity in order to establish very high FELIX power levels, hopefully reaching saturation conditions.

A detector system must be employed that is sensitive to the r.f. modulation added to the probe laser by the FEL action. The necessary resolution ~1 ns is at the limit of present detector technology but can be achieved with low temperature photoconductive devices; improved detectors might become available after development work, but a gain per pass of 5% should be easily measured, and ~ 1% with more difficulty.

Other useful diagnostic equipment would be available from the infra-red beam line of the SRS, including an interferometer.

Good electron beam diagnostics will also be necessary. Momentum spectra should be readily available with the electron spectrometer at the exit of the interaction region, but it will be necessary to develop techniques to measure the peak current within the 60 ps electron bunch.

OUTLINE OF EXPERIMENTAL PROGRAMME

Tunability

Both the magnetic field and electron beam energy can be varied to allow wide tuning of FELIX.

High Gain Regime

The predicted gains are sufficiently high to invalidate the assumption of a constant radiation field throughout the passage of each electron.

High Power and Saturation

Investigation of saturation conditions should be possible, at least when an external seed laser is used. The practical problems of dealing with the high peak and average output powers will also be of interest!

Tolerances

The effect of imperfect quality of both electron beam and periodic magnet can be explored under variable conditions.

Pulse Behaviour

Variable pulse length and slippage should allow a comprehensive investigation of this topic.

Harmonic Operation

A demonstration of FEL operation at a higher harmonic should be possible. It is interesting to note that for K = 2 most of the spontaneous power is NOT emitted into the fundamental line.

Electron Momentum Spectra

Detailed investigation under low gain, high gain and saturation conditions will be possible, and the effectiveness of proposed techniques to overcome the electron beam "heating" could be examined.

FELOKIX

A preliminary study has shown that FELIX could be converted to an optical klystron with a significant increase in the gain per pass. Only a simple reconnection of the periodic magnet coils is necessary, and a very large additional range of experiments would then be possible.

Design Changes

Major changes to the periodic magnet should be possible at some future stage, giving even greater scope to the experimental programme. Such changes include gradient fields and tapering of the magnet characteristics.

Operating Experience

The extensive programme of work will produce a wide experience of the operational problems of FELs that are often omitted from the theories. An example would be the problem of optimising the coupling between the electron beam and the radiation mode. Diagnostic techniques will also be established and may require the development of new equipment.

FELIX AS A FAR INFRA-RED SOURCE FOR RESEARCH

Although FEL investigations must be the major programme of FELIX experiments there is already interest in its use as a source of infra-red radiation, and it is certain that some exploratory work would be undertaken if FELIX were constructed. There is certainly a scientific case for a tunable coherent source with an output extending over the whole of the infra-red region of the spectrum; no such source exists at present in this region.

The tuning range of FELIX can be extended well beyond the 75-150 μm so far specified. The linac can certainly operate over a range 20-50 MeV, with some reduction in current only at the highest energies. At 50 MeV with a value K=1 a wavelength as short as 15 μm is attainable, but the gain per pass is only 5%; at this wavelength the radiation mode area is also less than that of the electron beam. The maximum output wavelength is not set by electron energy or magnetic field but by diffraction losses that are prohibitive above 200 μm. The useful range of the initial FELIX apparatus is therefore likely to be about 50-200 μm. At some future date this could be extended by further modification to the electron beam properties or adoption of a new magnet design.

Any infra-red facility would be complementary to the SRS at Daresbury. Areas for research application have already been identified and include the following

Electronic structure of semiconductors
Generation of coherent phonons
Spectroscopy of adsorbed molecules
Low frequency vibrations in amorphous materials
Molecular motion in liquids and polymers
Superconductivity studies

REFERENCES

1. M.W. Poole and R.P. Walker DL/SCI/P215A. To be published in Nuclear Instruments and Methods.

2. R. Barbini and G. Vignola LNF 80/12 (R) 10 March 1980.

3. J.M.J. Madey Private communication.

THE FEL - MICROTRON ACTIVITY AT THE C.N.E.N. FRASCATI CENTER

Giuseppe Dattoli, Angelo Marino and Alberto Renieri

Comitato Nazionale Energia Nucleare (CNEN)
Centro di Frascati
C.P. 65
00044 Frascati, Rome, Italy

1. INTRODUCTION

The first operation of a Free Electron Laser (FEL), i.e., the production of coherent electromagnetic radiation of short wavelength from relativistic electrons moving in a periodic magnetic field, was demonstrated at Stanford University[1].

After these pioneering experiments a noticeable amount of work has been undertaken, aimed essentially towards the realization of efficient high power FEL's.

FEL's have been proposed operating with variable parameter wigglers[2], gain expanded wigglers[3], and systems which utilize the continous electron beam (e.b.) generated by an electrostatic accelerator, where the efficiency enhancement is provided by the e.b. energy and change recovery after the FEL interaction[4].

Furthermore, various e.b. sources have been proposed which can be grouped in two main categories: recirculated e.b. (Storage Ring; for a review see Refs. 5,6) and those operating with single pass machines (e.g. Linac, microtrons and so on).

Although the characteristics of the Stanford Superconducting Linac were ideally suited for the study of the most important operating characteristics of the FEL, there are a number of practical considerations (e.g., the high cost of construction and maintenance, limitation on the cu-rent and criticality of operation) which suggest looking for other e.b. sources such as conventional linacs[7] or microtrons[8].

It is the purpose of the present note to describe the program at the C.N.E.N. center in Frascati which is directed towards the realization of a FEL operating with a circular microtron as the electron source.

Let us briefly summarize the practical considerations which have dictated us the choice of a microtron:

a) The existence at the center of a 12 MeV microtron[9].
b) The possibility of reaching working energies up to 30 or 40 MeV.
c) The fact that the characteristics of the e.b. microtron are better than those from a Linac of the same energy (in particular, the energy spread is lower).
d) The possibility in principle, of reaching, high e.b. power.

The plan of this note is as follows: In Sect. 2 we will, briefly, deal with the main characteristics of a microtron and in particular with the Frascati microtron; in Sect. 3 the general FEL - microtron characteristics are reported; and finally in Sect. 4 the principal laser beam (l.b.) features are analyzed.

2. THE MICROTRON

A) General Features

The original concept of the microtron was due to Veksler[10] during the last war years (1944). Indeed, Veksler proposed a new type of accelerator, which he called an "electron cyclotron:; unlike conventional cyclotrons, this device would be suitable for accelerating ultrarelativistic particles.

The microtron is thus a device which is particularly useful for accelerating light particles (e.g. electrons and positrons[11]).

In the microtron the electrons are accelerated by an alternating electric field of constant frequency in a constant magnetic field. In the vacuum chamber the electrons follow circular paths with a common tangent point (Fig. 1) At this point is placed the cavity which supplies the high frequency electric field needed to accelerate the electrons. The cavity is excited by a hig frequency powerful source (e.g. a pulsed magnetron with a power of at least hundreds of kilowatts). After each pass through the cavity, the electron gains a certain amount of energy Δu and passes to the next orbit. When the electrons reach the last orbit they pass through a magnetic shield and are extracted.

Let us briefly discuss the methods for particle injection in a microtron.

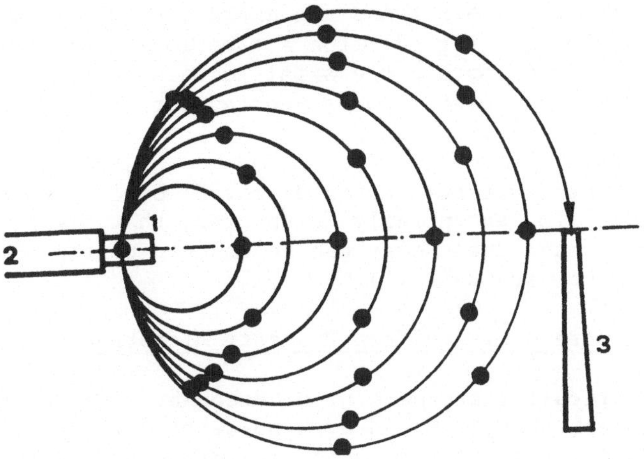

Figure 1. Microtron Layout

1) Accelerating Cavity
2) Waveguide
3) Shielding magnetic channel for electron extraction. The dots represent the electron bunches.

The first microtron (e.g. the Canadian one[12,13]) used electron injection by field emission from the cavity gap; even if high currents can be injected, only low currents are accelerated. It was just this limitation on the current which impeded for a long time the development and the wide introduction of microtrons.

After a number of intermediate proposals (see Refs. 4), a mechanism for injection was suggested which transformed the microtron into a practical device.

The new injection method was proposed by O. Wernholm[15]; the idea was that of injecting the electrons from a small gun outside the cavity (see Fig. 2). This method increased the accelerated current by an order of magnitude. Short afterwards two new methods of injection were proposed by Kapitza[16] and Melekin[17]. In

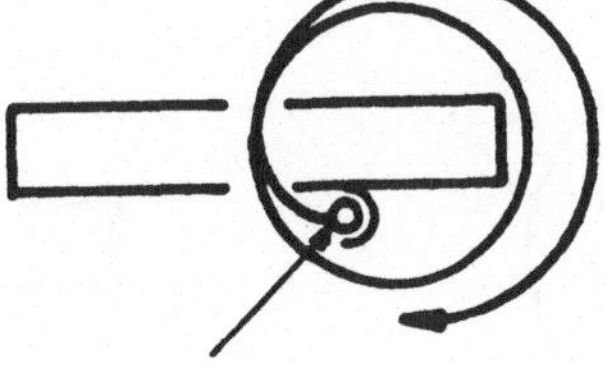

Figure 2. Wernholm Injection

both the methods the electrons are extracted by the radio frequency field (see Fig. 3) from a thermionic cathode in the cavity wall[16]. To improve the axial focusing it was also proposed that the cavity used should have rectangular, rather than circular, transit holes[17].

The advantage of the latter approved 16, 17 compared with previous ones lies essentially in the fact that higher energies can be reached and the energy can be changed continuously (for further comments see Ref. 11).

B) Main Characteristics of the Frascati Microtron

Let us recall that the e.b. produced by a circular microtron is, in general, characterized by a pulse of duration (τ_M) of few usec and by a microstructure which is formed by electron bunches of the duration (τ_b) of several tens of psec with a bunch-to-bunch time interval ($\Delta\tau_b$) of about several hundreds of psec.

In particular, the Frascati microtron, which operates S band, has

$$\tau_M \sim 2 \text{ to } 4 \mu\text{sec}, \quad \tau_b \sim 20 \text{ psec}, \quad \Delta\tau_b \sim 300 \text{ psec}$$

$$\text{Working energy} = \begin{cases} 12 \text{ MeV (Wernholm injection)} \\ 20 \text{ MeV (Kapitza injection)} \end{cases} \tag{2.1}$$

Average current (at 20 MeV) ≅ 35 mA
Peak current (at 20 MeV) ≅ 650 mA

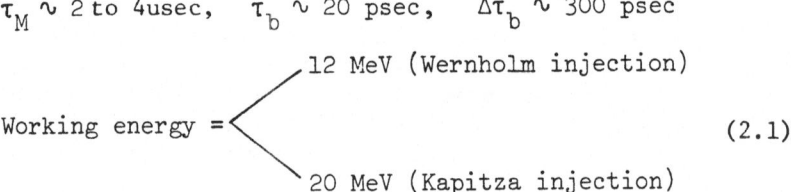

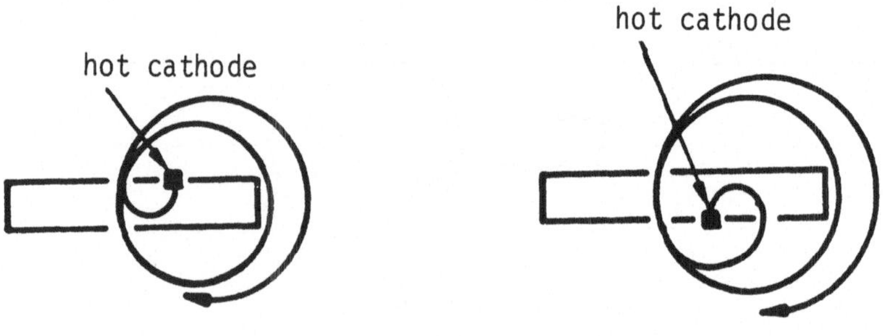

Figure 3. Kapitza and Melekin Injection

3. FEL - MICROTRON CHARACTERISTICS

A) Gain Parameters

It has been shown by us in a recent theoretical analysis of single pass FEL operation[18] that the gain per pass is given by

$$G = g_o \mathbb{R}e\, q_{\gamma}^{1}(\Theta,\mu_c) \tag{3.1}$$

where

$$g_o = 2\pi \left(\frac{2\lambda}{\lambda_q}\right)^{1/2} \frac{L_w\lambda}{\Sigma_L} \frac{I}{I_o} \frac{K^2}{(1+K^2)^{3/2}} \left(\frac{\Delta\omega}{\omega}\right)_o^{-2} \tag{3.2}$$

$$\Theta = -\frac{2}{\pi}\left(\frac{\Delta\omega}{\omega}\right)_o \frac{\omega\delta t}{g_o} \qquad (\delta t = T_c - T_e) \tag{3.3}$$

$$\mu_c = \frac{\lambda/2\sigma_z}{(\Delta\omega/\omega)_o} = \text{coupling parameter ;} \tag{3.4}$$

the meanings of the symbols are summarized in Table I (for further details see Refs. 18, 19).

Without entering into the details of the quantity let us recall that it is in general a function of Θ and the "coupling - parameter" μc; anyway in our particular experimental conditions it can be treated as a parameter ranging between $\sim 0.4 \div 0.8$ (for further comments and physical insight see Refs. 18,19). To achieve laser action we thus have to optimize the fundamental parameters of the apparatus to have enough gain per pass to reach saturation in a time small with respect to the e.b. pulse duration.

Furthermore, recalling that the net gain of the system is given by [18,19]

$$\alpha = (1-\gamma_T)G - \gamma_T, \tag{3.5}$$

(γ_T being the total cavity losses including the output mirror transmissivity) we must minimize the diffraction losses, in other words we must have a large Fresnel number N.

TABLE I

PHYSICAL CONSTANTS

E	= electron charge
m_o	= electron rest mass
c	= light velocity
r_o	= e^2/m_oc^2 = classical electron radius
I_o	= ec/r_o = Alfvén current

ELECTRON BEAM PARAMETERS

E	≡ energy
I	≡ peak current
σ_z	≡ r.m.s. bunch length
T_e	= bunch to bunch time interval

FEL PARAMETERS

λ_q, L_w	= wiggler wavelength and length
$\left(\frac{\Delta\omega}{\omega}\right)_o$	= $\lambda_q/2L_w$ = homogeneous bandwidth
λ	= $\lambda_q/2(1 + K^2)(m_oc^2/E)^2$ = laser wavelength, $\omega = 2\pi c/\lambda$
K	= $eB\lambda_q/\sqrt{2} \cdot 2\pi m_oc^2$, B ≡ peak wiggler magnetic field
Σ_L	= laser beam cross section
T_c	= cavity round trip period
γ_T	= total optical cavity losses
N	= $(h/2)^2/\lambda L_w$ = Fresnel number
h	= wiggler gap

B) e.b. Parameters

Before discussing the e.b. parameters let us recall that the above formulas, in particular that defining the gain, hold in general if the inhomogeneous broadening due to the energy spread and emittance is negligible with respect to the homogeneous one[18]. The Frascati microtron e.b. fulfills quite well the above requirements[20].

The electron beam characteristics suitable for laser action in the wavelength region we are interested in (see below) are summarized in Table II.

It is to be noticed that the above parameters will be obtained by further modification of the present e.b. Frascati microtron[9]. The main modifications are the substitution of the actual 2 MW, 2 to 4 usec S-band magnetron with a klystron with the

TABLE II

	(a)	(b)
e.b. Energy	20 MeV	⩾ 30 MeV
Average Current	∿ 350 mA	∿ 250 mA
Peak Current	∿ 6.5 A	∿ 4.5 A
Pulse Length	12 μsec	12 μsec
Energy Spread	< 0.20%	< 0.20%
Emittance (vertical)	5 mm.mrad	< 5 mm.mrad
Klystron Peak Power	P_P = 15 MW	P_P = 15 MW
Klystron Average Power	P_M = 30 KW	P_M = 30 KW
Magnet Diameter	80 cm	150 cm
Injection	Kapitza	Kapitza
Bunch Length	∿ 6 to 8 mm	∿ 6 to 8 mm
Repetition Frequency	∿ 1 Hz	∿ 150 Hz

characteristics shown in Table II and, at a later time, the replacement of the present 80 cm magnet with a new 150 cm one.

C) Wiggler and Transport Channel Performance

We will now briefly discuss the main wiggler and transport channel characteristics.

Permanent magnet technology has been a growing field of research which has recently received a high degree of development owing to the introduction of REC materials[21] having a remanent field which, for $SmCo_5$, reaches about 0.95T. By means of these magnets it is possible to realize small pass large gap high field wigglers. In particular for our purposes, for e.b. energies ranging from 15 to 35 MeV, we can operate at λ = 10 to 50 um with the following parameters (h = magnet gap).

$$\lambda_q = 5 \text{ cm}, \quad L_W = 2.25 \text{ m}, \quad \left(\frac{\Delta\omega}{\omega}\right)_0 \sim 1.1 \times 10^{-2}$$
$$h = 2.4 \text{ cm}, \quad B \sim 3 \text{ kG}, \quad K \cong 1 . \tag{3.6}$$

On the other hand we have also, considered the possibility of utilizing superconducting magnets which would be necessary to operate an open cavity in the far infrared region. The feasibility analysis for such a superconducting wiggler has been carried out[22] and the fundamental parameters are

$$\lambda_q = 12 \text{ cm}, \quad L_W = 5 \text{ m}, \quad \left(\frac{\Delta\omega}{\omega}\right)_0 \simeq 1.2 \times 10^{-2}$$
$$h = 8 \text{ cm}, \quad B \sim 1.0 \text{ to } 3.5 \text{ kG} . \tag{3.7}$$

The transport channel has been designed to modify the optical e.b. characteristic at the microtron output to minimize the inhomogeneous spread in the wiggler magnet (see Ref. 5, Sect. 3.1). To this aim two pairs of independent quadrupoles (F-D) have been inserted. The bending structure is provided by two identical dipole magnets (B). Furthermore, to obtain an achromatic channel, a focusing quadrupole in the radial plane has been inserted between the dipoles. The general layout is sketched in Fig. 4, while in Fig. 5 we have plotted as an example, the optical functions (βx, βy)[24] for the permanent magnet configuration.

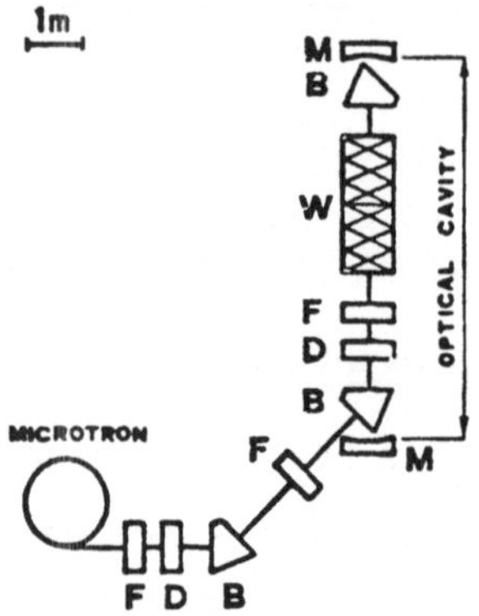

Figure 4. FEL Microtron Experimental Layout

F-D ≡ radial and vertical focusing quadrupole magnets
B ≡ bending magnet
W ≡ wiggler magnet
M ≡ optical cavity mirror

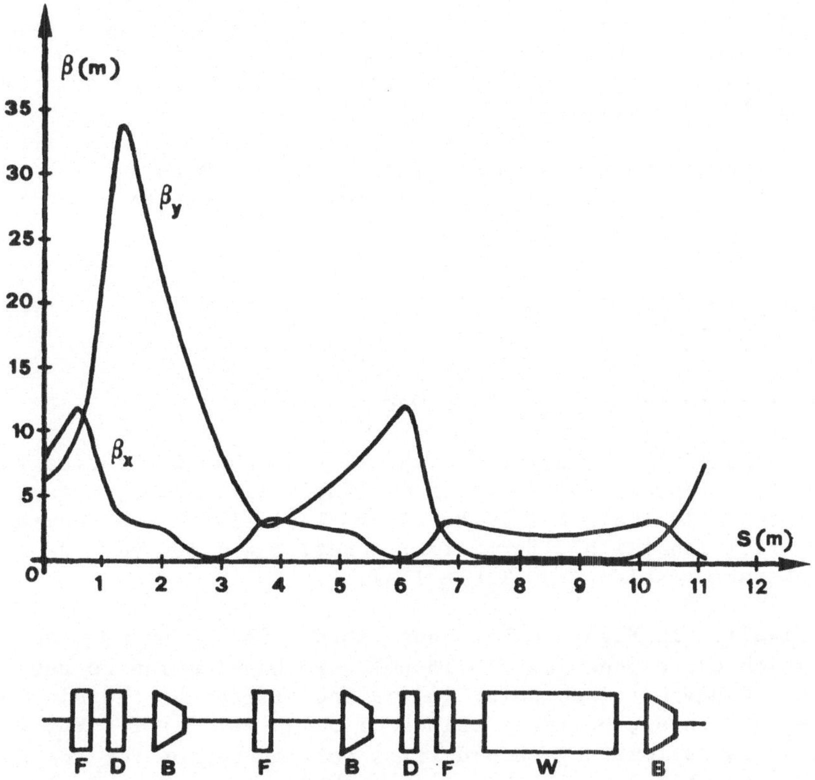

Figure 5. Transport Channel Optical Functions

β_x, β_y ≡ radial and vertical beta functions
(for the symbols see Fig. 4)

4. LASER BEAM CHARACTERISTICS

In Figure 6 we have plotted the operating wavelength λ, the gain per unit current G/I and the Fresnel number N vs the e.b. energy for the permanent magnet wiggler (3.6), while in Fig. 7 the same parameters are plotted vs the wiggler parameter K ($\propto$ B, see Table I) for the superconducting one (3.7). The two different kinds of plots are motivated by the fact that, in the case of permanent magnet wiggler, the magnetic field cannot be easily changed*, and then we must change the e.b. energy to tune the wavelength. In the case of the superconduction wiggler, on the other hand, the magnetic field is straightforwardly tunable by varying the coil current.

Before giving the details of the l.b. parameters let us recall that at saturation the l.b. power is linked to the e.b. power by the relationship[25].

$$P_L \sim \left(\frac{\Delta\omega}{\omega}\right)_o P_E \tag{4.1}$$

where $P_{L,E}$ are the l.b., e.b. power respectively.

In Table III we summarize typical l.b. characteristics for the working regions $\lambda \cong 25$ to 35 μm (Table IIIa, e.b. characteristics from Table IIa), $\lambda \cong 10$ to 20 μm (Table IIIb, e.b. characteristics from Table IIb), far infrared region $\lambda \sim 100$ μm (Table IIIc, e.b. characteristics from Table IIa).

Finally, in Figure 8 has been plotted $\mathrm{Re}\, q_\lambda^1$ vs Θ (at $\mu_c = 0.5$, which corresponds to $\lambda \sim 16$ μm). We have, owing to the lethargy behaviour (see Refs. 18, 19 and references therein for further comments), positive values of $\mathrm{Re}\, q_\lambda^1$ for $0 < \Theta < \Theta_M$; for lossless operation, this would fix a working region for the cavity length variation (recall that Θ is linked to $\delta t = T_c - T_e$, Eq. (3.3)). Furthermore including the losses the working region is narrowed, and typically we find for losses of about 3% a maximum cavity length variation admissible if we are to have laser action:

$$\delta L = c\delta t = 150 \ \mu\text{m}.$$

Analogous considerations hold for the far infrared region.

---

*The only method available is mechanically changing the magnet gap

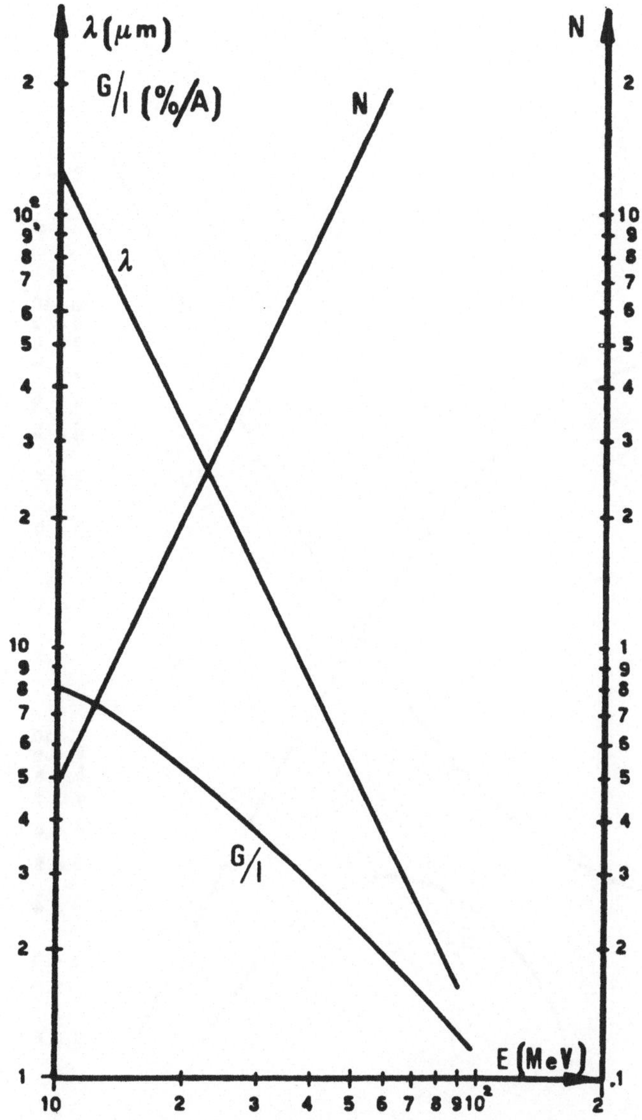

Figure 6. λ, G/I and N vs E (Permanent Magnet (3.6))

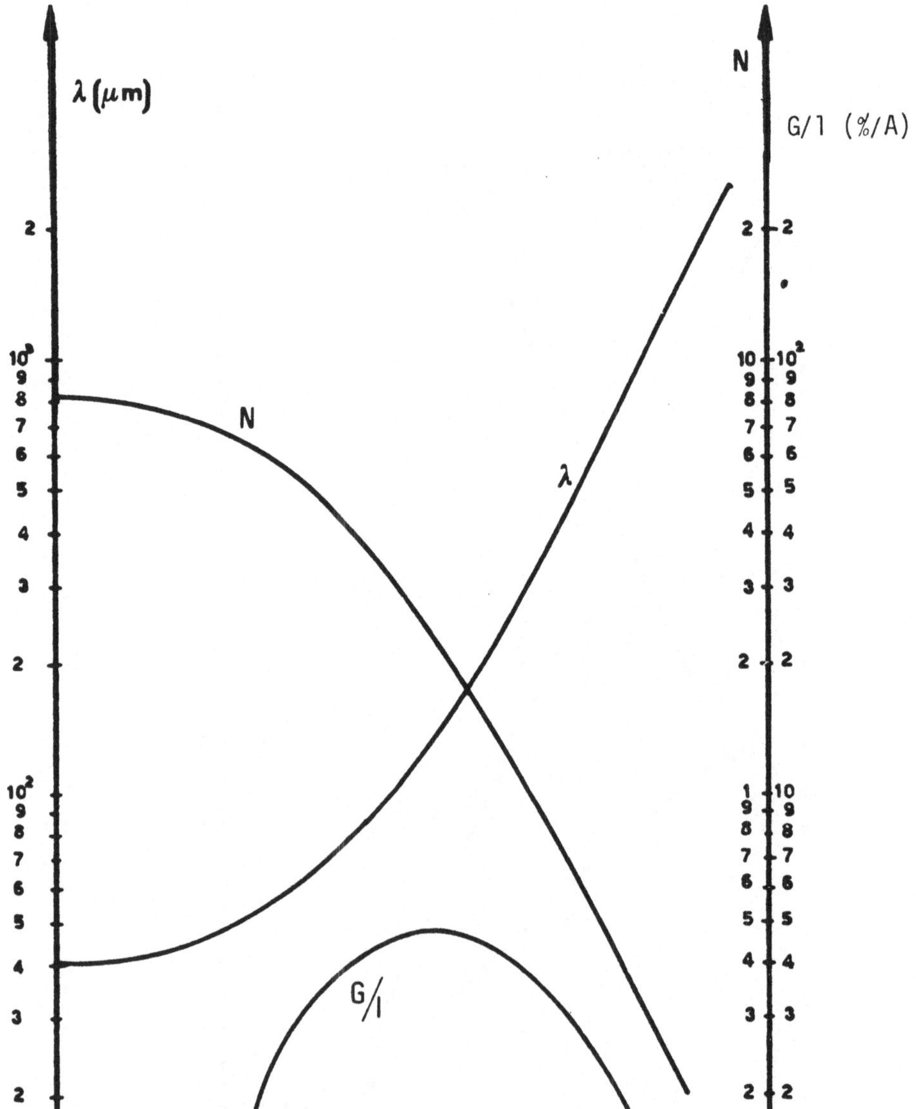

Figure 7. λ, G/I and N vs K (Superconducting Magnet (3.7))

TABLE III

		(a)	(b)	(c)
Wavelength	$\lambda(\mu m)$	32.6	16	120
e.b. energy	$E_{e.b.}(MeV)$	20	28.5	20
Gain	G(%)	33	18	32
Cavity losses	$\gamma_T(\%)$	5	3	5
Net gain	$\alpha(\%)$	28	15	27
Pulse rise time	$\tau(\mu sec)$	2.5	5	4.4
Energy per pulse	$\mathcal{E}(J)$	0.7	0.5	0.6
Average power (at 150 Hz)	$P_L(W)$	100	75	90
Wiggler magnet		Permanent magnet		Superconducting

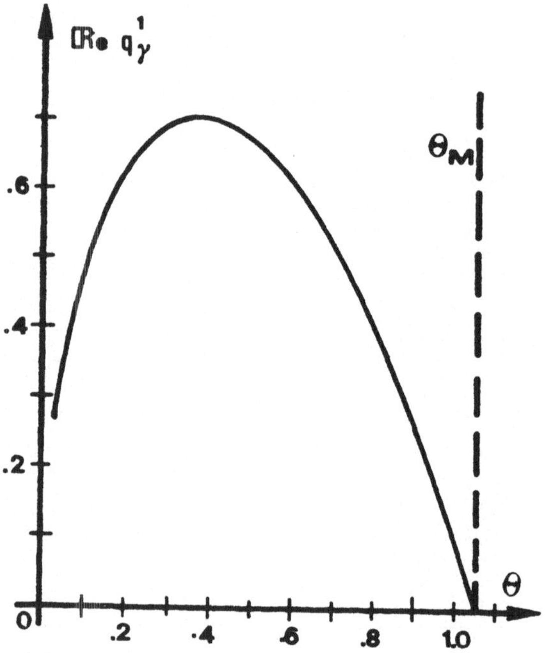

Figure 8. $\mathbb{R}e\ q^1_\lambda$ vs Θ (μ_c = 0.5)

ACKNOWLEDGEMENTS

During the compilation of this note we benefit of the assistance of many collegues, in particular we are grateful to U. Bizzarri and A. Vignati for all information concerning the Frascati microtron. F. Ciocci for having carried out all the calculations relevant to the optimization of the transport channel; G. Pasotti and M. Ricci for the continuous assistance in the problems connected with the superconducting and permanent magnet wigglers.

Finally we express greatly our gratitude to E. Fiorentino and T. Letardi for comments, criticism, kind interest and critical analysis.

REFERENCES

1. L.R. Elias, W.M. Fairbank, J.M.J. Madey, H.A. Schwettman and T.I. Smith, Phys. Rev. Lett. 36, 717 (1976)
   D.A.G. Deacon, L.R. Elias, J.M.J. Madey, G.J. Ramian, H.A. Schwettman and T.I. Smith, Phys. Rev. Lett. 38, 892 (1977)
2. N.M. Kroll, P. Morton, M.M. Rosenbluth, JRI Technical Report JSR-79-01, SRI International, 1980
3. J.M.J. Madey and R. Tabler, Physics of Quantum Electronics (Addison-Wesley Publishing Co.) Vol. 7, Chap. IV, 774 (1980)
4. L.R. Elias, Phys. Rev. Lett. 42, 977 (1979), L.R Elias "The UCSB FEL Experimental Program" these proceedings
5. G. Dattoli, A. Marino and A. Renieri "Storage Ring Operation of the FEL" these proceedings
6. C. Pellegrini, I.E.E.E. Trans. Nuc. Sci. NS-26, 3791 (1979)
7. M.W. Poole "FELIX a Proposed Experimental Facility for FEL Investigations" these proceedings
8. E.B. Shaw and C.K.N. Patell "Optimization of a FEL for Far Infrared" these proceedings
9. U. Bizzarri and A. Vignati, Nuovo Cimento 68A, 513 (1970)
10. V.I. Veksler, Proc. USSR Acad. Sci. 43, 346 (1944), J. Phys. USSR, 9, 153 (1945)
11. S.P. Kapitza and V.N. Melekin "The Microtron" Harwood Academic Publishers, London-Chur (1978)
12. W.J. Henderson, H. Le Caine and R. Montalbetti, Nature 162, 699 (1948)
13. P.A. Readhead, H. Le Caine, and W.J. Henderson, Canad. J. Res. A28, 73 (1950)
14. D.K. Aitken, F.F. Keymann, R.E. Jennings, P.I.P. Kalmus, Proc. Phys. Soc. 77, 769 (1961)
15. O. Wernholm, Arkiv. Fys. 26, 527 (1964)
16. S.P. Kapitza, V.P. Bykov and V.N. Melekin, Ž. Ėksp. Teor. Fiz. (JETP) 39, 997 (1960)
    Id. ibid. 41, 368 (1961)
17. V.N. Melekin, ibid. 42, 622 (1962)
18. G. Dattoli, A. Marino and A. Renieri, C.N.E.N. Report 80.29/p, Centro di Frascati, Frascati, Rome, Italy to appear in Opt. Commun.
19. G. Dattoli, A. Marino and A. Renieri "Analysis of the Single Pass FEL: the Multimode Small Signal Regime" these proceedings
20. U. Bizzarri and T. Letardi, private communication and Ref. 9
21. K. Halbach, I.E.E.E. Trans. Nucl. Sci. NS-26, 3882 (1979)
    H. Winick and J.E. Spencer, Nucl. Instrum. and Methods 172, 45 (1980)
22. G. Pasotti and M. Ricci, private communication
23. F. Ciocci, A. Marino and A. Renieri, private communication
24. See for example Ref. 5, Sect. 2.2
25. A. Renieri, Proc. of the Workshop in "The Possible Impact of FELS on Spectroscopy and Chemistry" Ed. by G. Scoles, Riva del Garda, Italy, 28 (1979)

OPTIMIZATION OF A FREE ELECTRON LASER FOR FAR INFRARED

E. D. Shaw and C. K. N. Patel

Bell Laboratories

Murray Hill, New Jersey 07974

ABSTRACT

The interaction distance for relativistic electrons and optical radiation in a mode-locked free-electron laser is often limited by the electron bunch length. For a far-infrared oscillator, it is indicated that the round-trip gain and thus the output coupling efficiency can be increased by the introduction of an intracavity filter composed of four metal meshes.

I. INTRODUCTION

In a free-electron laser, FEL, relativistic electrons travel through a static helical magnetic field and generate coherent radiation.[1,2] Motz recently noted[3] the electronic duality of this developing technology to the traveling wave tube technology pioneered by R. Kompfner, J. R. Pierce, and associates more than three decades ago. The earlier development of this microwave source paralleled the then growing understanding of electron interaction along distributed circuits. An electron beam was injected along the axis of a loaded transmission line designed to propagate a mode whose field configuration bunched the electrons into microbunches with spatial periodicity equal to the microwave wavelength. For the average electron velocity slightly greater than the phase velocity of the propagating mode, the electron beam is decelerated and the microwave signal is amplified.[4]

In the FEL we consider a relativistic electron beam bunched at the accelerator rf cavity wavelength (having experienced the opposite interaction) is injected into a far-infrared cavity on the axis of a static helical magnet. The amplified optical wave front outdistances the electron bunch in a fraction of the magnet length. A wave guide to reduce the optical pulse velocity to the electron beam velocity has an estimated diameter (∿0.5 cm) less than the expected amplitude of transverse beta oscillations associated with the transport of relativistic electrons in a static helical magnetic field.[5] For far-infrared radiation, FIR, we show that the interaction distance may be increased by the use of an intracavity filter composed of lump-constant elements.

II. FEL COMPONENTS FOR FIR GENERATION

We recently proposed a FEL oscillator tunable from 100-400 μm with potential optical peak power density output more than ten orders of magnitude greater than that for thermal sources currently used in FIR spectrometers.[6] The principal components of this system are an energy tunable (10-20 MeV) microtron accelerator[7] and a 10 m helical magnet with a bore radius of 6.3 cm and period of 20 cm. The high peak current (∿5 amps) from a microtron is composed of electron bunches of bunch length $L_b \gtrsim 5.5$ mm spaced at the rf cavity field wavelength $\lambda_{rf} \simeq 10$ cm, with an electron beam emittance $\varepsilon \simeq 1$ cm mr. At the 10 MeV output of the microtron, the 400 μm wavelength operation of the FEL is not anticipated to be optimum because the generated FIR traverses the electron bunch in a fraction of the magnet length. This limits the potential round-trip gain[8] and forces undercoupling of the laser. This may be the case in Ref. 1 where the first FEL achieved an efficiency of ∿0.01% with an output mirror with 1.5% transmission. The electron beam pulse length was ≃ 1 mm and several authors believe that the short pulse length influenced the laser performance.[8,9]

In principle, narrow band intracavity filters can be designed as time delay elements to resynchronize the FIR pulse with the electron beam pulse as well as to give a prescribed phase shift of the optical pulse field relative to the electron beam microbunches.[10] Low loss metal mesh Fabry-Perot filters with absorbance as low as 1% have been used as variable output FIR laser couplers.[11,12] Also, various aspects of multi-element metal mesh filters have been discussed in the literature.[13,14] In Section III, we review the optical properties of single metal meshes and use these properties to estimate the FIR optical properties of a tunable narrow band filter composed of four identical metal meshes that are washer-shaped. In Section IV, we review the FEL interaction to estimate the round-trip gain improvement expected with the use of the intracavity narrow band filter.

III. METAL MESH FILTERS FOR FIR FREQUENCIES

The optical properties of metal meshes are independent of polarization because the metal meshes have square symmetry. The properties of the single reflector are determined by the physical

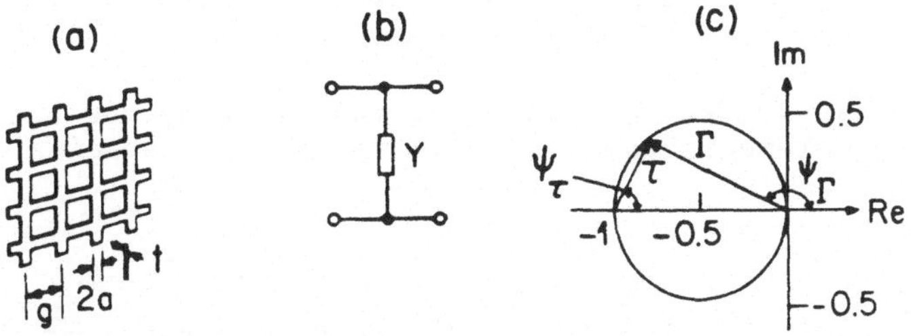

Fig. 1. (a) Two-dimensional metal mesh grid. (b) Equivalent circuit of a thin grid in free space for $g/\lambda < 1$. (c) The locus of the amplitude reflection (Γ) and transmission (τ) coefficients of a thin metal mesh.

dimensions of the reflector grid [see Fig. 1(a)]. The dimensions of the metal mesh studied in Ref. 12 are listed in Table 1 and these parameters are used with the considerations presented here to calculate the listed optical constants for 400 μm radiation. In the long wavelength limit ($g/\lambda < 1$), the optical properties are conveniently represented by an equivalent electrical circuit [see Fig. 1(b)]. The equivalent circuit is a uniform transmission line representing free space that is shunted by a single admittance Y representing the metal mesh. Using transmission line theory, the amplitude reflection coefficient of the grid can be related to its physical parameters by the following equations,[11]

$$\Gamma = \frac{-Y/2}{1+Y/2} \tag{1a}$$

$$Y = -\partial(2/Z)\ (\lambda/g) \tag{1b}$$

$$Z = 2\ \ell n\ \mathrm{cosec}\ (\pi a_o/g) \tag{1c}$$

$$a_o = a+t/2\pi[1+\ell n(8\pi a/t)]\ . \tag{1d}$$

Because the admittance is proportional to λ/g, the metal mesh reflectance in the FIR can be quite large. Equation (1) is used to calculate $|\Gamma| = 0.99$ for the metal mesh parameters listed in Table 1. The Q for a Fabry-Perot filter used in first order at 400 μm constructed from this metal mesh is $\gtrsim 110$. This would give a pulse delay time for a Fabry-Perot filter of

$$t_d = \frac{QT}{2\pi} \quad \text{where } Q = \frac{\pi|\Gamma|}{1-|\Gamma|^2}\ , \tag{2}$$

and T is the period of the 400 μm radiation. The results reported in Ref. 12 were at 170 μm and 118 μm where the expected Q is less because of the wavelength dependence of Γ. However, the observed loss of 1% is noteworthy because ohmic losses are smaller at 400 μm.

For the four-element filter designed as a narrow band filter, with adjacent elements being equally spaced, the spacing d can be[14]

$$kd = \psi_\Gamma , \tag{3}$$

where ψ_Γ is the reflection phase of the grid and $k = \frac{2\pi}{\lambda}$ is the wave vector for wavelength λ. The value of ψ_Γ can be approximated by considering the locus of Γ(λ) in the complex plane. The complex amplitude transmission coefficient is related to Γ by the relation $\tau = 1+\Gamma$. We see (from Fig. 1(c)) that for large reflection grids $\psi_\Gamma \gtrsim \pi$ and $\psi_\tau \gtrsim \frac{\pi}{2}$. Since $d \leq \lambda/2$, ohmic losses are further minimized because the intrafilter fields at the grids are smallest at $d = \lambda/2$. For narrow band four-element filters, the phase shift of the transmitted electric field is $4\psi_\tau \gtrsim 2\pi$ relative to the transmitted wave for the filter removed. The total filter length is $\gtrsim 3\lambda/2$.

The optical constants are now estimated for a four-element narrow band filter with washer-shaped metal meshes designed to transmit the electron beam without scattering (see Fig. 2). The reflectance of a washer Γ_w can be calculated from the mesh reflectance through the relation,

$$|\Gamma_w| = \left(1 - \frac{A_b}{A_{opt}}\right) |\Gamma| , \tag{4}$$

where A_b is the electron beam area and A_{opt} is the optical mode area. Since the electron beam waist is an order of magnitude larger than the 400 μm wavelength, diffraction effects will not affect the FIR performance of the filter. We use the locus of Γ (the circle in Fig. 1(c)) to calculate the reflectance and transmission phases in Table 1 for $A_b = 1.2\ cm^2$ and $A_{opt} = 60\ cm^2$. We also use Eq. (1a) to calculate the admittance Y_w from $|\Gamma_w|$. Although most of these numbers are derived from small differences of large numbers, one optical parameter can be used to calculate the other parameters. It is clear from the parameters listed in Table 1 that the washer shape does not destroy the high reflectivity of metal meshes in the FIR for reasonable values of A_b and A_{opt}.

Transmission theory is used to calculate the finesse of the four-element filter from the admittance Y_w derived above. We consider the four-element filter as a combination of two two-grid subfilters that are separated by the central spacer and that act as

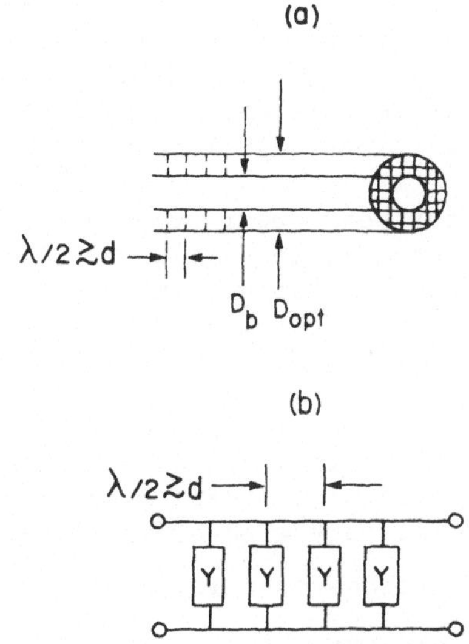

Fig. 2. (a) A four-element narrow band filter composed of washer-shaped metal meshes with diameters D_b and D_{opt}, and space separation $d \gtrsim \lambda/2$. λ is the cavity resonant wavelength. (b) Equivalent circuit of four-element filter.

the reflectors of the main interference filter.[14] The subfilters have high reflectivities in their stop band, resulting in a high finesse of the main filter (for example, at λ not satisfying Eq. (3)). The Q for the filter is larger than that for the subfilters. Because the spacers are spaced $\approx \frac{\lambda}{2}$ apart, the admittance of the subfilters seen from the central region looking back through the subfilters is the addition of the admittance of the subfilter meshes, that is $Y_f \cong 2Y_w$.[11] Thus, we use $\Gamma_f \cong \frac{Y_w}{1+Y_w}$ and Eq. (2) to calculate $Q_f \cong 175$ for the quality factor of the four-element filter giving a transient time response $t_d \approx \frac{Q_f}{2\pi} T$.

IV. FEL CAVITY WITH INTRACAVITY FILTER

We consider a FIR cavity consisting of a totally reflecting curved metal mirror, M, a multi-element wire mesh filter, F, and an output mirror R with power reflectance R, see Fig. 3. R may be a wire mesh reflector or a variable metal mesh Fabry-Perot coupler. Studies of variable reflectance output Fabry-Perot laser couplers in the FIR have demonstrated that there is an optimum reflectance for which the output laser power is maximized.[11,12] Since the laser threshold reflectance R is defined by the relation $Re^{\Gamma_s L} \simeq 1$, where Γ_s is the power gain per unit length, the threshold and optimum coupling reflectance is decreased with increased round-trip gain given by $\Gamma_s L$. The output peak power proportional to 1-R can be a strong function of the round-trip gain when $\Gamma_s L$ is small. Here the conditions for increasing $\Gamma_s L$ through the use of time delay filters are determined for FEL's operated in a parameter regime where the round-trip gain is limited by the electron beam pulse length.[9]

In Fig. 3, we consider an electron beam pulse of width L_b deflected into the FIR cavity near mirror M. The electron beam experiences a helical field of strength B and period $\lambda_m = \frac{2\pi}{k_m}$ that determines a helical trajectory with period λ_m for each electron. Each electron receives a transverse momentum P such that $\frac{P}{mc} = \frac{eB\lambda_m}{2\pi mc^2}$. The optimum magnetic field strength for FEL operation corresponds to $\frac{P}{mc} = 0.7$ which gives a value of the magnetic field of 380 gauss for this study.[15] The cavity is geometrically arranged so that the leading edge of the cavity optical pulse of pulse length L_p and wavelength λ overlaps the electron bunch and performs work on the electron beam by way of coupling to the transverse current. The

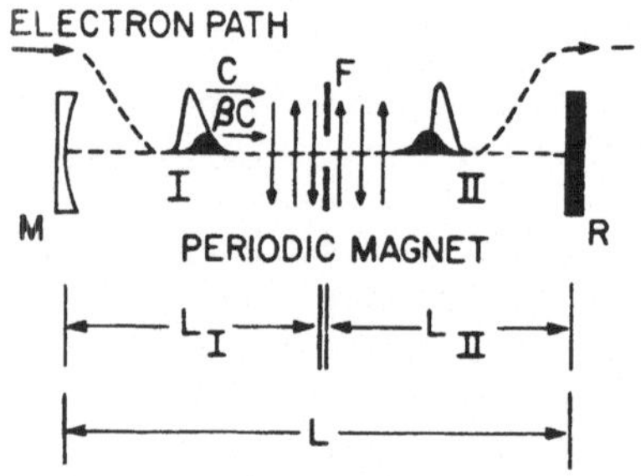

Fig. 3. The electron pulse is injected into a resonator with velocity βc. The electron pulse is removed after a single pass while the amplified optical pulse is stored in the cavity.

optical pulse travels at velocity c which is greater than the electron beam velocity βc and traverse the electron pulse before arriving at filter F. The electrons are initially randomly distributed along the magnet axis so that the optical pulse (circularly polarized) is attenuated by some electrons and amplified by others (dependent on initial phase between the optical field and the electron traverse motion). This combined static field-electromagnetic field interaction bunches the electrons along the magnet axis with the spacial periodicity λ which satisfies the resonant condition,

$$\frac{k}{k+k_m} = \beta \, . \tag{5}$$

For an electron beam with energy $E = \gamma mc^2$, Eq. (5) can be written $\lambda = \left(\lambda_m/2\gamma^2\right)[1+\left(\frac{P}{mc}\right)^2]$. For infinite interaction length at the magnetic field strength of this study, this coherent microbunching mechanism is referred to as the high pump single particle scattering regime and gives a gain identical to that for traveling wave tubes.[8,16] The estimated power gain coefficient for the parameters given here is $\Gamma_s \approx 2.2\times10^{-3}$/cm.

In the ultrashort pulse regime discussed here, this mechanism leads to a gain that grows exponentially (with the microbunch component) along the magnet axis to the traveling wave tube value at large optical fields.[3,8,17] Also the trailing edge of the optical pulse experiences greater gain than the leading edge.[9] These two considerations suggest a significant advantage in delaying the optical pulse at filter F so that the trailing edge of the optical pulse near the electron beam input in region I can be amplified at the end of region II as well as at the end of region I. In order to maintain the microbunch growth and maximize the gain, it is advantageous for filter F to maintain the same phase between the optical pulse and the microbunch in regions I and II. Thus the four-element filter which was shown in Section III to approximate a 2π phase shifter can potentially impact on these considerations.

The interaction length and the optical pulse width can be derived from Eq. (5). The resonance condition corresponds to a wave front traversing a distance $\lambda_m+\lambda$ for the corresponding beam travel of λ_m. Thus $\lambda_m+\lambda$ is the approximate interaction length for the wave front with a microbunch. For the electron bunch of bunch length L_b, the total interaction distance is given approximately as

$$L_{eff} = \frac{L_b}{\lambda} (\lambda_m+\lambda) \, . \tag{6}$$

For this study $L_{eff} \approx 3$ m. Similarly for $L_I \cong L_{II} \approx \frac{L}{2}$, $L_P = \frac{L_I\lambda}{(\lambda_m+\lambda)}$.

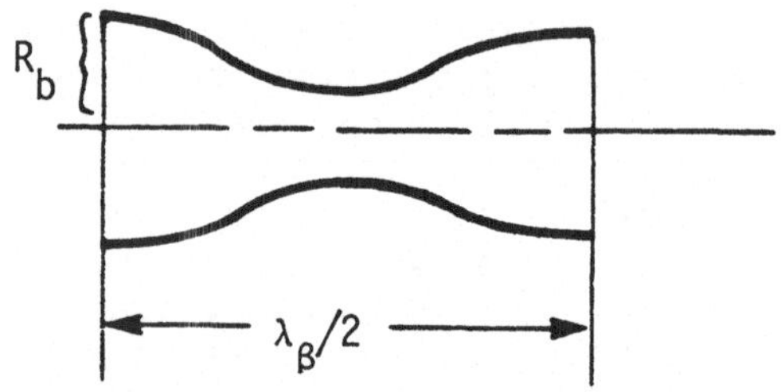

Fig. 4. The beam waist oscillations for a relativistic electron beam in a helical static magnetic field.

The criterion for the pulse delay time t_d is that t_d be short enough to allow the intrafilter optical field to grow to a value comparable to the pulse field in regions I and II, but sufficiently long to allow the electron pulse to pass through F with the optical field still stored in F. This is written

$$\frac{L_p}{c} > t_d > \frac{L_b}{\beta c}$$

or

$$\frac{k_m L}{2} > Q_F > kL_b \ . \tag{7}$$

This gives $160 > Q_F > 85$. This criterion is consistent with the technological potential for high Q metal meshes in the FIR as discussed in Section III. Because the gain is at the electron bunch, $L_I \cong L_{II} \cong n\beta\lambda_{rf}$ where n is a large integer.[9]

In Fig. 2, the metal mesh was designed to transmit an electron beam of finite beam waist. The beam waist in the magnet is determined by an interplay between the initial preparation of the beam (emittance = divergence of the beam times the beam waist) and the weak focusing properties of the helical magnet field.[5] The helical magnetic field is a minimum on the magnet axis increasing radially toward the bore. In this field the electron trajectory is one of a helix whose axis slowly oscillates about the magnet axis with a period $\lambda_\beta \cong 2\gamma\lambda_m$. The amplitude of these oscillations (beta oscillations) are determined by the initial conditions at the entrance of the magnet. The emittance of the beam is approximately constant in the magnet so that we can calculate an approximate equilibrium radius. The beam waist will oscillate in a manner illustrated in Fig. 4. We take $\frac{R_b}{\lambda_\beta/4}$ to represent the divergence at a beam waist maximum which may be written $\varepsilon/2R_b$ by the definition of the emittance ε. Thus we have for the beam radius

$$R_b \cong \sqrt{\frac{\varepsilon\lambda_\beta}{8}} \ . \tag{8}$$

Table 1

Four-Element Copper Metal Mesh Filter Parameters

g (μm)	29 (μm)	t (μm)	$\|Z\|$	$\|\Gamma\|$	Γ_w	$\|\Gamma_w\|^2$	ψ_τ	ψ_Γ	$\|Y_w\|$	$\|Y_F\|$	$\|\Gamma_F\|$	$\|\Gamma_F\|^2$	Q_F
50	12	3	1.38	.99	.97	.93	.42π	.92π	7.43	14.85	.99	.98	175

This gives $A_b \cong \pi R_b^2 = 0.3\ cm^2$ and $\lambda_\beta = 8$ m for this study. Also we estimate the optical mode area by the relation $A_{opt} = \lambda L$ where we take $L \approx 15$ m. The larger value we chose for A_b for the calculation in Section III permits less marginal design.

V. CONCLUSION

We have proposed that an intracavity four-element narrow band filter can be designed at FIR wavelengths for FEL's. The realizable round-trip gain of the laser discussed here $2\Gamma_s L_{eff} \cong 130\%$ is more than twice that previously described by us and this will have a marked improvement on the output efficiency. This approach of using the high peak current output of rf accelerators with intracavity filters may be more fruitful in developing more efficient FEL lasers than approaches that depend on the klystron analogy. This latter approach necessitates drift distances comparable to the interaction distance and depends sometimes on debunching of the microbunches within the FEL cavity.[18,19] The high repetition rate (∿100 Hz) and high peak power potential of the FIR oscillator discussed here offers the potential for unprecedented opportunity for new nonlinear and linear spectroscopic studies in the FIR.

REFERENCES

1. D. A. G. Deacon, L. R. Elias, J. M. J. Madey, G. J. Ramian, H. A. Schwettman, and T. I. Smith, Phys. Rev. Lett. 38, 892 (1977).
2. D. B. McDermott, T. C. Marshall, S. P. Schlesinger, R. K. Parker, and V. L. Granatstein, Phys. Rev. Lett. 41, 1368 (1978).
3. H. Motz, Phys. Lett. 71A, 41 (1979).
4. M. Chodorow and S. Susskind, Fundamentals of Microwave Electronics, (McGraw Hill, New York, 1964).
5. J. P. Blewett and R. Chasman, J. Appl. Phys. 48, 2692 (1977).
6. E. D. Shaw and C. K. N. Patel, Free Electron Generators of Coherent Radiation, S. F. Jacobs, H. S. Pilloff, M. Sargent III, M. D. Scully, and R. Spitzer, eds. (Addison-Wesley Publishing Company, London, 1980), p. 665.
7. S. P. Kapitza and V. N. Melekhin, The Microtron, ed. by Ednor M. Rowe (Harwood Academic Publishers, London-Chur, 1978).
8. N. M. Kroll, Novel Sources of Coherent Radiation (Addison-Wesley, 1978), p. 115.
9. F. A. Hopf, T. G. Kuper, G. T. Moore, M. O. Scully, Free-Electron Generators of Coherent Radiation, S. F. Jacobs, H. S. Pilloff, M. Sargent III, M. D. Scully, and R. Spitzer, eds. (Addison-Wesley Publishing Company, London, 1980), p. 31; and W. B. Colson and S. K. Ride, ibid., p. 377.
10. G. L. Matthaei, L. Young, and E. M. T. Jones, Microwave Filters, Impedance-Matching Networks, and Coupling Structures (McGraw Hill Book Company, Inc., New York, 1964).
11. R. Ulrich, T. J. Bridges, and M. A. Pollack, Appl. Opt. 9, 2511 (1970).
12. C. O. Weiss, Appl. Phys. 13, 383 (1977).
13. R. Ulrich, Infrared Phys. 7, 37 (1967).
14. R. Ulrich, Appl. Opt. 7, 1987 (1968).
15. B. M. Kincaid, J. Appl. Phys. 48, 2684 (1977).
16. P. Sprangle, R. A. Smith, and V. L. Granatstein, NRL Memorandum Report 3911.
17. W. B. Colson, Phys. Lett. 64A, 190 (1977).
18. C. Shif and A. Yariv, Opt. Lett. 5, 76 (1980).
19. R. H. Pantell, W. D. Kumura, J. A. Edighofter, and M. A. Piestrup, Free-Electron Generators of Coherent Radiation, S. F. Jacobs H. S. Piloff, M. Sargent III, M. D. Scully, and R. Spitzer, eds. (Addison-Wesley Publishing Company, London, 1980). p. 15.

THE UCSB FREE ELECTRON LASER EXPERIMENTAL PROGRAM

Luis R. Elias

Quantum Institute
University of California
Santa Barbara, CA 93106

INTRODUCTION

The University of California at Santa Barbara (UCSB) free electron laser (FEL) program was initiated in January of 1980 at the Quantum Institute to study the operation of electrostatic accelerator free electron lasers. A detailed discussion of the operation of these devices can be found elsewhere in this book under the title "Electrostatic Accelerator Free Electron Lasers" and also in reference [1].

The main goal in this paper is threefold: 1) to present a summary of the UCSB experimental FEL program to date, 2) to discuss future FEL device development objectives, and 3) to talk briefly about the future of applied research at UCSB.

There is presently in existence a large number of high energy electron accelerator machines that can be readily used as sources of electron beams suitable to demonstrate the basic operation of free electron lasers. This is the case, for example, with the superconducting electron accelerator at Stanford University where the first operation of the free electron laser was demonstrated in 1975, and where the potential operating characteristics of FEL's such as high power, high efficiency and broadband continuous tunability were clearly delineated. To realize all of the above predicted operating capabilities of free electron lasers it is, however, necessary to develop suitable electron beam sources.

Electrostatic accelerators appear to offer great promise for generating the electron beams needed by free electron lasers. Their conventional technology is well understood and there are good reasons to believe that this technology can be used with free electron lasers.

The UCSB FEL experimental program is based on the use of electrostatic accelerators in conjunction with the ideas developed by Elias [1] and Madey [2].

DESIGN OF THE UCSB FEL

A conceptual layout of the UCSB FEL machine is illustrated in Figure 1. The major components of the system are:

Accelerator high pressure tank
HV generator
Electron gun
Electron accelerator column
FEL periodic magnet and optical resonator
Electron decelerator column
Electron collector
Ancilliary electron beam optics and control system

The accelerator shown in Figure 1 is being modified for FEL operations at the National Electrostatic Corporation factory and will be ready for initial electron beam tests sometime before the end of this year.

Accelerator High Pressure Tank The high pressure tank contains the major components of the electron beam generation system. It is filled with high pressure sulfur hexafloride gas to electrically insulate the high voltage components. The tank stands 24 feet tall.

HV Generator The high-voltage generator is not shown in Figure 1, but it is located inside the vertical accelerator steel tank shown. It consists of two charging chains (pelletrons) capable of delivering a total current of 500 microamperes to the high-voltage terminal. The high voltage terminal has been designed initially to withstand an initial maximum negative potential of 3 megavolts. With a larger high-pressure tank and larger high-voltage terminal, the maximum holding voltage of the device can be extended to 6 MV.

Electron Gun The electrical design of the electron gun was carried out using the Herrmannsfeldt [2] computer code. It was designed to operate at a total electron current of 2 amperes at 60 kV. A schematic diagram of the gun is illustrated in Figure 2. The thermoionic cathode shown is 15 millimeters in diameter and operates at a current density of 1.13 Amp/cm^2. The intermediate electrode between the cathode and the anode is used to modulate the current in the gun. At +10 kV with respect to cathode potential, the modulating anode turns the gun on. At -2 kV the gun current is completely

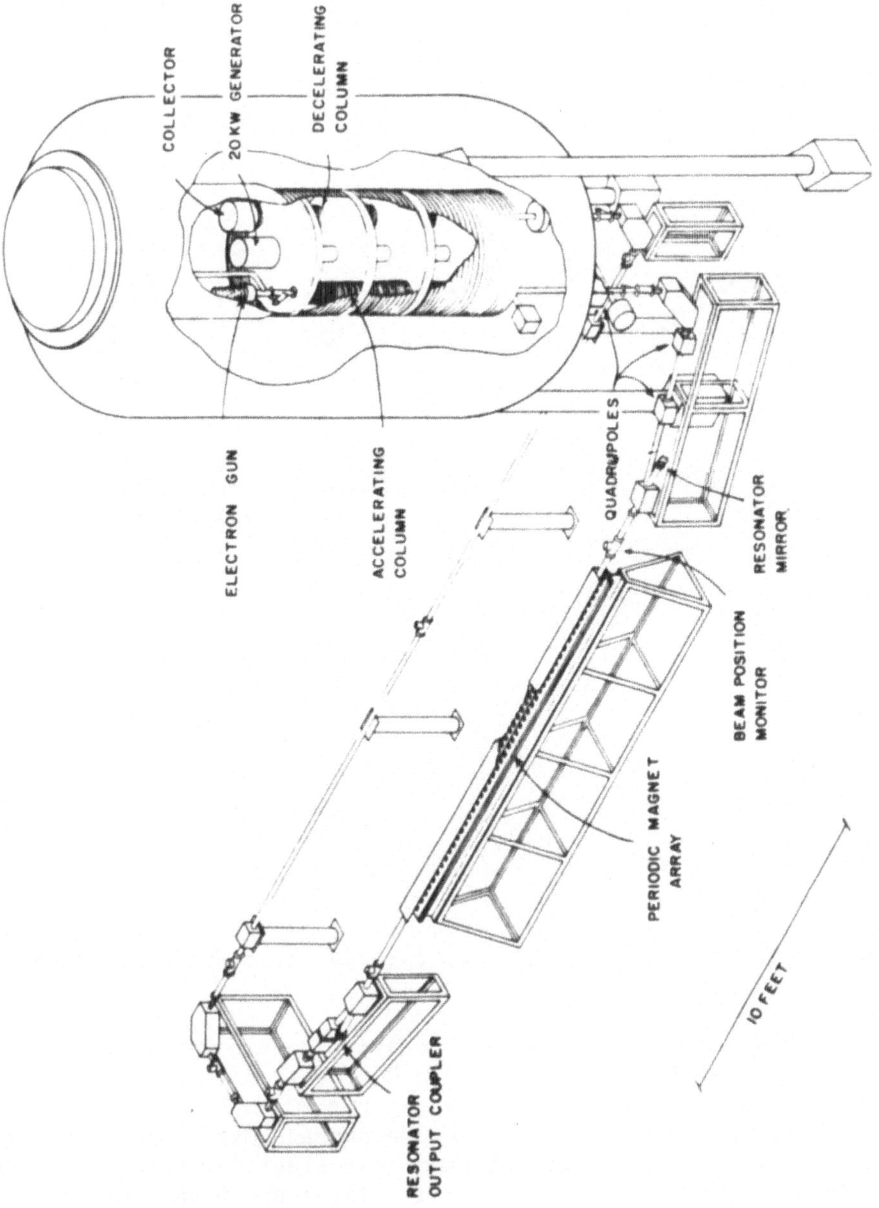

Figure 1. Conceptual layout of the UCSB free electron laser apparatus.

turned off. The electron beam trajectories inside the gun are all very nearly parallel to the gun axis. The emerging electron beam has an approximate radius of 6.2 mm and a diverging exit angle of 30 milliradians. In conjunction with its ancilliary electronics, the gun can produce continuously or on a pulsed basis an electron beam with a minimum pulse length of 100 microseconds.

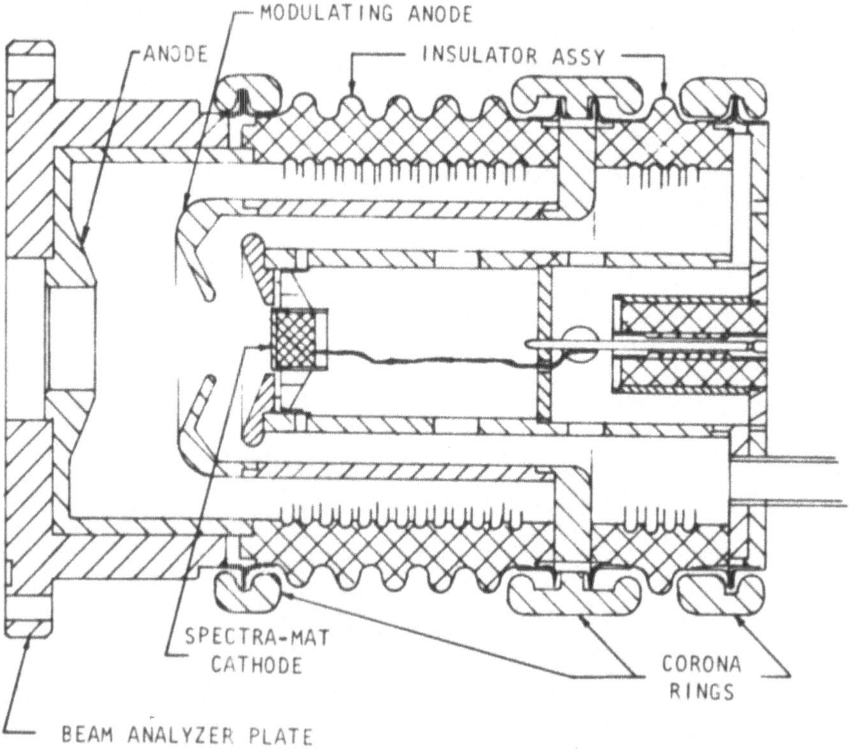

Figure 2. Electron gun design. a) cathode, b) modulating anode, c) anode.

<u>Accelerating Column</u> The accelerating column was designed and constructed by National Electrostatic Corporation in Middleton, Wisconsin. It consists of 9 modules nearly 20 cm long, each producing a constant electric field on axis whose average magnitude is about 15 kV/cm. The longitudinal variations of the electric field along the axis gives rise to transverse electric fields which have a focusing effect on the electron beam. A program has been developed at UCSB to follow the trajectories of electrons inside the electrostatic accelerator. It includes the focusing effect of the accelerating column and the defocusing effect due to space charge forces. The following paraxial equations of motions were numerically integrated:

a) longitudinal equations of motion

$$\frac{d\gamma\beta_z}{dz} = - \frac{q}{mc^2\beta_z} \frac{d\Phi}{dz} \tag{1}$$

b) radial equations of motion

$$\frac{d}{dz}\left(\gamma\beta_z \frac{dR}{dz}\right) = \frac{q}{mc^2}\left[\frac{R\Phi''}{2\beta_z} + \frac{I\mu_o C}{2\pi R}\left(\frac{1}{\beta_z^2} - 1\right)\right]$$

where: (MKS)

γmc^2 = electron energy

I = electron beam current

β_z = longitudinal electron velocity

R = electron beam radius

Φ = electrostatic potential along the axis of accelerator

z = position along the axis of the accelerator

$\Phi'' = \frac{d^2\phi}{dz^2}$

Figure 3 shows the variations of electron beam radius with positions along the accelerator volumn. The electron beam shown has a maximum radius of 9 mm. The aperture of the accelerator column is 25mm diameter. For higher current beams it will be necessary to include additional magnetic focusing elements along the accelerator tube.

FEL Periodic Magnet The FEL wiggler has not been constructed at this time. However, a permanent magnet wiggler will be built. Table 1 lists some of the possible wiggler designs being considered. Also, corresponding single-stage operating FEL characteristics are included.

Decelerating Column A computer code has also been developed to analyze the trajectories of electrons inside the decelerating column shown in Figure 1. Electron decelerating trajectories have been generated for a variety of input kinetic energies to simulate the electron beam energy spread created by the FEL. The trajectories obtained did not change significantly with initial energy. The optimum trajectories appeared very much like the one shown in Figure 3, seen from right to left.

Electron Beam Collector The electron beam collector was designed by R. Hechtel and its layout is shown in Figure 4. It is capable of collecting an electron beam having an energy spread of 10 kV and a mean energy of from 50 kV to 60 kV. The refocusing coil shown in the figure attaches directly to the end of the electron

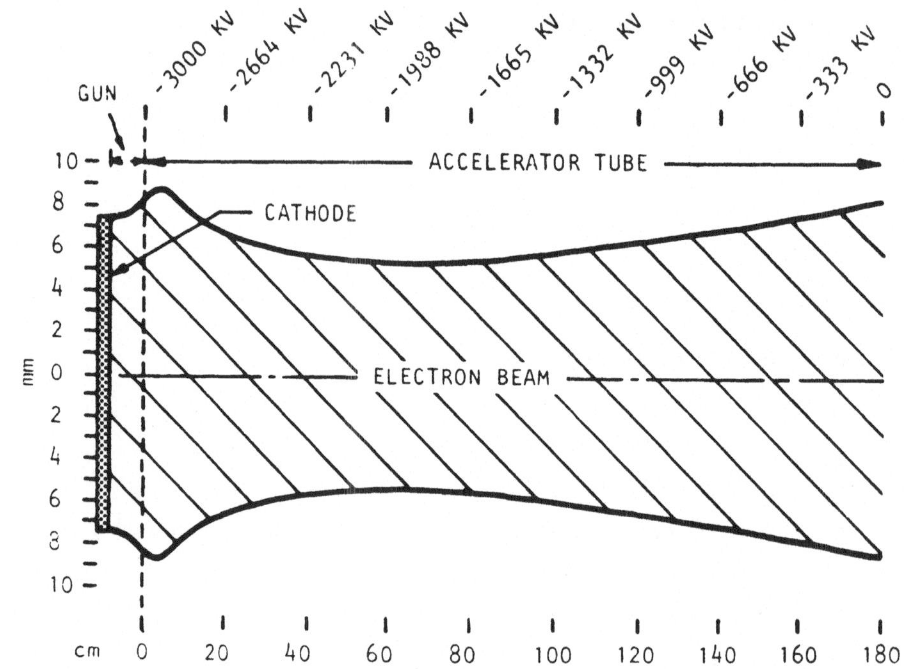

Figure 3. Electron beam radius as a function of position along the accelerating column. The vertical dashed line separates the gun from the accelerator column. The electron beam has cylindrical symmetry.

Table 1. Possible UCSB FEL design parameters. The optical mode assumed is a TEMoo circularly polarized. The current density in the interaction region is $16A/cm^2$.

	I	II	III	IV
Mag. period (cm)	1	1.5	2	3
# of mag. periods	300	200	200	100
Elect. energy (MeV)	3	3	3	3
Elect. beam radius (mm)	2	2	2	2
$K_{MAX} = (e\ \lambda o\ B_{MAX}/2\pi mc)$	0.03	0.058	0.08	0.175
MAX B-field (Gauss)	328	413	428	624
Wavelength (μm)	106	153	205	360
Opt. beam waist radius (cm)	0.54	0.65	0.87	1.0
Small signal gain (Amp^{-1})	0.16	0.24	0.47	0.60
Power output (kWatts/amp)	5	7.5	7.5	15

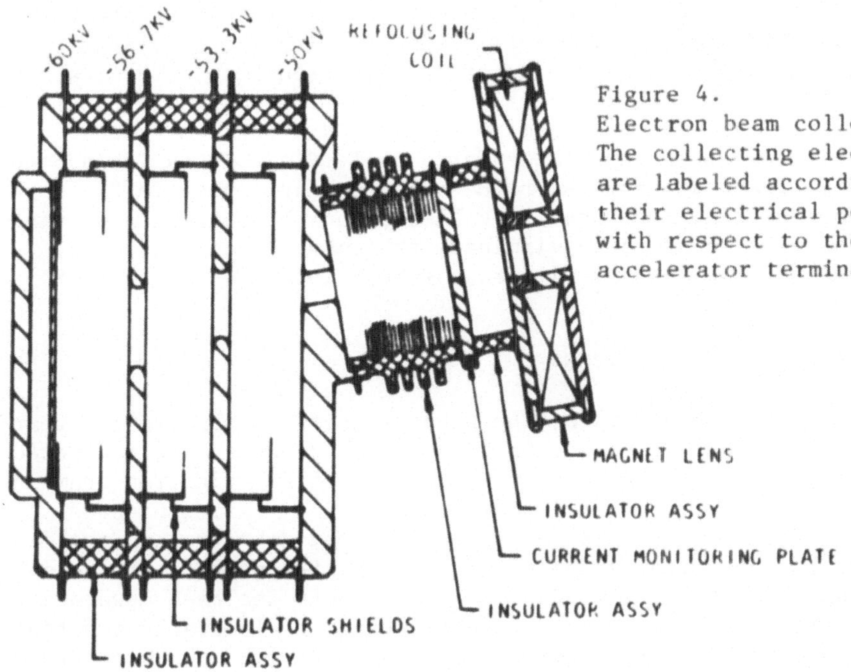

Figure 4.
Electron beam collector. The collecting electrodes are labeled according to their electrical potential with respect to the H.V. accelerator terminal.

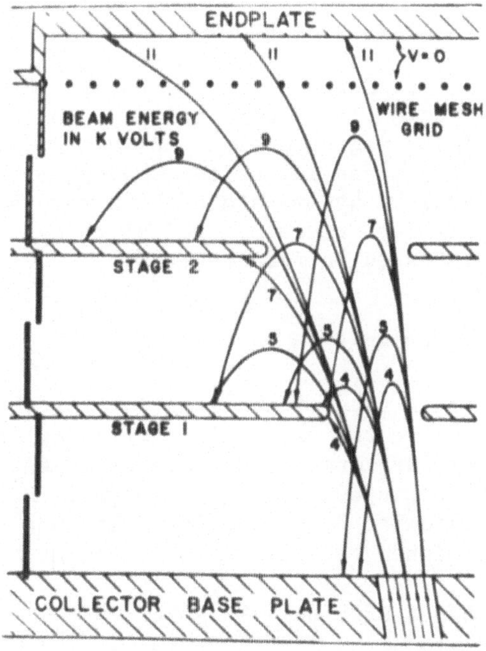

Figure 5.
Electron beam trajectories inside the electron beam collector. The numbers on the trajectories are associated with the electron's entrance energy.

decelerating tube and focuses the electron beam into the entrance of the actual collector. The electric field inside the insulator assembly removes the mean energy of the electron beam. Thus, at the entrance to the collector assembly, the kinetic energy of the electron beam can have a minimum value of a few eV and a maximum value of 10 keV. There are altogether four charge collecting plates inside the collector. The trajectories of the electrons are shown in Figure 5. The reason why the electrons are collected on the back side of the collecting electrodes is to eliminate the possibility that secondary emitted electron will back stream into the decelerator section.

IMMEDIATE PROGRAM OBJECTIVES

The primary objective of the UCSB project for the present year is to show that a 2 ampere electron beam can be produced and recovered using 3 MV electrostatic accelerator. Some of the electron beamline components needed to complete these abbreviated tests are shown in Figure 6. The electron beam emerging from the accelerating section is indicated by a vertical arrow at the top right-hand side of the illustration. The two 90° bending magnets return the beam upward into the decelerator section as shown in the figure. Initially the tests will be conducted using a pulsed electron beam. The pulse length of the beam will be later increased as the charge collection process becomes more efficient. Ideally, if only 500 μA of beam current is lost then the electron beam can circulate in a continuous manner because 500 μA is also the maximum current provided by the HV generator.

During the second year of research a complete single-stage free electron laser will be operated in the submillimeter region with one of the magnet designs shown in Table 1. Also, if the electron beam quality and intensity produced by the UCSB machine is as good as that required by two-stage FEL's then an attempt will be made to generate infrared or visible radiation using the two-stage FEL techniques discussed in reference [1].

FUTURE PROGRAM OBJECTIVES

A natural continuation for the UCSB experimental program is to explore the possibility of using very high voltage electrostatic accelerators to generate efficient single-stage FEL operations with electron beam energy recovery. Electrostatic accelerators have been tested at 32 MV and there is some optimism that these machines can be designed to operate in the 50 MV range.

FUTURE APPLICATIONS RESEARCH

The UCSB free electron laser will be able to provide scientists with a unique source of submillimeter radiation during its initial

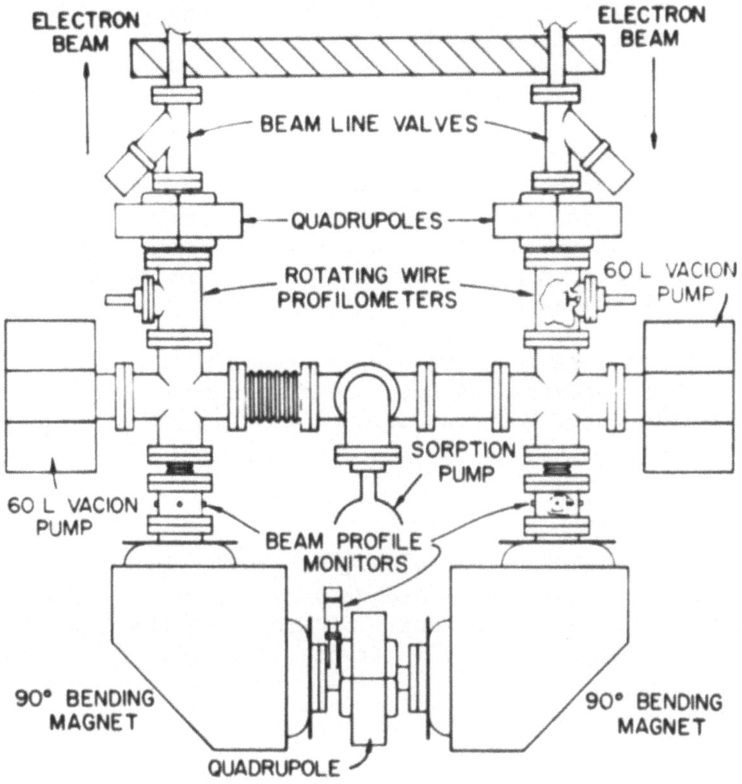

Figure 6. Electron beamline components for the first abbreviated UCSB tests. The accelerator tank is located above the figure.

stages of development. As is well known, there are no intense sources of coherent radiation in the submillimeter region. The availability of a strong monochromatic source of radiation in this region will allow experimenters to study, for example: 1) energy band gaps in superconductors, 2) rotational, torsional and orientational molecular states, 3) small electronic energy differences in semiconductors, and 4) localized heating of confined plasma. A review of the possible research applications of this device can be found in the report that evolved from the 1979 Trento workshop [3].

There is great interest at UCSB to established a User's Facility to provide scientists with the badly needed monochromatic source of FIR radiation. The UCSB machine is ideally suited to fill this need and a serious attempt is being made to procure the funds necessary to achieve this goal.

ACKNOWLEDGEMENTS

The author wishes to acknowledge the contributions of Gerald Ramian to the technical design of the UCSB apparatus. Also, the author wishes to acknowledge the cooperation and very useful discussions with Messrs. Ray Herb, James Ferry and Walid Mourad of National Electrostatics Corporation.

Funds for the project have been provided by the Office of Naval Research. This program has been undertaken in cooperation with National Electrostatic Corporation.

REFERENCES

1. L.R. Elias, Phys. Rev. Letts. 42, 977(1979).
2. J.M.J. Madey, Thesis, Stanford University, P. 150(1970), unpubl.
3. G. Scoles, "The Possible Impact of Free Electron Lasers on Spectroscopy and Chemistry". Trento, Italy(1974).

STORAGE RING OPERATION OF THE FREE ELECTRON LASER

Giuseppe Dattoli, Angelo Marino and Alberto Renieri

Comitato Nazionale Energia Nucleare
Centro di Frascati
C.P. 65
00044 Frascati, Rome, Italy

1. INTRODUCTION

These lectures are concerned with the steady state operation mode of a free electron laser (FEL), operating in a storage ring (SR). In Section 2 the equations of motion of an electron in a "standard" SR are derived. Section 3 is devoted to the study of the optimization of the electron beam (e.b.) parameters for FEL operation. Finally in Sect. 4 the amplifier and oscillator behavior of an FEL in a storage ring is investigated.

In Table I we list the symbols most frequently used throughout this paper. For the sake of conciseness and to share a common notation with the quoted references, we prefer to work with symbols generally used in machine theory (even if some of the symbols may be confused with other well known physical quantities e.g. $\alpha \equiv$ coefficient, $\alpha_c \equiv$ momentum compaction and $\alpha \equiv e^2/h_c \equiv$ fine structure constant). However, the meaning of the symbols will be clarified within the text.

2. SINGLE PARTICLE DYNAMICS IN STORAGE RING

It is beyond the scope of this paper to give a complete and rigorous treatment of the electron dynamics in a SR. Many standard books and excellent review articles are devoted to the subject. [1,2,3,4,5] We limit ourselves to the study of those features of the e.b. motion which are relevant to the FEL operation. We follow, mainly, the approach and the notations of Ref. 1.

TABLE I

PHYSICAL CONSTANTS

e	= electron charge,
m_0	= rest electron ma s,
c	= light velocity,
r_0	= e^2/m_0c^2 = classical electron radius,
I_0	= ec/r_0 = Alfvén current,
$\bar{\lambda}_e$	= $\hbar(m_0c)$ = reduced Compton wavelength of the electron,
$\hbar$	= reduced Planck constant.

STORAGE RING PARAMETERS

L	= orbit length,
R	= $L/2\pi$ = mean radius,
ω_0	= c/R = revolution frequency,
T	= $2\pi/\omega_0$ = revolution period,
E_0	= nominal working energy
E	= actual electron energy,
ε	= $(E - E_0)/E_0$,
ρ	= magnetic bending radius,
α,β,γ	= Twiss parameters,
$\nu_{x,y}$	= betatron tune,
$\mu_{x,y}$	= $2\pi\nu_{x,y}$,
W	= $\gamma u^2 = 2\alpha uu' + \beta u'^2$ = Courant-Snyder invariant,
A	= transverse emittance,
η	= off-energy function,
P_γ	= synchrotron radiation emitted power (per electron),
U_0	= total energy radiated per machine turn and per electron,
ω_{RF}	= frequency of the r.f. accelerating system,
h	= ω_{RF}/ω_0 = harmonic number,
ω_s	= synchrotron frequency,
$\tau_i(i=s,x,y,)$	= damping time,
J_i	= damping partition numbers,
α_c	= momentum compaction,
N	= number of electrons per bunch,
I	= $(eN/2\pi)\omega_0$ = mean electron current per bunch,
I_p	= $eNc/\sqrt{2\pi}\,\sigma_z$ = peak electron current,
σ_ε	= r.m.s. energy spread,
σ_z	= r.m.s. bunch length,
P_s	= $I.U_0/e$ = total average power radiated by the whole bunch.

FEL PARAMETERS

ω_i, W_i, i	= frequency, energy density and phase of the i-th longitudinal laser mode,
k_i	= ω_i/c = i-th longitudinal laser mode wave-number,
λ_o	= resonant wavelength,
λ_q	= wiggler wavelength,
L_w	= wiggler length,
B_o	= wiggler magnetic field,
L_c	= optical cavity length,
T_c	= $2L_c/c$,
$(\Delta\omega/\omega)_o$	= homogeneous frequency width,
$(\Delta\omega/\omega)_c$	= coupling width,
$(\Delta\omega/\omega)_L$	= laser bondwidth,
μ_c	= $(\Delta\omega/\omega)_c/(\Delta\omega/\omega)_o$ = coupling parameter,
σ_L	= r.m.s. laser beam length,
Σ_L	= laser beam cross section,
ω_L	= average laser frequency,
dP_L/dS	= intracivity laser power density,
$\mathcal{E}$	= intracivity laser energy,
P_L	= output laser power,
γ_T	= total losses of the optical cavity (during T_e),
γ_M	= mirror transmission.

In Section 2.1 a typical SR layout is described. The linear electron motion is investigated in Sects 2.2 - 2.5. Finally Sects 2.6 - 2.7 are devoted to the study of the effects of the stochastic properties of the synchrotron radiation emission on the beam dynamics.

2.1 Storage Ring Layout

In a circular accelerating machine such as a SR the stored particles oscillate around a closed equilibrium orbit. Assuming that the machine has a median plane of symmetry, the closed orbit lies just in this plane. The electron motion will be described by means of the following curvilinear coordinates (see Fig. 1)

s : running along the closed orbit

x,y: lying in the plane orthogonal to the orbit, x being in the orbit plane and y orthogonal to it

x and y will be referred to as the radial and vertical coordinates respectively.

The guide field is supplied by bending magnets keeping the electrons in a given closed orbit and by focusing magnetic elements, the quadrupoles (q-poles), stabilizing the transverse motion. A q-pole acts on the electrons as a lens. It is worthwile to note that if a q-pole is focusing in the radial direction, it is defocusing in the vertical one and viceversa. We use the following notations for the magnetic elements

B = bending magnet,

F = radial focusing (vertical defocusing) q-pole,

D = radial defocusing (vertical focusing) q-pole,

In general the bending magnets too have focusing properties. However, if we use a magnetic field configuration constant in space (or with small radial gradient) we can neglect this effect with respect to the q-pole focusing. In this condition we say that the SR has "separate functions", i.e., the bending function is separated from the focusing one.

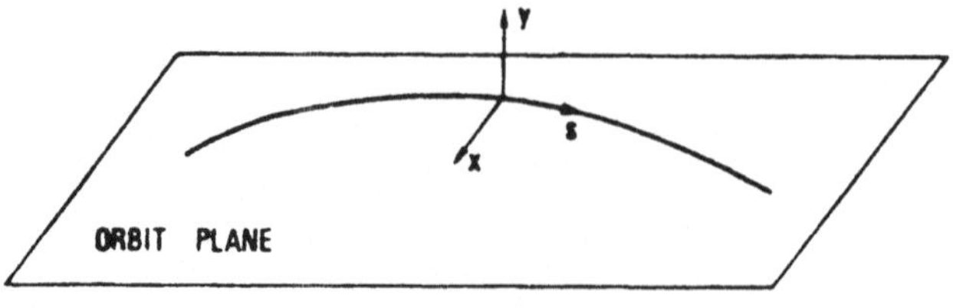

Fig. 1. Moving coordinate system

Other magnetic elements can be used for stabilizing the e.b. motion, such as sextupoles (S) and octupoles. In these lectures we shall not investigate the effects of these non linear elements. For details on these topics see, for example, Ref. 4, pp. 111 and 265.

A new magnetic element ("Wiggler Magnet", (WM)) is now being widely utilized[6]. In a WM the magnetic field varies periodically along the longitudinal direction. Accompanying this longitudinal variation there is a transverse gradient, which causes e.b. focusing. Let us consider, for example, the case of a linearly polarized WM, in which the magnetic field on the axis equals

$$\underline{B} \equiv \left(0,\ B_w \cos\left(\frac{2\pi}{\lambda_q} s\right),\ 0\right),$$

where λq is the WM "wavelength" (spatial periodicity). In this condition it is easy to show that the WM is focusing along the vertical direction and behaves like a free field straight section along the radial direction (see App. A). The WM are utilized for many purposes, such as e.b. stabilization and dimensional control,[7] generation of special synchrotron radiation[8] and, finally, for the FEL operation[9] we are interested.

Finally, one or more radio-frequency (r.f.) cavities must be arranged in a SR, for supplying the energy lost by the electrons, via synchrotron radiation, in the bending and wiggler magnets. (see Sect. 2.4).

As an example, we show in Fig. 2 the layout of the SR LEDA-F1[10]. It is interesting to analyze the parameters of this machine (even if it is, up to now, only at the level of a preliminary project study) because its design has been optimized just for FEL operation. LEDA-F1 has symmetry of order 2. This is a very important point. Indeed it is possible to show that lack of symmetry may generate many problems in the SR operation, in particular when the SR contains non linear magnetic elements such as sextupoles and octupoles (see e.g., Ref. 5, Chapter 3). For this reason, two WM are symmetrically inserted in the layout, but only one will be utilized (at least initially) for FEL operation. The machine has separate functions, the focusing being supplied by an alternate distribution of F and D q-poles. Finally we have, for each quarter of machine, two bending magnets and two sextupoles (whose utilization will be clarified in Sect. 2.3). The available free space is utilized for the insertion of r.f. cavities, for e.b. injection and for diagnostic tools. It is possible to modify the structure of LEDA-F1 in order to insert four high field WM for enhancing the synchrotron radiation emmission. The layout of this new machine (LEDA-F2[10]) is shown in Fig. 3. The desire for very large synchrotron radiation emission is related to FEL operation and will be discussed in Sects 4.3 and 4.6.

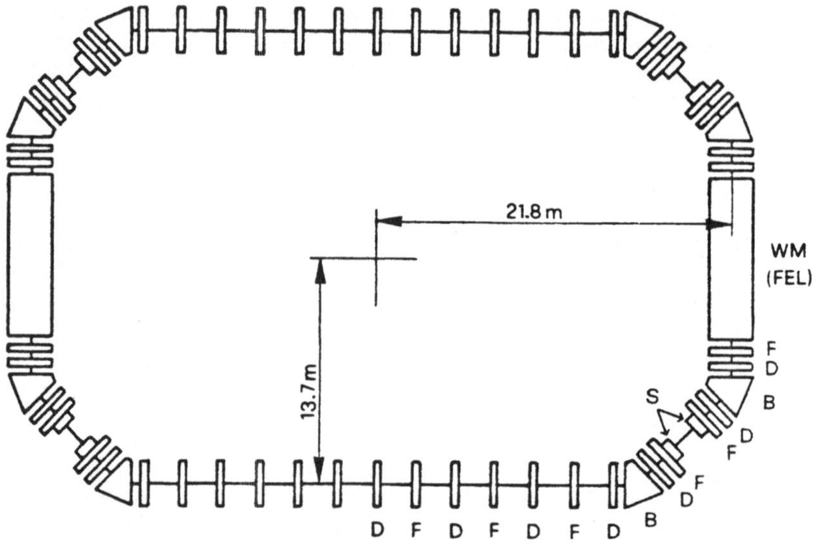

Fig. 2. LEDA-F1 layout
B = bending magnet
F = radial focusing q-pole
D = radial defocusing q-pole
S = sextupole magnet
WM = wiggler magnet

2.2 Transverse Motion for a Monoenergetic Electron Beam

In this section we investigate the electron motion in the transverse plane (x,y). As the independent variable we use the azimuth s instead of the time t. Indeed, for ultrarelativistic electrons, the velocity is practically independent of the energy ($ds/dt \sim c$). In this connection, assuming no coupling between x and y, the linearized equations of motion, for an electron of energy equal to the nominal one, read

$$\begin{cases} \dfrac{d^2x}{ds^2} = -K_x(s) \cdot x \\ \\ \dfrac{d^2y}{ds^2} = -K_y(s) \cdot y \end{cases} \tag{2.2}$$

Equations (2.2) describe the so-called "betatron motion". The coefficients $K_{x,y}$ (s) ("elastic constants") depend on the magnetic con-

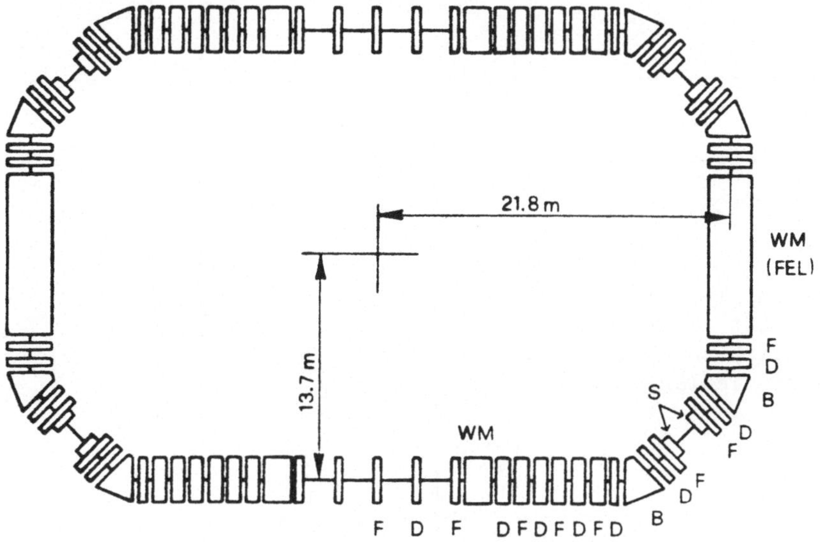

Fig. 3. LEDA-F2 layout
B = bending magnet
F = radial focusing q-pole
D = radial defocusing q-pole
S = sextupole magnet
WM = wiggler magnet

figuration of the guide field. In Table II we have listed $K_{x,y}$ for various elements (for the notations, see Table I). Note that, as previously pointed out, for the q-poles we have $K_x = - K_y$.

The "elastic constants" are periodic functions with the same periodicity of the machine, i.e.

$$K_{x,y}(s + L) = K_{x,y}(s), \tag{2.3}$$

where L is the orbit length. The Eqs (2.2), with the condition (2.3), describe the motion of a particle in a periodic potential. From the Floquet theorem (see, e.g., Ref. 3, Chapter V) we know that there are always two independent particular solutions of the form (u stands for x or y and K for K_x or K_y)

TABLE II

ELEMENT	K_x	K_y
FREE SPACE	0	0
BENDING MAGNET (without radial gradient and with the pole faces perpendicular to the orbit)	$\left(\frac{e\ B_y}{E}\right)^2$	0
q-pole	$\frac{e}{E} \cdot \frac{\partial B_y}{\partial x}$	$-\frac{e}{E} \cdot \frac{\partial B_y}{\partial x}$
WIGGLER MAGNET ($\underline{B} \parallel \hat{y}$)	0	$\frac{1}{2}\left(\frac{e\ B_y}{E}\right)^2$

$$\begin{cases} u_1(s) = \exp\left(j\ \mu\ \frac{s}{L}\right) \cdot p_1(s) \\ \qquad\qquad\qquad\qquad\qquad (j = \sqrt{-1}) \\ u_2(s) = \exp\left(-\ j\ \mu\ \frac{s}{L}\right) \cdot p_2(s), \end{cases} \tag{2.4}$$

where p_1 and p_2 have the same periodicity of K

$$p_{1,2}(s + L) = p_{1,2}(s). \tag{2.5}$$

The stability of the motion depends on μ, called the "characteristic coefficient" (if μ is real the motion is stable, if μ is complex it is unstable). The phase advance per machine turn of the "pseudo-oscillation" (2.4) is just μ. We have indeed (see Eqs (2.4) and (2.5))

$$u_{1,2}(s + L) = \exp(j\mu)u_{1,2}(s). \tag{2.6}$$

The number of "betatron oscillations" per turn is called "betatron tune" and it is given by

$$\nu = \frac{\mu}{2\pi}\ . \tag{2.7}$$

Assuming μ real (stable condition) we can write down the general solution of the Eq. (2.2) in the real form

$$u(s) = I(s) \cdot \cos(\varphi(s)). \tag{2.8}$$

with

$$I(s + L) = I(s) \tag{2.9}$$

and

$$\varphi(s + L) = \varphi(s) + \mu. \tag{2.10}$$

The motion described by the Eq. (2.8) is "pseudo-harmonic", i.e. it has a sinusoidal behaviour with an amplitude and phase periodic modulation. In the simple case $K(s) = K_o$ = const., we obtain a purely harmonic motion

$$I(s) = \text{const.}, \ \varphi(s) = \sqrt{K_o} \cdot s + \varphi_o \Rightarrow \mu = \sqrt{K_o} \cdot L. \tag{2.11}$$

The functions $I(s)$ and $\varphi(s)$, being not (in general) independent, are expressible in terms of a single function $\beta(s)$, having the machine periodicity:

$$I(s) = I_o\sqrt{\beta(s)}, \tag{2.12}$$

$$\varphi(s) = \int_o^s \frac{ds'}{\beta(s')} + \varphi_o. \tag{2.13}$$

From Eqs (2.10) and (2.13) we derive μ as a function of β

$$\mu = \int_o^L \frac{ds'}{\beta(s')} . \tag{2.14}$$

The methods for finding $\beta(s)$ (for a given $K(s)$) are described in the text books previously quoted, and we shall not deal with them here. The physical meaning of β is straightforward. It may be regarded as the "local reduced betatron wavelength". Where β is low the e-beam is more focused (Eq. (2.12)) and the phase advances more rapidly (Eq. (2.13)). As example, we show in Fig. 4 the $\beta_{x,y}$ functions for a quarter of the machines LEDA-F1,2.

Let us define now a motion invariant of Eq. (2.2) which plays the role of the energy invariant for a purely harmonic oscillator,

$$W = \gamma(s)u^2 + 2\alpha(s)uu' + \beta(s)u'^2 = \text{const.} \tag{2.15}$$

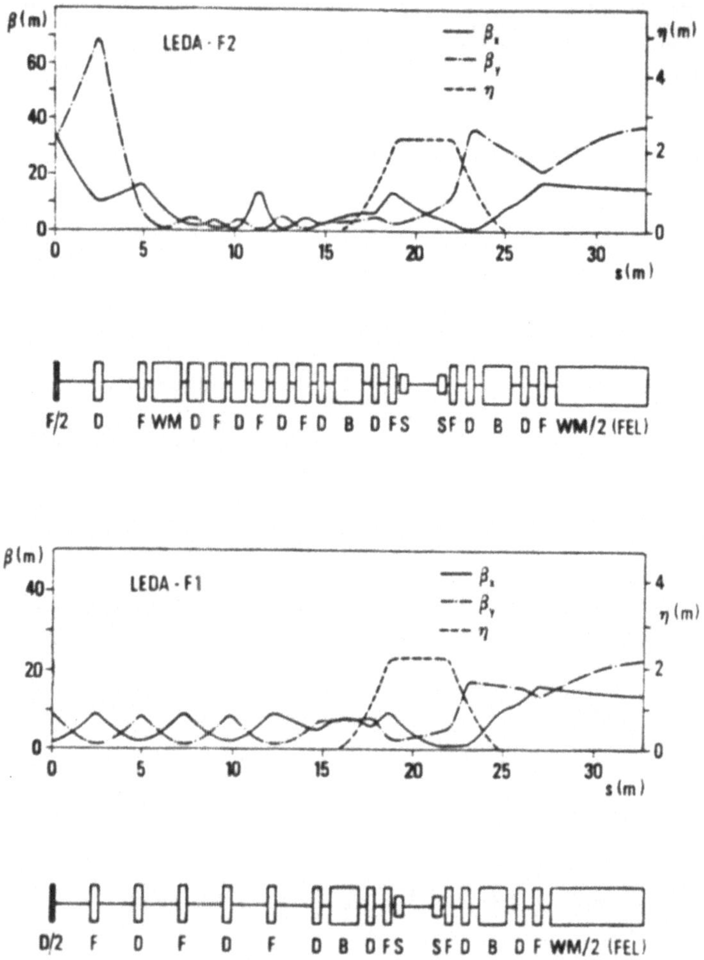

Fig. 4. Optical functions in LEDA-F1 and LEDA-F2
B = bending magnet
F = radial focusing q-pole
D = radial defocusing q-pole
S = sextupole magnet
WM = wiggler magnet

W is called "Courant-Snyder" invariant.[2] In Eq. (2.15) β is the function previously defined, α and γ depend on β:

$$\alpha(s) = -\frac{1}{2}\beta', \tag{2.16}$$

$$\gamma(s) = \frac{1 + \alpha^2}{\beta} . \tag{2.17}$$

The functions α, β and γ are generally called "Twiss coefficients".

If we follow the evolution of u and its derivative u', turn by turn, at a given azimuth s_0, we see that they describe an ellipse, according to Eq. (2.15), whose area is given by

$$A = \pi W. \tag{2.18}$$

A is called "beam emittance". The ellipse described by Eq. (2.15) changes throughout the machine, but its area is a constant. The emittance is then an e-beam intrinsic property and it is independent of the guide field parameters (Liouville theorem). From the Eqs (2.15) and (2.18) we derive the beam envelope σ and divergence σ', defined as the r.m.s. values of u and u' over many betatron oscillations (at fixed s)

$$\sigma(s) = \sqrt{\langle u^2 \rangle} = \sqrt{\frac{A \cdot \beta(s)}{2\pi}} , \tag{2.19}$$

$$\sigma'(s) = \sqrt{\langle u'^2 \rangle} = \sqrt{\frac{A \cdot \gamma(s)}{2\pi}} . \tag{2.20}$$

2.3 Off-energy Transverse Motion

In the previous section we have assumed an electron energy exactly equal to the nominal one. A particle with a different energy sees different $K_{x,y}$ (see Table II), which leads to a change in the betatron tune $\nu_{x,y}$. We call the chromaticity of the machine the variation of ν with the particle energy:

$$C_{x,y} = \frac{E_0}{\nu_{x,y}} \cdot \frac{\Delta \nu_{x,y}}{\Delta E} . \tag{2.21}$$

In a well designed machine this parameter must be kept as small as possible, in order to avoid too large a frequency spread in the beam. It is possible to correct the chromaticity by utilizing sextupole magnets (see, e.g., Ref. 4, p. 120). Indeed just for this reason these elements have been inserted in the LEDA-F1,2 layout (see Sect. 2.1).

Another effect is connected with the change of the magnetic bending radius ρ:

$$\rho = \frac{E}{e B_y} . \tag{2.22}$$

As a consequence, in the radial equation of motion we have a nonhomogeneous term. The first of Eqs (2.2) becomes (see, e.g., Ref. 4, p. 11),

$$\frac{d^2x}{ds^2} + K_x(s)x = \frac{\varepsilon}{\rho} , \qquad (2.23)$$

where

$$\varepsilon = \frac{E - E_o}{E_o} , \qquad (2.24)$$

E = actual electron energy,
E_o = nominal machine energy.

The general solution of Eq. (2.23) may be written as the sum of the general solution of the homogeneous one (Eq. (2.8)), which describes the "free betatron oscillation", and a particular solution of the inhomogeneous equation. That is, we can write

$$x(s) = x_\beta(s) + \eta(s) \cdot \varepsilon, \qquad (2.25)$$

with

$$\frac{d^2x_\beta}{ds^2} + K_x(s)x_\beta = 0, \qquad (2.26)$$

and

$$\frac{d^2\eta}{ds^2} + K_x(s)\eta = \frac{1}{\rho} . \qquad (2.27)$$

It is convenient to choose the particular solution of Eq. (2.27) which has the machine periodicity

$$\eta(s + L) = \eta(s). \qquad (2.28)$$

The function η is called the "off-energy function". The physical meaning is straightforward, i.e. $\eta \cdot \varepsilon$ is the new reference orbit around which a particle with off-energy ε oscillates.

In Figure 4 is plotted, as an example, $\eta(s)$ for LEDA-F1,2. It is interesting to note that this function is null through all the machine, except in the region between two neighbouring bending magnets and in the WM, where it is so small that it is not resolved in Fig. 4 (see App.B).

The reason for this choice is that, when the FEL is operating, the e.b. energy spread may be very high (see Sects 4.3 and 4.9). Then we must limit, as much as possible, the magnitude of the off-energy orbit around the machine. Furthermore the low η in WM minimize the transverse e.b. dimensions (see Sect. 2.7) and the inhomogeneous synchrotron emission bandwidth (see Sect. 3.1).

2.4 Longitudinal Motion

The longitudinal motion (usually called "synchrotron motion") of an electron in a SR is strongly affected by the synchrotron radiation emission.

The power P_γ radiated by an ultrarelativistic particle with energy E, mass m_o and charge e in a magnetic field B, orthogonal to the particle velocity, is given by [11]

$$P_\gamma = \frac{2}{3}\frac{e^4 \cdot c}{(m_o c^2)^4}(B \cdot E)^2. \tag{2.29}$$

The power radiated in the bending magnet can be written in the form (see Eq. (2.22), r_o = classical electron radius)

$$P_\gamma = \frac{2}{3}\frac{r_o c}{(m_o c^2)^3} \cdot \frac{E^4}{\rho^2}, \tag{2.30}$$

while in a WM it is (see Eq. (2.1))

$$P_\gamma = \frac{2}{3}\frac{r_o c}{(m_o c^2)^3}\left[e\, B_w \cos\left(\frac{2\pi}{\lambda_q} s\right) E\right]^2. \tag{2.31}$$

The total energy radiated per machine turn is then given by

$$U = \oint P_\gamma dt = \frac{r_o}{3}\left(m_o c^2\right)\left\{\frac{4\pi}{\rho}\left(\frac{E}{m_o c^2}\right)^4 + \left(\frac{E}{m_o c^2}\right)^2 \sum_i L_w^i \left(\frac{eB_w^i}{m_o c^2}\right)^2\right\}, \tag{2.32}$$

where the summation is extended over all the WM in the SR (L_w^i, $B_w^i \equiv$ length and field of the i-th WM).

Table III summarizes the maximum working energy, the bending radius ρ, the mean machine radius R (= $L/2\pi$) and the energy radiated per turn for some of the existing SR and for LEDA-F1,2. We can see that U increases very rapidly with the working energy. In particular let us note the considerable increase in energy radiated due

TABLE III

	E	ρ	R	U
	GeV	m	m	
ACO	0.54	1.11	3.5	6.8 keV
ADONE	1.5	5	16.6	89.6 keV
SPEAR	4.5	12.7	37.3	2.86 MeV
LEDA-F1	0.75	2	20.8	15.5 keV
LEDA-F2	0.75	2	20.8	72.6 keV

to the insertion of the high field WM ($\sim$ 50 kG) in LEDA-F2, with respect to LEDA-F1. However, in all the cases illustrated in Table III, U is a small fraction ($\sim 10^{-4} \div 10^{-5}$) of the electron energy E. In this connection, in dealing with the longitudinal motion, we are allowed to make the following simplifying assumptions:

a) An electron which starts a given revolution with energy E continuously loses energy and changes its orbit and instantaneous radiated power $P_\gamma(E)$. We assume that, in one revolution, it moves along the designed orbit and radiates the energy U corresponding to the starting energy E. We will take into account, however, the cumulative multiturn effects.
b) We neglect betatron oscillations. They affect the instantaneous emitted power and not its average (there are generally several betatron oscillations per turn).

In order to maintain the electron distribution in the machine, we must supply (on the average) the lost energy. We cannot use static electric fields, because we must have a net energy gain over a closed path. For this reason a time varying electric field is used. This field must be periodic in time with particle revolution period, so that a typical electron sees the same accelerating voltage turn by turn. Then the frequency ω_{RF} of the radio-frequency (r.f.) accelerating system must be equal or multiple of the electron revolution frequency ω_o

$$\omega_{RF} = h\,\omega_o, \quad h = 1, 2, 3, \ldots \tag{2.33}$$

where

$$\omega_o = \frac{2\pi}{T} = \frac{2\pi c}{L} = \frac{c}{R} , \tag{2.34}$$

h is called "harmonic number of the r.f. system".

Let us assume we have only one r.f. resonating cavity in the machine. In Figure 5 we have the plot of the energy supplied by the accelerating system to the electrons versus the particle time passage through the r.f. cavity (in the case h = 3).

The accelerating voltage makes h complete oscillations in one period machine (T). A particle with energy E_o which crosses the cavity at the time t_i' or t_i'' (with i = 1, 2...h), gains exactly the energy U_o lost by synchrotron radiation (see Fig. 5). In the next turn, the particle crosses the cavity at the time $t_i' + T$ (or $t_i'' + T$) and gains the same energy U_o. Thus, this particle will move along the designed orbit indefinitely.

In general we have 2h "synchronous phases"

$$(t_i', t_i''), \qquad i = 1, 2, \ldots, h \tag{2.35}$$

with (see Fig. 5)

$$\begin{cases} t_i' = t_1' + \frac{i}{h} T \\ \qquad\qquad\qquad i = 1, 2, \ldots, h \\ t_i'' = t_1' + \frac{i}{h} T \end{cases} \tag{2.36}$$

Let us consider now the case of an electron with $E \neq E_o$ and whose time passage t_j (at the j-th turn) throughout the r.f. cavity is different from the synchronous one and reads,

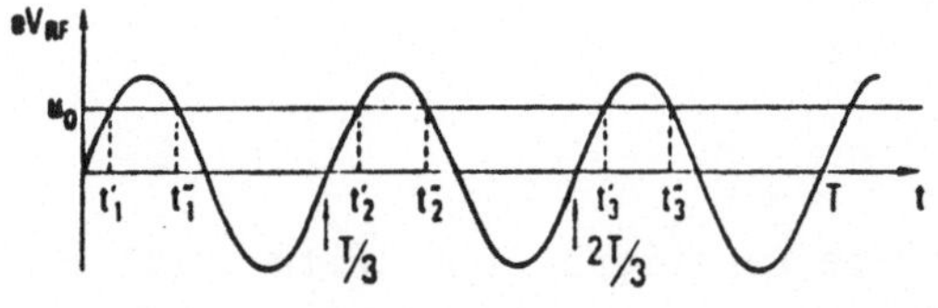

Fig. 5. R.F. System supplied energy

$$t_j = t_o + jT - \frac{z_j}{c}\,, \tag{2.37}$$

where t_o is the "synchronous phase" (i.e. t_o may be t_i' or t_i''), and z_j is the longitudinal displacement of the particle from the synchronous one. If $z_j > 0$ the electron is traveling ahead of the bunch. The energy gain is given by

$$eV_j = eV_{RF}\left(t_o + \frac{i}{h}T + jT - \frac{z_j}{c}\right). \tag{2.38}$$

The r.f. field has period T/h (see Eq. (2.33)). Then Eq. (2.38) becomes

$$eV_j = eV_{RF}\left(t_o - \frac{z_j}{c}\right). \tag{2.39}$$

The total particle energy variation (including the radiation losses) at the jth turn is

$$\Delta\varepsilon_j = \varepsilon_{j+1} - \varepsilon_j = \frac{eV_j - U}{E_o}\,. \tag{2.40}$$

We are interested in studying the small amplitude synchrotron motion (for the large amplitude motion, which is relevant to the investigation of the mean life of the stored e.b., see, e.g. Ref. 1, Sect. 3.6).

Expanding U and V to first order in ε and z, we obtain

$$\frac{U}{E_o} \sim \frac{U_o}{E_o} + \frac{U_o'}{E_o}\varepsilon \qquad \left(U_o' = \left(\frac{\partial U}{\partial\varepsilon}\right)_{\varepsilon=o}\right), \tag{2.41}$$

$$eV_j \sim eV_{RF}(t_o) - \frac{e}{c}V_{RF}' \cdot z_j = U_o - \frac{e}{c}V_{RF}' \cdot z_j \tag{2.42}$$

$$\left(V'_{RF} = \left(dV_{RF}/dt\right)_{t=t_o}\right).$$

For a SR with separate functions (i.e. with small radial gradient in the bending magnets), we derive the following rough estimate

$$U_o' \sim 2U_o \tag{2.43}$$

obtained by differentiating Eq. (2.29) with B $\sim$ const. and neglecting the variation of the orbit length (in the next section we shall give the general expression of U_0'). Equation (2.40) becomes

$$\Delta\varepsilon_j = -\frac{e}{cE_0} V'_{RF} \cdot z_j + 2\frac{U_0}{E_0}\varepsilon_j. \tag{2.44}$$

The change in z is connected with ε by the so called "momentum compaction" α_c, measuring the variation of the orbit length with the particle energy. This parameter depends on η(s). We have indeed (see Ref. 1, Part III),

$$\alpha_c = \frac{\Delta L/L}{\Delta E/E} = \frac{1}{L}\oint \frac{\eta(s)}{\rho(s)}\, ds, \tag{2.45}$$

where ρ(s) is the local bending radius (including the WM). The variation on z is then given by (for ultrarelativistic electrons, i.e. v $\sim$ c)

$$\Delta z_j = z_{j+1} - z_j = -L\alpha_c\,\varepsilon_{j+1}. \tag{2.46}$$

The recurrence Eqs (2.44) and (2.46) describe completely the linear synchrotron motion. The variations Δε and Δz are usually very small, so that we can approximate the differences by differencials

$$\Delta\varepsilon \sim \frac{d\varepsilon}{dt}\cdot T, \quad \Delta z \sim \frac{dz}{dt}\cdot T. \tag{2.47}$$

This leads to

$$\begin{cases} \dfrac{dz}{dt} = -(c\alpha_c)\cdot\varepsilon \\[2ex] \dfrac{d\varepsilon}{dt} = -\dfrac{eV'_{RF}}{cT\,E_0}\cdot z - \dfrac{2\,U_0}{T\,E_0}\varepsilon. \end{cases} \tag{2.48}$$

Equations (2.48) describe a damped harmonic motion of frequency given by

$$\omega_s = \left(-\frac{\alpha_c}{E_0 T}\cdot eV'_{RF}\right)^{1/2}, \tag{2.49}$$

and damping time

$$\tau_s = \frac{T\,E_o}{U_o} = \frac{3}{4\pi} \cdot \frac{(m_o c^2)^3}{r_o} \cdot \frac{\rho}{E_o} \cdot T, \tag{2.50}$$

(In deriving Eq. (2.49) we have neglected $1/\tau_s$ with respect to ω_s). From Eq. (2.49) we realize that the stable synchronous phases lie on the negative slope sides of the r.f. voltage (i.e. t_1'', t_2'', ..., t_h'' in Fig. 5). In the neighborhood of these h "stable synchronous phases" we have the h possible "bunches" of the machine. Generally we have

$$T \ll \frac{2\pi}{\omega_s} \ll \tau_s, \tag{2.51}$$

i.e. the synchrotron motion is slow with respect to the revolution rate, and the damping is a small perturbation to the synchrotron motion. The recurrence Eqs (2.44) and (2.46), written in terms of ω_s and τ_s, read,

$$\begin{cases} \Delta z_j = -\left(c\alpha_c\ \varepsilon_{j+1}\right)T \\ \Delta\varepsilon_j = \left(\dfrac{\omega_s^2}{c\alpha_c}\ z_j - \dfrac{2}{\tau_s}\ \varepsilon_j\right)T. \end{cases} \tag{2.52}$$

2.5 The Damping--General Formulation

Taking into account the radial gradient in the bending magnet and the variation of the closed orbit length, the quantity U_o' (see Eq. (2.41)) reads (see Ref. 1, Part IV)

$$U_o' = U_o(2 + \Delta), \tag{2.53}$$

where Δ is given by

$$\Delta = \frac{1}{cU_o} \oint \left\{ P_\gamma \left(\frac{1}{\rho} + \frac{2}{B} \cdot \frac{dB}{dx} \right) \eta \right\}_{\varepsilon=o} ds; \tag{2.54}$$

for dB/dx = 0, Δ reduces to (for a SR without WM)

$$\Delta = \frac{1}{cU_o} \oint P_\gamma \frac{\eta}{\rho} ds = \frac{R}{\rho} \alpha_c . \qquad (2.55)$$

Typically we have

$$\frac{R}{\rho} \sim 3, \ \alpha_c \sim 10^{-1} \Rightarrow \Delta \sim 3 \times 10^{-1} \ (\ll 2) \ ;$$

then Eq. (2.43) is a very good estimate for U_o'.

The damping constant per turn may be written in the general form

$$\frac{T}{\tau_s} = J_s \cdot \frac{U_o}{2E_o}, \qquad (2.56)$$

where

$$J_s = 2 + \Delta. \qquad (2.57)$$

It can be shown that the betatron oscillations too are damped. The damping mechanism arises again from synchrotron emission.

The damping constants per turn for the three modes of oscillations can be written in the general form

$$\frac{T}{\tau_i} = J_i \frac{U_o}{2E_o} \quad (i = x, y, s) \qquad (2.58)$$

where J_i ("damping partition numbers") are given by

$$J_s = 2 + \Delta, \quad J_x = 1 - \Delta, \quad J_y = 1. \qquad (2.59)$$

For the case i=s, Eq. (2.59) reduces to Eq. (2.57.)
The quantities J_i satisfy the equations

$$J_s + J_x + J_y = 4 \qquad (2.60)$$

$$J_s + J_x = 3, \quad J_y = 1. \qquad (2.61)$$

Equation (2.60) holds in general (Robinson's theorem[12]), while

Eqs (2.61) hold only for plane machines.

2.6 Synchrotron radiation excitation

In Section 2.4 we have assumed that the synchrotron power (Eq. (2.29)) is radiated continuously in the bending and wiggler magnets. We know, however, that the radiation is emitted in quanta of discrete energy. The emission of a given quantum is statistically independent of the other ones. In addition the energy change is a very small fraction of the electron energy. Thus we may consider this phenomenon as a purely random process. These many (and small) discontinuous disturbances generate a kind of "noise". As a consequence the electrons undergo a diffusion process enlarging the oscillation amplitudes, until this radiation excitation is balanced by the damping.

In this section we shall investigate in detail the effect of this stochastic phenomenon on the synchrotron motion, while in the next one, we shall describe, in a qualitative way, the effects on the betatron motion.

The synchrotron motion recurrence equations (Eqs (2.52)), with the quantum noise, read

$$\begin{cases} \Delta z_j = -\left(c \cdot \alpha_c \, \varepsilon_{j+1}\right) \cdot T \\ \Delta \varepsilon_j = \left(\frac{\omega_s^2}{c\alpha_c} z_j - \frac{2}{\tau_s} \varepsilon_j\right) \cdot T + \frac{u}{E_o} ; \end{cases} \tag{2.62}$$

u describes the stochastic properties of the emitted radiation. The average values of u and u^2, taken over a time Δt, which is small with respect to the synchrotron period, but big enough to have a high number of elementary photon emission processes, are given by (see Ref. 1, Part V)

$$< u > = 0$$

$$< u^2 > = \frac{55}{16\sqrt{3}} \frac{\lambda\!\!\!^{-}_e}{\rho} \cdot E_o \, P_\gamma \cdot \left(\frac{E_o}{m_o c^2}\right)^2 \Delta t, \tag{2.63}$$

where $\lambda\!\!\!^{-}_e$ is the reduced Compton wavelength of the electron, P_γ is given by Eq. (2.29) and we have assumed that only bending magnets are operating in the SR. At the end of this section we shall consider the general case (i.e. with WM inserted in the SR).

The electron distribution function describing the synchrotron motion may be derived from a FOKKER-PLANCK equation (FPE), where the damping mechanism arises from the mean emitted power and the diffusion from the statistical behaviour of the radiation process. We have[13]

$$\frac{\partial f}{\partial t} = -\frac{\partial}{\partial z}\left(\frac{\langle \Delta z\rangle}{\Delta t} f\right) - \frac{\partial}{\partial \varepsilon}\left(\frac{\langle \Delta \varepsilon\rangle}{\Delta t} f\right) + \frac{1}{2}\left[\frac{\partial^2}{\partial z^2}\left(\frac{\langle \Delta z^2\rangle}{\Delta t} f\right) + \frac{\partial^2}{\partial \varepsilon^2}\left(\frac{\langle \Delta \varepsilon^2\rangle}{\Delta t} f\right) + 2\frac{\partial^2}{\partial z \partial \varepsilon}\left(\frac{\langle \Delta z \Delta \varepsilon\rangle}{\Delta t} f\right)\right], \tag{2.64}$$

where Δt is the previously defined averaging time. We can choose (see Sect. 2.4) Δt just equal to the revolution period T. From the Eqs (2.62) and (2.63) we obtain, to first order in Δt (i.e. in T),

$$\langle \Delta z\rangle = -(c\alpha_c\, \varepsilon)\cdot T, \quad \langle \Delta z^2\rangle = 0, \quad \langle \Delta z \Delta \varepsilon\rangle = 0$$

$$\langle \Delta \varepsilon\rangle = \left(\frac{\omega_s^2}{c\alpha_c} z - \frac{2}{\tau_s}\varepsilon\right)\cdot T, \quad \langle \Delta \varepsilon^2\rangle = \frac{\langle u^2\rangle}{E_o^2}. \tag{2.65}$$

The FPE (2.64) becomes

$$\frac{\partial f}{\partial t} = \omega_s\left(\left(\frac{c\alpha_c}{\omega_s}\right)\varepsilon\frac{\partial f}{\partial z} - \left(\frac{\omega_s}{c\alpha_c}\right) z \frac{\partial f}{\partial \varepsilon}\right) + \frac{\partial}{\partial \varepsilon}\left(2\varepsilon\frac{f}{\tau_s} + D\frac{\partial f}{\partial \varepsilon}\right), \tag{2.66}$$

where D ("diffusion coefficient") is given by (see Eq.(2.63))

$$D = \frac{55}{32\sqrt{3}}\cdot\frac{\lambda\!\!\!^{-}_e}{\rho}\,\frac{P_\gamma}{E_o}\left(\frac{E_o}{m_o c^2}\right)^2. \tag{2.67}$$

The steady state distribution ($\partial f_o/\partial t = 0$) has the following guassian form, which is a consequence of the central limit theorem:

$$f_o(z,\varepsilon) = \frac{1}{2\pi\,\sigma_z\,\sigma_\varepsilon}\exp\left[-\frac{1}{2}\left(\frac{z^2}{\sigma_z^2}+\frac{\varepsilon^2}{\sigma_\varepsilon^2}\right)\right], \tag{2.68}$$

where

$$\sigma_\varepsilon = \sqrt{\frac{D\cdot\tau_s}{2}} = \text{r.m.s. energy spread,} \tag{2.69}$$

$$\sigma_z = \frac{c\alpha_c}{\omega_s}\sigma_\varepsilon = \text{r.m.s. bunch length.} \tag{2.70}$$

By using Eqs (2.56) and (2.67), σ_ε can be written in the form

$$\sigma_\varepsilon = \sqrt{C_q\cdot\frac{1}{J_s\rho}}\cdot\frac{E_o}{m_oc^2}, \tag{2.71}$$

where we have defined

$$C_q = \frac{55}{32\sqrt{3}}\lambda_e. \tag{2.72}$$

Equation (2.71) holds for "isomagnetic machines", i.e. without WM. In the general case σ_ε reads (see Ref. 14 and Eq. (2.1))

$$\sigma_\varepsilon = \sqrt{C_q\frac{1}{J_s\rho}}\cdot\sqrt{\frac{1+(2/3\pi^2)\sum_i\frac{L_w^i}{\rho}\left(\frac{B_w^i}{B_b}\right)^3}{1+(1/4\pi)\sum_i\frac{L_w^i}{\rho}\left(\frac{B_w^i}{B_b}\right)^2}}\cdot\left(\frac{E_o}{m_oc^2}\right) \tag{2.73}$$

where L_w^i, B_w^i are the length and the field of the i-th WM and B_b is the field in the bending magnets.

To give the relative orders of magnitude, we report in Table IV σ_ε and $\sigma_\varepsilon \cdot E_o$ for the machines previously described in Table III. The bunch length depends (Eq. (2.70)), on σ_ε, α_c and ω_s. Typically we have

TABLE IV

	E_o(GeV)	σ_ε	$\sigma_\varepsilon \cdot E_o$
ACO	.54	4.4×10^{-4}	240 keV
ADONE	1.5	5.7×10^{-4}	860 keV
SPEAR	4.5	1.1×10^{-3}	4.9 MeV
LEDA-F1	.75	$4\ \times 10^{-4}$	300 keV
LEDA-F2	.75	$8\ \times 10^{-4}$	600 keV

$$\alpha_c \sim 10^{-2} - 10^{-1},\ \omega_s \sim 10^5\ \mathrm{sec}^{-1},\ \sigma_\varepsilon \sim 10^{-3} \Rightarrow \sigma_z \sim 3 \text{ to } 30 \text{ cm}$$

In deriving the e-beam energy spread, two effects, generating additional noise, have been neglected. They are
1) Multiple Toushek effect (see Ref. 3, Chapter XXXI)
2) Anomalous lengthening (see Ref. 4, p. 172)

The former is due to inter-particle multiple scattering processes, while the latter is caused by the interaction of the whole beam with machine elements (cavities, electrodes, etc.) resonating at wavelengths of the order of the bunch length. These effects depend on the energy and on the beam current. However they may be neglected if we work at high energy ($E_0 \gtrsim 300 \div 400$ MeV) in a "smooth machine" (i.e. without spurious high frequency resonating cavities).

2.7 Transverse Beam Dimension

Let us consider the radial motion. It is described by the Eqs (see Eq. (2.25))

$$\begin{cases} x(s) = x_\beta(s) + \eta(s)\cdot\varepsilon \\ x'(s) = x'_\beta(s) + \eta'(s)\cdot\varepsilon. \end{cases} \qquad (2.74)$$

The emission of a quantum of synchrotron radiation produces a sudden change in the reference orbit, generating an amplitude variation of the free betatron oscillation (see Fig. 6). Then we have

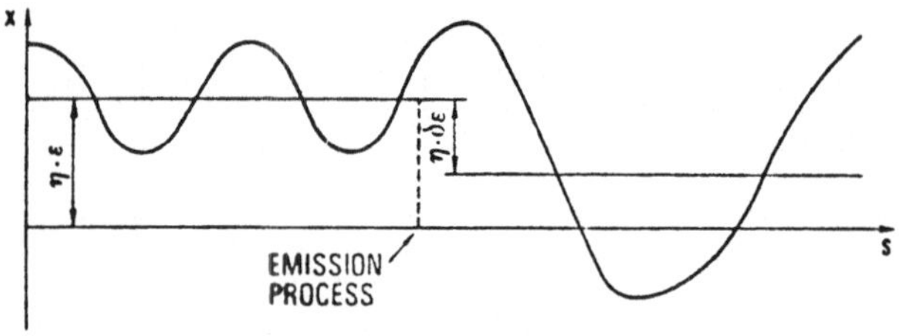

Fig. 6. Betatron motion amplitude variation due to a synchrotron emission process

$$\delta x_\beta = -\eta \cdot \Delta\varepsilon$$

$$\delta x'_\beta = -\eta' \cdot \Delta\varepsilon. \tag{2.75}$$

Equation (2.75) does not account for the emission angle of the photon. The change in x'_β for a photon emitted at angle Θ_x is

$$\delta x'_\beta = \Theta_x \cdot \Delta\varepsilon. \tag{2.76}$$

This effect is to be compared with Eq. (2.75), where typically $\eta' \sim 1$:

$$\left| \frac{\left(\delta x'_\beta\right)_{\Theta_x}}{\left(\delta x'_\beta\right)_{\eta'}} \right| \sim \Theta_x. \tag{2.77}$$

The value of Θ_x, for ultrarelativistic electrons, is very low. Indeed, the emitted radiation lies in a cone (around the e-beam trajectory) whose half angle is

$$\Theta_x \sim \frac{m_o c^2}{E_o} \ll 1. \tag{2.78}$$

As a consequence we can neglect this effect and only treat the variations due to the off-energy closed orbit.

The stochastic behaviour of the emitted radiation (see previous section) generates a diffusion process, which will be counteracted only by the damping mechanism.

In this section we shall derive in a simple, but not rigorous, way, the radial emittance for a SR without WM (however, the results we find are correct). A rigorous treatment may be found in Ref. 1, Sect. 5.5. The general case (SR with WM) will be discussed at the end of this section.

The change of the radial invariant W (Eq. (2.15)) due to the process (2.75) is given by

$$\delta W = (\Delta\varepsilon)^2(\gamma\eta^2 + 2\alpha\eta\eta' + \beta\eta'^2) -$$

$$- 2\Delta\varepsilon\left[\gamma x_\beta \cdot \eta + \alpha\left(x_\beta\eta' + \eta\, x'_\beta\right) + \beta \cdot x'_\beta \cdot \eta'\right]. \tag{2.79}$$

There is no correlation between the phase of the oscillation and the time of the photon emission. An averaging over all the initial phases gives

$$\overline{\delta W} = (\Delta\varepsilon)^2 \cdot H(s), \tag{2.80}$$

where

$$H(s) = \gamma\eta^2 + 2\alpha\eta\eta' + \beta\eta'^2. \tag{2.81}$$

It is to be noted that, since $\eta(s)$ is not a free betatron oscillation, $H(s)$ is not a constant. In Eq. (2.80) we can approximate $H(s)$ with its mean value in the bending magnets (H_M). We obtain

$$\overline{\delta W} \approx (\Delta\varepsilon)^2 H_M. \tag{2.82}$$

The total change per unit time in W is connected with the stochastic properties of the emitted radiation. We have (see Eqs (2.65) and (2.67))

$$\frac{<\overline{\delta W}>}{\Delta t} = \frac{<\Delta\varepsilon^2>}{\Delta t} H_M = 2D \cdot H_M. \tag{2.83}$$

By taking into account the damping, Eq. (2.83) reads

$$\frac{<\overline{\delta W}>}{\Delta t} = 2DH_M - \frac{2}{\tau_x} <W>. \tag{2.84}$$

The equilibrium invariant $\langle W\rangle_o$ is then given by

$$\langle W\rangle_o = \tau_x \cdot D \cdot H_M . \tag{2.85}$$

From Eq. (2.85) and (2.18) we derive the radial emittance

$$A = \pi \langle W\rangle_o = \pi \tau_x D H_M. \tag{2.86}$$

The electron distribution function becomes (again from the central limit theorem)

$$f(x,x') = \frac{1}{A} \exp\left(- \frac{\pi W}{A}\right) = \frac{1}{A} \exp\left(- \frac{\pi}{A}\left(\gamma x_\beta^2 + 2\alpha x_\beta x_\beta' + \beta x_\beta'^2\right)\right). \tag{2.87}$$

We can write A as a function of σ_ε. We obtain (see Eqs (2.69) and (2.58))

$$A = 2\pi \frac{J_s}{J_x} H_M \cdot \sigma_\varepsilon^2. \tag{2.88}$$

The beam envelope and divergence are (see Eqs (2.19) and (2.20)),

$$\sigma_{x\beta}^2 = \frac{J_s}{J_x} H_M \sigma_\varepsilon^2 \beta(s),$$

$$\sigma_{x\beta}'^2 = \frac{J_s}{J_x} H_M \sigma_\varepsilon^2 \gamma(s), \tag{2.89}$$

where the suffix β stands for free betatron motion.

Let us consider now the radial spread that arises from energy oscillations. We have,

$$x_\varepsilon = \eta\varepsilon, \quad x_\varepsilon' = \eta'\varepsilon. \tag{2.90}$$

We obtain

$$\sigma_{x\varepsilon}^2 = \eta^2\sigma_\varepsilon^2, \ \sigma_{x\varepsilon}'^2 = \eta'^2\sigma_\varepsilon^2 . \tag{2.91}$$

The two widths (2.89) and (2.91) arise from the same stochastic effect, the quantum emission. However, the betatron frequency is very different from the synchrotron one. Thus we may consider the two contributions as statistically independent. Then the total widths will be given by

$$\sigma_x^2 = \sigma_{x\beta}^2 + \sigma_{x\varepsilon}^2, \quad \sigma_x'^2 = \sigma_{x\beta}'^2 + \sigma_{x\varepsilon}'^2. \tag{2.92}$$

From Eqs (2.89) and (2.91) we obtain

$$\sigma_x = \sigma_\varepsilon \sqrt{\eta^2(s) + \frac{J_s}{J_x} H_M \, \beta(s)}.$$

$$\sigma_x' = \sigma_\varepsilon \sqrt{\eta'^2(s) + \frac{J_s}{J_x} H_M \, \gamma(s)}. \tag{2.93}$$

Let us consider now the vertical motion. Plane machines do not give rise to a vertical off energy orbit, and the only effect generating noise is related to the emission angle (Eq. (2.76)). Then we can expect that the vertical dimension will be $\sim (m_0c^2/E_0)$ times smaller than the radial one. E.g., for $E_0 \sim 500$ MeV, we have a factor $\sim 10^{-3}$.

However, perfectly plane orbits do not exist. The guide field imperfections give always coupling between radial and vertical motion.

In general, the radial and vertical emittance may be written in the form (see Ref. 1, Sect. 5.6)

$$A_x = \frac{1}{1+\delta} \cdot A_0, \quad A_y = \frac{\delta}{1+\delta} \cdot A_0, \tag{2.94}$$

where A_0 is the radial emittance without coupling (Eq. (2.88)) and δ is the "coupling coefficient", which, in principle, is a number in the range

$$0 \leqslant \delta \leqslant 1. \tag{2.95}$$

However the minimum δ (experimental) is of the order of $\sim 10^{-2}$, which corresponds to a beam height of the order of $\sim 10^{-1}$ of the beam width (and not $\sim 10^{-3}$). The upper limit ($\delta = 1$) may be achieved using skewed q-poles or operating the machine near to the coupling resonance (see Ref. 5, p. 91)

$$\nu_x = \nu_y. \tag{2.96}$$

In this condition we have $\delta \sim 1$ and

$$A_x = A_y = A_0/2. \tag{2.97}$$

When WM are operating in a SR, we cannot average H(s) and $\Delta\varepsilon^2$ separately in Eq. (2.80). As a consequence the relationship between the emittance and the energy spread is no longer as simple as in Eq. (2.88). The general equations, defining σ_ε and A for an arbitrary magnetic structure in a SR, are reported in Ref. 14.

To give a numerical example, in Table V we present the emittance and the transverse dimensions for full coupling (eq. (2.97)) in the FEL region, for LEDA-F1,2. It is interesting to note the strong reduction of the emittance due to the insertion of a high field WM in a region where the off-energy orbit is minimum, due merely to the dispersive properties of the WM itself (see App.B).

3. FREE ELECTRON LASER AND ELECTRON BEAM PARAMETERS

In these lectures use will be made of the single particle FEL theory developed in Ref. 15 and summarized in Ref. 16.

The following Sects 3.1 and 3.2 are devoted to a very short review of the main features of the spontaneous and stimulated (i.e. FEL) synchrotron radiation emission in a WM.

3.1 Spontaneous Synchrotron Radiation Emission in a Wiggler Magnet

A detailed study of the synchrotron emission in a WM can be found in Ref. 8. In this section we describe in a pictorial, but substantially correct, way, the properties of the "zero angle emission" which are relevant to the FEL process.

Let us consider an electron passing throughout a linearly polarized WM, whose wavelength is λ_q (see Eq. (2.1)). It can be shown[17] that the electron "sees" this field as a running wave of a radiation field, traveling oppositely to its motion with a wavelength given by

$$\lambda_{PRF} = 2\lambda_q . \tag{3.1}$$

TABLE V

	$A_x = A_y$	σ_x	σ_y
	mm × mrad	mm	mm
LEDA-F1	2.2	2.3	2.5
LEDA-F2	0.4	1.1	1.5

We use the label PRF to indicate that the effective field is not a true radiation field but a "Pseudo Radiation Field". This approximation is valid only for ultrarelativistic electrons, within a small region around the wiggler axis, because the magnetic field (2.1), with B = const, does not satisfy the Maxwell equations. Indeed, a field varying along one direction (e.g.,s) must vary also in the normal plane (e.g., x,y), (see App. A). In the reference frame where the longitudinal electron velocity is zero, the Doppler shifted wavelength of the ultrarelativistic electrons is

$$\lambda'_{PRF} = \lambda_{PRF}/(2\gamma^{\star}). \tag{3.2}$$

We denote with a prime all quantities evaluated in the electron frame. In Equation (3.2) $\gamma^{\star}$ is given by

$$\gamma^{\star} = 1/\sqrt{1 - (v_s/c)^2}, \tag{3.3}$$

where v_s is the electron longitudinal velocity in the lab frame. The spatial wavelength of the static magnetic field, λ_q, is of the order of several centimeters. In the electrons' frame, with $\gamma \sim 100$, the effective wavelength of the PRF is of the order of several tenths of millimeters. Therefore, in that frame, Thomson scattering of the PRF occurs. Let us focus our attention on the backscattered radiation.

In the lab frame the field scattered in this direction has a wavelength given by

$$\lambda = \frac{\lambda'_{PRF}}{2\gamma^{\star}} = \frac{\lambda_q}{2\gamma^{\star 2}} . \tag{3.4}$$

Therefore, with $\lambda_q \sim 20$ cm, $\gamma \sim 100$ to 1000, we obtain radiation with a wavelength

$\lambda \sim 10$ μm to 0.1 μm.

Let us look at the properties of the scattered radiation. Its bandwidth is limited by two factors. There is a "homogeneous" bandwidth, due to the finite length of the wiggler, that is to the finite flight time (interaction time) of the electron through the wiggler. This type of broadening is called homogeneous because it affects all the scattering processes, independently of the particles' velocity and position. The bandwidth is given by[8]

$$\left(\frac{\Delta\omega}{\omega}\right)_o = \frac{\lambda_q}{2L_w} , \qquad (3.5)$$

where L_w is the wiggler length.

There is also an inhomogeneous broadening of the scattered radiation. This comes from the distribution of the velocities of the scattering particles and their displacement from the wiggler axis. Indeed, the scattered radiation depends on γ^*, and therefore its spectrum reflects the distribution of the velocities. Moreover, the magnetic field B_W experienced by the electrons depends on their transverse displacements, as we have noted earlier. A change in the magnetic field amplitude effects the longitudinal motion of the particles (i.e. the value of γ^*).

A detailed study of the inhomogeneous broadening can be found, for example, in Ref. 18 and 19, from which we find that γ^* can be written in the form (for a linearly polarized WM).

$$\frac{1}{\gamma^*} = \frac{1}{\gamma} (1 + K^2 + (\gamma \cdot \delta\Theta)^2)^{1/2}, \qquad (3.6)$$

where we have put

$$\gamma = \frac{E}{m_o c^2} , \qquad (3.7)$$

$$K = \frac{e B_w \lambda_q}{2\pi\sqrt{2}\, m_o c^2} \qquad (3.8)$$

$$\delta\Theta^2 = \left(\frac{dx}{ds}\right)^2 + \left(\frac{dy}{ds}\right)^2 = \text{mean angle between the electron motion direction and the WM axis.} \qquad (3.9)$$

The wavelength of the back-scattered radiation is then given by (see Eqs (3.4) and (3.6))

$$\lambda = \frac{\lambda_q}{2\gamma^2} (1 + K^2 + (\gamma \cdot \delta\Theta)^2). \qquad (3.10)$$

From Eq. (3.10) we derive the frequency shifts due to the energy, magnetic field and angular deviation dependence.

We have

$$\left(\frac{\delta\omega}{\omega}\right)_\gamma = 2\,\frac{\delta\gamma}{\gamma} = 2\,\varepsilon \quad \left(\varepsilon = \frac{\delta E}{E}\right), \tag{3.11}$$

$$\left(\frac{\delta\omega}{\omega}\right)_B = -\frac{2K^2}{1+K^2}\cdot\frac{\delta B}{B}, \tag{3.12}$$

$$\left(\frac{\delta\omega}{\omega}\right)_\Theta = -\frac{1}{1+K^2}\,\gamma^2(\delta\Theta)^2. \tag{3.13}$$

In App. A (Eq. (A.6)) $\delta B/B$ is derived as a function of the vertical displacement y, yielding

$$\frac{\delta B}{B} = 2\left(\frac{\pi}{\lambda_q}\right)^2 y^2. \tag{3.14}$$

The total inhomogeneous detuning reads ($x' = dx/ds$, etc.),

$$\frac{\delta\omega}{\omega} = 2\varepsilon - \frac{4K^2}{1+K^2}\left(\frac{\pi}{\lambda_q}\right)^2 y^2 - \frac{\gamma^2}{1+K^2}\left(x'^2 + y'^2\right), \tag{3.15}$$

and the inhomogeneous broadening is given by

$$\left(\frac{\delta\omega}{\omega}\right)_i^2 = \left\langle\left(\frac{\delta\omega}{\omega}\right)^2\right\rangle - \left\langle\frac{\delta\omega}{\omega}\right\rangle^2, \tag{3.16}$$

where the average < > is performed over the e-beam distribution function, which may be written in the form (see Eqs (2.68) and (2.87)),

$$f(\varepsilon,x,x',y,y') = \frac{1}{\sqrt{2\pi}\,A_x A_y \sigma_\varepsilon}\exp\left\{-\frac{\varepsilon^2}{2\sigma_\varepsilon^2} - \pi\left(\frac{W_x}{A_x} + \frac{W_y}{A_y}\right)\right\}, \tag{3.17}$$

where σ_ε is the energy spread, $A_{x,y}$ the emittances and $W_{x,y}$ the Courant-Snyder invariants of the beam (Eq. (2.15)).

From the Eq. (3.15) we see that the detuning is affected by the divergence $x'^2 + y'^2$ and by the vertical beam dimension y. For a fixed vertical emittance, if we reduce one parameter we increase the other one. As a consequence, if we went to minimize the detuning we must optimize the vertical e-beam focusing parameters, and minimize the radial divergence x', which is obtained by requiring very large β in the WM (see Eqs (2.17) and (2.20)). Obviously we have a limit on the maximum β, namely the e.b. radial dimension must not exceed the WM radial one.

Furthermore, we need the e-beam transport channel to be achromatic. In a SR we obtain this result by matching the spatial insertion to the zero off-energy closed orbit ($\eta = 0$, $\eta' = 0$) as done in LEDA-F1,2 machines (see Fig. 4). In this condition x and y are just the free betatron coordinates.

From Equations (3.15), (3.16) and (3.17) we obtain

$$\left(\frac{\delta\omega}{\omega}\right)_i^2 = 4\sigma_\varepsilon^2 + \frac{1}{2\pi^2}\left(\left(A_x\gamma_x G\right)^2 + A_y^2\left(4\left(F\beta_y\right)^2 + 4FG\alpha_y^2 + \left(G\gamma_y\right)^2\right)\right) \tag{3.18}$$

where $\alpha_{x,y}$, $\beta_{x,y}$ and $\gamma_{x,y}$ are the Twiss coefficients (see Eqs (2.16) and (2.17)) and F and G are given by

$$F = \frac{2K^2}{1+K^2}\left(\frac{\pi}{\lambda_q}\right)^2, \quad G = \frac{1}{1+K^2}\left(\frac{E}{m_oc^2}\right)^2. \tag{3.19}$$

The minimum of the inhomogeneous bandwidth due to the vertical emittance is obtained for the following values of the Twiss coefficients (as can be easily verified by differentiating Eq. (3.18))

$$\begin{cases} \alpha_y^\star = 0 \\ \beta_y^\star = \dfrac{\lambda_q}{2\pi K} \cdot \dfrac{E}{m_o c^2} = \dfrac{\sqrt{2}\,E}{e\,B_w} \\ \gamma_y^\star = 1/\beta_y^\star . \end{cases} \tag{3.20}$$

There is a very lucky occurrence in Eq. (3.20). We note that, in general, the Twiss coefficients depend on the SR azimuth s (see Sect. 2.2). However, if we enter the WM with α_y, β_y and γ_y defined by Eqs (3.20), they remain constant (see App. A, Eq. (A.13)). As a consequence $(\delta\omega/\omega)_i$ is minimized throughout all the WM. As previously pointed out, in order to minimize $(\delta\omega/\omega)_i$ with respect to the radial motion, we must choice γ_x as small as possible. However it is sufficient to require that this broadening will be negligible with respect to the vertical one. In this connection we have (see Eqs (3.18) and (3.20))

$$\gamma_x^\star \ll \frac{\sqrt{2}}{\beta_y^\star}\left(\frac{A_y}{A_x}\right). \tag{3.21}$$

Note that γ_x is a constant throughout the WM (see App. A, Eq. (A.14)), thus Eq. (3.21) applies everywhere in the WM. For plane SR Eq. (3.21) reads

$$\gamma_x^\star \ll \frac{\sqrt{2}}{\beta_y^\star}\,\delta \tag{3.22}$$

where δ is the "coupling coefficient" (Eq. (2.94)).

Under this condition, Eq. (3.18) yields

$$\left(\frac{\delta\omega}{\omega}\right)_i^2 = 4\sigma_\varepsilon^2 + \left[\frac{2A_y}{\lambda_q}\left(\frac{E}{m_oc^2}\right)\frac{K}{1+K^2}\right]^2. \tag{3.23}$$

From Equation (3.23) we can derive the limitations on σ_ε and A_y required to have purely homogeneously broadened back-scattered radiation. Namely, we must require

$$\left(\frac{\Delta\omega}{\omega}\right)_i \ll \left(\frac{\Delta\omega}{\omega}\right)_o, \tag{3.24}$$

which implies

$$\sigma_\varepsilon \ll \sigma_\varepsilon^\star = \frac{1}{2}\left(\frac{\Delta\omega}{\omega}\right)_o = \frac{\lambda_q}{4L_w}, \tag{3.25}$$

$$A_y \ll A_y^{\star} = \frac{1 + K^2}{4K} \frac{\lambda_q^2}{L_w} \cdot \left(\frac{m_o c^2}{E}\right). \tag{3.26}$$

Conditions (3.25) and (3.26) are generally largely fulfilled for typical SR and WM parameters. As example, in Table VI we report the LEDA-F1,2 cases. In conclusion we can say that the spontaneous synchrotron radiation spectrum in WM is homogeneously broadened.

This situation changes when we are dealing with the stimulated process (the FEL one). Indeed we shall see that the FEL interaction modifies the "natural" energy spread and emittance of the e.b.. It is possibile to show that, with $\eta(s) = 0$ in the WM, the e.b. emittance is very weakly affected by the FEL interaction (see Ref. 19, App. D). As a consequence the condition (3.26) can be, in principle, always fulfilled. This is not true for the energy spread, whose increase with working energy generates an inhomogeneous broadening of the same order of the homogeneous one (see Sect. 4.9). For this reason in the following we shall consider only the effect of the energy spread.

3.2 Stimulated Synchrotron Radiation Emission in a Wiggler Magnet

In the previous section we have investigated some features of spontaneous backward Thomson scattering. Let us consider now the situation in which one or more of the modes of the electromagnetic (e.m.) field is occupied by a large number n of photons. In this condition stimulated scattering in these modes occurs, consisting of the scattering cross section enhanced by a factor n. The variations of the energy densities and phases of the e.m. modes in the "small signal regime" (Ref. 22) and with homogeneous broadening due to the e.b. energy spread (see previous section), have been derived in Ref. 16 (Eqs (A.6)), with the following result:

TABLE VI

	σ_ε	$\sigma_\varepsilon^{\star}$	A_y (mm × mrad)	$A_y^{\star}(K = 1)$ (mm × mrad)
LEDA-F1	4×10^{-4}	10^{-2}	2.2	6
LEDA-F2	8×10^{-4}	10^{-2}	0.4	6

$$\begin{cases} \Delta W_i = -A \sum_r \left(W_i W_r\right)^{1/2} \left(B^c_{i,r} \cos\left(\varphi_i - \varphi_r\right) - B^s_{i,r} \sin\left(\varphi_i - \varphi_r\right)\right) \\ \Delta\varphi_i = \omega_i \Delta t + \frac{A}{2} \sum_r \left(\frac{W_r}{W_i}\right)^{1/2} \left(B^s_{i,r} \cos\left(\varphi_i - \varphi_r\right) + B^c_{i,r} \sin\left(\varphi_i - \varphi_r\right)\right) \end{cases}, \tag{3.27}$$

where

W_i ≡ i-th e.m. mode energy density

φ_i ≡ i-th e.m. mode phase

ω_i ≡ i-th e.m. mode frequency

Δt ≡ interaction time = L_w/c

$$A \equiv 4\pi^2 r_o \sqrt{\frac{2\lambda_o}{\lambda_q}}\, L_w \lambda_o n_e \frac{K^2}{(1+K^2)^{3/2}} \left(\frac{\Delta\omega}{\omega}\right)_o^{-2}$$

$$\lambda_o = \frac{2\pi c}{\omega_o} \equiv \frac{\lambda_q}{2}\left(1 + K^2\right)\left(\frac{m_o c^2}{E_o}\right)^2$$

L_w, λ_q ≡ wiggler length and wavelength

n_e ≡ electron density (3.28)

$(\Delta\omega/\omega)_o$ and K are defined by Eqs (3.5) and (3.8) respectively and, finally ($j = \sqrt{-1}$),

$$B^c_{i,r} = \mathrm{Re}\, B_{i,r},\quad B^s_{i,r} = -\mathrm{Im}\, B_{i,r}, \tag{3.29}$$

$$B_{i,r} = \int f(z,\varepsilon)\, dz\, d\varepsilon\, E\left[\eta_i(\varepsilon), \eta_r(\varepsilon)\right] \exp\left(+j\left(k_i - k_r\right)z\right), \tag{3.30}$$

where $f(z,\varepsilon)$ is the synchrotron e.b. distribution introduced in Sects 2.4 and 2.6, and

$$k_i = \frac{\omega_i}{c} = \text{i-th e.m. mode wavenumber},$$

$$\eta_i(\varepsilon) = \pi\left(\frac{\Delta\omega}{\omega}\right)_o^{-1}\left\{2\varepsilon - \frac{\omega_i - \omega_o}{\omega_o}\right\}, \tag{3.31}$$

$$E\left(\eta_i,\eta_r\right) = \exp\left(j\eta_i\right)\left\{\frac{1}{\eta_i - \eta_r}\left(\frac{1 - \exp(-j\eta_r)}{\eta_r^2} - \frac{1 - \exp(-j\eta_i)}{\eta_i^2}\right) - \frac{j}{\eta_i\eta_r}\right\}. \tag{3.32}$$

A detailed discussion of the physical meaning and the validity range of Eqs (3.27) is reported in Ref. 16.

4. FREE ELECTRON LASER OPERATION IN A STORAGE RING (INTRODUCTION)

The analysis of FEL operation in a SR has been carried out in Ref. 20 for an amplifier and in Refs. 19, 21 and 22 for an oscillator. These papers concern the operation of a "constant parameters" and "zero gradient" WM and a "standard" SR. New schemas have been proposed in more recent years in order to improve the FEL performances in SR[23]. In this paper we shall not deal with this topic (other lectures in this school are devoted to this subject).

We shall follow, essentially, Refs. 20 and 22, where a FOKKER-PLANCK formalism has been utilized for studying the e.b. evolution (a "Monte Carlo" approach for computing FEL operation in SR has been developed in Ref. 24).

Before describing the details of our investigation of SR FEL operation, let us make some qualitative observations.

The main experimental problems connected with the practical realization of FEL lie with the high quality required of the e.b., i.e. high peak current, low energy spread and emittance. The characteristics of an e.b. in a SR are generally very good, as they affect FEL operation, (as pointed out in Sect. 3.1). Unfortunately in such machines the e.b. is continuously recirculated throughout the FEL interaction region. As a consequence the energy spread and emittance increase and, in turn, the FEL amplification decreases. By properly choosing the machine parameters (see Sect. 3.1) the emittance is only very weakly affected by the FEL interaction and it remains practically unchanged. On the other hand, we suffer a large enhancement of the energy spread, which is counteracted only by the "damping mechanism" due to the synchrotron radiation emission (see Sect. 2.5). The stationary regime is reached when the FEL amplification, lowered by the large e.b. energy spread, becomes equal to the cavity losses, and the "FEL noise" is balanced by radiation damping.

In Section 4.1 we shall describe the operating conditions of the SR FEL. Section 4.2 is devoted to the study of the e.b. longitudinal motion. The FEL amplifier mode of operation is investigated in Sect. 4.3. Finally, in Sects 4.4 - 4.9 the FEL oscillator regime is considered.

4.1 Operating Conditions

The schematic machine layout is sketched in Fig. 7.

In Part I of Fig. 7 is depicted the FEL region (i.e. wiggler magnet and optical cavity), in Part II a conventional circular electron machine (SR).

In dealing with the electron dynamics we begin by neglecting the transverse motion, which follows the usual betatron equations, and concern ourselves only with the longitudinal motion. The approximation just described relies on the following two conditions:

(i) inhomogeneous frequency broadening, due to the transverse emittance, must be negligible with respect to the homogeneous broadening (see Sect. 3.1),

(ii) laser cross section greated than the e.b. cross section.

The longitudinal electron motion will be described by means of the "synchrotron variables" (z,ε), introduced in Sects 2.3 and 2.4. The laser field is described in terms of the energy densities W_i and phases φ_i of the longitudinal modes of the optical cavity. In

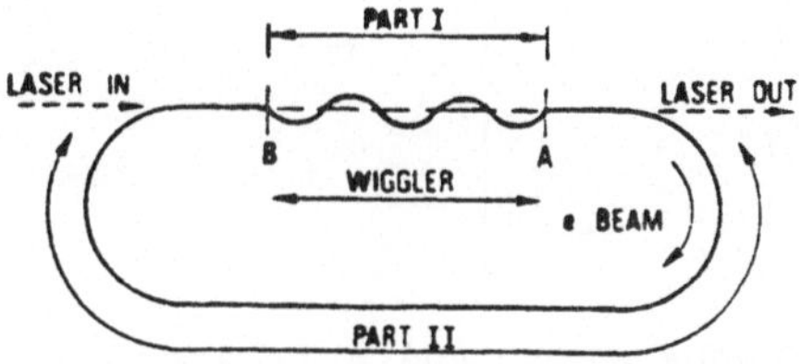

Fig. 7. Machine layout

the absence of an interaction of the fields with the electrons, the time history of W_i and φ_i obeys the equations

$$\begin{cases} \dfrac{dW_i}{dt} = 0 \\ \dfrac{d\varphi_i}{dt} = \omega_i, \end{cases} \tag{4.1}$$

where ω_i is the i-th mode frequency :

$$\omega_i = \frac{2\pi c}{L_c} \cdot n_i, \tag{4.2}$$

where n_i is an integer, c is the velocity of light and L_c is the optical cavity length.

We assume the WM and the laser field to be circularly polarized. This choice is different from that adopted in the previous sections. We prefer to work with this assumption because the formalism is simplified; however, the results we obtain are fully valid for the linearly polarized case too.

The laser vector potential, written in terms of W_i and φ_i is given by

$$\begin{cases} A_x = \sqrt{4\pi} \sum_i \dfrac{1}{k_i} \sqrt{W_i} \sin(\varphi_i - k_i s) \\ A_y = \sqrt{4\pi} \sum_i \dfrac{1}{k_i} \sqrt{W_i} \cos(\varphi_i - k_i s) \\ A_z = 0. \end{cases} \tag{4.3}$$

where x and y are the transverse coordinates, s is the longitudinal coordinate and k_i ($\equiv \omega_i/c$) is the i-th mode wavenumber.

We can derive from (4.3) the intracavity power density:

$$\frac{dP_L}{dS} = c\left|\sum_i \sqrt{W_i} \exp\left(j\left(\varphi_i - k_i s\right)\right)\right|^2 . \tag{4.4}$$

The average laser frequency ω_L and the laser bandwidth $(\Delta\omega/\omega)_L$ are

$$\omega_L = \sum_i \omega_i W_i / \sum_i W_i \tag{4.5}$$

and

$$\left(\frac{\Delta\omega}{\omega}\right)_L = \left\{\sum_i \left(\frac{\omega_i - \omega_L}{\omega_L}\right)^2 W_i / \sum_i W_i\right\}^{1/2} . \tag{4.6}$$

We assume that we are working in the "small signal regime"[16] (see App. C).

4.2 Equations of Motion for Electrons

The synchrotron recurrence equations of motion, as perturbed by the FEL interaction, are (see Eqs (2.62))

$$\begin{cases} \Delta z_j = -\left(c\alpha_c \, \varepsilon_{j+1}\right) \cdot T \\ \Delta\varepsilon_j = \left(\dfrac{\omega_s^2}{c\alpha_c} z_j - \dfrac{2}{\tau_s}\varepsilon_j\right) \cdot T + \dfrac{u}{E_o} + \delta\varepsilon, \end{cases} \tag{4.7}$$

where $\delta\varepsilon$ is the relative energy change due to the FEL interaction.

Let us investigate the features of this perturbation. In Ref. 16 (Sect. 4) it was shown that $\delta\varepsilon$ is a rapidly varying function of the phase differences between the laser modes and the wiggler at the time the electron enters the FEL region. For simplicity let us consider only one laser mode in the cavity. If we call φ_o the phase difference relative to the synchronous particle ($z = 0$, $\varepsilon = 0$), for an electron displaced by z we have

$$\varphi = \varphi_o - 2\pi \frac{z}{\lambda_o} \quad (\text{mod } 2\pi) \tag{4.8}$$

where λ_o is the laser wavelength. If the average variation of z in one turn is very large with respect to λ_o, the phase φ jumps (turn by turn) discontinuously in the interval $[0, 2\pi]$ and behaves like a random variable. In Appendix B of the first of Ref. 20 this problem was investigated and a "stochasticity condition" was derived. Under this condition we can assume that φ is really a "random phase" and, as a consequence, $\delta\varepsilon$ may be regarded as an additional noise perturbing the synchrotron motion.

In this connection the electron distribution function describing the longitudinal motion may be derived from a FOKKER-PLANCK equation in which the diffusion mechanism arises from the radiation noise and the FEL interaction. In particular the e.b. equilibrium distribution function $f(z,\varepsilon)$ $(\partial f/\partial t = 0)$ is a solution of the following FPE (see Eqs (2.65) and (2.66))

$$\omega_s\left(\left(\frac{c\alpha_c}{\omega_s}\right)\left(\varepsilon\,\frac{\partial f}{\partial z}\right) - \left(\frac{\omega_s}{c\alpha_c}\right)\left(z\,\frac{\partial f}{\partial \varepsilon}\right)\right) +$$

$$+ \frac{1}{T}\frac{\partial}{\partial \varepsilon}\left(\left(\frac{2\varepsilon}{\tau_s} - \frac{1}{T}\left(<\delta\varepsilon> - \frac{1}{2}\frac{\partial}{\partial\varepsilon}<\delta\varepsilon^2>\right)\right)f + \left(D + \frac{1}{2}\frac{<\delta\varepsilon^2>}{T}\right)\frac{\partial f}{\partial\varepsilon}\right), \tag{4.9}$$

where $<\delta\varepsilon>$ and $<\delta\varepsilon^2>$ are averaged over the "fast running phase" (4.8) and are given by (see Ref. 16 Eqs (A.16) and (A.17))

$$<\delta\varepsilon^2> = \left(\frac{2\pi r_o}{m_o c^2}\right)\lambda_o^2\left(\frac{\Delta\omega}{\omega}\right)_o^{-2}\frac{K^2}{(1+K^2)^2}\,\cdot$$

$$\cdot\,\left|\sum_i\left(W_i\right)^{1/2}\exp\left\{j\left(k_i z - \varphi_i + \frac{\eta_i}{2}\right)\right\}\frac{\sin(\eta_i/2)}{(\eta_i/2)}\right|^2, \tag{4.10}$$

$$<\delta\varepsilon> = \frac{1}{2}\frac{\partial}{\partial\varepsilon}<\delta\varepsilon^2>. \tag{4.11}$$

For the meaning of the symbols, see Sect. 3.2. The only difference required involves K (Eq. (3.8)), which for a circularly polarized WM is given by

$$K = \frac{e\,B_w\,\lambda_q}{2\pi\,m_o c^2}\,.$$

From Equation (4.9) we find that the FEL interaction generates the following terms

$$\frac{1}{2}\frac{<\delta\varepsilon^2>}{T}, \tag{4.12}$$

$$<\delta\varepsilon> - \frac{1}{2}\frac{\partial}{\partial\varepsilon}<\delta\varepsilon^2>, \tag{4.13}$$

which can be interpreted as a diffusion and damping term respectively.

In the small signal approximation, however, (4.13) is identically zero (see Eq. (4.11)); then the diffusion term (4.12) is counteracted only by the usual damping generated by the synchrotron emission in the bending magnets.

Eventually we can insert in the machine a special high field wiggler in order to enhance the synchrotron emission and limit the energy spread, as done in the LEDA-F2 project[10].

In this case Eq. (4.9) reads

$$\omega_s \left[\left(\frac{c\alpha_c}{\omega_s}\right)\left(\varepsilon \frac{\partial f}{\partial z}\right) - \left(\frac{\omega_s}{c\alpha_c}\right)\left(z \frac{\partial f}{\partial \varepsilon}\right) \right] +$$

$$+ \frac{1}{T}\frac{\partial}{\partial \varepsilon}\left[2\frac{T}{\tau}(\varepsilon\, f) + \left(DT + \left(\frac{<\delta\varepsilon^2>}{2}\right)\right)\frac{\partial f}{\partial t}\right] = 0. \tag{4.14}$$

In order to simplify the notation let us introduce the following "normalized variables", which relate respectively, to the electrons and to the i-th laser mode frequency:

A) Electron variables

$$\begin{cases} \tilde{\varepsilon} = 2\left(\dfrac{\Delta\omega}{\omega}\right)_o^{-1} \varepsilon \\ \tilde{z} = 2\left(\dfrac{\Delta\omega}{\omega}\right)_o^{-1}\left(\dfrac{\omega_s}{c\alpha_c}\right) z, \end{cases} \tag{4.15}$$

B) i-th laser mode frequency

$$\nu_i = \pi\left(\frac{\Delta\omega}{\omega}\right)_o^{-1} \cdot \frac{(\omega_o - \omega_i)}{\omega_o} . \tag{4.16}$$

By means of the above new variables, η_i (see Eq. (3.31)) can be rewritten as

$$\eta_i = \pi\tilde{\varepsilon} + \nu_i . \tag{4.17}$$

Using these new variables the FPE (4.14) becomes

$$\tilde{\varepsilon}\,\frac{\partial f(\tilde{z},\tilde{\varepsilon})}{\partial \tilde{z}} - \tilde{z}\,\frac{\partial f(\tilde{z},\tilde{\varepsilon})}{\partial \tilde{\varepsilon}} + \frac{2}{\omega_s \tau_s}$$

$$\frac{\partial}{\partial \tilde{\varepsilon}}\left\{\tilde{\varepsilon} f(\tilde{z},\tilde{\varepsilon}) + \left(\tilde{\sigma}_o^2 + G(\tilde{z},\tilde{\varepsilon})\right)\frac{\partial f(\tilde{z},\tilde{\varepsilon})}{\partial \tilde{\varepsilon}}\right\} = 0, \tag{4.18}$$

where we have taken

$$\tilde{\sigma}_o = 2\left(\frac{\Delta\omega}{\omega}\right)_o^{-1} \sigma_\varepsilon^o, \tag{4.19}$$

σ_ε^o accounts for the e.b. energy spread due to the synchrotron radiation noise (natural energy spread), as given by Eq. (2.69). It is interesting to note that $\tilde{\sigma}_o$ is just the ratio between the inhomogeneous linewidth due to the energy spread and the homogeneous linewidth (see Sect. 3.1). Furthermore G $(\tilde{z},\tilde{\varepsilon})$, which is proportional to $\langle\delta\varepsilon^2\rangle$, may be expressed

$$G(\tilde{z},\tilde{\varepsilon}) = G_o\left|\sum_i \left(W_i\right)^{1/2} \exp\left\{j\left[\nu_i\left(\frac{\tilde{z}}{\beta} - \frac{1}{2}\right) + \varphi_i\right]\right\}\frac{\sin \eta_i/2}{\eta_i/2}\right|^2, \tag{4.20}$$

with

$$G_o = \frac{\tau_s}{T}\,\frac{2\pi r_o}{m_o c^2}\cdot \lambda_o^2 \left(\frac{\Delta\omega}{\omega}\right)_o^{-4} \cdot \frac{K^2}{(1+K^2)^2}, \tag{4.21}$$

and

$$\beta = \left(\frac{\Delta\omega}{\omega}\right)_o^{-2} \lambda_o \left(\frac{\omega_s}{c\alpha_c}\right). \tag{4.22}$$

(Do not confuse this β with the Twiss coefficient introduced in Sect. 2.2).

We can easily integrate Eq. (4.18) if the damping time τ_s is very large with respect to the synchrotron period, i.e. (see Eq. (2.51)),

$$\tau_s \gg \frac{2\pi}{\omega_s}. \tag{4.23}$$

Under this condition it can be shown (see Ref. 3, Chapter XXVI) that f is simply a function of the "synchrotron energy" defined by

$$H = \frac{1}{2}\left(\tilde{\varepsilon}^2 + \tilde{z}^2\right); \tag{4.24}$$

i.e., we have

$$f(\tilde{z},\tilde{\varepsilon}) = h(H), \tag{4.25}$$

where h satisfies

$$\frac{d}{dH}\left\{H\left[h(H) + \left(\tilde{\sigma}_o^2 + R(H)\right)\frac{dh(H)}{dH}\right]\right\} = 0, \tag{4.26}$$

with

$$R(H) = \frac{2}{\pi}\int_0^{\pi}\cos^2\vartheta\, G\left(\sqrt{2H}\sin\vartheta,\ \sqrt{2H}\cos\vartheta\right)d\vartheta. \tag{4.27}$$

In deriving Eq. (4.26) use has been made of the following polar transformation

$$\begin{cases}\tilde{z} = \sqrt{2H}\sin\vartheta \\ \tilde{\varepsilon} = \sqrt{2H}\cos\vartheta,\end{cases} \tag{4.28}$$

and the equation has been averaged over the phase ϑ.

It is straightforward to integrate Eq. (4.26).

$$h(H) = h_o \exp\left(-\int_0^H \frac{dH'}{\tilde{\sigma}_o^2 + R(H')}\right), \tag{4.29}$$

$$h_o = \left\{2\pi\int_0^{\infty} dH \exp\left[-\int_0^H \frac{dH'}{\tilde{\sigma}_o^2 + R(H')}\right]\right\}^{-1} \tag{4.30}$$

The distribution (4.29) describes the stationary state of the e.b. interacting with a multimode laser and wiggler fields. We can distinguish two different regimes, which will be investigated in the following sections:

(i) Amplifier regime: only one e.m. mode (generated by an external laser) is present (Sect. 4.3)

(ii) Oscillator regime: many e.m. modes are lasing, whose energy densities and phases must be evaluated, in a self-consistent way, by coupling Eq. (4.29) with Eqs (3.27) (Sects 4.4 - 4.9).

4.3 The FEL Amplifier Regime in a Storage Ring

If only one e.m. mode, generated by an external laser, is present, Eq. (4.20) becomes

$$G(\tilde{z},\tilde{\varepsilon}) = \tilde{W}\left(\frac{\sin\left[\frac{1}{2}(\pi\tilde{\varepsilon} + \nu)\right]}{\frac{1}{2}\,(\pi\tilde{\varepsilon} + \nu)}\right)^2, \tag{4.31}$$

where we have defined the dimensionless laser power density

$$\tilde{W} = G_o \cdot W. \tag{4.32}$$

By inserting Eq. (4.31) in Eqs (4.27) and (4.29), we obtain the e.b. stationary distribution as a function of the following dimensionless quantities

$\tilde{\sigma}_o$ = normalized natural e.b. energy spread (Eq. (4.19))
ν = normalized laser frequency (Eq. (4.16))
$\tilde{W}$ = normalized laser power density (Eq. (4.32))

The variances of $\tilde{z}$ and $\tilde{\varepsilon}$ can be easily derived from Eqs (4.25) and (4.28):

$$\tilde{\sigma}^2\left(\tilde{W},\ \tilde{\sigma}_o,\ \nu\right) = \langle\tilde{z}^2\rangle = \langle\tilde{\varepsilon}^2\rangle = 2\pi\int_0^\infty Hh\left(H|\tilde{W},\ \tilde{\sigma}_o,\ \nu\right)dH. \tag{4.33}$$

In Figure 8 $\tilde{\sigma}$ is plotted vs. $\tilde{W}$ in the limit case $\tilde{\sigma}_o = 0$, i.e. for negligible synchrotron radiation noise. For large $\tilde{W}$, $\tilde{\sigma}$ has the following behaviour:

$$\tilde{\sigma}(\tilde{W},\ 0,\ \nu) \propto \tilde{W}^{1/4} \qquad (W \gg 1). \tag{4.34}$$

For $\tilde{\sigma}_o \neq 0$, $\tilde{\sigma}$ starts from the unperturbed value $\tilde{\sigma}_o(\tilde{W} = 0)$ and, as $\tilde{W}$ increases, approaches asymptotically the $\tilde{\sigma}_o = 0$ case, i.e.

$$\tilde{\sigma}(\tilde{W},\ \tilde{\sigma}_o,\ \nu) \xrightarrow[\tilde{W}\to\infty]{} \sigma(\tilde{W},\ 0,\ \nu). \tag{4.35}$$

As example we have plotted $\tilde{\sigma}$ vs. $\tilde{W}$ in Fig. 9 for $\nu = 2.51$ (which corresponds, roughly, to the maximum of the small signal gain for a monoenergetic beam (see Fig. 2 curve a in Ref. 16), and $\tilde{\sigma}_o = 0,\ 0.5,\ 1$.

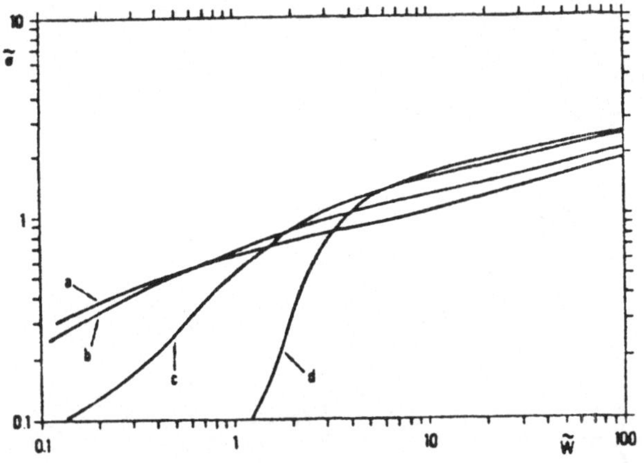

Fig. 8. $\tilde{\sigma}$ vs. $\tilde{W}$ for $\tilde{\sigma}_o = 0$:
a) $\nu = 0$, b) $\nu = 2.51$, c) $\nu = 5.03$, d) $\nu = 6.28$

In Figure 10 has been plotted the e.b. distribution function normalized to $\tilde{\sigma}^2$, in the case $\tilde{\sigma}_o = 0.5$, $\nu = 2.5$ and $\tilde{W} = 0, 1, 5, 10$. For $\tilde{W} = 0$ we have the unperturbed distribution,

$$h\left(H \,\middle|\, 0,\ \tilde{\sigma}_o,\ \nu\right) = \frac{1}{2\pi\tilde{\sigma}_o^2} \exp\left(-\frac{H}{\tilde{\sigma}_o^2}\right). \tag{4.36}$$

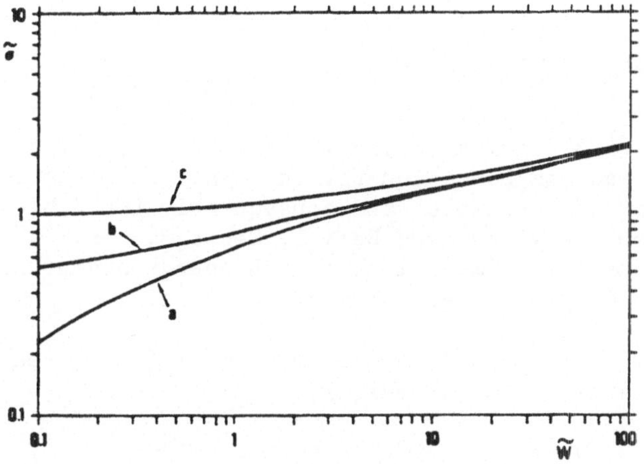

Fig. 9. $\tilde{\sigma}$ vs. $\tilde{W}$ for $\nu = 2.51$
a) $\tilde{\sigma}_o = 0$, b) $\tilde{\sigma}_o = 0.5$, c) $\tilde{\sigma}_o = 1.0$

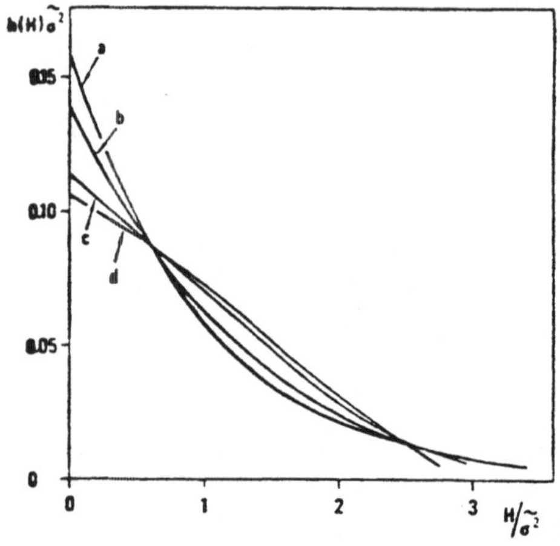

Fig. 10. e.b. distribution function, $\nu = 2.5$, $\tilde{\sigma}_0 = 0.5$
a) $\tilde{W} = 0$, b) $\tilde{W} = 1$, c) $\tilde{W} = 5$, d) $\tilde{W} = 10$.

Let us return to the old variables ε and z. The energy spread and the bunch length are given by (see Eqs (4.15))

$$\sqrt{\langle\varepsilon^2\rangle} = \frac{1}{2}\left(\frac{\Delta\omega}{\omega}\right)_o \tilde{\sigma},$$
$$\sqrt{\langle z^2\rangle} = \frac{1}{2}\left(\frac{\Delta\omega}{\omega}\right)_0 \frac{c\alpha_c}{\omega_s}\tilde{\sigma}. \tag{4.37}$$

Equations (4.37) set an upper limit to $(\Delta\omega/\omega)_0$. Indeed for high energy spread and bunch length, the synchrotron motion becomes highly nonlinear and unstable, and consequently the e.b. mean life in the SR becomes very low (see Ref. 1, Sects 3.6 and 5.8). In order to overcome this limitation, the SR energy acceptance (see Ref. 1, Sect. 3.6) must be as large as possible.

The energy variation of the laser beam is just given by the total energy change of the e-beam, i.e.

$$\Delta E = -N \cdot E_0 \int \langle\delta\varepsilon\rangle f(\varepsilon,z)\, d\varepsilon\, dz, \tag{4.38}$$

where N is the number of electrons in the bunch.

Equations (4.38) may be written in the form (see Eqs (4.11), (4.31) and (4.32))

$$\Delta E = - E_o N \frac{T}{\tau_s} \tilde{W}\left(\frac{\Delta\omega}{\omega}\right)_o \int d\tilde{\varepsilon}\, d\tilde{z}\, h\left(\frac{\tilde{z}^2 + \tilde{\varepsilon}^2}{2}\bigg|\tilde{W},\tilde{\sigma}_o,\nu\right)\frac{\partial}{\partial\tilde{\varepsilon}}\left[\frac{\sin\left(\frac{(\pi\tilde{\varepsilon}+\nu)}{2}\right)}{(\pi\tilde{\varepsilon}+\nu)/2}\right]^2 . \tag{4.39}$$

For a SR with "separate functions", the damping time is roughly given by (see Eq. (2.56))

$$\tau_s = \simeq T \frac{E_o}{U_o} . \tag{4.40}$$

Inserting Eq. (4.40) in Eq. (4.39) we obtain

$$\Delta \underline{E} = \left(\frac{\Delta\omega}{\omega}\right)_o U_T \cdot \chi\left(\tilde{W}, \tilde{\sigma}_o, \nu\right) \tag{4.41}$$

where we have defined

$$U_T = NU_o = \text{total synchrotron energy radiated per turn by the whole beam} \tag{4.42}$$

and

$$\chi\left(\tilde{W}, \tilde{\sigma}_o, \nu\right) = - W\int d\tilde{\varepsilon}\, d\tilde{z}\, h\left(\frac{\tilde{z}^2 + \tilde{\varepsilon}^2}{2}\bigg|\tilde{W},\tilde{\sigma}_o,\nu\right)\frac{\partial}{\partial\tilde{\varepsilon}}\left[\frac{\sin\left(\frac{(\pi\tilde{\varepsilon}+\nu)}{2}\right)}{(\pi\tilde{\varepsilon}+\nu)/2}\right]^2 . \tag{4.43}$$

In Equation (4.41) we find that the energy ΔE radiated by the e-beam into the laser mode is proportional to the total energy emitted in one turn via synchrotron radiation (U_T) and to the homogeneous linewidth $(\Delta\omega/\omega)_o$.

The dependences on the laser power density $\tilde{W}$, the e-beam width $\tilde{\sigma}_o$ and the laser frequency ν are all contained in the function χ.

The function χ is antisymmetric with respect to ν, i.e. (see Eq. (4.43))

$$\chi(\tilde{W}, \tilde{\sigma}_o, \nu) = - \chi(\tilde{W}, \tilde{\sigma}_o, - \nu). \tag{4.44}$$

In Figure 11 χ is plotted vs. $\tilde{W}$ in the case $\tilde{\sigma}_o = 0$. It is possible to distinguish two limiting regimes, namely
i) $\tilde{W} \ll 1$; in this case χ follows the small signal gain profile corresponding to a monoenergetic e.b. (see Ref. 16, Sect. 4)

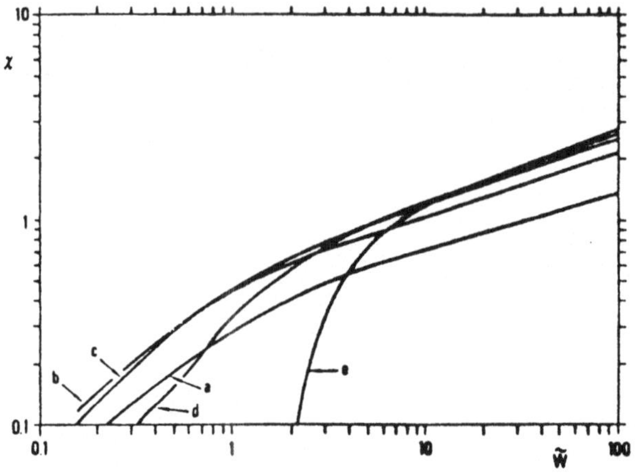

Fig. 11. χ vs. $\widetilde{W}$ for $\widetilde{\sigma}_0 = 0$:
a) $\nu = 1.26$, b) $\nu = 2.51$, c) $\nu = 3.77$, d) $\nu = 5.03$,
e) $\nu = 6.28$

$$\chi(\widetilde{W} \ll 1, 0, \nu) \propto \widetilde{W} \frac{\partial}{\partial \nu} \left(\frac{\sin \nu/2}{\nu/2}\right)^2 \tag{4.45}$$

ii) $\widetilde{W} \gg 1$; in this condition we have a large broadening of the e-beam energy distribution, which lowers the FEL gain. The dependence of χ on $\widetilde{W}$ is weaker than in the case i). We have, roughly,

$$\chi(\widetilde{W} \gg 1, 0, \nu) \propto \widetilde{W}^{1/3} . \tag{4.46}$$

The behaviour of χ for $\widetilde{\sigma}_0 \neq 0$ is similar to the $\widetilde{\sigma}_0 = 0$ case; there is merely a lowering of χ. However, for sufficiently large W, i.e. when we can neglect the synchrotron radiation noise with respect to the FEL one, χ approaches the $\widetilde{\sigma}_0 = 0$ case.

From Equation (4.41) we derive the "efficiency" of the FEL amplifier, defined as the ratio between the energy lost by the e-beam in the laser mode (ΔE) and the total energy radiated ($\Delta E + U_T$)

$$\eta = \frac{\Delta E}{\Delta E + U_T} = \frac{(\Delta\omega/\omega)_0\, \chi}{(\Delta\omega/\omega)_0 \chi + 1} . \tag{4.47}$$

In the range investigated ($W \lesssim 100$), the maximum χ is of the order of unity. As a consequence Eq. (4.47) may be written in the form (recall that $(\Delta\omega/\omega)_o \ll 1$)

$$\eta \sim \left(\frac{\Delta\omega}{\omega}\right)_o , \tag{4.48}$$

that is, the efficiency has the order of magnitude of the homogeneous linewidth.

4.4 FEL Oscillator in SR: Evolution Equations for the Laser Modes

A first analysis of a FEL oscillator in a SR has been carried out in Refs. 19 and 21 in the framework of a "single mode theory" and assuming "smooth varying laser envelope". However, the assumption of only one laser mode is unrealistic. We are aware that a consistent analysis requires a multimode approach. Indeed, the bunched structure of the e.b. in a SR gives rise to a strong coupling between the longitudinal modes of the optical cavity, which generates a kind of "mode locking" and, as a consequence, a bunching in the laser field itself.

In these lectures we shall follow, essentially, the theory developed in Ref. 22.

The energy density and phase variations of the i-th laser mode, during its time of passage between two adjacent electron bunches, can be easily derived from Eqs (3.27) (see also App. A in Ref. 16):

$$\begin{cases} \Delta W_i = -W_i\gamma_T - A\left(1-\gamma_T\right)\sum_r \left(W_i W_r\right)^{1/2}\left[B^c_{i,r}\cos\left(\varphi_i-\varphi_r\right) - B^s_{i,r}\sin\left(\varphi_i-\varphi_r\right)\right] \\ \Delta\varphi_i = -\omega_i\delta t + \frac{1}{2}A\sum_r \left(\frac{W_r}{W_i}\right)^{1/2}\left[B^s_{i,r}\cos\left(\varphi_i-\varphi_r\right) + B^c_{i,r}\sin\left(\varphi_i-\varphi_r\right)\right], \end{cases} \tag{4.49}$$

where we have taken

$$\delta t = T_c - T_e \tag{4.50}$$

T_e ≡ time distance between two adjacent electron bunches
T_c ≡ optical cavity round trip period
γ_T ≡ total optical cavity losses (during T_e).

The "gain coefficient" A, which is defined in Eq. (3.28), can be written in terms of the average electron current

$$I = \frac{Ne}{T_e} \qquad (N \equiv \text{number of electron per bunch}), \tag{4.51}$$

giving

$$A = 4\pi^2 \left(\frac{2\lambda_o}{\lambda_q}\right)^{1/2} \frac{L_w \lambda_o}{\Sigma_L} \left(\frac{I}{I_o}\right) \frac{K^2}{\left(1 + K^2\right)^{3/2}} \left(\frac{\Delta\omega}{\omega}\right)_o^{-2}, \tag{4.52}$$

where I_o is the Alfvén current, which is defined by the equation

$$I_o = \frac{ec}{r_o} .$$

From Equations (4.49) and (3.30) we can see that the strength of the coupling between two laser modes depends primarily on the bunch length σ_z. In fact, as it has been pointed out in Ref. 16 (Sect. 4), the coupling coefficient $B_{i,r}$ goes rapidly to zero once

$$\left|k_i - k_r\right| \gg \frac{1}{\sigma_z} . \tag{4.53}$$

The preceding statement relies on the fact that $B_{i,r}$ depends on $k_i - k_r$ via the Fourier transform of the longitudinal electron distribution (see Eq. (3.30)). For example, for a Gaussian beam whose r.m.s. length is σ_z, we have

$$\left|B_{i,r}\right| \propto \exp\left(-\frac{1}{2}\left(k_i - k_r\right)^2 \sigma_z^2\right)^2 . \tag{4.54}$$

Furthermore Equation (4.53) can be used to obtain a definition of "coupling width". In fact by defining such a "coupling width" as

$$\left(\frac{\Delta\omega}{\omega}\right)_c = \frac{\lambda_o}{2\sigma_z} , \tag{4.55}$$

we can distinguish two different operating regimes, depending on the magnitude of $(\Delta\omega/\omega)_c$ with respect to the emission line homogeneous width $(\Delta\omega/\omega)_o$:

A) Small coupling width regime

$$\left(\frac{\Delta\omega}{\omega}\right)_c \ll \left(\frac{\Delta\omega}{\omega}\right)_o, \tag{4.56}$$

B) Large coupling width regime

$$\left(\frac{\Delta\omega}{\omega}\right)_c \gtrsim \left(\frac{\Delta\omega}{\omega}\right)_o. \tag{4.57}$$

Once the condition (4.56) is satisfied, one laser mode couples only with a small number of modes in the gain curve (whose bandwidth is just given by $(\Delta\omega/\omega)_o$). In this connection the function E in Eq. (3.30) can be approximated as follows

$$E\left(\eta_i, \eta_r\right) \approx E\left(\eta_i, \eta_i\right), \tag{4.58}$$

that is, the function E reduces to that of the single mode theory (see Ref. 16, Eq. (4.25)), becoming

$$E(\eta, \eta) = E(\eta) = \frac{d}{d\eta}\left(\frac{1 - \exp(+j\eta) + j\eta}{\eta^2}\right). \tag{4.59}$$

On the other hand, in the regime of Eq. (4.57) all the modes in the gain curve are coupled and we cannot perform any simplifying approximation.

In the following we assume that we work in the regime of Eq. (4.56), which is relevant for a FEL operating in a Sr. This last observation is supported by the fact that in a practical device we have, typically,

$$\lambda_o \sim 1\ \mu m \quad \sigma_z \sim 1\ cm,$$

from which it follows that the "coupling width" is given by

$$\left(\frac{\Delta\omega}{\omega}\right)_c \sim 5 \times 10^{-5},$$

which is very small when compared with a typical homogeneous width

$$\left(\frac{\Delta\omega}{\omega}\right)_o \sim 10^{-2}.$$

Using the coupling width, we can define a new parameter:

$$\mu_c = \left(\frac{\Delta\omega}{\omega}\right)_c \Big/ \left(\frac{\Delta\omega}{\omega}\right)_o = \text{coupling parameter.} \tag{4.60}$$

Condition (4.56) can now be written

$$\mu_c \ll 1. \tag{4.61}$$

Equation (4.61) has a particularly interesting physical meaning which relates to the coupling width. Let us write Eq. (4.61) in the form (see Eq. (3.5))

$$\sigma_z \gg \frac{\lambda_o}{\lambda_q} L_w. \tag{4.62}$$

From Equations (3.4) and (3.3) we find that the longitudinal electron velocity, in the WM, is

$$v_s \approx c\left(1 - \frac{\lambda_o}{\lambda_q}\right). \tag{4.63}$$

From Equations (4.62) and (4.63) we obtain

$$\sigma_z \gg \left(1 - \frac{v_s}{c}\right) L_w. \tag{4.64}$$

The r.h.s. of the Eq. (4.64) is just the "vacuum phase slippage" between the electrons and the laser field during one wiggler passage. This slippage is due to the difference in velocity between electrons and photons (in the vacuum). In conclusion the "small coupling width" condition is equivalent to requiring that the e.b. length σ_z is very large compared with the electron-photon "vacuum phase slippage" in one wiggler passage.

We emphasize that this slippage must not be confused with the "group slippage" between electron and photon bunches due to the difference between the electron velocity and the laser pulse group velocity, which depends on the FEL interaction. In a later section we will show that the photon bunch is pushed back towards the trailing edge of the e.b.. As a consequence, in order to maintain synchronism between photons and electrons, the optical round trip period T_c must be lower than the electron bunch time distance

T_e ($\delta t < 0$ as follows from Eq. (4.50)). This phenomenon (known as FEL lethargy) was first derived and investigated in the study of the single pass FEL oscillator[25], in the framework of the coupled Maxwell and quasi-Bloch equations[26].

4.5 FEL Oscillator in SR: Steady State Equations

The equations describing the steady state laser regime read

$$\begin{cases} \Delta W_i = 0 \\ \Delta \varphi_i = \psi, \end{cases} \tag{4.65}$$

where ΔW_i and $\Delta \varphi_i$ are given by Eqs (4.49) and ψ is a phase advance identical for all the laser modes. The "coupling coefficient" $B_{i,r}$ in Eqs(4.49), to be evaluated by means of the stationary electron distribution function (4.29), depends on (W,φ). Thus the set of Eqs (4.65) is complete with respect to the laser variables (W, φ), defining completely the stationary FEL regime.

Equations (4.65) can be rewritten in a more compact and convenient form using complex notation, i.e.,

$$x_i = \left(W_i\right)^{1/2} \exp\left(j \varphi_i\right). \tag{4.66}$$

Inserting (4.65) and (4.66) in (4.49) one easily finds

$$\left[\frac{\gamma_T}{1 - \gamma_T} + 2j\left(\psi + \omega_i \delta t\right)\right] x_i = - A \sum_r B_{i,r} x_r \tag{4.67}$$

where owing to the condition of small coupling width condition, $B_{i,r}$ is given by (see Eqs (3.30), (4.29), (4.59), (4.15), (4.16) and (4.17))

$$B_{i,r} = \int h\left(\frac{\tilde{z}^2 + \tilde{\varepsilon}^2}{2}\right) E\left(\pi\tilde{\varepsilon} + \nu_i\right) \cos\left(\frac{\nu_i - \nu_r}{\beta} \tilde{z}\right) d\tilde{z}\, d\tilde{\varepsilon}. \tag{4.68}$$

The number of modes involved in the Eq. (4.67) is generally very high. Indeed, we find that the spacing between two adjacent modes is given by

$$\omega_{i+1} - \omega_i = \frac{2\pi c}{L_c}, \tag{4.69}$$

where L_c is the optical cavity length. The number of modes N_c contained in a "coupling width" is given by

$$N_c = \left(\frac{\Delta\omega}{\omega}\right)_c \bigg/ \left(\frac{\omega_{i+1} - \omega_i}{\omega_o}\right) = \frac{L_c}{2\sigma_z} . \tag{4.70}$$

We have typically

$$L_c \sim 20 \text{ m}, \quad \sigma_z \sim 1 \text{ cm} \Rightarrow N_c \sim 10^3 .$$

Since the number of modes is so large, the discrete spectrum we are analyzing can be approximated by a continuous one. In this approximation Eq. (4.67) becomes

$$\left[\frac{1}{\xi} + j(\Psi + \theta\nu)\right] x(\nu) = \int K(\{x\}|\nu,\nu')x(\nu')d\nu' , \tag{4.71}$$

where

$$\xi = \frac{1 - \gamma_T}{\pi\gamma_T} \frac{\omega_s T}{\alpha_c} \left(\frac{\Delta\omega}{\omega}\right)_o^{-1} A, \tag{4.72}$$

$$\Psi = \frac{2\omega_o\delta t + 2\psi}{\xi} \cdot \frac{1 - \gamma_T}{\gamma_T} , \tag{4.73}$$

$$\theta = -\frac{2}{\pi}\left(\omega_o\delta t\right)\left(\frac{\Delta\omega}{\omega}\right)_o \frac{1}{\xi} \cdot \frac{1 - \gamma_T}{\gamma_T} . \tag{4.74}$$

The Kernel $K(\{x\}|\nu'\nu')$ is

$$K(\{x\}|\nu, \nu') = -\frac{B(\{x\}|\nu, \nu')}{\beta} , \tag{4.75}$$

where β has been defined by Eq. (4.22).

The nonlinear integral equation (4.71) accounts completely for the stationary state of a FEL operating in a SR. The parameters ξ and θ are fixed by the experimental conditions. Indeed, ε is proportional to the ratio between the "gain coefficient" A and the cavity losses γ_T, while θ is proportional to the "time-delay" δt between electrons and photons. The parameter ψ depends on the phase advance ψ (which is idantical for all the modes) and will be evaluated selfconsistently from Eq. (4.71) itself.

It is easy to integrate Eq. (4.71) under the hypothesis that the laser bandwidth $(\Delta\omega/\omega)_L$ satisfies the conditions

$$\left(\frac{\Delta\omega}{\omega}\right)_o \gg \left(\frac{\Delta\omega}{\omega}\right)_L \gg \left(\frac{\Delta\omega}{\omega}\right)_c \tag{4.76}$$

and that the laser pulse length σ_L is very short with respect to the e.b. one σ_z, i.e.,

$$\sigma_L \ll \sigma_z . \tag{4.77}$$

It is appearent that the conditions (4.76) can be fulfilled only in the "small coupling width" regime (see Eq. (4.61)). However, we shall find later in this paper that Eq. (4.61) is not only necessary but also sufficient to fulfill (4.76) and (4.77), for typical FEL parameters.

The conditions above allow us to carry out a solution of (4.71) as follows (the details can be found in Appendix B of Ref. 22):

$$x(\nu) = \left(\frac{y_3}{2\pi^3\mu_c}\right)^{1/4}\left(\frac{y_o}{\beta\, G_o}\right)^{1/2} \cdot$$

$$\cdot \exp\left\{\frac{1}{\mu_c}\left[j\, y_1\left(\nu - \nu_L\right) - \left(y_3 + j\, y_2\right)\left(\nu - \nu_L\right)^2\right]\right\}, \tag{4.78}$$

where μ_c is the coupling parameter, G_o and β are defined by Eqs (4.21) and (4.22), and finally, $y_i (i = 0, 1, 2, 3)$ and ν_L satisfy the following equations

$$\begin{cases} F_1(y_i, \nu_L|\tilde{\sigma}_o) = -\frac{1}{\xi} \\ F_2(y_i, \nu_L|\tilde{\sigma}_o) = 0 \\ F_3(y_i, \nu_L|\tilde{\sigma}_o) = -\theta \\ F_4(y_i, \nu_L|\tilde{\sigma}_o) = 0 \\ F_5(y_i, \nu_L|\tilde{\sigma}_o) = 0 \end{cases} \tag{4.79}$$

where F_i are rather complicated functions of the arguments and are reported in Appendix B (Eqs B.12) of Ref. 22

In addition to Eqs (4.79) we have the condition

$$y_3 > 0 \tag{4.80}$$

which is related to the requirement that $x(\nu)$ will be a normed function. Eqs (4.80), (4.76) and (4.77) define the region of validity of the solution (4.78).

4.6 FEL Oscillator in SR: Laser Beam Parameters

A) Laser spectrum

One can easily derive from Eq. (4.78) that the laser spectrum has a Gaussian profile, i.e.

$$g(\nu) = \frac{|x(\nu)|^2}{\int |x(\nu)|^2 d\nu} = \frac{1}{\sqrt{2\pi}\,\sigma_\nu} \exp\left\{-\frac{(\nu - \nu_L)^2}{\sigma_\nu^2}\right\}, \tag{4.81}$$

where σ_ν and ν_L are the "normalized" bandwidth and central frequency respectively; σ_ν is defined in terms of μ_c and y:

$$\sigma_\nu = \frac{1}{2}\left(\mu_c / y_3\right)^{1/2}. \tag{4.82}$$

The laser bandwith $(\Delta\omega/\omega)_L$ is thus given by

$$\left(\frac{\Delta\omega}{\omega}\right)_L = \frac{1}{\pi}\left(\frac{\Delta\omega}{\omega}\right)_o \sigma_\nu = \frac{1}{2\pi} q_\omega\left(\xi, \theta, \tilde{\sigma}_o\right)\left[\left(\frac{\Delta\omega}{\omega}\right)_c \left(\frac{\Delta\omega}{\omega}\right)_o\right]^{1/2}, \tag{4.83}$$

where $q_\omega(\xi, \theta, \tilde{\sigma}_o)$ is simply

$$q_\omega\left(\xi, \theta, \tilde{\sigma}_o\right) = \frac{1}{\sqrt{y_3(\xi, \theta, \tilde{\sigma}_o)}}. \tag{4.84}$$

Finally, the central laser frequency ω_L is related to ν_L by the equation (see Eq. (4.16))

$$\omega_L = \omega_o\left[1 - \frac{1}{\pi}\left(\frac{\Delta\omega}{\omega}\right)_o \nu_L\right]. \tag{4.85}$$

We show schematically in Fig. 12 the laser spectrum and the small signal single mode (SSSM) gain profile, which as its maximum value for

$$\nu = \nu^{\star} \cong 2.61 .$$

B) Laser power density (at the starting time for the FEL interaction)

From Eqs (4.4), (4.21),(4.66) and (4.78) we find that the laser beam (l.b.) has a gaussian profile, with power density

$$\frac{dP_L}{dS} = \frac{\mathscr{E}}{\Sigma_L} \frac{c}{\sqrt{2\pi}\,\sigma_L} \exp\left\{ - \frac{\left(s + y_1\,\sigma_z\right)^2}{2\sigma_L^2} \right\}, \tag{4.86}$$

where $\mathscr{E}$ indicates the intracavity laser pulse energy:

$$\mathscr{E} = \frac{1 - \gamma_T}{\gamma_T}\, P_s T \left(\frac{\Delta\omega}{\omega}\right)_o \chi\left(\xi,\ \theta,\ \tilde{\sigma}_o\right); \tag{4.87}$$

P_s is the average power emitted by the whole e.b. via synchrotron radiation, explicitly given by

$$P_s = \frac{I \cdot U_o}{e}\ , \tag{4.88}$$

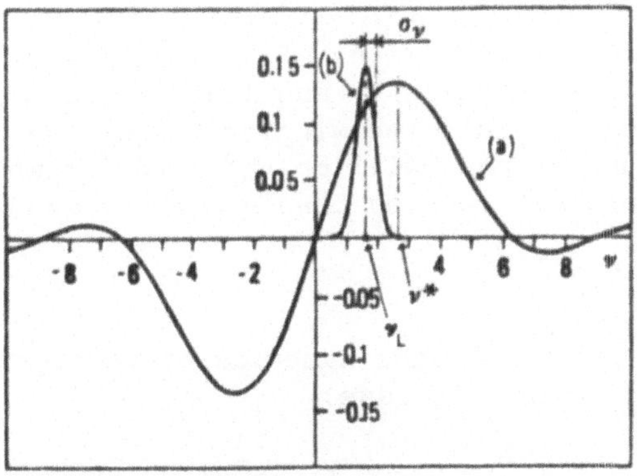

Fig. 12. Small signal single mode gain profile (curve (a)) and laser spectrum (curve (b))

U_o being the energy radiated per turn and per electron (Eq. (2.32)). The function χ is defined by the equation

$$\chi\left(\xi,\ \theta,\ \tilde{\sigma}_o\right) = \frac{y_o\left(\xi,\ \theta,\ \tilde{\sigma}_o\right)}{\xi} \ . \tag{4.89}$$

In deriving Eq. (4.87) we have assumed that the damping time τ_s is given by Eq. (4.40) (separate functions SR). Finally Σ_L is the l.b. cross-section and σ_L is the r.m.s. l.b. length,

$$\sigma_L = q_\sigma\left(\xi,\ \theta,\ \tilde{\sigma}_o\right)\sigma_z\left[\left(\frac{\Delta\omega}{\omega}\right)_c \left(\frac{\Delta\omega}{\omega}\right)_o^{-1}\right]^{1/2} , \tag{4.90}$$

where

$$q_\sigma\left(\xi,\ \theta,\ \tilde{\sigma}_o\right) = \left[\frac{y_2^2\left(\xi,\ \theta,\ \tilde{\sigma}_o\right) + y_3^2\left(\xi,\ \theta,\ \tilde{\sigma}_o\right)}{y_3\left(\xi,\ \theta,\ \tilde{\sigma}_o\right)}\right]^{1/2} . \tag{4.91}$$

From Equation (4.86) we see that at the time of entrance of the e.b. into the WM, the l.b. is displaced from the e.b. center of mass by the distance

$$\delta s = - y_1\sigma_z . \tag{4.92}$$

To better understand these notations, Fig. 13 illustrates schematically the longitudinal e.b. and l.b. distributions at the entrance time of the e.b. into the wiggler. What one expects is that optimum laser performance will be obtained for maximum overlapping between laser and electron bunches, i.e., when

$$\delta s = 0 \Rightarrow y_1 = 0. \tag{4.93}$$

Section 4.9 is devoted to the numerical investigation of this particular, but very important, case.

The average output laser power P_L is given by

$$P_L = \frac{\mathscr{E}}{T_e}\gamma_M = \gamma_M\left(\frac{1-\gamma_T}{\gamma_T}\right)P_s\left(\frac{\Delta\omega}{\omega}\right)_o \chi\left(\xi,\ \theta,\ \tilde{\sigma}_o\right) , \tag{4.94}$$

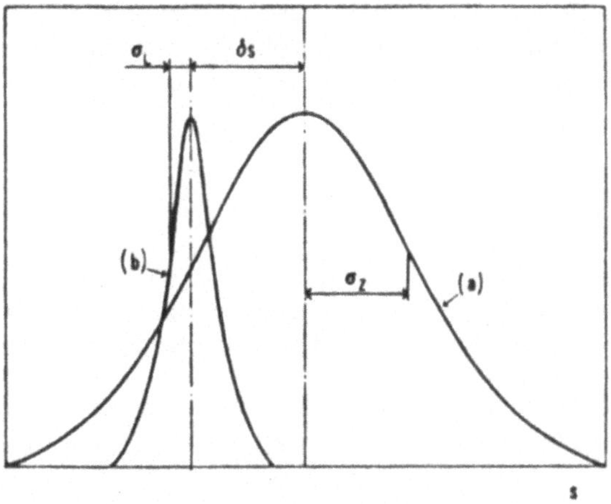

Fig. 13. Curve (a): electron beam longitudinal distribution
Curve (b): laser beam longitudinal distribution

where γ_M is the mirror transmission.

Let us investigate further Eq. (4.94). We can see that the average output laser power is proportional to the synchrotron radiation power P_s. The same relationship has been obtained in Sect. 4.2 between the energy radiated by the e.b. into the laser mode and the synchrotron radiation energy (Eq. (4.41)).

The physical meaning of this dependence is quite clear. The damping time τ_s, counteracting the diffusion due to the FEL interaction, depends simply on the synchrotron radiation energy emitted per machine turn U_o. The damping mechanism becomes more efficient as U_o (and then P_s) increases, and, as a consequence, the energy which can be emitted by the e.b. into the laser modes is enhanced.

We can define (as has been done in Sect. 4.3) the efficiency for a FEL operating in a SR, as

$$\eta = \frac{P_L}{P_L + P_s} . \tag{4.95}$$

If we assume small cavity losses ($\gamma_T \ll 1$) mainly due to the mirror transmission ($\gamma_M \sim \gamma_T$), Eq. (4.95) becomes

$$\eta \sim \frac{(\Delta\omega/\omega)_o \, \chi}{(\Delta\omega/\omega)_o \, \chi + 1} . \tag{4.96}$$

In Section 4.9 we shall find that the maximum χ we can obtain is of the order of the unity. As a consequence the maximum efficiency we can have is roughly given by

$$\eta \sim \left(\frac{\Delta\omega}{\omega}\right)_0 \tag{4.97}$$

as previously obtained for the amplifier case (Sect. 4.3). It is worthwhile to note that it is possible to derive that Eq. (4.97) holds in general, by using purely thermodynamic arguments[27].

Finally, let us rewrite the condition for validity of the solution (4.78) in terms of μ_c, q_ω (Eq. (4.84)) and $q\sigma$ (Eq. (4.91)). Eqs (4.76) become

$$\sqrt{\mu_c} \ll \frac{q_\omega}{2\pi} \ll \frac{1}{\sqrt{\mu_c}} , \tag{4.98}$$

while Eq. (4.77) reads

$$q_\sigma \ll \frac{1}{\sqrt{\mu_c}} . \tag{4.99}$$

In the following sections we shall analyze two particular, but interesting, cases:
i) Threshold regime ($y_0 \to 0$) (Sects 4.7 and 4.8)
ii) Optimum operating regime ($y_1 = 0$) (Sect. 4.9)

4.7 FEL Oscillator in SR: Threshold Regime

The theory developed uo to this point does not enable us to treat consistently the case of a very weak laser field, owing to the presence of quantum fluctuations which cannot be taken into account in any way in the present classical formalism. Furthermore, the "stochasticity condition" (Ref. 20) which has allowed us to utilize a FOKKER-PLANCK formalism for the e.b. distribution, might be not satisfied. A "threshold regime" can be defined as follows:
i) The laser field is sufficiently small that one can neglect the effect of the FEL interaction on the e.b. compared with the noise due to the synchrotron emission process. In this case the e.b. distribution is simply the unperturbed one and it is given by the Eq. (4.36).
ii) The laser field is sufficiently strong that we may neglect any quantum effects, allowing a completely classical treatment.

In Appendix C of Ref. 22 we have discussed the conditions under which i) and ii) can be satisfied simultaneously. We have found that, for typical FEL and SR parameters, it is always possible to define such an operating regime.

Equations (4.79) are considerably simplified in the limit $y_0 \to 0$. In this case the e.b. distribution which appears in the definition of the functions F_i does not depend on the e.b. parameters $\{y\}$ and ν_L.

The first of Equations (4.79) yields the "threshold value" for ξ, while the remaining four give y_i (i = 1, 2, 3) and ν_L as function of θ and $\tilde{\sigma}_0$.

Furthermore we can assume we have a homogeneously broadened emission line (see Sect. 3.1). As a consequence we have (see Eqs (3.25) and (4.19))

$$\tilde{\sigma}_0 \ll 1. \tag{4.100}$$

In this connection the dependence of y_i and ν_L on θ and $\tilde{\sigma}_0$ has the following simple form:

$$\begin{cases} y_i = y_i\left(\sqrt{2\pi}\,\tilde{\sigma}_0\,\theta\right) \\ \nu_L = \nu_L\left(\sqrt{2\pi}\,\tilde{\sigma}_0\,\theta\right), \end{cases} \tag{4.101}$$

and the threshold value reads

$$\xi_{th} = \sqrt{2\pi}\,\tilde{\sigma}_0/q_\gamma\left(\sqrt{2\pi}\,\tilde{\sigma}_0\,\theta\right), \tag{4.102}$$

where q_γ is a function which must be evaluated numerically. From Equations (4.101) and (4.102) it follows that all the features of the "threshold regime" depend only on the parameter

$$\text{Ⓗ} = \sqrt{2\pi}\,\tilde{\sigma}_0\,\theta, \tag{4.103}$$

which can be rewritten in a more obvious form as (see Eqs (4.19) and (4.74)),

$$\text{Ⓗ} = -\frac{2}{\pi}\left(\frac{\Delta\omega}{\omega}\right)_0 \frac{\omega_0\delta t}{g_0} \tag{4.104}$$

where g_0 is given by

$$g_0 = 2\pi \left(\frac{2\lambda_0}{\lambda_q}\right)^{1/2} \frac{L_w \lambda_0}{\Sigma_L} \left(\frac{I_p}{I_0}\right) \frac{K^2}{\left(1 + K^2\right)^{3/2}} \left(\frac{\Delta\omega}{\omega}\right)_0^{-2}, \qquad (4.105)$$

$$I_p = \frac{c\, T_e\, I}{\sqrt{2\pi}\, \sigma_z^0} \equiv \text{peak electron current}, \qquad (4.106)$$

where σ_z^0 is the r.m.s. "natural e.b. length" (see Eq. (2.70)).

The range of validity of the solution (4.78) is defined by Eqs (4.98) and (4.99) and (4.80). Furthermore we have the new condition,

$$q_\gamma(\text{Ⓗ}) > 0 \qquad (4.107)$$

which is due to the requirement of positive gain ($\xi_{th} > 0$, see Eq. (4.102)).

4.8 FEL Oscillator in SR: Numerical Results for the Threshold Regime

In Figure 14 we have plotted q_γ and y_1 versus Ⓗ, while in Fig. 15 q_ω and q_σ, describing the bandwidth and pulse length respectively, are plotted. We have found that condition (4.80) is satisfied only for Ⓗ > 0. As a consequence the optical cavity round trip period must be shorter than the time distance between two adjacent electron bunches ($\delta t < 0$, see Eqs (4.104)). This effect is due to

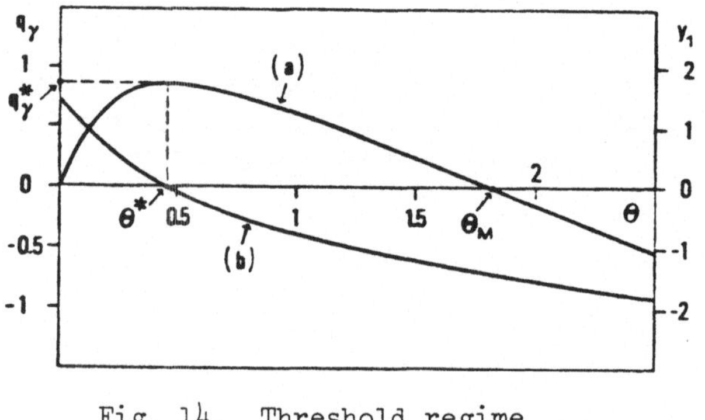

Fig. 14. Threshold regime
Curve (a): q_γ vs Ⓗ
Curve (b): y_1 vs Ⓗ

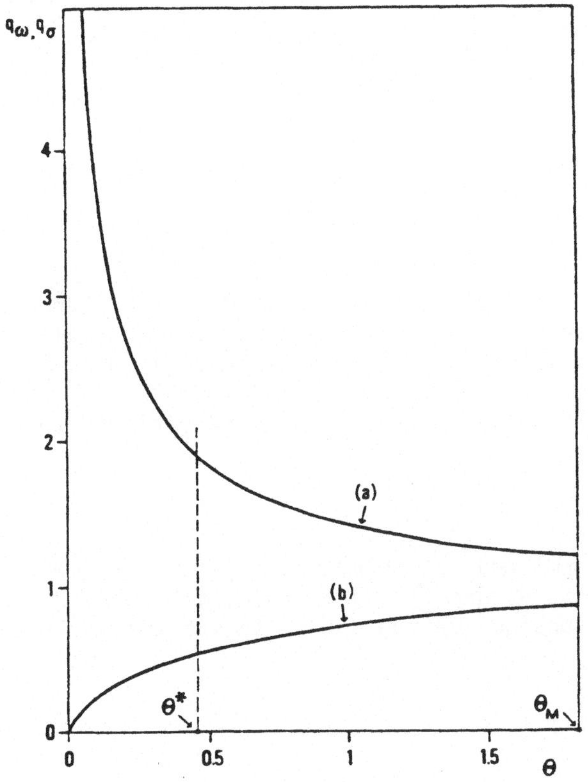

Fig. 15. Threshold regime
Curve (a): q_ω vs Ⓗ
Curve (b): q_σ vs Ⓗ

the so-called "FEL lethargy"[25] we briefly discussed in Sect. 4.4.

The condition (4.107) is satisfied (as is shown in Fig. 14, curve a)) in the region

$$0 < Ⓗ < Ⓗ_M, \tag{4.108}$$

where $Ⓗ_M$ has the value

$$Ⓗ_M \sim 1.82.$$

Figure 15 shows that q_σ satisfies the condition (4.99) in the full range (for $\mu_c \ll 1$) while q_ω is divergent when Ⓗ $\rightarrow 0$. Thus, in the neighborhood of Ⓗ ~ 0, the solution is not valid, the second inequality in Eqs (4.98) not being satisfied. The range of validity of our solution is dependent on μ_c. For example, for

LEDA-F1[10] we have

$$\mu_c \sim 3 \times 10^{-3}$$

corresponding to

$$q_\omega(\textcircled{H}) \ll 100.$$

From Figure 15 one derives that for $\textcircled{H} \gtrsim 5 \times 10^{-2}$ one has $q_\omega \lesssim 5$ ($\ll 100$), which amounts to saying that, apart from a small region around $\textcircled{H} \sim 0$, our solution is valid in the full range (4.108).

Let us now discuss in further detail the various l.b. parameters.

A) Bandwidth and pulse length

As previously pointed out, almost in the full range (4.108) (apart a small region around $\textcircled{H} \sim 0$) we have, to an order of magnitude, $q_\omega \sim q_\sigma \sim 1$. As a consequence the bandwidth and the pulse length read

$$\left(\frac{\Delta\omega}{\omega}\right)_L \cong \frac{1}{4\pi}\left(\frac{\lambda_q\ \lambda_o}{L_w\ \sigma_z^o}\right)^{1/2}, \tag{4.109}$$

$$\sigma_L \cong \sigma_z^o\left(\frac{\lambda_o\ L_w}{\sigma_z^o\ \lambda_q}\right)^{1/2}. \tag{4.110}$$

In Figure 16 has been plotted ν_L vs. $\textcircled{H}$. For $y_1 = 0$, ν_L lies just on the maximum of the SSSM gain ($\nu_L = \nu$). For $\textcircled{H} < \textcircled{H}$ we have $\nu_L < \nu^\star$, while for $\textcircled{H} > \textcircled{H}^\star$, $\nu_L \sim \nu^\star$.

B) Threshold cavity loss γ_T^{th}

The total cavity loss, which corresponds to the threshold, is given by (see Eqs (4.72), (4.102) and (4.105))

$$\gamma_T^{th} = \frac{g_o\ q_\gamma(\textcircled{H})}{1 + g_o\ q_\gamma(\textcircled{H})}. \tag{4.111}$$

In order to have laser action, the overall gain must be positive. As a consequence γ_T must satisfy the following condition:

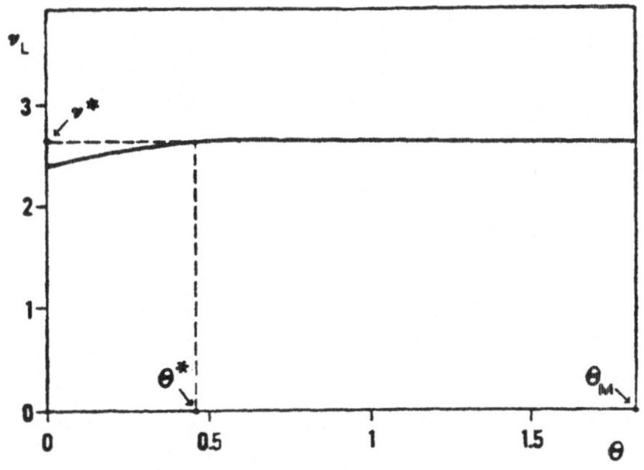

Fig. 16. Threshold regime: ν_L vs Ⓗ

$$\gamma_T < \gamma_T^{th}. \tag{4.112}$$

From Figure 14 we see that the maximum q_γ is obtained for Ⓗ = Ⓗ$^\bigstar \sim .46$, for which we have $q_\gamma^\bigstar \sim .85$. The minimum g_0 value for lasing is then given by

$$g_0^{min} \sim 1.2\ \gamma_T \Big/ \left(1 - \gamma_T\right). \tag{4.113}$$

C) e.b. - l.b. displacement

In Figure 14 (curve b) has been plotted y_1, which is equal to the ratio between the e.b. - l.b. displacement δs and the e.b. pulse length σ_z° (see Eq. (4.92)). It is worthwhile noticing that the maximum for q_γ corresponds to $y_1 = 0$, i.e. the maximum overlapping between e.b. and l.b.. This configuration will be called, henceforth, "optimum operating regime".

4.9 FEL Oscillator in SR Optimum Operating Regime

The "optimum operating regime" is defined by the requirement for maximum overlapping between the l.b. and the e.b., i.e., by the equation

$$y_1 = 0. \tag{4.114}$$

Equations (4.79) admit a solution only for

$$\xi > \xi_{th}, \tag{4.115}$$

where ξ_{th} has been defined in Sect. 4.7. For $\tilde{\sigma}_o \ll 1$, $y_1 = 0$, ξ_{th} becomes (see Eq. (4.102))

$$\xi_{th} = \sqrt{2\pi}\,\frac{\tilde{\sigma}_o}{q_\gamma^{\star}} . \tag{4.116}$$

In general (for $y_1 = 0$ but for arbitrary σ_o) ξ_{th} can be derived from the first of Eqs (4.79)

$$y_o\left(\xi_{th}, \tilde{\sigma}_o\right) = 0 \Rightarrow \xi_{th} = \xi_{th}\left(\tilde{\sigma}_o\right), \tag{4.117}$$

In Figure 17 ξ_{th} has been plotted versus $\tilde{\sigma}_o$ in the range $0 < \tilde{\sigma}_o < 1$.

Furthermore in Figure 18 q_ω and q_σ have been plotted vs ξ for various $\tilde{\sigma}_o$ values. We can see that the dependence on $\tilde{\sigma}_o$ is very weak, apart from the different thresholds. This behaviour is general for all the e.b. and l.b. parameters. In other words, there is very

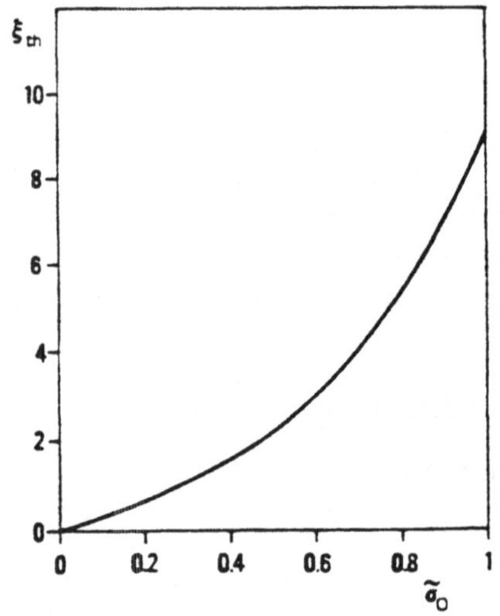

Fig. 17. Optimum operating regime: ξ_{th} vs $\tilde{\sigma}_o$

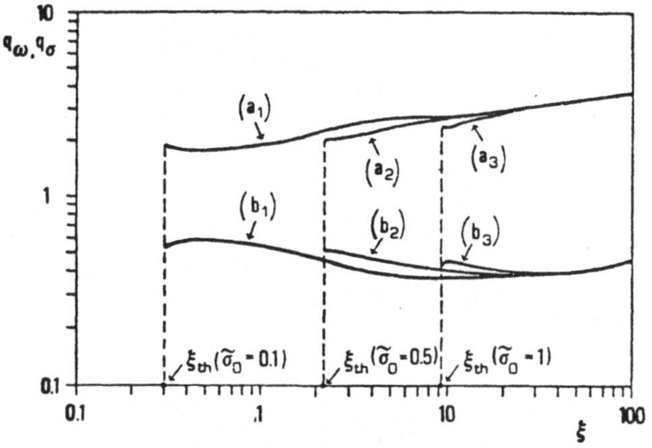

Fig. 18. Optimum operating regime:
Curves (a_1), (a_2), (a_3): q_ω vs ξ
Curves (b_1), (b_2), (b_3): q_σ vs ξ
(a_1), (b_1) $\Rightarrow \tilde{\sigma}_0 = 0.1$
(a_2), (b_2) $\Rightarrow \tilde{\sigma}_0 = 0.5$
(a_3), (b_3) $\Rightarrow \tilde{\sigma}_0 = 1.0$

little difference between the curves corresponding to different $\tilde{\sigma}_0$ (in the range $0 < \tilde{\sigma}_0 < 1$), and this difference approaches zero for large ξ values. Figure 18 shows also that, up to $\xi \sim 100$, q_ω and q_σ are roughly constant, given by

$$q_\omega \sim 2 \text{ to } 3, \; q_\sigma \sim .4 \text{ to } .5.$$

Thus conditions (4.98) and (4.99) are largely fulfilled; as a consequence, in the range investigated, the solution we find is valid.

Let us now consider in some more detail the e.b. and l.b. parameters.

A) Electron distribution function

The r.m.s. e.b. "normalized" dimension $\tilde{\sigma}$ is plotted vs ξ in Fig. 19, for various $\tilde{\sigma}_0$ values. It is shown that $\tilde{\sigma}$ starts from the unperturbed $\tilde{\sigma}_0$ value ($\xi = \xi_{th}$) and, for large ξ, saturates towards $\tilde{\sigma} \sim 1$ (in this region it is practically independent of $\tilde{\sigma}_0$). The physical meaning of this last observation is that for large ξ, the inhomogeneous broadening $(\Delta\omega/\omega)_i$ due to the energy spread is of the same order of magnitude as the homogeneous one $(\Delta\omega/\omega)_0$ (see Sect. 3.1).

The e.b. density profile, defined as

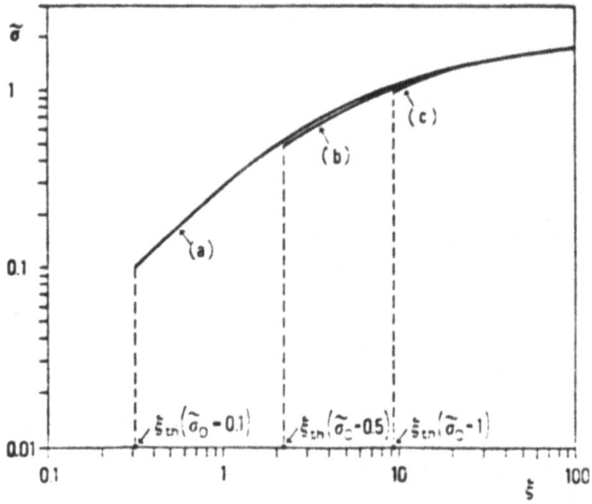

Fig. 19. Optimum operating regime: $\tilde{\sigma}$ vs ξ
Curve (a_1): $\tilde{\sigma}_0 = 0.1$
Curve (a_2): $\tilde{\sigma}_0 = 0.5$
Curve (a_3): $\tilde{\sigma}_0 = 1.0$

$$f(\tilde{z}) = \int h\left(\frac{\tilde{\varepsilon}^2 + \tilde{z}^2}{2}\right) d\tilde{\varepsilon}, \tag{4.118}$$

is plotted in Fig. 20 for different ξ values. Curve (a) corresponds to the threshold condition (unperturbed e.b.), curve (c) to the maximum output laser power (see the following paragraph in this section) and, finally, curve (b) is related to an intermediate ξ value.

B) Laser beam parameters

i) Output power

The laser average output power P_L is given by Eqs (4.94), where χ is the "efficiency function" defined by Eq. (4.89). In Figure 21 the function χ has been plotted versus ξ for various values of $\tilde{\sigma}_0$. The curve (a) defines the limiting case corresponding to vanishingly small initial energy spread. Figure 21 shows that there is, above the threshold, very small difference between the cases (a) ($\tilde{\sigma}_0 \to 0$) and (b) ($\tilde{\sigma}_0 = 0.1$). Let us note that this range ($\tilde{\sigma}_0 \ll 1$) is just the operating range for FEL and SR of practical interest.

Furthermore, a different behaviour arises for larger $\tilde{\sigma}_0$: χ decreases and the maximum is pushed towards larger ξ. The maximum χ is obtained for $\xi \sim 10$, and it is of the order of

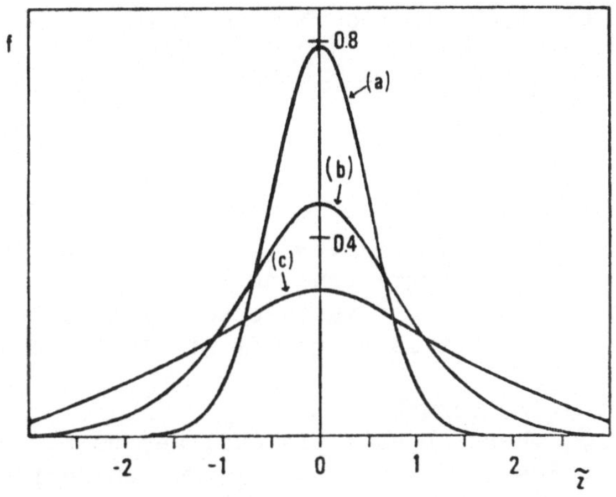

Fig. 20. Optimum operating regime: longitudinal electron beam distribution ($\tilde{\sigma}_0 = 0.5$)
Curve (a): $\xi = \xi_{th}$
Curve (b): $\xi = 6$
Curve (c): $\xi = \xi_M = 20$

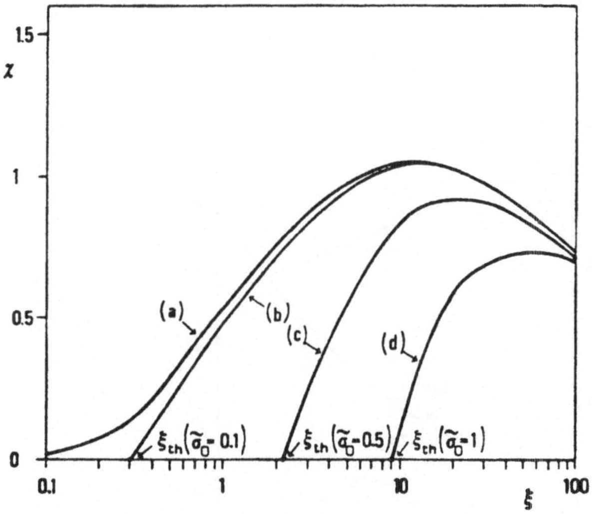

Fig. 21. Optimum operating regime: efficiency function χ vs ξ
Curve (a): $\tilde{\sigma}_0 = 0.0$
Curve (b): $\tilde{\sigma}_0 = 0.1$
Curve (c): $\tilde{\sigma}_0 = 0.5$
Curve (d): $\tilde{\sigma}_0 = 1.0$

unity. More precisely, for $\sigma_o \to 0$ we have

$$\xi\Big|_{\chi_{MAX}} \cong 12 \qquad \chi_{MAX} \cong 1.05. \tag{4.119}$$

In Figure 22 we have plotted the maximum of the efficiency function (χ_M) and the corresponding ξ value (ξ_M) versus $\tilde{\sigma}_o$. We can see that up to $\tilde{\sigma}_o \lesssim .2$, χ_M is practically constant, while for higher $\tilde{\sigma}_o$ values it is roughly linearly decreasing.

ii) Bandwidth and pulse length
From Figure 18, where we have plotted, versus ξ, the parameters q_ω and q_σ, and Eqs (4.83) and (4.90), we can derive the laser bandwidth and the pulse length. However, we must take into account that the "coupling width" $(\Delta\omega/\omega)_c$ depends on the e.b. length σ_z, that is on $\tilde{\sigma}$. It is then convenient to rewrite Eqs (4.83) and (4.90) in the following more compact form

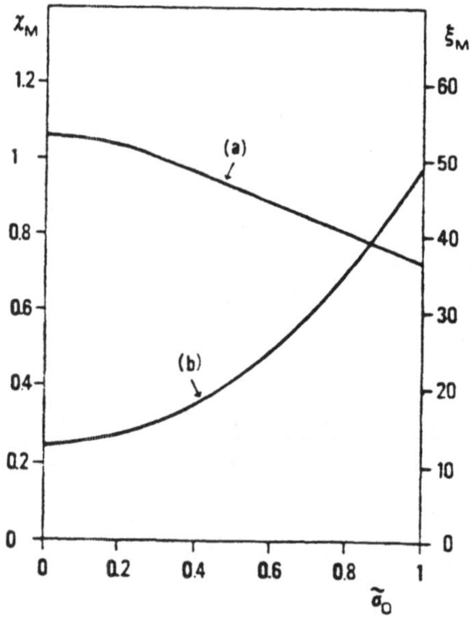

Fig. 22. Optimum operating regime:
Curve (a): χ_M vs $\tilde{\sigma}_o$
Curve (b): ξ_M vs $\tilde{\sigma}_o$

$$\left(\frac{\Delta\omega}{\omega}\right)_L = Q_\omega\left(\xi,\ \tilde{\sigma}_o\right)\sqrt{\frac{\lambda_o}{\ell^\star}}\ , \tag{4.120}$$

$$\sigma_L = Q_\sigma\left(\xi,\ \tilde{\sigma}_o\right)\sqrt{\lambda_o\ell^\star}\ , \tag{4.121}$$

where we have put

$$\ell^\star = \frac{c\alpha_c}{\omega_s}\ , \tag{4.122}$$

$$Q_\omega\left(\xi,\ \tilde{\sigma}_o\right) = \frac{1}{2\pi}\frac{q_\omega\left(\xi,\ \tilde{\sigma}_o\right)}{\sqrt{\tilde{\sigma}\left(\xi,\ \tilde{\sigma}_o\right)}}\ , \tag{4.123}$$

$$Q_\sigma\left(\xi,\ \tilde{\sigma}_o\right) = \frac{1}{2}\,q_\sigma\left(\xi,\ \tilde{\sigma}_o\right)\sqrt{\tilde{\sigma}\left(\xi,\ \tilde{\sigma}_o\right)}\ . \tag{4.124}$$

The functions Q_ω and Q_σ are plotted in Fig. 23. We can see that they are practically independent of $\tilde{\sigma}_o$, (apart from the threshold) and, for $\xi \gtrsim 1$, are roughly constant:

$$Q_\omega \sim .4 \text{ to } .5, \qquad Q_\sigma \sim .15 \text{ to } .25.$$

iii) Average laser frequency ν_L
The "normalized" average laser frequency ν_L is plotted in Fig. 24 as a function of ξ for $\tilde{\sigma}_o = 0.1$, 0.5 and 1.0. As pointed out previously, it corresponds to the maximum of the SSSM inhomogeneous broadened gain profile.

iv) Timing between e.b. and l.b.
The difference δt between the optical cavity round trip period T_c and the time distance T_e between two adjacent electron bunches is a function of θ (see Eq. (4.74)), which has been derived from the condition (4.114). Let us call $\theta^\star$ the value of θ which satisfies Eq. (4.114); we have

$$\omega_o\delta t = -\,Q_\theta\left(\xi,\ \tilde{\sigma}_o\right)\frac{\gamma_T}{1-\gamma_T}\left(\frac{\Delta\omega}{\omega}\right)_o^{-1}\ , \tag{4.125}$$

where we have put

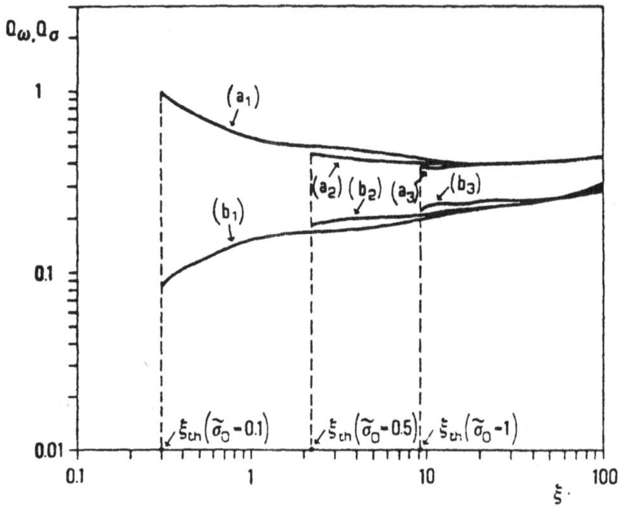

Fig. 23. Optimum operating regime:
Curves (a_1), (a_2), (a_3): Q_ω vs ξ
Curves (b_1), (b_2), (b_3): Q_σ vs ξ
(a_1), $(b_1) \Rightarrow \tilde{\sigma}_o = 0.1$
(a_2), $(b_2) \Rightarrow \tilde{\sigma}_o = 0.5$
(a_3), $(b_3) \Rightarrow \tilde{\sigma}_o = 1.0$

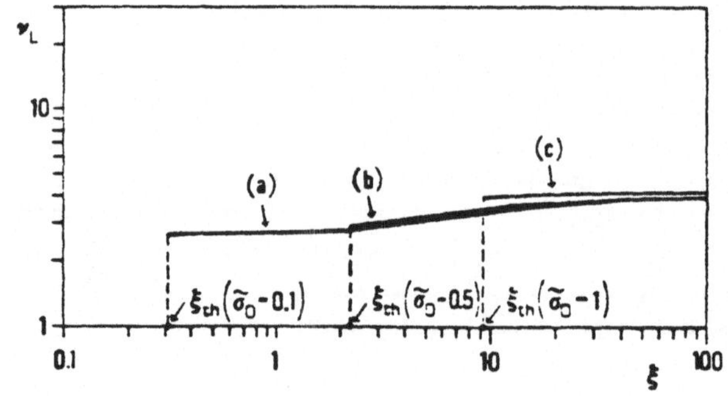

Fig. 24. Optimum operating regime: ν_L vs ξ
Curve (a): $\tilde{\sigma}_o = 0.1$
Curve (b): $\tilde{\sigma}_o = 0.5$
Curve (c): $\tilde{\sigma}_o = 1.0$

$$Q_\theta\left(\xi,\ \tilde{\sigma}_o\right) = \frac{\pi}{2}\,\theta^\star\left(\xi,\ \tilde{\sigma}_o\right)\xi. \tag{4.126}$$

We have plotted Q_θ versus ξ for various $\tilde{\sigma}_o$ in Fig. 25. We can see that the dependence on $\tilde{\sigma}_o$ and ξ is very weak; for $\xi \gtrsim 1$ we have

$$Q_\theta \sim 0.8 \text{ to } 0.4.$$

5. CONCLUSIONS

From Eq. (4.94) we see that the power P_L we can obtain from a FEL operating a SR is proportional to the spontaneous synchrotron radiation power P_S. Thus, if we want to increase P_L, we must enhance P_s with the insertion in the electron machine of special high magnetic field wigglers. This is a non trivial problem for the SR design. We indeed find that betatron tunes (Eq. (2.7)) are strongly affected by the high wiggler field (in LEDA-F2[10] this problem has been resolved with the insertion of six large q-poles near the WM, see Fig. 3).

The efficiency (see Eq. (4.97)) has the same order of magnitude of the single pass FEL. Unfortunately, the use of a recirculated

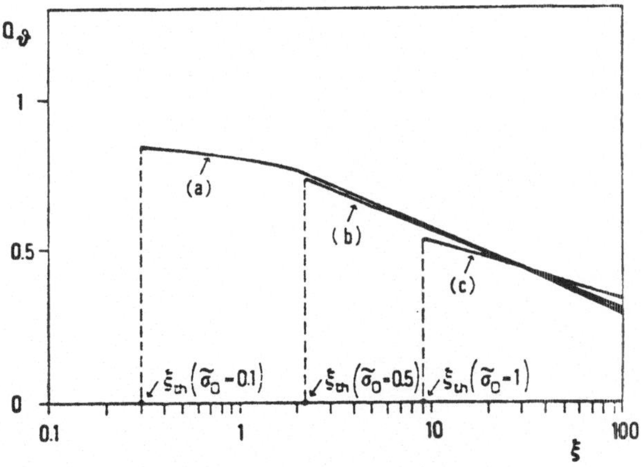

Fig. 25. Optimum operating regime: Q_θ vs ξ
Curve (a): $\tilde{\sigma}_o = 0.1$
Curve (b): $\tilde{\sigma}_o = 0.5$
Curve (c): $\tilde{\sigma}_o = 1.0$

e-beam does not increase this parameter (apart from "non-conventional" FEL and SR operation[23]).

However, in a SR it is possible to store very high currents, which allows us to obtain very high average power laser beams and very high gain per pass. This last feature can be very important, especially for operation in the UV region of the spectrum[28].

Finally, we want to point out that we have investigated only the steady state regime. The main problems which remain to be studied in order to have a complete picture of FEL behaviour in a storage ring are the following:

a) stability of the stationary solution.

b) Transient behaviour.

c) Single pulse operation (Q-switching).

APPENDIX A - WIGGLER MAGNET FOCUSING PROPERTIES

The magnetic field defined by (2.1) does not satisfy the Maxwell equations. Let us consider a more general magnetic field, given by

$$B_x = 0,\ B_y = F(y)\cos\left(\frac{2\pi}{\lambda_q} s\right),\ B_s = G(y)\sin\left(\frac{2\pi}{\lambda_q} s\right). \qquad (A.1)$$

If we require

$$F(0) = B_w,\ G(0) = 0, \qquad (A.2)$$

the field (A.1) reduces to the (2.1) in the median plane (y = 0).

In order to satisfy the Maxwell equations we must have

$$\begin{aligned} \mathrm{div}\underline{B} = 0 &\Rightarrow \frac{dF}{dy} = -\frac{2\pi}{\lambda_q} G(y) \\ \mathrm{rot}\underline{B} = 0 &\Rightarrow \frac{dG}{dy} = -\frac{2\pi}{\lambda_q} F(y). \end{aligned} \qquad (A.3)$$

The solution of Eqs (A.3), with the conditions (A.2), are

$$\begin{cases} F(y) = B_w \cosh\left(\dfrac{2\pi}{\lambda_q} y\right) \\ G(y) = -B_w \sinh\left(\dfrac{2\pi}{\lambda_q} y\right). \end{cases} \qquad (A.4)$$

To the lowest order in y, Eqs (A.4) yield

$$\begin{cases} F(y) \approx B_w\left(1 + 2\left(\dfrac{\pi y}{\lambda_q}\right)^2\right) \\ G(y) \approx -B_w\left(\dfrac{2\pi}{\lambda_q} y\right). \end{cases} \qquad (A.5)$$

From the first of Eqs (A.5) we derive the field variation along the vertical direction,

$$\frac{\delta B}{B} \approx 2 \left(\frac{\pi y}{\lambda_q}\right)^2 . \tag{A.6}$$

From Equations (A.1) we can derive the e.b. motion equations. We have (E = electron energy, $\dot{x} = dx/dt$, etc.)

$$\begin{cases} \ddot{x} = -\dfrac{ec}{E}\left[\dot{y}G(y)\sin\left(\dfrac{2\pi}{\lambda_q}s\right) + \dot{s}F(y)\cos\left(\dfrac{2\pi}{\lambda_q}s\right)\right] \\ \ddot{y} = -\dfrac{ec}{E}\dot{x}G(y)\sin\left(\dfrac{2\pi}{\lambda_q}s\right) \\ \ddot{s} = \dfrac{ec}{E}\dot{x}F(y)\cos\left(\dfrac{2\pi}{\lambda_q}s\right). \end{cases} \tag{A.7}$$

If we assume

$$\frac{e B_w \lambda_q}{2\pi E} \ll 1, \tag{A.8}$$

it is easy to integrate Eqs (A.7). The reference orbit along the WM is given by

$$\begin{cases} x_R(s) = \dfrac{e B_w}{E}\left(\dfrac{\lambda_q}{2\pi}\right)^2\left[\cos\left(\dfrac{2\pi}{\lambda_q}s\right) - 1\right] \\ y_R(s) = 0. \end{cases} \tag{A.9}$$

The motion around (x_R, y_R) is obtained by inserting Eqs (A.4) in Eqs (A.7). Up to the first order in x and y, we obtain ($x' = dx/ds$, etc.)

$$\begin{cases} x'' = 0 \\ y'' = -\left(\frac{e\,B_w}{E}\sin\left(\frac{2\pi}{\lambda_q}s\right)\right)^2 y. \end{cases} \tag{A.10}$$

That is, the WM behaves like a free field straight section in x and like a focusing lens in y. The average "elastic constant" along a WM wavelength is given by

$$K_y = \frac{1}{2}\left(\frac{e\,B_w}{E}\right)^2 \tag{A.11}$$

(this averaging procedure is allowed by Eq. (A.8)).

If the condition (A.8) does not hold, the integration of the Eqs (A.7) can be performed by using the Jacobian elliptic functions[29]. However this case is not relevant for typical SR and WM parameters.

Finally let us consider the evolution of the Twiss coefficients (α, β, γ) throughout a WM. The general methods for studying the propagation of these parameters can be found in the previously quoted Refs. 1-5 and in Ref. 30, where the most useful formulae on these topics are collected.

In this appendix we limit ourselves to the following particular cases:

a) Vertical motion

Let us consider an e.b. which enters the WM with arbitrary (α_y, β_y, γ_y). These coefficients (and then σ_y and σ'_y, Eqs (2.19), (2.20)) change throughout the WM. However, if we "match" the input Twiss coefficients to the WM focalization structure, (α_y, β_y, γ_y) do not change. This matching can be achieved by requiring that the input Courant-Snyder invariant (Eq. (2.15)) be equal to the harmonic oscillator invariant related to the second of the Eqs (A.10), which reads (see Eq. (A.11))

$$W_o = \frac{1}{\sqrt{2}}\left(\frac{e\,B_w}{E}\right)y^2 + \sqrt{2}\left(\frac{E}{e\,B_w}\right)y'^2. \tag{A.12}$$

From (2.15) and (A.12) we obtain

$$\alpha_y = 0, \; \beta_y = \sqrt{2}\left(\frac{E}{e\,B_w}\right), \; \gamma_y = \frac{1}{\beta_y} . \tag{A.13}$$

b) Radial motion

From the first of Eqs (A.10) we see that, along the x direction, the WM has no focalization properties. In this connection the e.b. divergence cannot change (there are no forces!). Thus we have the first result (see Eq. (2.20))

$$\gamma_x = \text{const.} \tag{A.14}$$

From Equations (2.17) and (2.16) we obtain the following equation for $\beta_x(s)$

$$\frac{1}{4}\,\beta_x'^2(s) + 1 = \beta_x(s) \cdot \gamma_x . \tag{A.15}$$

The solution of Eq. (A.15) can be easily found:

$$\beta_x(s) = \beta_x^o + \frac{\left(s - s_o\right)^2}{\beta_x^o} . \tag{A.16}$$

The minimum β_x^o is obtained in $s = s_o$, called the "beam waist". The e.b. envelope is given by (see Eq. (2.19))

$$\sigma^2(s) = \sigma^2(s_o)\left(1 + \left(A\,\frac{\left(s - s_o\right)}{2\pi\,\sigma^2(s_o)}\right)^2\right) . \tag{A.17}$$

It is interesting to note that Eq. (A.17) is identical to that describing the evolution of the envelope of a photon beam in an optical resonator[31].

APPENDIX B - OFF-ENERGY CLOSED ORBIT IN A WIGGLER MAGNET

From the first of Eqs (A.9) we can derive the off-energy closed orbit η in a WM,

$$\eta_{WM}(s) = \left.\frac{\partial x_R}{\partial \varepsilon}\right|_{\varepsilon=0} = \frac{e\,B_w}{E_o}\left(\frac{\lambda_q}{2\pi}\right)^2\left(1 - \cos\left(\frac{2\pi}{\lambda_q}\,s\right)\right). \tag{B.1}$$

The quantity (B.1) is generally very small compared with the typical η generated by the bending magnets. For example, in LEDA-F2 the maximum η is (see Fig. 4)

$$\eta_{MAX} \sim 2\ \text{m}$$

while the η generated by the high field WM is

$$\left.\eta_{WM}\right|_{MAX} \approx 1\ \text{cm}\ \left(B_w \sim 50\ \text{kG},\ E_o = 750\ \text{MeV},\ \lambda_q = 40\ \text{cm}\right).$$

REFERENCES

1. For a comprehensive review of problems relative to the electron storage ring, see M. Sands, The Physics of the Electron Storage Rings, Proc. Int. School "E. Fermi", Course XLVI, Ed. B. Touschek, Academic Press (1971)
2. E.D. Courant and H.S. Snyder, Ann. Phys. 3, 1 (1958)
3. H. Bruck, "Accélérateurs Circulaires de Particules", Press Universitaire de France, Paris (1966)
4. M.H. Blewett Ed. "Theoretical Aspects of the Behaviour of Beams in Accelerators and Storage Rings", CERN 77-13, Geneva (1977)
5. A.A. Kolomensky and A.N. Lebedev, "Theory of Cyclic Accelerators" North-Holland, Amsterdam (1966)
6. See, for example,"Wiggler Magnets"- Stanford Synchrotron Radiation Project, Report 77/05 - Stanford (California) (1977)
7. A. Hutton, Part. Accel. 7, 177 (1976)
8. B.M. Kincaid, J. Appl. Phys. 48, 2684 (1977)
9. J.M.J. Madey, J. Appl. Phys. 42, 1906 (1971)
L.R. Elias,W.M. Fairbank, J.M.J. Madey, H.A. Schwettman and T.I. Smith, Phys. Rev. Lett. 36, 717 (1976)
D.A.G. Deacon, L.R. Elias, J.M.J. Madey, G.J. Ramian, H.A. Schwettman and T.I. Smith, Phys. Rev. Lett. 38, 892 (1977)
10. R. Barbini, G. Dattoli, T. Letardi, A. Marino, A. Renieri and G. Vignola, IEEE Trans. Nucl. Sci. NS26, 3836 (1979)
11. L. Landau and E.Lifshitz "The Classical Theory of Fields", Pergamon Press, Oxford, Chapter 9 (1961)
12. K. Robinson, Phys. Rev. 111, 373 (1958)
13. See, for example, I. Prigogine "Non Equilibrium Statistical Mechanics", Interscience Publishers, 64 (1962)
14. R.H. Helm, M.J. Lee and P.L. Morton, Proc. of the 1973 Particle Accelerator Conference, San Francisco (California) (1973)
15. A. Bambini, A. Renieri, Proc. VII Course Int. School Quantum Electronics, Le Pianore 1977, Plenum Press (1977), Nuovo Cimento Lett. 21, 399 (1978)
A. Bambini, A. Renieri and S. Stenholm, Phys. Rev. A 19, 2013 (1979)
G. Dattoli and A. Renieri, Nuovo Cimento Lett. 24, 121 (1979)
G. Dattoli and A. Renieri,CNEN Report 79.37/p, Centro di Frascati, Frascati, (Italy)
16. G. Dattoli, A. Marino and A. Renieri, "On the Theory of the Free Electron Laser", these Proceedings
17. J.D. Jackson "Classical Electrodynamics", J. Wiley, New York, Chapter 15.5 (1975)
18. A. Renieri, CNEN Report 77.30, Centro di Frascati, Frascati (Italy) (1977)
19. A. Renieri "The Free Electron Laser: The Storage Ring Operation" Proc. Int. School of Phys. "E. Fermi" - 2nd Course 1978, Ed. by A. Marino and C. Pellegrini, Varenna 1978 - LXXIV SIF), to be be published

20. A. Renieri, Nuovo Cimento B 53, 160 (1979) and IEEE Trans. Nucl. Sci. NS-26, 3827 (1979)
21. A. Renieri, CNEN Report 77.33, Centro di Frascati, Frascati (Italy)
22. G. Dattoli and A. Renieri, Nuovo Cimento B 59, 1 (1980)
23. D.A.G. Deacon and J.M.J. Madey, Appl. Phys. 19, 295 (1979)
D.A.G. Deacon "Theory of the Isochronous Storage Ring Laser" Report HEPL 854 (1979), Stanford University, Stanford (California)
D.A.G. Deacon and J.M.J. Madey, Phys. Rev. Lett. 44, 449 (1980)
N.M. Kroll, P. Morton and M.N. Rosenbluth, SRI Technical Report JSR-79-01 (1980), SRI International, Arlington (Virginia)
C.A. Brau, IEEE J. Quant. Electron. QE-16, 335 (1980)
24. L.R. Elias, J.M.J. Madey and T.I. Smith, "One Dimensional Monte Carlo Analysis of a Free Electron Laser in a Storage Ring", Stanford High Energy Physics Laboratory Report HEPL 824 (1978)
25. H. Al-Abawi, F.A. Hopf, G.T. Moore and M.O. Scully, Opt. Commun. 30, 235 (1979)
26. F.A. Hopf, P. Meystre, M.O. Scully and W.L. Luisell, Opt. Commun. 18, 413 (1976) and Phys. Rev. Lett. 37, 1342 (1976)
H. Al-Abawi, F.A. Hopf and P. Meystre, Phys. Rev. A 16, 666 (1977)
27. S.A. Mani "A Study to Design a Free Electron Laser Experiment-Volume IV - FEL Theory", WJSA-FTR-79-135, W.J. Schafer Associates, Inc., Wakefield (Massachusetts) (1979)
28. C. Pellegrini, Proc. of the Workshop in "The Possible Impact of Free Electron Lasers on Spectroscopy and Chemistry" Ed. by G. Scoles, Riva del Garda (Italy), 1 (1979)
A. Renieri, ibidem, 28
29. See, for example, M. Abramowitz and I.A. Stegun, Handbook of Mathematical Functions, Dover, New York, 567 (1970)
30. C. Bovet, G.S. Cohen, R. Gouiran, I. Gumowski, K.H. Reich and W.M. Saxon, "A Selection of Formulae and Data Useful for the Design of A.G. Synchrotrons" MPS-SI (Int. DL/68-3) CERN, Geneva (1968)
31. H. Kogelnick and T.Li, Proc. IEEE 54, 1312, Sect. 3.2 (1966)
32. G.M. Zaslavskii and B.V. Chirikov, Sov. Phys. Usp. 14, 549 (1972)
33. J.M.J. Madey and D.A.G. Deacon, in Cooperative Effects in Matter and Radiation, Plenum Press, New York, N.Y., 313 (1977)

GAIN EXPANDED FREE ELECTRON LASERS*

James N. Eckstein

W. W. Hansen Laboratories of Physics
Stanford University
Stanford, California 94305 USA

Gain expansion is a technique that was proposed to make the operation of an FEL insensitive to energy spread in the electron beam (1,2). To see how this can be done, we note that the basic synchronism condition

$$\lambda_r = \frac{\lambda_w}{2\lambda^2} \left(1 + \frac{e^2 B_w^2 \lambda_w^2}{(2\pi\ mc^2)^2}\right) \equiv \frac{\lambda_w}{2\gamma^2} \left(1 + \alpha^2\ B_w^2\right) \tag{1}$$

can be satisfied even for electrons of differing energies if the value of the magnetic field seen by any electron, B_w, is correlated with its energy, γ. This leads to energy independent gain. Gain expanded magnets accomplish this correlation by incorporating a transverse gradient to the wiggler field and dispersing the electrons injected in the FEL to match the intrinsic dispersion of the wiggler.

We will investigate the mechanism for gain in these devices and will find that it differs fundamentally from the gain mechanism for standard FELs. It is necessary in gain expanded FELs to couple the transverse motion of the electrons to their phase evolution. We will present new scaling relations which incorporate new laser parameters that characterize the transverse dynamics. In addition, we will show that a second regime of operation exists at high optical power intensities which exhibits electron trapping and allows for high single pass energy extraction.

---

*Work supported in part by the Office of Naval Research under contract N00014-78-C-0403.

If we consider variations in γ and B, we see that

$$d\lambda_r = -\frac{2}{\gamma}\frac{\lambda w}{2\gamma^2}(1+\alpha^2 B_w^2)\,d\gamma + \frac{2}{B}\frac{\lambda w}{2\gamma^2}\alpha^2 B_w^2\,dB_w\,. \tag{2}$$

Demanding that this be zero and dividing through by dx, we have

$$\frac{1}{\gamma}\frac{d\gamma}{dx} = \frac{\alpha^2 B_w^2}{1+\alpha^2 B_w^2}\;\frac{1}{B_w}\frac{dB_w}{dx}\,. \tag{3}$$

If we introduce η as the energy dispersion coefficient and k as the magnetic field dispersion coefficient,

$$\delta(x) = \gamma_0\,(1+\eta x) \quad B_w(x) = B_w\,(1+kx)$$

for energy independent gain we need

$$k = \eta\frac{(1+\alpha^2 B_w^2)}{\alpha^2 B_w^2} \tag{4}$$

from Eqn. 3. When this relationship holds electrons are properly dispersed in the transverse coordinate, x, as a function of energy to match the transverse dispersion of the magnet. All electrons remain synchronized to radiate at the same wavelength throughout the interaction.

The transverse gradient also introduces higher order effects, most notably it tends to deflect the beam from the region of high field to low field. This can be remedied by adding a constant, non-periodic transverse field that deflects the beam back on axis. A gradient, s, can be included in this field as a convenient way to adjust the net dispersion of the wiggler. We write the wiggler field as

$$B_w(x,z) = \hat{y}\,\left|B_w(1+kx)\,\sin k_w z + B_c(1+sx)\right| \tag{5}$$

where

$$B_c = \alpha^2 B_w^2\,\frac{\pi^2 k^2}{2\gamma}\,\frac{mc^2}{e}$$

and γ_0 is the energy at which an electron injected at $x = x' = 0$ will travel a straight line through the magnet. Likewise when "s" is chosen properly, electrons with an energy offset $\delta\gamma$ and injected with no transverse momentum at a position $x = (k-s)^{-1}\,\delta\gamma/\gamma$ will also travel straight through as long as no energy is exchanged with the optical field. Electrons in inappropriate transverse positions, for example after having exchanged energy with the optical field, will execute transverse "betatron" oscillations around their ideal transverse position determined by their instantaneous value of $\delta\gamma$.

Detailed microscopic equations can be written for the transverse and longitudinal coordinates including the interaction with and the dynamics of the light field. The ensuing motion occurs on two scales. On the smallest scale electrons are jiggled back and forth by the magnetic field at the optical frequency. This motion is superimposed on the slower betatron oscillations and phase slippage between the electron's transverse velocity and the electric field's phase. The rapid variations can be averaged out, and in a way reminiscent of renormalization theory, equations exclusively for the larger scale motion are then obtained with microscopic parameters appearing in the coupling constants. If the resulting nonlinear equations are linearized in small transverse displacements and energy offsets, we get (2):

$$x'' = -\frac{k}{2}\left(\frac{\lambda_w}{2\pi\rho_o}\right)^2 \left|(k - s)\, x - \delta\right| \tag{6}$$

$$\delta' = \frac{eE}{\gamma mc^2}\,\frac{\lambda_w}{2\pi\rho_o}\,\sin\psi\,\left|J_0(\xi) - J_1(\xi)\right| \tag{7}$$

$$\psi' = \frac{4\pi}{\lambda_w}\,\frac{\gamma_o}{\gamma_{r_o}}\left|\frac{\gamma_o - \gamma_{r_o}}{\gamma_o} + \delta - (k - s)x\right| \tag{8}$$

where the following definitions are used:

γ_o = injection energy at $x = 0$.

γ_{r_o} = resonance energy at $x = 0$; at this energy the optical wave slips exactly one period past an electron while the electron traverses one wiggler period.

$\delta = (\gamma - \gamma_o) / \gamma_o$.

$\rho_o = \gamma_o mc^2/eB_w$.

$\xi = \alpha^2 B_w^{\ 2}/(4(1 + \alpha^2 B_w^{\ 2}))$.

J_0, J_1 are cylindrical Bessel functions.

These equations describe the dynamics of the electron in the presence of an assumed carrier wave. The angle, ψ, as usual is the angle between the electric field vector and the electrons transverse velocity.

From the structure of the equations, we can see that the dynamics of a particle entering at $x = 0$ with $\gamma = \gamma_0$, $(\delta = 0)$, is identical to that of a particle entering at $x = x_{off}$ with $\delta = (k - s)x_{off}$. This explicitly shows that the dynamics are independent of the initial electron energy.

In analogy with conventional wigglers we now construct a quasi-pendulum equation. The crucial point to note is that terms on the right hand side of Eqn. 6 and Eqn. 8 are identical. If we substitute from Eqn. 6 into Eqn. 8 and integrate from 0 to z, we get

$$x'(z) - x'(0) = \frac{\lambda_w}{4\pi} \; \frac{\gamma v_0}{\gamma_0} \; \frac{k}{2} \left(\frac{\lambda_w}{2\pi\rho_0}\right)^2 |\psi(z) - \psi_0 - q_z| \qquad (9)$$

where we have written $\Psi'(0) = q$, and qL is the total optical phase slip. For perfect injection, $x'(0) = 0$ and Eqn. 9 can be used to eliminate $x'(z)$ in a second order differential equation for Ψ'' obtained by taking the spatial derivative of Eqn. 8. What one gets is:

$$\psi''(z) = \frac{2e^2B_wE}{(\gamma mc^2)^2} \; (J) \sin\psi - \frac{k(k - s)}{2} \; \left(\frac{\lambda_w}{2\pi\rho_0}\right)(\psi - \psi_0 - qz) \qquad (10)$$

where (J) represents the Bessel function term in Eqn. 7. This is always near 1 and for convenience we ignore it from now on. This equation clearly implies the two regimes of operation; we can most easily see this if we normalize the interaction length to 2π. To do this, we multiply both sides by $(L/2\pi)^2$ and use the substitution $\mu = 2\pi\ z/L$, where μ is now our length variable. Writing derivatives with respect to μ with primes, we arrive at

$$\psi'' = \frac{E}{E_s} \sin\psi - N^2 \left(\psi - \psi_0 - \frac{qL}{2\pi}\mu\right) \qquad (11)$$

where

$$E_s = \frac{(2\pi\gamma mc^2)^2}{2e^2B_wL^2}$$

is a field amplitude characteristic of saturated operation of standard FELs and

$$N^2 \equiv \left(\frac{\tilde{\Lambda}L}{2\pi}\right)^2 = \frac{L^2}{(2\pi)^2} \; \frac{k(k - s)}{2} \; \left(\frac{\lambda_w}{2\pi\rho_0}\right)^2$$

is the square of the number of transverse betatron oscillations an electron executes during the interaction. The first term in the dimensionless equation causes motion characteristic of standard FELs and is responsible for energy exchange with the optical field. The second term comes from introducing the transverse gradient to the field and it causes the transverse betatron motion to occur.

The two regimes are determined by the relative magnitudes of the coefficients in Eqn. 11. Fig. 1 shows this graphically. In the case where $E/E_s < N^2$ the motion is dominated by the second term. In this regime the gain mechanism requires phase matching not only between the optical wave and electron bunches, but also requires phase matching with the betatron oscillations. Gain and saturated output power are related to results for conventional FELs by simple scaling relations involving E/E_s and N. In the case where $E/E_s >> N^2$ electrons are trapped in pondermotive potential wells and high energy extraction is possible. The transverse motion is no longer periodic for the trapped electrons. Rather they adiabatically are correlated in transverse position with their energy loss. The transverse gradient plays the role that tapering the wiggler does in other high energy extraction devices.

NORMAL REGIME ($E/E_s \lesssim N^2$)

To see how gain and saturation occur, we have numerically integrated the equations of motion. Gain curves are obtained using Eqns. 7 and 11 by integrating a set of electrons, averaging over ψ_0 from 0 to 2π. For gain expanded magnets, gain curves are most conveniently plotted versus optical phase slip, qL, which is just the "extra" amount of electron phase slippage during the interaction beyond what occurs exactly at resonance. That is, for a magnet of "n" periods the number of optical phases that slip past an electron is n - qL. Regular FELs have maximum small signal gain at qL - 2.6. A "gain" curve for a gain expanded magnet is shown in Fig. 2. Here $\langle\delta\gamma\rangle$ is plotted, in arbitrary units, versus $qL/2\pi$. For this example N, the number of betatron oscillation periods, was chosen to be 10. This result contrasts sharply with what is obtained for regular FELs. First, maximum gain occurs when $qL/2\pi = N$. The gain curve is also symmetric about the point of maximum gain, at least in that neighborhood, although it is still antisymmetric around resonance, qL = 0. This leads to considerably different frequency pulling behavior at saturation. To see how small signal gain scales with betatron phase advance, we integrated gain curves for many different values of "N". The magnitude of the small signal gain is proportional to 1/N. In fact comparing to regular FELs, the maximum small signal gain is given by

$$G_{ss}(N) = 0.293\, G_{ss},\ \text{reg.} \cdot \frac{1}{N} \qquad (12)$$

It is also interesting to look at how the gain saturates with intensity. For values of $E/E_s \simeq N$, maximum gain still occurs at

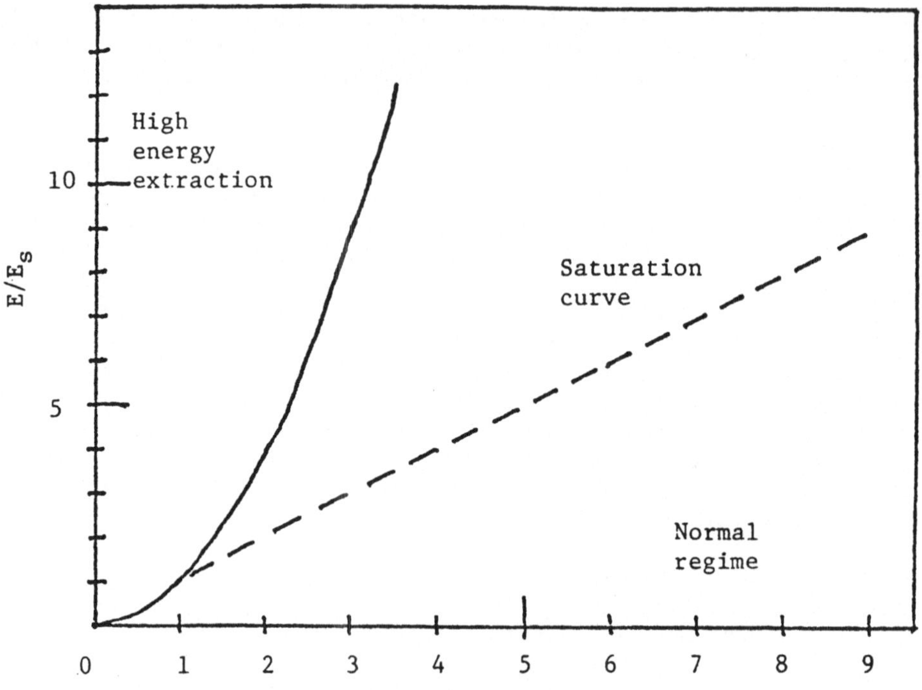

Figure 1

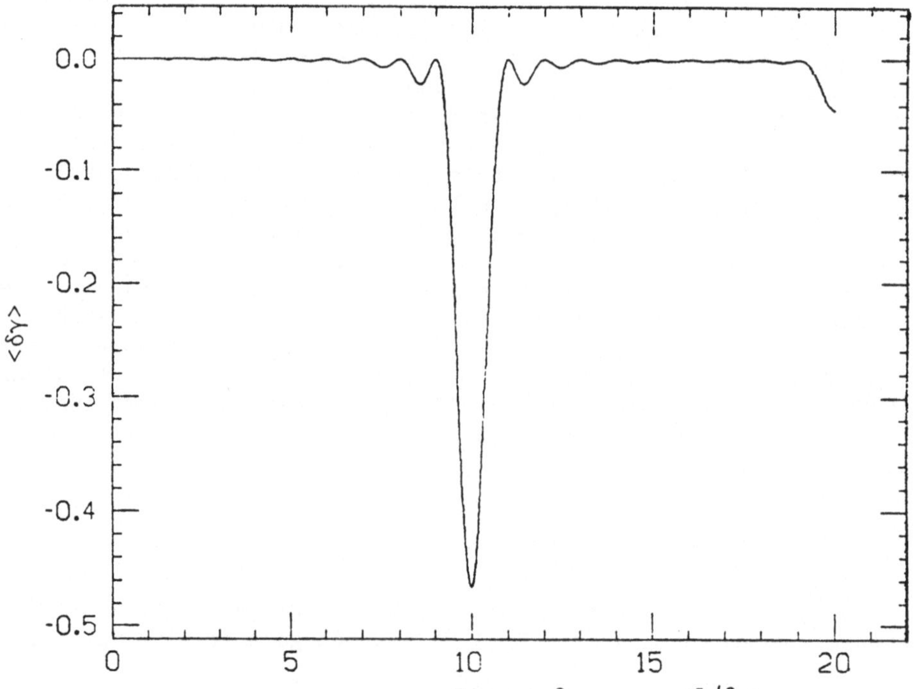

Figure 2

$qL/2\pi = N$, though the saturated gain is given by

$$G(N,I) = G_{ss}(N) \cdot \frac{1}{1 + I/I_s} \tag{13}$$

where

$$I_s(N) = \frac{c}{8\pi} E_s^{\,2} \cdot N^2 \tag{14}$$

It is easy to show that the maximum intensity that can be coupled out is then given by

$$I_{out}(N) = G_{ss}(N) I_s(N) = 0.293\ G_{ss},\ \text{reg}\ I_s \cdot N \tag{15}$$

These relationships, coupled with the scaling behavior of standard FELs, show how gain and saturated output power scale with all of the parameters describing a device, as well as how operation of a standard FEL compares with a gain expanded FEL in the normal operating regime.

Gain curves tell us about how well we can expect gain expanded FELs to work and how their characteristics scale, but do not tell us how they work. Indeed, it is almost paradoxical that maximum gain occurs for qL greater than 2π. It would seem that no net energy would be given to the optical field since the electrons would spend as much time absorbing light as radiating light. This paradox can be resolved by forming a physical picture of the radiation process which we can obtain by looking at electron trajectories in the ψ, x "phase space".

In Fig. 3 a histogram is plotted which shows the relative amount of time different values of ψ are populated during the interaction. Here qL has been chosen to equal $2\pi N$ for maximum gain. Electrons initially populate the phases between 0 and 2π. The electrons slip to near $\psi = qL$ by the end, but along the way they preferentially populate those values of ψ which radiate in phase into the optical field.

To see how this happens, consider Fig. 4. Here we plot $x(k-s)n$, where "n" is the number of magnet periods, versus ψ. Again electrons are initially distributed between 0 and 2π. Electron phase evolution is, as. Eqn. 8 shows, coupled to transverse position. Physically, when the electrons move into a region of higher field they are bent more sharply, reducing their longitudinal velocity. This serves to increase the bunching of the beam, to momentarily slow the rate of optical phase slippage so that electrons spend an increased amount of time phased to radiate into the field, and thirdly, to increase

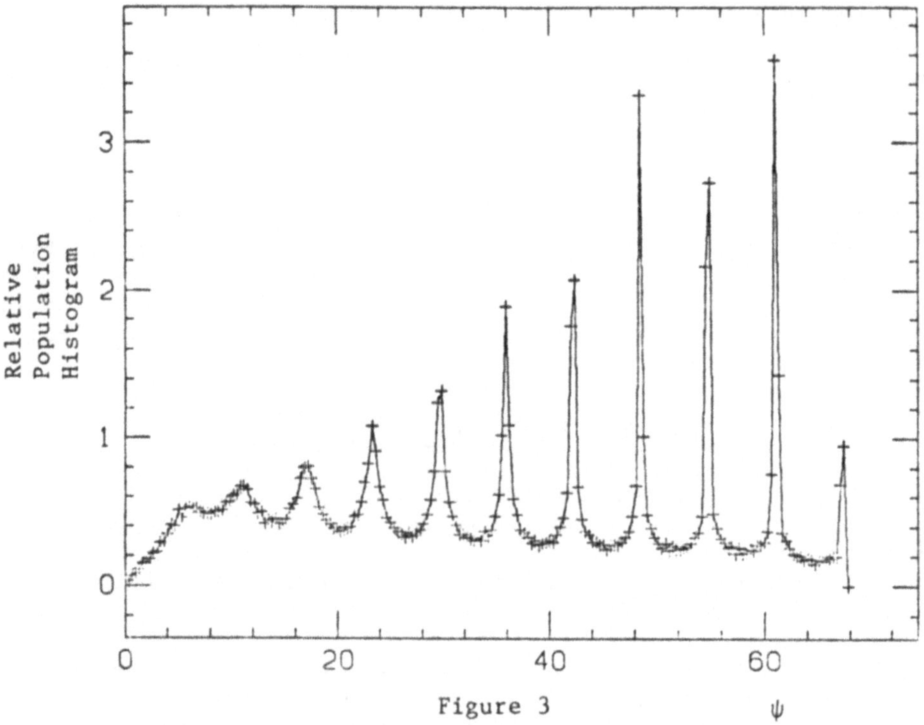

Figure 3

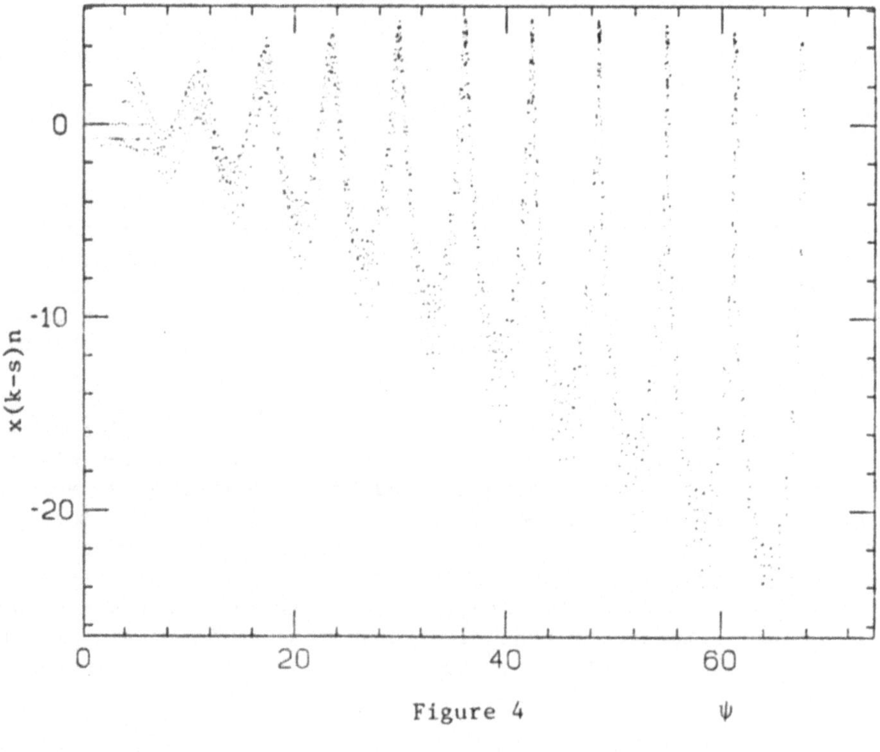

Figure 4

the rate at which they radiate. Conversely, when electrons reach the low field regions their longitudinal velocity increases decreasing the bunching, optical phase slippage occurs more rapidly causing the electrons to spend less time at phases that absorb light and since they experience a weaker periodic magnetic field, they also radiate less. It is the phase matching between the betatron motion and the optical phase slippage, expressed by

$$qL = 2\pi N$$

that causes the bunches to radiate into the field.

HIGH ENERGY EXTRACTION REGIME ($E/E_s >> N^2$)

When the light field becomes very intense, the electron dynamics no longer occur as discussed above. Optical traps contain a fraction of the electrons which then are decellerated giving a substantial fraction of their energy to the optical field. This is shown in Fig. 5, where as before electrons are initially distributed between 0 and 2π. This time "N" was chosen to be 2 and E/E_s was taken to be 256. If the electric field were less than or about equal to "N", the electrons would populate phases out to $2\pi N$. Here, however, some of the electrons are trapped near their initial phase. Figures 6 and 7 indicate what happens to these electrons. In Fig. 6 we follow the transverse displacement as a function of time for several of the electrons. When the optical field is such that the device is operating in the "normal" regime, similar graphs show the transverse betatron oscillations around the transverse position correlated with the instantaneous average value of the electrons' energy. In this case, electrons which are trapped to not oscillate. Instead they adibatically move to transverse positions which maintain synchronism with the optical field, depending on their energy loss. This can be seen to be closely analogous to the electron dynamics in tapered wigglers where synchronism is maintained by changing magnet parameters as a function of longitudinal position. In this case, though, the actual trajectory followed in phase space is not fixed a priori by magnet design. Rather the electrons adiabatically choose the appropriate trajectory.

The author would like to thank the following individuals for many useful conversations: J. Madey, N. Kroll, P. Morton and M. Rosenbluth. Two publications now in preparation with these individuals as well as the present author will present this material in more detail as well as provide analytic solutions to the equations of motion for gain expanded magnets.

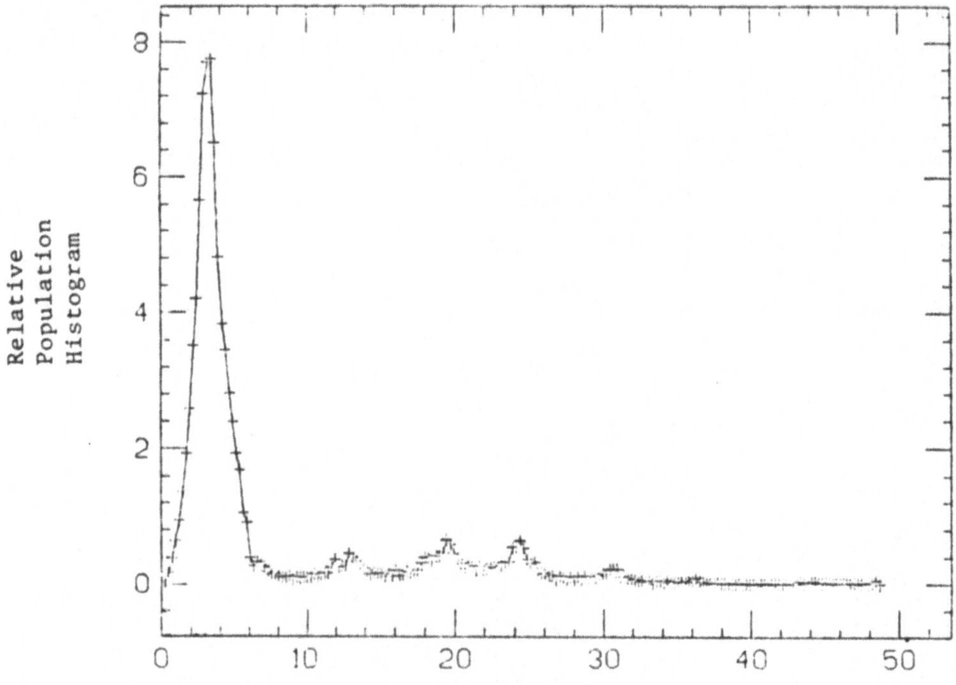

Figure 5

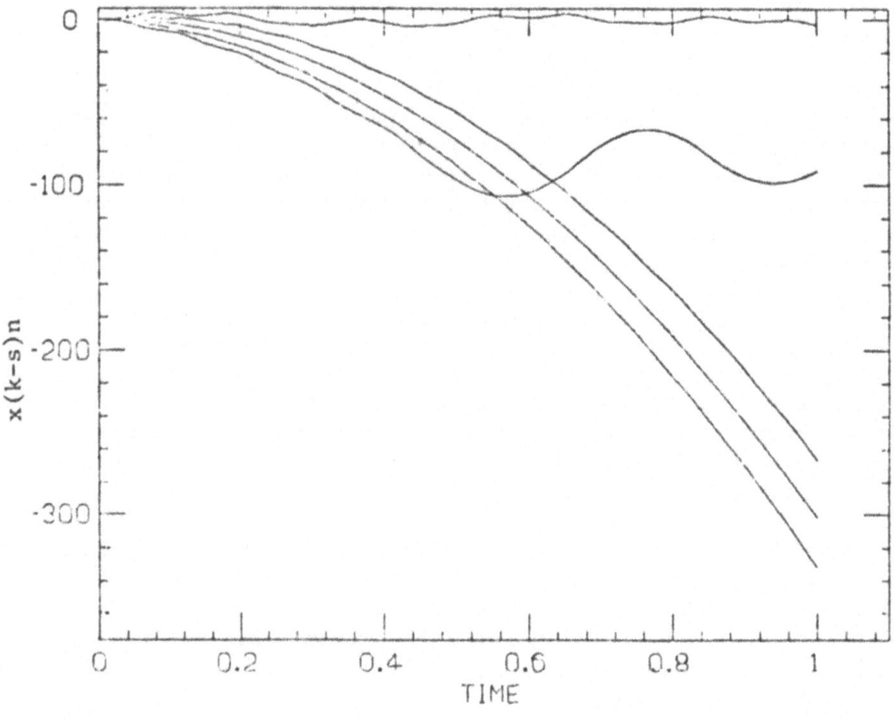

Figure 6

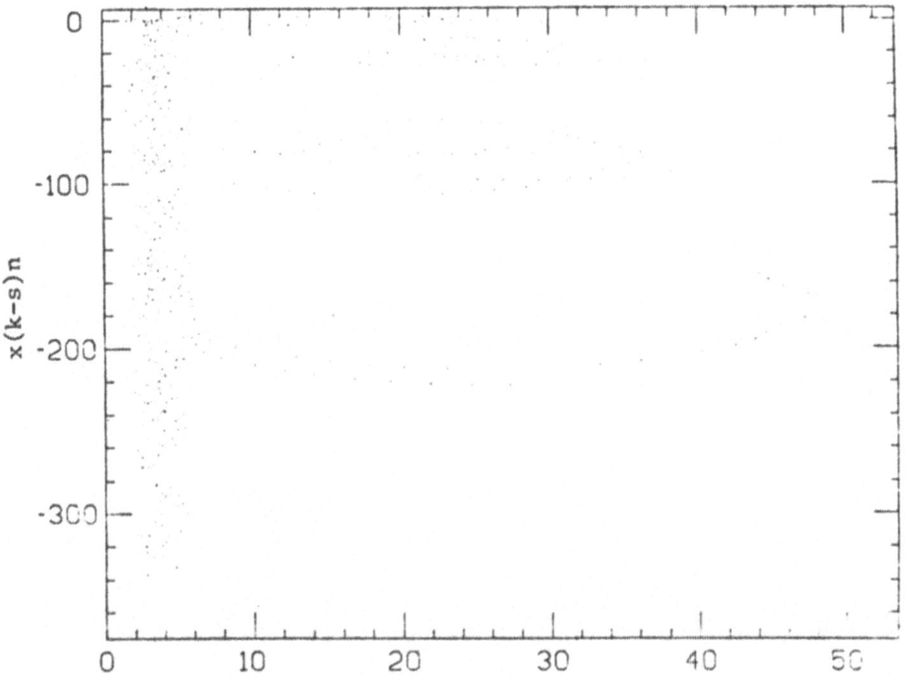

Figure 7

REFERENCES

1. T. I. Smith, J. M. J. Madey, L. R. Elias and D. A. G. Deacon J. Appl. Phys. 50, 4580 (1979).
2. J. M. J. Madey and R. C. Taber, in Physics of Quantum Electronics, V. 7, ed. S. Jacobs, H. Pilloff, M. Sargent, M. Scully and R. Spitzer, Addison-Wesley (1980).

CANCELLATION OF TRANSVERSE EXCITATION IN GAIN EXPANDED FREE ELECTRON LASERS

John M. J. Madey

Physics Department & High Energy Physics Laboratory
Stanford University
Stanford, CA 94305

ABSTRACT

We define and analyze emittance growth in gain expanded free electron lasers, develop criteria for the minimization of this phenomenon, and present some examples of the application of these criteria.

I. INTRODUCTION

Smith, Madey, Elias and Deacon have proposed the use of a periodic magnetic field with a transverse gradient to reduce the sensitivity of free electron lasers to variations in electron energy.[1] The realization of energy independent gain in a storage ring free electron laser would permit a substantial improvement in laser power output and efficiency.[2] The present analysis considers the effect of gain expansion on the transverse coordinates of the electrons, and the means by which these effects can be minimized.

Smith's discussion of gain expansion assumed the correlation of the electrons' energy and transverse position. In a storage ring, the nominal transverse position of the electrons is determined by the dispersion function $\eta(s)$:

$$x(s) = \eta(s)\cdot\delta \tag{1.1}$$

when δ is the fractional deviation of the electron energy from the design energy γ_0:

$$\delta \equiv (\gamma - \gamma_0)/\gamma_0 \tag{1.2}$$

and x(s) is the transverse displacement of the electron measured relative to the design orbit.[3] But Eq. (1.1) predicts only the nominal transverse position; the electrons will oscillate about this position with an amplitude determined by the quantum fluctuations in the synchrotron radiation, the excitation due to laser operation, and the fields and gradients in the magnet lattice. This "betatron motion" tends to spoil the correlation of x and δ, and can significantly reduce the effectiveness of the gain expansion scheme.

The means to reduce the betatron motion in a storage ring due to quantum fluctuations in the synchrotron radiation are well known[4], and it would appear that the loss of correlation due to these fluctuations would be quite acceptable for a gain expanded system. This paper extends the analysis to include the emittance growth due to the laser, defines the conditions which must be satisfied to minimize the laser-induced emittance growth, and demonstrates the application and effectiveness of these guidelines with numerical data for some selected examples.

The analysis makes use of the equations of motion developed by Madey and Taber.[5] Succeeding papers in this series describe the general effect of the electrons' transverse excitation on gain,[6] the use of scaling relations in the analysis of laser performance,[7] and the optimization of magnet and laser parameters.[8]

II. NATURE AND DESCRIPTION OF TRANSVERSE EXCITATION

If the nominal transverse position of an electron with energy deviation δ is $\eta\delta$, and the actual transverse displacement is x, the betatron excitation amplitude x_β can be defined as:

$$x_\beta = x - \eta\delta \; . \tag{2.1}$$

Similarly, the slope x_β' can be defined as:

$$x_\beta' = dx_\beta/dS = x' - \eta'\delta - \eta\delta' \; . \tag{2.2}$$

Even if δ' is constant, the values of x_β and x_β' will change with position about the ring due to the oscillatory character of the betatron motion and the variation of η with position. The values of x_β and x_β' taken individually, are therefore not particularly useful measures of the degree of transverse excitation.

If we were to follow the motion of an electron moving off-axis through many orbits around the ring, the coordinates (x_β, x_β') measured at a given point on the circumference of the ring would move around the perimeter of an ellipse. Although the aspect ratio and orientation of the ellipse would depend on the point at which x_β and x_β' are measured, the area of the ellipse is nearly invariant[9]:

$$A = \pi[\frac{x_\beta^2}{\beta} + \beta(x_\beta' - \frac{\beta'}{2\beta} x_\beta)^2] . \tag{2.3}$$

Computed for an individual electron, the value of A defines the degree of transverse excitation of that electron. The average of A over the electron distribution defines the contribution of the betatron motion to the emittance of the electrons.

The function β in Eq. (2.3) must be computed from the magnitude of the magnetic fields and gradients in the ring. Specifically, β is the solution to the equation[10]:

$$\frac{d\beta^2}{dS^2} - \frac{1}{2\beta} \frac{d\beta}{dS}^2 = 2[K_x(S)\beta + \frac{1}{\beta}] \tag{2.4}$$

where

$$K_x(S) = - \frac{1}{\rho_0} [\frac{1}{\rho_0} + \frac{dB/dx}{B}] \tag{2.5}$$

ρ_0 = nominal gyro radius

= $p_0c/eB_0(S)$

$dB/dx \equiv$ transverse gradient

$S \equiv$ azimuthal coordinate.

The dispersion function η is also determined by ring parameters.[11]

$$\eta'' = K_x(S) + \frac{1}{\rho_0} . \tag{2.6}$$

As noted, the ellipse area A is only approximately invariant. We can observe that the adiabatic changes in energy associated with the synchrotron motion result in the periodic variation of the ellipse area, that damping due to synchrotron radiation causes the ellipse area to shrink, and that the non-adiabatic quantum fluctuations in the synchrotron radiation randomly excite both the transverse and longitudinal coordinates. The changes in A due to quantum fluctuations and radiation damping are of particular importance, since they determine the equilibrium magnitude of A.[12] The changes in A due to the laser will determine the magnitude of the laser-induced emittance growth.

The perturbation due to the laser interaction arises both from the change in position x and slope x' during the interaction, and the change in energy deviation δ. Some insight into the problem can be gained by considering the case in which the electrons enter the laser moving on-axis. If an ensemble of electrons uniformly distributed in x were to enter the interaction region with $x = x' = \delta = 0$, as shown in Fig. 1, each would in general emerge with

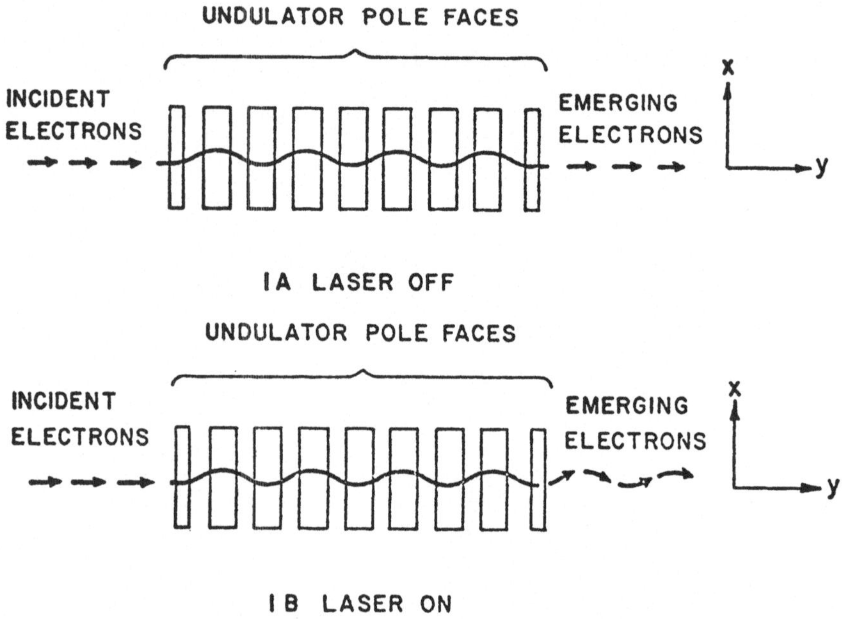

Fig. 1. In Fig. 1A, we indicate the trajectories of an ensemble of electrons moving through the FEL insertion when the laser is off. If the electrons enter on-axis, with no transverse momentum, they emerge in the same state. The motion of these electrons defines the "nominal" electron trajectory. In Fig. 1B, the laser has been turned on. Electrons entering on-axis now deviate in both position and transverse momentum from the nominal trajectory when they leave the insertion.

transverse and energy coordinates which differed from zero, resulting in an emittance equal to:

$$\langle \delta A \rangle = \langle \frac{x_\beta^{\ 2}}{\beta} + \beta \left(x_\beta' - \frac{\beta'}{2\beta} x_\beta \right)^2 \rangle$$

$$x_\beta \equiv x - \eta \cdot \delta$$

$$x_\beta' = x' - \eta' \delta \ . \qquad (2.7)$$

(In defining x_β', we assume that the distribution is analyzed just after is leaves the interaction region, so that $\delta' = 0$.)

In a non-gain expanded FEL, this emittance growth can be controlled by designing the ring and the laser to maintain $\eta = \eta' = 0$ at the ends of the interaction.[13]

This choice is not possible in a gain expanded system which requires a finite value of η for operation. For gain expanded systems, suppression of the emittance growth requires a magnet design for which $x = \eta\delta$ and $x' = \eta'\delta$ at the end of the laser magnet. Surprisingly enough, this is not particularly difficult to accomplish. An analysis of the problem and the constraints on magnet design is given in the next section.

The definition of emittance growth for off-axis initial motion can be developed as a generalization of the result for on-axis initial motion. If we assume that an ensemble of electrons enters the interaction region distributed uniformly in the axial coordinate x, with betatron coordinates (x_β, x_β') distributed uniformly around the perimeter of the phase space ellipse, A, the electrons emerging from the interaction will be distributed on the perimeter of a deformed ellipse A' (Fig. 2). The emittance growth can then be defined as:

$$\langle \delta A \rangle = \pi\left\langle \frac{x_\beta^2}{\beta} + \beta\left(x_\beta' - \frac{\beta'}{2\beta}x_\beta\right)^2 \right\rangle - A_i$$

$$= \pi\left\langle \frac{x_\beta^2}{\beta} + \beta\left(x_\beta' - \frac{\beta'}{2\beta}x_\beta\right)^2 - A_i \right\rangle \tag{2.8}$$

where A_i is the area of the initial ellipse and the average is taken over the members of the ensemble.[14] (Note that since x_β and x_β' can change substantially during the laser interaction, it may not be appropriate in computing this average to approximate δx_β^2 by $2x_\beta\delta x_\beta$ as is done to evaluate the excitation due to the quantum fluctuations in synchrotron radiation.)

There does not appear to be any general way to cancel the emittance growth for off-axis initial motion. To the contrary, the linearized equations of motion suggest that non-zero values for δA are inevitable for off-axis motion, even if $\delta A = 0$ for the on-axis case. The quantitative evaluation of the effects of off-axis motion requires the integration of the full equations of motion. This analysis is carried out in Sec. V, and it is found that the emittance growth is tolerable if the magnet parameters are chosen with care.

Note that δA in Eq. (2.8) has no immediate relation to the difference in the areas enclosed by the initial ellipse A and the deformed ellipse A'. The areas enclosed by these two curves will usually be nearly identical, differing only in proportion to the

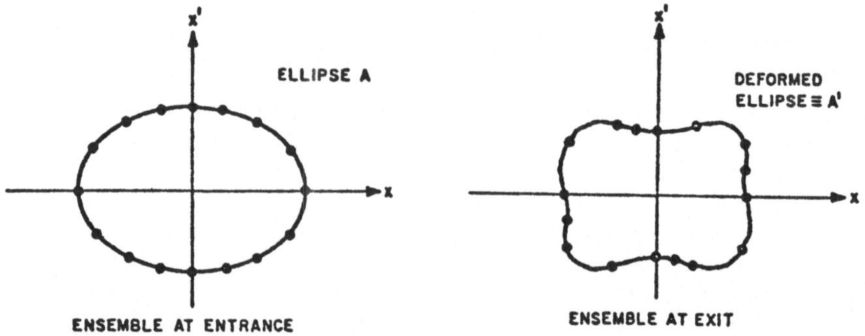

Fig. 2. The transverse excitation induced by the laser can be computed by considering the change in the mean value of the transverse invariant for an ensemble of electrons moving through the laser. At the entrance to the FEL insertion, the members of the ensemble are assumed to be uniformly distributed (in phase) around the perimeter of the ellipse generated by Eq. (2.15) (left figure). The right figure indicates the configuration of the ensemble at the end of the insertion. In general, both the shape of the volume enclosed by the ensemble, and the electrons' distribution around that shape will be altered by the interaction.

change in electron energy lost during the interaction.[15,16] The definition in Eq. (2.8) involves the distribution of electrons around the circumference of A' as well as the form of A'. The sign of δA can be negative even if the area enclosed by A' exceeds the area of the initial ellipse.

Note also that δA specifies only the shift of the centroid of the transverse distribution during the interaction. As discussed in the Appendix, to fully specify the electrons' distribution in the transverse coordinates, the higher moments δA^n must also be computed:

$$\delta A^n = \pi^n \left\langle \left[\frac{x_\beta^{\,2}}{\beta} + \beta (x_\beta{}' - \frac{\beta'}{2\beta} x_\beta)^2 - A_i \right]^n \right\rangle \; . \tag{2.9}$$

The Laser Insertion and the Storage Ring

Formally, the functions β and η are defined only if all the fields and gradients around the storage ring are known [Eq. (2.4)-(2.6)]. But the analysis of the excitation problem requires knowledge of β and η only within the laser insertion. To simplify the present analysis, we will fix by assumption the value of η within the insertion, and the values of β and β' at the start of the insertion, leaving to future studies the question of how these values can best be realized. We will assume that the dispersion function η is chosen so that the averaged dispersion in the insertion equals the intrinsic dispersion (1/k-s) of the periodic array[17], and will typically choose β and β' so that the trajectory in the coordinates (x,x') traced out by an electron as it moves through the interaction region will have the form of an erect ellipse.

In this approach, motion through the storage ring laser is divided into two parts: motion through the laser insertion, and motion through the remainder of the ring. In the laser insertion, the electrons interact with the optical field, losing (or gaining) energy with an associated displacement in phase and the transverse coordinates. The storage ring accepts the beam at the end of the insertion and transports it back to the beginning.

The change per turn in the electrons' distribution is then the sum of the changes due to the laser and the storage ring. The effects of the laser are discussed in the following sections. The effects of motion in the storage ring are well known. The focussing magnets in the ring advance the betatron phase, and energy is lost as synchrotron radiation. Also, the electrons are accelerated by the fields in the rf cavity, and drift in longitudinal position in proportion to their deviation from the synchronous energy.

The betatron phase advanced/turn is typically quite large. If the tune of the ring is ν, electrons move around the phase space ellipse at the angular frequency $\nu \cdot \omega_0$,[18] where ω_0 is the orbit frequency. The value of ν typically lies between 1 and 10. Physically, we rely on this rapid periodic circulation around the phase space ellipse to justify the hypothesis that the evolution of the electrons distribution in the transverse coordinates (x,x') will be governed by the ensemble average of the change in the invariant A.

As a consequence of synchrotron radiation the amplitude of the transverse motion is damped on each orbit. This synchrotron damping is necessary to limit the growth of the electrons' distribution, but represents a dissipation mechanism which must be minimized to optimize the efficiency. Quantum fluctuations in the synchrotron radiation also contribute to the excitation of the transverse (and longitudinal) coordinates, and set a lower limit to the attainable emittance.

The rf cavity in the storage ring is, of course, required to replace the energy lost by the electrons as laser and synchrotron radiation. The variation of RF voltage with phase, and the energy dependence of the phase also lead to slow oscillations of the individual electron's energy and phase, the so-called synchrotron oscillations.

We assume the synchrotron motion plays no role in the evolution of the transverse coordinates. We also assume that the phase difference between the electrons' transverse velocity and the optical electric field at the start of the interaction changes in a random way from one orbit to the next. The assumption is justified by noting that, unless special measures are taken, the energy spread acquired by the electrons during the laser interaction, and due to quantum fluctuations in the synchrotron radiation, will lead to a variation in the time required for electrons to travel around the ring and a resulting spread in the phase of the optical field acting on the fastest and slowest electrons in the next orbit. Although that portion of the variation in the transit time around the ring which can be attributed to laser operation is formally deterministic, the magnitude of the variation in optical phase at the start of the next interaction due to the laser induced energy spread and the sensitivity of the laser interaction to the initial optical phase indicates that there would be little sign of systematic or correlated motion on successive orbits. Of course, the energy and phase spread induced by the quantum fluctuations in synchrotron radiation would be purely random.

The criteria for the random phase approximation are well satisfied at infrared and visible wavelengths for time dilation factors $\alpha \gtrsim 10^{-2}$. Note, however, that correlated motion can be achieved if the storage ring lattice can be designed to reduce α to a sufficiently small value.[19]

Assuming that the criteria for the random phase approximation have been satisfied, the only storage ring parameters required to analyze the transverse excitation due to the laser are the damping time, or equivalently, the number of turns to damp, and the natural emittance due to quantum fluctuations. Although the details of motion in the storage ring are complex in their own right, these scalar parameters, together with the equations of motion in the laser insertion, are sufficient to define the equilibrium distribution of the electrons in the system.

Standard, Instantaneous, and Averaged Coordinates

The published general analyses[3,20,21] of particle motion in storage rings treat the evolution of the electrons' instantaneous coordinates, referenced to the ideal trajectory, with the path length S along the ideal trajectory taken as the independent variable. In contrast, the coordinates we will use to analyze motion in the laser insertion refer to the electrons absolute position in cartesian coordinates (x,x',y,y',z,δ) with z as the independent variable (the z-axis is aligned with the nominal direction of motion of electrons in the insertion). Furthermore, our equations of motion deal with the averaged, rather than the instantaneous, values of the coordinates.

While the use of averaged cartesian coordinates substantially simplifies the form of the equations of motion in the insertion, some care must be taken to define how the two coordinate systems are related.

Formally, if we wish to follow the motion of an electron around the system with different coordinate systems in the laser and the ring, we would have to perform a coordinate transformation each time we entered and left the laser insertion. But, fortunately, we can obtain the required data on the change in the invariant A without transforming from the averaged coordinates. The proof follows in straightforward fashion from the properties of the transformation equations for x and β.

For present purposes, we can assume that the transformation is made at a point in the orbit at which $d\underset{\sim}{S} \parallel \hat{z}$, corresponding to an extremum of the electrons' short-period motion through the static periodic field. At these points, the normal storage ring and cartesian coordinates systems are identical, and the transformation matrix is the unity matrix. The transformation from the instantaneous to the average cartesian coordinates was derived in Eq. (20) of Ref. 5. In matrix form:

$$\begin{pmatrix} x(z) \\ x'(z) \end{pmatrix} \approx \begin{pmatrix} [1-\frac{k}{\rho}\left(\frac{\lambda_q}{2\pi}\right)^2 \cos(\frac{2\pi}{\lambda_q} z)] & [\frac{2k}{\rho}\left(\frac{\lambda_q}{2\pi}\right)^3 \sin(\frac{2\pi}{\lambda_q} z)] \\ [\frac{k}{\rho}\left(\frac{\lambda_q}{2\pi}\right) \sin(\frac{2\pi}{\lambda_q} z)] & [1+\frac{k}{\rho}\left(\frac{\lambda_q}{2\pi}\right)^2 \cos(\frac{2\pi}{\lambda_q} z)] \end{pmatrix}$$

$$\times \begin{pmatrix} \bar{x}(z) \\ \bar{x}'(z) \end{pmatrix} + \begin{pmatrix} -\frac{1}{\rho}\left(\frac{\lambda_q}{2\pi}\right)^2 \cos(\frac{2\pi}{\lambda_q} z) \\ +\frac{1}{\rho}\left(\frac{\lambda_q}{2\pi}\right)^2 \sin(\frac{2\pi}{\lambda_q} z) \end{pmatrix} \qquad (2.10)$$

Note that, to order $[\frac{k}{\rho}(\lambda_q/2\pi)^2]^2$, the transformation is area preserving, e.g.,

$$\mathrm{Det} \begin{pmatrix} [1-\frac{k}{\rho}\left(\frac{\lambda_q}{2\pi}\right)^2 \cos(\frac{2\pi}{\lambda_q} z)] & [\frac{2k}{\rho}\left(\frac{\lambda_q}{2\pi}\right)^3 \sin(\frac{2\pi}{\lambda_q} z)] \\ [\frac{k}{\rho}\left(\frac{\lambda_q}{2\pi}\right) \sin(\frac{2\pi}{\lambda_q} z)] & [1+\frac{k}{\rho}\left(\frac{\lambda_q}{2\pi}\right)^2 \cos(\frac{2\pi}{\lambda_q} z)] \end{pmatrix} \approx 1 \ . \qquad (2.11)$$

The transformation equation implies that the values of η, β and β' appropriate for the averaged coordinates $\bar{x}$ and $\bar{x}'$ will differ from the values appropriate for the local coordinates x and x'. The transformation relates x to $\bar{x}$. But $x = x_\beta + \eta\delta$, and we may similarly define $\bar{x} = \bar{x}_\beta + \bar{\eta}\delta$, where $\bar{x}$ and $\bar{\eta}$ are computed using the averaging procedure in Ref. 6. Assuming that the transformation from $(\bar{x},\bar{x}')$

to (x,x') is made at an extremum of the short period motion so that $\cos\left(\frac{2\pi}{\lambda_q} z\right) = 1$, the betatron coordinates (x_β, x_β'), η, and η' in the two systems are related as:

$$x_\beta + \eta\delta \approx [1 - \frac{k}{\rho_0}\left(\frac{\lambda_q}{2\pi}\right)^2] (\bar{x}_\beta + \bar{\eta}\delta) + \frac{1}{\rho_0}\left(\frac{\lambda_q}{2\pi}\right)^2 \delta$$

$$x_\beta' + \eta'\delta \approx [1 + \frac{k}{\rho_0}\left(\frac{\lambda_q}{2\pi}\right)^2] (\bar{x}_\beta' + \bar{\eta}'\delta) \tag{2.12}$$

implying:

$$x_\beta \approx [1 - \frac{k}{\rho_0}\left(\frac{\lambda_q}{2\pi}\right)^2]\bar{x}_\beta$$

$$x_\beta' \approx [1 + \frac{k}{\rho}\left(\frac{\lambda_q}{2\pi}\right)^2]\bar{x}_\beta'$$

$$\eta \approx [1 - \frac{k}{\rho_0}\left(\frac{\lambda_q}{2\pi}\right)^2]\bar{\eta} + \frac{1}{\rho_0}\cdot\left(\frac{\lambda_q}{2\pi}\right)^2$$

$$\eta' \approx [1 + \frac{k}{\rho}\left(\frac{\lambda_q}{2\pi}\right)^2]\bar{\eta}' . \tag{2.13}$$

If the averaged dispersion function $\bar{\eta}$ is assumed equal to (1/k-s), the intrinsic dispersion of the periodic magnet as computed in Ref. 5, then $\bar{\eta}' = \eta' = 0$, and:

$$\eta = [1 - \frac{s}{\rho_0}\left(\frac{\lambda_q}{2\pi}\right)^2]\bar{\eta} . \tag{2.14}$$

To determine how β and β' should be defined in the averaged coordinate system, we compare the electrons' averaged and instantaneous motion in the betatron coordinates (x_β, x_β'). As previously noted, followed over a number of orbits, the coordinates (x_β, x_β') at fixed x (or z) of an electron in the ring will trace out the perimeter of an ellipse. The aspect ratio and orientation of the ellipse are determined by β and β'. Specifically,

$$x = \sqrt{\frac{A\beta}{\pi\left(1 + \frac{\beta'^2}{4}\right)}} \{\cos\theta + \frac{1}{2}\beta' \sin\theta\}$$

$$x' = \sqrt{\frac{A\; 1+ \frac{\beta'^2}{4}}{\pi\beta}} \quad \sin\theta \tag{2.15}$$

where θ is the betatron phase. Assuming again that the transformation from (x,x') to $(\overline{x},\overline{x}')$ is made at an extremum of the short period motion, the corresponding values of $\overline{x}$ and $\overline{x}'$ are:

$$\overline{x} \approx [1+\frac{k}{\rho}\left(\frac{\lambda_q}{2\pi}\right)^2] \sqrt{\frac{A\beta}{\pi\left(1+ \frac{\beta'^2}{4}\right)}} \quad \{\cos\theta + \frac{1}{2}\beta' \sin\theta\}$$

$$\overline{x}' \approx [1-\frac{k}{\rho}\left(\frac{\lambda_q}{2\pi}\right)^2] \sqrt{\frac{A\left(1+ \frac{\beta'^2}{4}\right)}{\pi\beta}} \quad \sin\theta \; . \tag{2.16}$$

The transformation is area-preserving, so that A in Eq. (2.16) is the same as in Eq. (2.15). Requiring that β and β' be defined to reduce Eq. (2.16) to the same form as Eq. (2.15), or alternatively, to permit the area of the ellipse to be computed by substitution of the averaged variables $\overline{x}_\beta$, $\overline{x}_\beta'$, $\overline{\beta}$ and $\overline{\beta}'$ in Eq. (2.3), we find that:

$$\overline{\beta} \approx [1- \frac{2k}{\rho}\left(\frac{\lambda_q}{2\pi}\right)^2]\beta \tag{2.17}$$

in the averaged coordinate system. The function β' is the same in both systems.

In general, the function β and β' change with z. Starting with the assumed values of β and β' at the beginning of the laser insertion, and the corresponding values for β and β' in the averaged coordinates, we can follow the evolution of these functions by following the evolution of the phase space ellipse as determined by the motion of the electron ensemble in the laser-off state. Having defined the value of β and β' at the end of the insertion by these means, the magnitude of the moments of the change A in the transverse invariant can be computed as outlined earlier in this section.

The conclusion that all calculations can be done in the averaged coordinate system follows from the properties of the transformation equations and the functional form of δA^n [Eq. (2.8)]. Specifically, the value of $\langle \delta A^n \rangle$ is not altered by the transformation from averaged to instantaneous coordinates since the quantities A, $(x_\beta{}^2/\beta)$,

$(\beta x_\beta{}'^2)$, and $(\beta' x_\beta x_\beta')$ which determine δA are each left unchanged by the transformation.

III. SMALL SIGNAL ANALYSIS

Linearized Equations of Motion for $\bar{x}$, $\bar{\delta}$, and $\bar{\psi}$

The equations of motion for an electron moving through the interaction region in a free electron laser magnet with a transverse gradient, and with constant period and amplitude, have been derived in Ref. 5. If these equations are linearized, they can be used to construct and generate a series solution for $\bar{x}$, $\bar{\delta}$, and $\bar{\psi}$ in powers of the optical electric field. From Madey and Taber:

$$\frac{d^2}{dz^2}\,\bar{x}(z) \approx -\frac{k}{2}\left(\frac{1}{\rho_0}\right)^2\left(\frac{\lambda_q}{2\pi}\right)^2 \left[(k-s)\bar{x}-\bar{\delta}(1+s\bar{x})\right](1-2\bar{\delta})-2\bar{x}'\bar{\delta}'$$

$$\frac{d}{dz}\,\bar{\gamma}(z) \approx \frac{e|\mathcal{E}|}{mc^2}\,\frac{\lambda_q}{2\pi\rho_0}\,(1+k\bar{x})(1-\delta)\,\sin[\psi_0(z)]$$

$$\left[J_{\left(\frac{N-1}{2}\right)}(N\xi) - J_{\left(\frac{N+1}{2}\right)}(N\xi)\right]$$

$$\frac{d}{dz}\,[\psi_0(z)] \approx N\left\{\frac{4\pi}{\lambda_q}\left[\frac{\gamma-\gamma_{r_0}}{\gamma_{r_0}} - \frac{\gamma}{\gamma_{r_0}}\,\frac{\left(\frac{\gamma_0^2}{2}\right)\left(\frac{\lambda_q}{2\pi\rho_0}\right)^2}{1+\left(\frac{\gamma_0^2}{2}\right)\left(\frac{\lambda_q}{2\pi\rho_0}\right)^2}\,k\bar{x}\right]\right.$$
$$\left.-\frac{\omega_0}{2c}\left(\frac{d\bar{x}}{dz}\right)^2\right\} \qquad (3.1)$$

where $\bar{x}(z) \equiv$ average transverse position

$\bar{\gamma}(z) \equiv$ average energy

$\bar{\delta}(z) \equiv$ average fractional deviation from design energy

$= \left(\bar{\gamma}(\bar{z}) - \gamma_0\right)/\gamma_0$

$\bar{\psi}_0(z) = \bar{\psi}_1(z) + (N-1)\left(\frac{2\pi}{\lambda_q}\right) z$, where $\bar{\psi}_1(z)$ is the average phase difference between the electron transverse velocity and the optical electric field

$\gamma_0 \equiv$ design energy

γ_{r_0} ≡ resonance energy at $\bar{x} = 0$

λ_q ≡ magnet period

ω_0 ≡ fundamental operating frequency $\approx 2\gamma_{r_0}^2(2\pi c/\lambda_q)$

N ≡ harmonic number of operating wavelength (N = 1,3,5,...)

$\xi \equiv \frac{\omega_0}{4c}\left(\frac{\lambda_q}{2\pi\rho_0}\right)^2\left(\frac{\lambda_q}{4\pi}\right)(1-2\delta)(1+2k\bar{x})$

$\tilde{\mathcal{E}}$ ≡ optical electric field

$\tilde{B}$ = static magnetic field ≡ $[B_0(1+kx)\cos\left(\frac{2\pi}{\lambda_q}z\right) + B_c(1+sx)]\hat{y}$

k,s ≡ gradients of periodic and constant components of static field

ρ_0 ≡ characteristic gyro radius = $(\gamma_0 mc^2/eB_0)$

$J_{\left(\frac{N-1}{2}\right)}, J_{\left(\frac{N+1}{2}\right)},\ldots$ ≡ Bessel functions of order $\left(\frac{N-1}{2}\right), \left(\frac{N+1}{2}\right)\ldots$

These equations can be linearized by dropping the cross products and powers of the assumed small quantities x, x', $\bar{\delta}$, and $\bar{\delta}'x$ and setting $\xi = \xi_0 \equiv (\omega_0/c)(1/\rho_0)^2(\lambda_q/4\pi)^3$. We obtain:

$$\frac{d^2\bar{x}}{dz^2} \approx -\frac{\tilde{\Lambda}^2}{k-s}\left[(k-s)\bar{x} - \bar{\delta}\right]$$

$$\frac{d\bar{\delta}}{dz} = \frac{e|\mathcal{E}|}{\gamma_0 mc^2}\frac{\lambda_q}{2\pi\rho_0}(1+k\bar{x}-\bar{\delta})\sin[\psi_0(z)]\{J_{\left(\frac{N-1}{2}\right)}(N\xi_0) - J_{\left(\frac{N+1}{2}\right)}(N\xi_0)\}$$

$$\frac{d\bar{\psi}_0}{dz} = N\frac{4\pi}{\lambda_q}\left\{\left(\frac{\gamma_0-\gamma_{r_0}}{\gamma_{r_0}}\right) + \frac{\gamma_0}{\gamma_{r_0}}\left[\bar{\delta} - \frac{\left(\frac{\gamma_0^2}{2}\right)\left(\frac{\lambda_q}{2\pi\rho_0}\right)^2}{1+\left(\frac{\gamma_0^2}{2}\right)\left(\frac{\lambda_q}{2\pi\rho_0}\right)^2}k\bar{x}\right]\right\}. \qquad (3.2)$$

Note that we have introduced the parameter $\tilde{\Lambda} \equiv \sqrt{k(k-s)/2}\ (\lambda_q/2\pi\rho_0)$ in (3.2). This key parameter defines the focussing strength of the magnet array and appears with considerable frequency in the analysis of electron motion in the insertion.

The phase evolution equation can be simplified if the discussion is restricted to gain expanded magnets in which the variation of gain with energy is exactly cancelled by the transverse gradient ("infinite gain expansion"). For this case,

$$\frac{\left(\frac{\gamma_0^2}{2}\right)\left(\frac{\lambda_q}{2\pi\rho_0}\right)^2}{1+\left(\frac{\gamma_0^2}{2}\right)\left(\frac{\lambda_q}{2\pi\rho_0}\right)^2}\cdot k = \frac{1}{\overline{\eta}} . \tag{3.3}$$

We assume, further, that the storage ring has been designed to yield an averaged dispersion equal to the intrinsic dispersion of laser magnet, $\overline{\eta} = (1/k\text{-}s)$.[17] Setting $\overline{\gamma} = \gamma_0(1+\overline{\delta})$ in the phase evolution equation, we obtain:

$$\frac{d\overline{\psi}_0}{dz} = N\,\frac{4\pi}{\lambda_q}\left[\left(\frac{\gamma_0-\gamma_{r_0}}{\gamma_{r_0}}\right) + \left(\frac{\gamma_0}{\gamma_{r_0}}\right)\left(\overline{\delta} - (k\text{-}s)\overline{x}\right)\right] . \tag{3.4}$$

Our objective is to solve the three equations in (3.3) given the initial electron coordinates $(\overline{x},\overline{x}')$, energy deviation δ, and relative phase ψ_0. We will combine the three equations to generate a single equation for $\overline{x}$ which relates this coordinate and its derivatives to the optical electric field. This equation can then be expanded in powers of the electric field to obtain the terms responsible for energy spread, transverse excitation and amplification in the small signal limit.

From Eq. (3.2), the energy deviation can be expressed in terms of $\overline{x}$ as:

$$\overline{\delta} = (k\text{-}s)\left(\frac{1}{\tilde{\Lambda}^2}\,\frac{d^2\overline{x}}{dz^2} + \overline{x}\right) . \tag{3.5}$$

This can be used to express the phase evolution equation as:

$$\frac{d\overline{\psi}_0}{dz} = N\left(\frac{4\pi}{\lambda_q}\right)\left[\left(\frac{\gamma_0-\gamma_{r_0}}{\gamma_{r_0}}\right) + \frac{\gamma_0}{\gamma_{r_0}}\left(\frac{k\text{-}s}{\tilde{\Lambda}^2}\right)\frac{d^2\overline{x}}{dz^2}\right] . \tag{3.6}$$

Equation (3.6) can be formally integrated to obtain $\psi_0(z)$ in terms of $\overline{x}'$ and $\int\overline{x}dx$:

$$\overline{\psi}_0(z) = \overline{\psi}_0\left(-\frac{L}{2}\right) + \int_{-L/2}^{z} dz'\,\frac{d\psi_0}{dz}$$

$$= \overline{\psi}_0\left(-\frac{L}{2}\right) + q\left(z+\frac{L}{2}\right) + N\,\frac{4\pi}{\lambda_q}\,\frac{\gamma_0}{\gamma_{r_0}}\left(\frac{k\text{-}s}{\tilde{\Lambda}^2}\right)\frac{d\overline{x}}{dz}\bigg|_{-L/2}^{z} \tag{3.7}$$

where we have assumed the interaction begins at $z = -L/2$ and continues to z = +L/2. The quantity $q \equiv N(4\pi/\lambda_q)(\gamma_0-\gamma_{r_0}/\gamma_{r_0})$ is the nominal phase shift per unit length of the optical electric field relative to the electron transverse velocity, and $\psi_0(-L/2)$ is the initial phase. For subsequent purposes, it is convenient to define

$$\Delta\psi \approx N\left(\frac{4\pi}{\lambda_q}\right)\int_{-L/2}^{z} \frac{\gamma_0}{\gamma_{r_0}} [\bar{\delta} - (k-s)\bar{x}]$$

$$\approx N\left(\frac{4\pi}{\lambda_q}\right)\frac{\gamma_0}{\gamma_{r_0}}\left(\frac{k-s}{\tilde{\Lambda}^2}\right)\frac{d\bar{x}}{dz}\bigg|_{-L/2}^{z} \tag{3.8}$$

with the definition:

$$\psi_0(z) = \psi_0\left(-\frac{L}{2}\right) + q\left(z+\frac{L}{2}\right) + \Delta\psi \tag{3.9}$$

and the effect of transverse motion is contained entirely in $\Delta\psi$. We will further define $[J_{\left(\frac{N-1}{2}\right)}(N\xi_0) - J_{\left(\frac{N+1}{2}\right)}(N\xi_0)] \equiv J(N\xi_0)$ in Eq. (3.2). With these conventions, the energy evolution equation from (3.2) can be combined with Eq. (3.5) for $\bar{\delta}(z)$ to yield:

$$\frac{d}{dz}\bar{\delta}(z) = \frac{d}{dz}(k-s)\left(\frac{1}{\tilde{\Lambda}^2}\frac{d^2\bar{x}}{dz^2} + \bar{x}\right)$$

$$= \frac{e|\mathcal{E}|}{\gamma_0 mc^2}\left(\frac{\lambda_q}{2\pi\rho_0}\right)\left[1 - \frac{k-s}{\tilde{\Lambda}^2}\frac{d^2x}{dz^2} + s\bar{x}\right]$$

$$\cdot\ J(N\xi_0)\sin\left[\psi_0\left(-\frac{L}{2}\right) + q\left(z+\frac{L}{2}\right) + \Delta\psi\right] . \tag{3.10}$$

If $\bar{x}$ is expanded as a power series in $\mathcal{E}$, the amplitude of the optical electric field:

$$\bar{x} = x_0 + x_1\mathcal{E} + x_2\mathcal{E}^2 + \ldots , \tag{3.11}$$

Eq. (3.10) can be used to solve for x_0 and to relate each subsequent term in the series to the terms preceding it. The zeroth order term must satisfy the equations of motion in the absence of the optical field:

$$\frac{d}{dz}(k-s)\left(\frac{1}{\tilde{\Lambda}^2}x_0'' + x_0\right) = 0 . \tag{3.12}$$

These solutions will have the form:

$$x_0 = x^0 + a\cos(\tilde{\Lambda}z + \phi_0) \tag{3.13}$$

where a and ϕ_0 are constants of integration determined by the initial conditions. The zeroth order energy deviation δ_0 is simply:

$$\delta_0 = x^0(k-s) \tag{3.14}$$

while the phase difference $\Delta\psi_0$ is:

$$\Delta\psi_0 = N\left(\frac{4\pi}{\lambda_q}\right)\frac{\gamma_0}{\gamma_{r_0}}\left(\frac{k-s}{\tilde{\Lambda}}\right)a[\sin(\tilde{\Lambda}z+\phi_0)+\sin\left(\frac{\tilde{\Lambda}L}{2}-\phi_0\right)] \; . \tag{3.15}$$

The first order term $x_1(z)$ must then satisfy:

$$\frac{d}{dz}(k-s)\left[\left(\frac{1}{\tilde{\Lambda}}\right)^2\frac{d^2x_1}{dz^2}+x_1\right]\mathcal{E}$$
$$=\frac{e|\mathcal{E}|}{\gamma_0 mc^2}\left(\frac{\lambda_q}{2\pi\rho_0}\right)J(N\xi_0)\,[1-\frac{k-s}{\tilde{\Lambda}^2}\frac{d^2x_0}{dz^2}+sx_0]$$
$$\cdot\sin[\psi_0\left(-\frac{L}{2}\right)+q\left(z+\frac{L}{2}\right)+\Delta\psi_0] \tag{3.16}$$

while the second order term $x_2(z)$ is the solution of:

$$\frac{d}{dz}(k-s)\left[\left(\frac{1}{\tilde{\Lambda}}\right)^2\frac{d^2x_2}{dz^2}+x_2\right]\mathcal{E}^2$$
$$=\frac{e|\mathcal{E}|}{\gamma_0 mc^2}\left(\frac{\lambda_q}{2\pi\rho_0}\right)J(N\xi_0)\left\{\left[-\frac{k-s}{\tilde{\Lambda}^2}\frac{d^2}{dz^2}(x_1\mathcal{E})+s(x_1\mathcal{E})\right]\right.$$
$$\cdot\sin[\psi_0\left(-\frac{L}{2}\right)+q\left(z+\frac{L}{2}\right)+\Delta\psi_0]$$
$$+\;[1-\frac{(k-s)}{\tilde{\Lambda}^2}\frac{d^2x_0}{dz^2}+sx_0]$$
$$\left.\cdot\;\psi_1\mathcal{E}\cos[\psi_0\left(-\frac{L}{2}\right)+q\left(z+\frac{L}{2}\right)+\Delta\psi_0]\right\} \; . \tag{3.17}$$

The first and second order contributions to δ and $\Delta\psi$ can be computed from Eqs. (3.5) and (3.8), respectively.

The zeroth order terms in $\overline{x}$, $\overline{\delta}$, and $\Delta\psi$ are required to satisfy the initial conditions at the start of the interaction, but they do not alter the energy of the electrons during the interaction. In the small signal regime, the first order terms dominate the evolution of the electrons' energy and emittance. Note, however, that the first order terms do not alter the average energy of an initially continuous electron beam: The average of $\psi_0(-L/2)$ of the first order change in electron energy during the interaction is zero,

$$\langle\overline{\delta}\left(z=\frac{L}{2}\right)-\overline{\delta}\left(z=-\frac{L}{2}\right)\rangle=\frac{1}{2\pi}\int_0^{2\pi}d\psi_0\left(-\frac{L}{2}\right)\int_{-L/2}^{L/2}dz\left(\frac{d\overline{\delta}}{dz}\right)$$

(cont'd)

(cont'd)

$$= \frac{1}{2\pi}\int_0^{2\pi} d\psi_0\left(-\frac{L}{2}\right)\int_{-L/2}^{L/2} dz \frac{e|\mathcal{E}|}{\gamma_0 mc^2}\left(\frac{\lambda_q}{2\pi\rho_0}\right) J(N\xi_0)\,[1+\frac{k-s}{\tilde{\Lambda}^2}\frac{d^2x_0}{dz^2}+sx_0]$$

$$\cdot \sin[\psi_0\left(-\frac{L}{2}\right)+q\left(z+\frac{L}{2}\right)+\Delta\psi_0]$$

$$= 0\ . \qquad (3.18)$$

As in the case of the one-dimensional free electron laser, the first order terms do not contribute to gain.[23] To estimate the gain, it is necessary to include the second order terms, while the onset of saturation involves terms of third order and higher.

Our present purpose is to estimate the growth in emittance during the interaction, and for this purpose it is sufficient to consider only the zeroth and first order terms. The second order terms are considered in detail in the discussion of gain and saturation.[6]

On Axis Initial Motion

We consider first the case in which the electrons enter the interaction region on-axis, e.g., with $\overline{x}(-L/2) = \overline{x}'(-L/2) = 0$. For this case, Eq. (3.16) reduces to

$$\frac{d}{dz}(k-s)\left[\left(\frac{1}{\tilde{\Lambda}}\right)^2\frac{d^2x_1}{dz^2}+x_1\right]$$

$$= \frac{e|\mathcal{E}|}{\gamma_0 mc^2}\left(\frac{\lambda_q}{2\pi\rho_0}\right) J(N\xi_0)\sin[\psi_0\left(-\frac{L}{2}\right)+q\left(z+\frac{L}{2}\right)]\ . \qquad (3.19)$$

This can be integrated once to obtain

$$(k-s)\left[\left(\frac{1}{\tilde{\Lambda}}\right)^2\frac{d^2x_1}{dz^2}+x_1(z)\right]$$

$$= \frac{e|\mathcal{E}|}{\gamma_0 mc^2}\left(\frac{\lambda_q}{2\pi\rho_0}\right) J(N\xi_0)\,\frac{1}{q}\{\cos[\psi_0\left(-\frac{L}{2}\right)+q\left(z+\frac{L}{2}\right)]$$

$$-\cos[\psi_0\left(-\frac{L}{2}\right)]\}\ , \qquad (3.20)$$

where we have assumed $x_1(-L/2) = x_1'(-L/2) = \delta_1(-L/2) = 0$, consistent with the assumption that the initial conditions would be matched using the zeroth order solutions.

To determine the effect of the interaction on the betatron coordinates we need to compute the values of x_1, x_1', and δ_1 at the end of the interaction at $z = +L/2$. The end of the interaction at

$z = +L/2$ will be defined by truncation of the optical or the periodic magnetic field. Formally, it is useful to compute x_β at $z = L/2+$ where $\delta' = 0$. Since each of the functions is continuous, $x_1(z = L/2+) = x_1(z = L/2)$, $x'(z = L/2+) = x'(z = L/2)$, and $\delta_1(z = L/2+) = \delta_1(z = L/2)$. Solving Eq. (3.20), we obtain:

$$x_1\left(z=\frac{L}{2}+\right) = -\left(\frac{1}{k-s}\right)\frac{e}{\gamma mc^2}\left(\frac{\lambda_q}{2\pi\rho_0}\right)J(N\xi_0)\left\{\frac{1}{q}\left[\cos(qL+\psi_0)-\cos\psi_0\right] - \frac{1}{q^2-\tilde{\Lambda}^2}\left[q\cos(\tilde{\Lambda}L+\psi_0)+(q-\tilde{\Lambda})\sin\psi_0\sin(\tilde{\Lambda}L) - q\cos(qL+\psi_0)\right]\right\}$$

$$x_1'\left(z=\frac{L}{2}+\right) = \left(\frac{1}{k-s}\right)\frac{e}{\gamma_0 mc^2}\left(\frac{\lambda_q}{2\pi\rho_0}\right)J(N\xi_0)\left[\frac{\tilde{\Lambda}}{q^2-\tilde{\Lambda}^2}\right] \cdot \left[q\sin(\tilde{\Lambda}L+\psi_0)-(q-\tilde{\Lambda})\sin\psi_0\cos(\tilde{\Lambda}L) - q\sin(qL+\psi_0)+(q-\tilde{\Lambda})\sin(qL+\psi_0)\right]$$

$$\delta_1\left(z=\frac{L}{2}+\right) = -\frac{e}{\gamma_0 mc^2}\left(\frac{\lambda_q}{2\pi\rho_0}\right)J(N\xi_0)\,\frac{1}{q}\left[\cos(qL+\psi_0)-\cos\psi_0\right], \tag{3.21}$$

where we have abbreviated $\psi_0\left(-\frac{L}{2}\right)$ as ψ_0. The betatron amplitudes $\overline{x}_\beta$ and $\overline{x}_\beta'$ are then:

$$\overline{x}_\beta\left(z=\frac{L}{2}+\right) \approx \left(\frac{1}{k-s}\right)\frac{e|\mathcal{E}|}{\gamma_0 mc^2}\left(\frac{\lambda_q}{2\pi\rho_0}\right)J(N\xi_0)\left(\frac{1}{q^2-\tilde{\Lambda}^2}\right) \cdot \left[q\cos(\tilde{\Lambda}L+\psi_0)-(q-\tilde{\Lambda})\sin\psi_0\sin(\tilde{\Lambda}L)-q\cos(qL+\psi_0)\right]$$

$$\overline{x}_\beta' \approx \left(\frac{1}{k-s}\right)\frac{e|\mathcal{E}|}{\gamma_0 mc^2}\left(\frac{\lambda_q}{2\pi\rho_0}\right)J(N\xi_0)\left(\frac{\tilde{\Lambda}}{q^2-\tilde{\Lambda}^2}\right) \cdot \left[q\sin(\tilde{\Lambda}L+\psi_0)+(q-\tilde{\Lambda})\sin\psi_0\cos(\tilde{\Lambda}L) - q\sin(qL+\psi_0)+(q-\tilde{\Lambda})\sin(qL+\psi_0)\right]. \tag{3.22}$$

From Eq. (3.22), it can be seen that we can suppress the betatron excitation if we use the appropriate transverse gradient and optical phase slip. Any choice of q and $\tilde{\Lambda}$ such that:

$$|qL - \tilde{\Lambda}L| = 2\pi J$$

$$|qL| = \pi K, \tag{3.23}$$

where J and K are integers, $J = 1,2,\ldots$, and $K = 0,1,2,3,\ldots$, will force $\overline{x}_\beta$ and $\overline{x}_\beta'$ to zero to first order in $\mathcal{E}$, independent of the optical phase at the start of the interaction. The first order change in energy for such a choice would be:

$$\delta_1\left(z=\frac{L}{2}\right) = \frac{e|\mathcal{E}|}{\gamma_0 mc^2} \frac{\lambda_q}{2\pi\rho_0} J(N\xi_0) \frac{1}{q} \left[(-1)^K - 1\right]\cos\psi_0 \ . \tag{3.24}$$

Even values of K reduce the first order change in energy to zero.

Given this special result, it is natural to inquire as to the effects of off-axis initial motion, and the effects of the higher order terms, which may not be cancelled by the choices of q and $\tilde{\Lambda}$ allowed in Eq. (3.23).

Numerical integration of the full equations of motion [Eq. (3.1)] indicates that the choices outlined in Eq. (3.23) remain useful for off-axis initial motion, and up to the onset of saturation. There is, however, some definite deterioration at saturation and with non-zero values of x_0.

Some qualitative insight into the problem of off-axis initial motion can be acquired from Eq. (3.16), which can be approximated for small x_0 and $\Delta\psi_0$ as:

$$\frac{d}{dz}(\dot{k}-s)\left[\left(\frac{1}{\tilde{\Lambda}}\right)^2 \frac{d^2x_1}{dz^2} + x_1\right] \approx \frac{e|\mathcal{E}|}{\gamma_0 mc^2}\left(\frac{\lambda_q}{2\pi\rho_0}\right) J(N\xi_0)$$
$$\left\{\left[1-\left(\frac{k-s}{\tilde{\Lambda}^2}\right)\frac{d^2x_0}{dz^2} + sx_0\right]\sin\left[\psi_0\left(-\frac{L}{2}\right)+q\left(z+\frac{L}{2}\right)\right]\right.$$
$$\left. + \Delta\psi_0 \cos\left[\psi_0\left(-\frac{L}{2}\right) \pm q\left(z+\frac{L}{2}\right)\right]\right\} \ . \tag{3.25}$$

The solution to Eq. (3.25) will be the sum of two components. One component is the solution to Eq. (3.20), which will, from the preceding arguments, result in zero transverse excitation if the gradient and optical phase slip are appropriately chosen. The second component is the solution to:

$$\frac{d}{dz}(k-s)\left[\left(\frac{1}{\tilde{\Lambda}}\right)^2 \frac{d^2x_1}{dz^2} + x_1\right] = \frac{e|\mathcal{E}|}{\gamma_0 mc^2}\left(\frac{\lambda_q}{2\pi\rho_0}\right) J(N\xi_0)$$
$$\left\{\left[-\left(\frac{k-s}{\tilde{\Lambda}^2}\right)\frac{d^2x_0}{dz^2} + sx_0\right]\sin\left[\psi_0\left(-\frac{L}{2}\right)+q\left(z+\frac{L}{2}\right)\right]\right.$$
$$\left. + \Delta\psi_0(z) \cos\left[\psi_0\left(-\frac{L}{2}\right)+q\left(z+\frac{L}{2}\right)\right]\right\} \ . \tag{3.26}$$

The solution to this equation will, in general, not result in $\overline{x}_\beta = 0$ or $\overline{x}_\beta' = 0$ at the end of the interaction. To the contrary, we note that the solution of this equation and the resultant value of $\overline{x}_\beta$ and

$\overline{x}_\beta'$ will be bilinear in $\mathcal{E}$ and x_0. The value of δA associated with this solution will therefore increase with both the optical power density and the initial emittance of the electron beam.

Of course, the actual excitation growth will not be monotonic; higher order terms, neglected in this section, typically change the sign of δA at large A. The practical question to be resolved is the extent to which the off-axis excitation growth which occurs in consequence of (3.26) spoils the correlation of x and δ. To answer this question we must integrate the full equations of motion, and compute the effect of off-axis excitation on the probability distribution. A comparison of the performance of some excitation cancelling and non-excitation cancelling systems is presented in Sec. VI.

IV. FUNCTIONAL CANCELLATION

The condition on q and $\tilde{\Lambda}$ derived in the last section are not the most general conditions for the cancellation of the transverse excitation if we consider laser magnets in which the amplitude and period of the field are functions of longitudinal position. From Eqs. (2.1) and (2.2), the conditions for zero excitation at the ends of the interaction are:

$$\overline{x}_\beta = \overline{x} - \overline{\eta}\,\overline{\delta} = 0$$

$$\overline{x}_\beta' = \overline{x}' - \overline{\eta}\,\overline{\delta}' = x' = 0 \qquad (4.1)$$

where, as in Sec. III, we assume that x, x', δ, and δ' are measured just after the interaction, so that $\delta' = 0$, and that the dispersion has been set equal to the intrinsic dispersion $\overline{\eta} = (1/k\text{-}s)$ with $\overline{\eta}' = 0$. Assuming, further, that $\overline{\delta}$ can be continued to be expressed in terms of $\overline{x}$ as $\overline{\delta} = (k\text{-}s)[(\overline{x}''/\tilde{\Lambda}^2)+\overline{x}]$, the first of the above equations imposes the condition:

$$\overline{x}_\beta = \overline{x} - \overline{\eta}\,\overline{\delta} = -(x''/\tilde{\Lambda}^2) = 0 \,. \qquad (4.2)$$

Equations (4.1) and (4.2) indicate that the condition $\overline{x}' = \overline{x}'' = 0$ at the end of the insertion is necessary and sufficient for the cancellation of the transverse excitation.

As with the constant period magnet, this condition cannot, in general, be satisfied if the electrons enter the laser off-axis. But if the electrons enter on-axis, that is, if $x' = x'' = 0$ at the beginning of the magnet, we can explicitly construct the magnet so that the derivatives x' and x'' are forced to zero at the end of the insertion. In such a magnet, the derivative of the first order solution, $x_1(z)$ will have the form:

$$x_1'(z) = \mathrm{Re}\{\left(z+\tfrac{L}{2}\right)^2\left(z-\tfrac{L}{2}\right)^2 f(z)\exp i[\psi_0\left(-\tfrac{L}{2}\right)+q\left(z+\tfrac{L}{2}\right)-\tfrac{\pi}{2}]\} \qquad (4.3)$$

where the interaction starts at $z = -L/2$ and ends at $z = +L/2$, and $f(z)$ can be (almost) any continuous function of z. The second order zero in $x_1'(z)$ at $z = L/2$ forces both x_1 and x_1' to zero, identically satisfying the excitation cancellation condition, independent of the optical phase. (The second order zero at $z = -L/2$ follows from the assumption of on-axis initial motion.)

Qualitatively, the idea with this class of magnets is to start with a functional form for the trajectory $x(z)$ which satisfies the conditions on x' and x'' in (4.1) and (4.2), then specify the amplitude and phase of the static field to match the amplitude and phase of the function obtained when the derivative x' is inserted into the left hand side of Eq. (3.19). This approach is made possible by the structure of the right hand side of (3.19). The magnitude and phase of the right hand side is fixed by the amplitude and period of the static field in a formal relation which can be inverted to express the parameters of the magnetic field in terms of the amplitude and phase of the left hand side of the equation. Quantitatively, the argument is somewhat more complicated, because $\tilde{\Lambda}$ is itself a function of the field's period and amplitude, and the attempt to change these quantities during the interaction alters the form of the relation between $\bar{x}$ and $\bar{\delta}$ unless additional measures are taken.

Modifications to Equations of Motion

To maintain the form of (3.19), it is necessary to include a second spatially periodic component in the static magnetic field. We consider the case in which the period $\lambda_q(z)$ and amplitude $B_0(z)$ of the principal component of the static field are allowed to vary with z. The second component of the field is assumed to have the period $\lambda_n(z)$ and amplitude $B_n(z)$, with λ_n harmonically related to λ_q:

$$\lambda_n(z) = \lambda_q(z)/n \quad . \tag{4.4}$$

The two components are assumed to have the same gradient k. With these modifications, the static magnetic field takes the form:

$$\begin{aligned} B_y(z) &= B_0(z)(1+kx) \cos\left[\int_{-L/2}^{z} \frac{2\pi dz'}{\lambda_q}\right] \\ &+ B_n(z)(1+kx) \cos\left[\int_{-L/2}^{z} \frac{2\pi dz'}{\lambda_n}\right] + B_c(1+sx) \quad . \end{aligned} \tag{4.5}$$

We assume $(d\lambda_q/dz) \ll 1$. The effects of these modifications on the equations of motion for $\bar{x}$ and $\bar{\delta}$ can be obtained by repeating the derivations in Ref. 5. The equation governing the evolution of $\bar{x}$ is straightforward. We obtain:

$$\frac{d^2\bar{x}}{dz^2} \approx -\frac{k}{2}\left[\left(\frac{\lambda_q}{2\pi\rho_0}\right)^2 + \left(\frac{\lambda_n}{2\pi\rho_{n0}}\right)^2\right] [(k-s)\bar{x} - \bar{\delta}(1+s\bar{x})](1-2\bar{\delta})$$

(cont'd)

(cont'd)

$$\rho_0 \equiv (\gamma_0 mc^2/eB_0)$$
$$\rho_{n0} \equiv (\gamma_0 mc^2/eB_n) \;, \tag{4.6}$$

where the average value of $\bar{x}$ is defined with respect to the <u>local</u> magnet period:

$$\bar{x} \equiv \left[\frac{1}{\lambda_q(z)}\right]^2 \int_{z-\lambda_q/2}^{z+\lambda_q/2} dz' \int_{z'-\lambda_q/2}^{z'+\lambda_q/2} dz''\, x(z'') \;. \tag{4.7}$$

Equation (4.6) has the same form as in the corresponding equation in (3.1), except that the parameter corresponding to $\tilde{\Lambda}$ now has the value:

$$\tilde{\Lambda} = \sqrt{\frac{k(k-s)}{2}} \left[\left(\frac{\lambda_q}{2\pi\rho_0}\right)^2 + \left(\frac{\lambda_n}{2\pi\rho_{n0}}\right)^2\right]^{1/2} \;. \tag{4.8}$$

In what follows, we will assume that ρ_{n0} is chosen to keep $\tilde{\Lambda}$ in (4.8) constant as λ_q, ρ_0, and λ_n are varied.

The derivation of the energy evolution equation is more complicated. As before, we know from first principles that $(d\delta/dz) \approx (e/\gamma_0 mc^2)\underset{\sim}{\mathcal{E}}\cdot\underset{\sim}{\beta}_\perp$, and can approximate $\beta_\perp$ as:

$$\begin{aligned} \beta_\perp &\approx x'(z) \approx \bar{x}'(z) + \left[\frac{1}{\rho}\left(\frac{\lambda_q}{2\pi}\right)\sin\left(\int\frac{2\pi dz}{\lambda_q}\right)\right. \\ &\qquad \left. + \frac{1}{\rho_n}\left(\frac{\lambda n}{2\pi}\right)\sin\left(\int\frac{2\pi dz}{\lambda_n}\right)\right](1+k\bar{x}) \\ &\qquad + \left[\frac{1}{\rho}\left(\frac{\lambda_q}{2\pi}\right)^2\cos\left(\int\frac{2\pi dz}{\lambda_q}\right) + \frac{1}{\rho_n}\left(\frac{\lambda_n}{2\pi}\right)^2\cos\left(\int\frac{2\pi dz}{\lambda_n}\right)\right]k\bar{x}' \\ &\approx \bar{x}'(z) + \left[\frac{1}{\rho}\left(\frac{\lambda_q}{2\pi}\right)\sin\left(\int\frac{2\pi dz}{\lambda_q}\right) + \frac{1}{\rho_n}\left(\frac{\lambda_q}{2\pi}\right)\sin\left(\int\frac{2\pi dz}{\lambda_n}\right)\right](1+k\bar{x}) \;, \end{aligned} \tag{4.9}$$

where $\rho = eB_0/\gamma mc^2$ and $\rho_n = eB_n/\gamma mc^2$. For $\mathcal{E} \approx |\mathcal{E}|\cos\frac{\omega}{c}(z-ct)$, this leads to the result:

$$\begin{aligned} \frac{d\delta}{dz} \approx \frac{e|\mathcal{E}|}{\gamma_0 mc^2} &\{\bar{x}'\cos\left(\frac{\omega}{c}(z-ct)+\phi_0\right) \\ &+\left(\frac{\lambda_q}{4\pi\rho}\right)(1+k\bar{x})\left[\sin\left(\int\frac{2\pi dz}{\lambda_q}+\frac{\omega}{c}(z-ct+\phi_0)\right)\right. \\ &\qquad \left. + \sin\left(\int\frac{2\pi dz}{\lambda_q}-\frac{\omega}{c}(z-ct-\phi_0)\right)\right] \end{aligned}$$

(cont'd)

(cont'd)

$$+\left(\frac{\lambda_n}{4\pi\rho_n}\right)(1+k\bar{x})\,[\sin\left(\int \frac{2\pi dz}{\lambda_n} + \frac{\omega}{c}(z-ct+\phi_0)\right)$$

$$+ \sin\left(\int \frac{2\pi dz}{\lambda_n} - \frac{\omega}{c}(z-ct-\phi_0)\right)]\} \ . \qquad (4.10)$$

Converting this into an equation for the average energy deviation $\bar{\delta}$, we are faced with the task of performing five integrals of a type familiar from Ref. 5:

$$\frac{1}{\lambda_q}\int_{z_0-\lambda_q/2}^{z_0+\lambda_q/2} dz\ \sin\left[\left(\int_{-L/2}^{z} \frac{2\pi dz'}{\lambda_q}\right) + \frac{\omega}{c}(z-ct)+\phi_0\right]$$

$$\equiv \frac{1}{\lambda_q}\int_{z_0-\lambda_q/2}^{z_0+\lambda_q/2} dz\ \sin\psi_1(z)$$

$$\frac{1}{\lambda_q}\int_{z_0-\lambda_q/2}^{z_0+\lambda_q/2} dz\ \sin\left[\left(\int_{-L/2}^{z} \frac{2\pi dz'}{\lambda_q}\right) - \frac{\omega}{c}(z-ct)-\phi_0\right]$$

$$\equiv \frac{1}{\lambda_q}\int_{z_0-\lambda_q/2}^{z_0+\lambda_q/2} dz\ \sin\psi_2(z)$$

$$\frac{1}{\lambda_q}\int_{z_0-\lambda_q/2}^{z_0+\lambda_q/2} dz\ \sin\left[\left(\int_{-L/2}^{z} \frac{2\pi dz'}{\lambda_q}\right) + \frac{\omega}{c}(z-ct)+\phi_0\right]$$

$$\equiv \frac{1}{\lambda_q}\int_{z_0-\lambda_q/2}^{z_0+\lambda_q/2} dz\ \sin\psi_3(z)$$

$$\frac{1}{\lambda_q}\int_{z_0-\lambda_q/2}^{z_0+\lambda_q/2} dz\ \sin\left[\left(\int_{-L/2}^{z} \frac{2\pi dz'}{\lambda_q}\right) - \frac{\omega}{c}(z-ct)-\phi_0\right]$$

$$\equiv \frac{1}{\lambda_q}\int_{z_0-\lambda_q/2}^{z_0+\lambda_q/2} dz\ \sin\psi_4(z)$$

$$\frac{1}{\lambda_q}\int_{z_0-\lambda_q/2}^{z_0+\lambda_q/2} dz\ \cos\left[\frac{\omega}{c}(z-ct)\right] \equiv \frac{1}{\lambda_q}\int_{z_0-\lambda_q/2}^{z_0+\lambda_q/2} dz\ \cos\psi_5(z) \ . \qquad (4.11)$$

As in Ref. 5, the form of $\psi_1(z),\ldots,\psi_5(z)$ is determined by the derivative dt/dz. As an example,

$$\frac{d\psi_1}{dz} = \frac{2\pi}{\lambda_q} + \frac{\omega}{c} - \omega\left(\frac{dt}{dz}\right) \approx \frac{2\pi}{\lambda_q} - \frac{\omega}{2c}\left(\frac{1}{\gamma^2} + \beta_\perp^{\,2}\right) . \qquad (4.12)$$

Using Eq. (4.9) to compute $\beta_\perp$, Eq. (4.12) can be integrated to approximate $\psi_1(z)$ in the vicinity of z_0:

$$\psi_1(z) \approx \psi_0(z_0) + (1-N)\int_{-L/2}^{z} \frac{2\pi dz'}{\lambda_q}$$
$$- \left\{ \frac{\omega}{2c}\left[\frac{1}{\gamma^2} + \overline{x'}^2 + \frac{1}{2}\left[\left(\frac{\lambda_q}{2\pi\rho}\right)^2 + \left(\frac{\lambda_n}{2\pi\rho_n}\right)^2\right](1+k\bar{x})^2\right] - N\left(\frac{2\pi}{\lambda_q}\right)\right\}(z-z_0)$$
$$+ \frac{\omega}{c}\left\{\left[\frac{1}{\rho}\left(\frac{\lambda_q}{2\pi}\right)^2 \cos\left(\int_{-L/2}^{z}\frac{2\pi dz'}{\lambda_q}\right) + \frac{1}{\rho_n}\left(\frac{\lambda_n}{2\pi}\right)^2 \cos\left(\int_{-L/2}^{z}\frac{2\pi dz'}{\lambda_n}\right)\right](1+k\bar{x})\overline{x'}\right.$$
$$+ \left(\frac{1}{\rho}\right)^2\left(\frac{\lambda_q}{4\pi}\right)^3 (1+k\bar{x})^2 \sin\left(\int_{-L/2}^{z}\frac{4\pi dz'}{\lambda_q}\right)$$
$$+ \left(\frac{1}{\rho_n}\right)^2\left(\frac{\lambda_n}{4\pi}\right)^3 (1+k\bar{x})^2 \sin\left(\int_{-L/2}^{z}\frac{4\pi dz'}{\lambda_n}\right)$$
$$- \left(\frac{\lambda_q}{4\pi\rho}\right)\left(\frac{\lambda_n}{4\pi\rho_n}\right)(1+k\bar{x})^2\left[\frac{1}{\left(\frac{2\pi}{\lambda_q}\right)-\left(\frac{2\pi}{\lambda_n}\right)} \sin\left(\int\frac{2\pi dz}{\lambda_q} - \int\frac{2\pi dz}{\lambda_n}\right)\right.$$
$$\left.\left. - \frac{1}{\left(\frac{2\pi}{\lambda_q}\right)+\left(\frac{2\pi}{\lambda_n}\right)} \sin\left(\int\frac{2\pi dz}{\lambda_q} + \int\frac{2\pi dz}{\lambda_n}\right)\right]\right\} \qquad (4.13)$$

where we assume operation on the Nth harmonic of the fundamental wavelength:

$$\frac{\omega}{2c}\left\{\frac{1}{\gamma_0^2} + \frac{\tilde{\Lambda}^2}{k(k-s)}\right\} \approx N\left(\frac{2\pi}{\lambda_q}\right) \qquad (4.14)$$

[see Eqs. (4.23)-(4.25) for a detailed discussion of the resonance condition]. Equation (4.11) is similar to the result derived in Ref. 5 for the constant period magnet in Eqs. (4.5) and (4.6), except for the larger number of oscillating terms in the present result. Carrying out the average of $\sin\psi_1(z)$, we obtain:

$$\frac{1}{\lambda_q}\int_{z_0-\lambda_q/2}^{z_0+\lambda_q/2} dz\, \sin\psi_1(z) \approx \sum_i \sum_j \sum_\ell \sum_m \sum_o \sum_p (-1)^o \sin[\psi_0(z_0) + (i+j)\frac{\pi}{2}]$$

$$i+2\ell+o+p+n(j+2m-o+p) = (N-1)$$

(cont'd)

(cont'd)

$$\cdot\ J_i\left[\frac{1}{\rho}\ \frac{\omega}{c}\left(\frac{\lambda_q}{2\pi}\right)^2\ (1+k\bar{x})\bar{x}'\right]$$

$$\cdot\ J_j\left[\frac{1}{\rho_n}\ \frac{\omega}{c}\left(\frac{\lambda_n}{2\pi}\right)^2\ (1+k\bar{x})\bar{x}'\right]$$

$$\cdot\ J_\ell\left[\frac{1}{\rho}^2\ \frac{\omega}{c}\left(\frac{\lambda_q}{4\pi}\right)^3\ (1+k\bar{x})^2\right]$$

$$\cdot\ J_m\left[\frac{1}{\rho_n}^2\ \frac{\omega}{c}\left(\frac{\lambda_n}{4\pi}\right)^3 (1+k\bar{x})^2\right]$$

$$\cdot\ J_o\left[\frac{\omega}{c}\left(\frac{\lambda_q}{4\pi\rho}\right)\left(\frac{\lambda_q}{4\pi\rho_n}\right)\frac{(1+k\bar{x})^2}{\left(\frac{2\pi}{\lambda_q}\right)-\left(\frac{2\pi}{\lambda_n}\right)}\right]$$

$$\cdot\ J_p\left[\frac{\omega}{c}\left(\frac{\lambda_q}{4\pi\rho}\right)\left(\frac{\lambda_q}{4\pi\rho_n}\right)\frac{(1+k\bar{x})^2}{\left(\frac{2\pi}{\lambda_q}\right)+\left(\frac{2\pi}{\lambda_n}\right)}\right] \tag{4.15}$$

where the sum is carried out over all indices satisfying the designated condition. The complexity of Eq. (4.15) is regrettable, although some simplifications appear possible. The arguments of the Bessel Functions indexed by ℓ, m, o, and p are of order unity:

$$\left(\frac{1}{\rho}\right)^2\ \frac{\omega}{c}\left(\frac{\lambda_q}{4\pi}\right)^3\ (1+k\bar{x})^2 \lesssim N/2$$

$$\left(\frac{1}{\rho_n}\right)^2\ \frac{\omega}{c}\left(\frac{\lambda_n}{4\pi}\right)^3\ (1+k\bar{x})^2 \lesssim N/2$$

$$\frac{\omega}{c}\left(\frac{\lambda_q}{4\pi\rho}\right)\left(\frac{\lambda_n}{4\pi\rho_n}\right)\frac{(1+k\bar{x})^2}{\left(\frac{2\pi}{\lambda_q}\right)\pm\left(\frac{2\pi}{\lambda_n}\right)} \lesssim N/2 \tag{4.16}$$

while the arguments of the functions indexed by i, j will typically be small compared to 1:

$$\frac{1}{\rho}\ \frac{\omega}{c}\ \left(\frac{\lambda_q}{2\pi}\right)^2\ (1+k\bar{x})\bar{x}' \leqslant 2N\ \frac{(\bar{x}')}{(\lambda_q/4\pi\rho)} \ll 1$$

$$\frac{1}{\rho_n}\ \frac{\omega}{c}\ \left(\frac{\lambda_n}{2\pi}\right)^2\ (1+k\bar{x})\bar{x}' \leqslant 2N\ \frac{(\bar{x}')}{(\lambda_n/4\pi\rho_n)} \ll 1\ . \tag{4.17}$$

Accordingly, only the terms with $i = j = 0$ in Eq. (4.15) can be expected to contribute to the sum, while the dominant contribution in the sum over the remaining indices will be made for small ℓ, m, o, and p. If $n \equiv \lambda_q/\lambda_n$ is large compared to 1, then some additional simplifications are possible. Assuming $i = j = 0$, the indices must satisfy the reduced equation:

$$2\ell = (N-1) + n(o-p-2m) - (o+p) \ . \tag{4.18}$$

So long as the harmonic number N remains of order unity, a non-zero value of (o-p-2m) implies that either or (o+p) must be large. But since the arguments of the Bessel functions are all of order unity, this would imply that at least one of J_ℓ, J_o or J_p would be small, suppressing the contribution of that term to the sum. Thus, as a practical matter, for large n, the sum needs only to be carried out over those values of o, p, and m for which o-p-2m = 0, significantly reducing the number of terms to be considered. Of course, in the limit of weak magnetic fields, the arguments of the Bessel functions all approach zero, and only the product i = j = ℓ = m = o = p = 0 for N = 1 makes a finite contribution.

Repetition of these procedures for $\sin\psi_2, \ldots, \sin\psi_5$ leads to the result:

$$\frac{d\overline{\delta}}{dz} = \frac{e|\mathcal{E}|}{\gamma_0 mc^2} \{\overline{x}'(z_0) \sum_i \sum_j \sum_\ell \sum_m \sum_o \sum_p (-1)^o \sin[\psi_0(z) + (i+j)\tfrac{\pi}{2}]$$

$$i+2\ell+o+p+n(j+2m-o+p) = N$$

$$\cdot J_i\left[\frac{\omega}{c}\,\frac{1}{\rho}\left(\frac{\lambda_q}{2\pi}\right)^2 (1+k\overline{x})\overline{x}'\right]$$

$$\cdot J_j\left[\frac{\omega}{c}\,\frac{1}{\rho_n}\left(\frac{\lambda_n}{2\pi}\right)^2 (1+k\overline{x})\overline{x}'\right]$$

$$\cdot J_\ell\left[\frac{\omega}{c}\left(\frac{1}{\rho}\right)^2\left(\frac{\lambda_q}{2\pi}\right)^3 (1+k\overline{x})^2\right]$$

$$\cdot J_m\left[\frac{\omega}{c}\left(\frac{1}{\rho_n}\right)^2\left(\frac{\lambda_n}{4\pi}\right)^3 (1+k\overline{x})^2\right]$$

$$\cdot J_o\left[\frac{\omega}{c}\left(\frac{\lambda_q}{4\pi\rho}\right)\left(\frac{\lambda_n}{4\pi\rho_n}\right)\frac{(1+k\overline{x})^2}{\left(\frac{2\pi}{\lambda_q}\right)-\left(\frac{2\pi}{\lambda_n}\right)}\right]$$

$$\cdot J_p\left[\frac{\omega}{c}\left(\frac{\lambda_q}{4\pi\rho}\right)\left(\frac{\lambda_n}{4\pi\rho_n}\right)\frac{(1+k\overline{x})^2}{\left(\frac{2\pi}{\lambda_q}\right)+\left(\frac{2\pi}{\lambda_n}\right)}\right]$$

$$+ \frac{\lambda_q}{4\pi\rho}(1+k\overline{x})\Big[\sum_i\sum_j\sum_\ell\sum_m\sum_o\sum_p(-1)^o\sin[\psi_0(z)+(i+j)\tfrac{\pi}{2}]J_iJ_jJ_\ell J_m J_o J_p$$

$$i+2\ell+o+p+n(j+2m-o+p) = (N-1)$$

$$- \sum_i\sum_j\sum_\ell\sum_m\sum_o\sum_p(-1)^o\sin[\psi_0(z)+(i+j)\tfrac{\pi}{2}]J_iJ_jJ_\ell J_m J_o J_p$$

$$i+2\ell+o+p+n(j+2m-o+p) = (N+1)$$

(cont'd)

(cont'd)

$$+ \frac{\lambda_n}{4\pi\rho_n} (1+k\bar{x}) [\sum_{\substack{i\,j\,\ell\,m\,o\,p \\ i+2\ell+o+p+n(j+2m-o+p) = (N-n)}} (-1)^o \sin[\psi_0(z)(i+j)\tfrac{\pi}{2}] J_i J_j J_\ell J_m J_o J_p$$

$$- \sum_{\substack{i\,j\,\ell\,m\,o\,p \\ i+2\ell+o+p+n(j+2m-o+p) = (N+n)}} (-1)^o \sin[\psi_0(z)+(i+j)\tfrac{\pi}{2}] J_i J_j J_\ell J_m J_o J_p] \qquad (4.19)$$

For $\bar{x}' \ll (\lambda_q/2\pi\rho)$, this can be reduced to:

$$\frac{d\bar{\delta}}{dz} = \frac{e|\mathcal{E}|}{\gamma_0 mc^2} \sin[\psi_0(z)] \left\{ \left(\frac{\lambda_q}{4\pi\rho}\right) (1+k\bar{x}) \left[\sum_{\substack{\ell\, m\, o\, p \\ 2\ell+o+p+n(2m-o+p) = (N-1)}} (-1)^o \right.\right.$$

$$\cdot J_\ell \left[\frac{\omega}{c} \left(\frac{1}{\rho}\right)^2 \left(\frac{\lambda_q}{4\pi}\right)^3 (1+k\bar{x})^2 \right]$$

$$\cdot J_m \left[\frac{\omega}{c} \left(\frac{1}{\rho}\right)^2 \left(\frac{\lambda_q}{4\pi}\right)^2 (1+k\bar{x})^2 \right]$$

$$\cdot J_o \left[\frac{\omega}{c} \left(\frac{\lambda_q}{4\pi\rho}\right) \left(\frac{\lambda_n}{4\pi\rho_n}\right) \frac{(1+k\bar{x})^2}{\left(\frac{2\pi}{\lambda_q}\right)-\left(\frac{2\pi}{\lambda_n}\right)} \right]$$

$$\cdot J_p \left[\frac{\omega}{c} \left(\frac{\lambda_q}{4\pi\rho}\right) \left(\frac{\lambda_n}{4\pi\rho_n}\right) \frac{(1+k\bar{x})^2}{\left(\frac{2\pi}{\lambda_q}\right)+\left(\frac{2\pi}{\lambda_n}\right)} \right]$$

$$\left. - \sum_{\substack{\ell\, m\, o\, p \\ 2\ell+o+p+n(2m-o+p) = (N+1)}} (-1)^o J_\ell J_m J_o J_p \right]$$

$$+ \left(\frac{\lambda_n}{4\pi\rho_n}\right) (1+k\bar{x}) [\sum_{\substack{\ell\, m\, o\, p \\ 2\ell+o+p+n(2m-o+p) = (N-n)}} (-1)^o J_\ell J_m J_o J_p$$

$$\left. - \sum_{\substack{\ell\, m\, o\, p \\ 2\ell+o+p+n(2m-o+p) = (N+n)}} (-1)^o J_\ell J_m J_o J_p] \right\} . \qquad (4.20)$$

In the limit of weak magnetic fields, only the terms $\ell = m = o = p = 0$ contribute, and only the choice $N = 1$, corresponding to operation at the fundamental wavelength, yields a non-zero derivative. In this limit, the result is simply:

$$\frac{d\bar{\delta}}{dz} = \frac{e|\mathcal{E}|}{\gamma_0 mc^2} \sin\psi_0(z) \left(\frac{\lambda_q}{4\pi\rho}\right) (1+k\bar{x}) . \qquad (4.21)$$

If we assume $\overline{x'} \ll (\lambda_q/2\pi\rho)$, we can retain the form of Eq. (4.21) in the strong field limit by writing $(d\overline{\delta}/dz)$ as:

$$\frac{d\overline{\delta}}{dz} = \frac{e|\mathcal{E}|}{\gamma_0 mc^2} \sin\psi_0(z) \left(\frac{\lambda_q}{4\pi\rho}\right) (1+k\overline{x})\ J(\lambda_q,\rho,n,N,\overline{x}) \ . \qquad (4.22)$$

Here, J is used as a shorthand symbol to denote the series representation in (4.20). Formally:

$$\begin{aligned} J(\lambda_q,\rho,n,N,\overline{x}) = & \Big[\sum_{\substack{\ell\ m\ o\ p \\ 2\ell+o+p+n(2m-o+p)\ =\ (N-1)}} (-1)^o J_\ell J_m J_o J_p \\ & - \sum_{\substack{\ell\ m\ o\ p \\ 2\Pi+o+p+n(2m-o+p)\ =\ (N+1)}} (-1)^o J_\ell J_m J_o J_p \Big] \\ + \frac{1}{n}(\rho/\rho_n) & \Big[\sum_{\substack{\ell\ m\ o\ p \\ 2\ell+o+p+n(2m-o+p)\ =\ (N-n)}} (-1)^o J_\ell J_m J_o J_p \\ & - \sum_{\substack{\ell\ m\ o\ p \\ 2\ell+o+p+n(2m-o+p)\ =\ (N+n)}} (-1)^o J_\ell J_m J_o J_p \Big] \end{aligned} \qquad (4.23)$$

The value of J approaches unity in the limit of small fields, but must be computed by explicit evaluation of the series at large fields.

The phase evolution equation can be constructed by noting that the value of $\overline{\psi}_1(z)$ must not change with z_0. The imposition of this condition fixes the variation of ψ_0 with z_0:

$$\frac{d}{dz_0}\psi_0(z_0) = -\frac{\omega}{2c}\left\{\frac{1}{\gamma^2} + \frac{1}{2}\left[\left(\frac{\lambda_q}{2\pi\rho}\right)^2 + \left(\frac{\lambda_n}{2\pi\rho_n}\right)^2\right] + \overline{x'}^2\right\} + N\frac{2\pi}{\lambda_q} \ . \qquad (4.24)$$

If we define λ_0 as the "nominal" period of the field, and express the resonance energy in terms of this period, we can reduce Eq. (4.24) to the form:

$$\frac{d}{dz}\psi_0(z) = N\left[\frac{2\pi}{\lambda_q} - \frac{2\pi}{\lambda_0}\right]$$

(cont'd)

(cont'd)

$$+ N\frac{4\pi}{\lambda_0}\left\{\frac{\gamma_0-\gamma_{r_0}}{\gamma_{r_0}} + \frac{\gamma_0}{\gamma_{r_0}}\left[\bar{\delta} - \frac{\left(\frac{\gamma_0^2}{k(k-s)}\tilde{\Lambda}^2\right)}{1+\left(\frac{\gamma_0^2}{k(k-s)}\tilde{\Lambda}^2\right)}k\bar{x}\right]\right\} - \frac{\omega}{2c}\bar{x}'^2 \qquad (4.25)$$

where:

$$\gamma_{r_0} \equiv \left(\frac{\lambda_0}{2\lambda}\right)^{1/2}\left[1 + \frac{\gamma_0^2}{k(k-s)}\tilde{\Lambda}^2\right]^{1/2} . \qquad (4.26)$$

We note that the definition of λ_0 is somewhat arbitrary. It is introduced primarily to facilitate the expression of the derivative of ψ_0 in terms of the small variables $\bar{x}$ and $\bar{\delta}$. As such, its values need to be selected to minimize the magnitude of the difference $|\lambda_q-\lambda_0|$ during the interaction. The value of λ_0 could, for example, be set equal either to the initial value of λ_q or its average over the interaction region.

The position, energy, and phase evolution equations [Eqs. (4.6), (4.22), and (4.25)] for this generalized magnet structure have a form remarkably similar to their counterparts for the constant period magnet:

$$\frac{d^2\bar{x}}{dz^2} = -\frac{\tilde{\Lambda}^2}{k-s}\left[(k-s)\bar{x} - \bar{\delta}(1+s\bar{x})\right](1-2\bar{\delta}) - 2\bar{x}'\bar{\delta}' \qquad (4.6)$$

$$\frac{d\bar{\delta}}{dz} = \frac{e|\mathcal{E}|}{\gamma_0 mc^2}\sin[\psi_0(z)]\left(\frac{\lambda_q}{4\pi\rho}\right)(1+k\bar{x})\ J(\lambda_q,\rho,n,N,\bar{x}) \qquad (4.22)$$

$$\frac{d\psi_0}{dz} = N\left[\frac{2\pi}{\lambda_q} - \frac{2\pi}{\lambda_0}\right] + N\frac{4\pi}{\lambda_0}\left\{\frac{\gamma_0-\gamma_{r_0}}{\gamma_{r_0}} + \frac{\gamma_0}{\gamma_{r_0}}\left[\bar{\delta} - \frac{\left(\frac{\gamma_0^2}{k(k-s)}\tilde{\Lambda}^2\right)}{1+\left(\frac{\gamma_0^2}{k(k-s)}\tilde{\Lambda}^2\right)}k\bar{x}\right]\right\}$$

$$- \frac{\omega}{2c}\bar{x}'^2 \qquad (4.25)$$

where $\tilde{\Lambda}^2$ is defined in Eq. (4.8), J in Eq. (4.23), and γ_{r_0} in Eq. (4.26). The principle changes in those equations, as compared to the constant period system, concern the strength of focussing, the nature of the function J, and the addition, to the phase evolution equation, of a term proportional to the deviation of the magnet period from the "nominal" period.

The use of a second component to the periodic field permits the amplitude and period of the fundamental component to be changed without altering the form of the equation governing the evolution of $\bar{x}$. As was described at the beginning of the section, this modification is a precondition to the implementation of the cancellation scheme. The change in the phase evolution equation is also necessary: the term $N(2\pi/\lambda_q - 2\pi/\lambda_q)$ permits the phase ψ_0 to be varied almost at will during the interaction by changing the local magnet period λ_q. The change in the definition of J is not necessary for excitation cancellation, but is apparently unavoidable given the requirement for a second periodic component to the field to stabilize the focussing strength.

The demonstration of the relation of the form of $x_1'(z)$ in Eq. (4.3) to the variation in amplitude and phase of the periodic field follows the lines of the linearized small signal analysis in Sec. III. As with the constant period magnet, the phase evolution equation can be simplified for gain expanded magnets with "infinite" gain expansion. For these structures:

$$\frac{\dfrac{\gamma_0^2\tilde{\Lambda}^2}{k(k-s)}}{1+\dfrac{\gamma_0^2\tilde{\Lambda}^2}{k(k-s)}}\,k = k - s\ . \tag{4.26}$$

It also remains convenient to define $\Delta\psi$ as in Eq. (3.8):

$$\Delta\psi = N\left(\frac{4\pi}{\lambda_0}\right)\int_{-L/2}^{z} dz'\ \frac{\gamma_0}{\gamma_{r_0}}\ [\bar{\delta} - (k-s)\bar{x}]\ . \tag{4.27}$$

Linearizing the equations and motion and setting $J = J(\lambda_q,\rho_0,n,N,\bar{x}=0)$, we obtain:

$$\frac{d^2\bar{x}}{dz^2} = -\frac{\tilde{\Lambda}^2}{(k-s)}\ [(k-s)\bar{x} - \bar{\delta}]$$

$$\frac{d\bar{\delta}}{dz} = \frac{e|\mathcal{E}|}{\gamma_0 mc^2}\left(\frac{\lambda_q}{4\pi\rho_0}\right)\ [1 - \frac{k-s}{\tilde{\Lambda}^2}\frac{d^2\bar{x}}{dz^2} + s\bar{x}]\ J(\lambda_q,\rho_0,n,N,\bar{x}=0)$$
$$\cdot\ \sin\left[\psi_0\left(-\frac{L}{2}\right) + q\left(z+\frac{L}{2}\right) + \Delta\psi + N\int_{-L/2}^{z} dz\left(\frac{2\pi}{\lambda_q} - \frac{2\pi}{\lambda_0}\right)\right]$$

$$\frac{d\Delta\psi}{dz} = N\left(\frac{4\pi}{\lambda_q}\right)\ \frac{\gamma_0}{\gamma_{r_0}}\ [\bar{\delta} - (k-s)\bar{x}]$$
$$= N\left(\frac{4\pi}{\lambda_q}\right)\ \frac{\gamma_0}{\gamma_{r_0}}\ \frac{k-s}{\tilde{\Lambda}^2}\ \bar{x}'' \tag{4.28}$$

where, as before, $q \equiv N(4\pi/\lambda_q)(\gamma_0 - \gamma_{r_0}/\gamma_{r_0})$. Representing $\bar{x}$ as a power series in $\mathcal{E}$ as in Eq. (3.10):

$$\bar{x} = x_0 + x_1 \mathcal{E} + x_2 \mathcal{E}^2 + \ . \ . \ . \tag{3.10}$$

we find that $x_0(z)$ obeys the equation:

$$\frac{d}{dz} \ (k-s)[\frac{1}{\tilde{\Lambda}^2} x_0'' + x_0 \ ] = 0 \tag{4.29}$$

while the first order term $x_1(z)$ satisfies:

$$\frac{d}{dz} \ (k-s)[\frac{1}{\tilde{\Lambda}^2} x_1'' + x_1 \ \mathcal{E}] = \frac{e|\mathcal{E}|}{\gamma_0 mc^2} \left(\frac{\lambda_q}{2\pi\rho_0}\right) J(\lambda_q, \rho_0, n, N, \bar{x}=0)$$

$$\cdot \left[1 - \frac{k-s}{\tilde{\Lambda}^2} \ \frac{d^2 x_0}{dz^2}\right] + sx_0 \ \sin\left[\psi_0(-\tfrac{L}{2}) + q(z+\tfrac{L}{2})\right.$$

$$\left. + \Delta\psi_0 + N\int_{-L/2}^{z} dz\left(\frac{2\pi}{\lambda_q} - \frac{2\pi}{\lambda_0}\right)\right] . \tag{4.30}$$

Assuming the initial motion is on-axis, $x_0 = \delta_0 = 0$, the equation for $x_1(z)$ is reduced to:

$$\frac{d}{dz} \ (k-s)[\frac{1}{\tilde{\Lambda}^2} x_1'' + x_1 \ \mathcal{E}] = \frac{e|\mathcal{E}|}{\gamma_0 mc^2} \left(\frac{\lambda_q}{2\pi\rho_0}\right) J(\lambda_q, \rho_0, \ n, N, \bar{x}=0)$$

$$\cdot \ \sin\left[\psi_0(-\tfrac{L}{2}) + q(z+\tfrac{L}{2}) + N \int_{-L/2}^{z} dz\left(\frac{2\pi}{\lambda_q} + \frac{2\pi}{\lambda_0}\right)\right] . \tag{4.31}$$

If $x_1'(z)$ is assumed to have the form hypothesized in Eq. (4.3), the left hand side of (4.31) becomes:

$$\frac{d}{dz} \ (k-s)[\frac{1}{\tilde{\Lambda}^2} x_1'' + x_1 \ \mathcal{E}] = \frac{k-s}{\tilde{\Lambda}^2} \ \mathcal{E} \cdot \mathrm{Re}\left[\exp \ i[\psi_0(-\tfrac{L}{2}) + q(z+\tfrac{L}{2}) - \tfrac{\pi}{2}]\right.$$

$$\cdot \ \{(z^2 - \tfrac{L}{4})^2 \ f''(z) + f'(z) [8z + i2q(z^2 - \tfrac{L^2}{4})] (z^2 - \tfrac{L^2}{4})$$

$$\left. + f(z) [(12z^2 - L^2) + i8qz(z^2 - \tfrac{L^2}{4}) + (\tilde{\Lambda}^2 - q^2)(z^2 - \tfrac{L^2}{4})^2]\}\right]$$

$$\equiv \frac{k-s}{\tilde{\Lambda}^2} \ \mathcal{E} \cdot \mathrm{Re}\{G(z) \ \exp \ i[\psi_0(-\tfrac{L}{2}) + q(z+\tfrac{L}{2}) - \tfrac{\pi}{2}]\} . \tag{4.32}$$

We have defined the quantity in brackets as $G(z)$, in recognition of its role as a "generating function" for the magnet structure.

To yield a trajectory of the form assumed in Eq. (4.3), the magnet period and amplitude must vary in such a way that the right side of Eq. (4.31) is equal, within a constant factor, to $G(z)$. Matching magnitudes and arguments:

$$N\int_{-L/2}^{z} dz\left(\frac{2\pi}{\lambda_q} - \frac{2\pi}{\lambda_0}\right) = \arg G(z)$$

$$\Rightarrow N\left(\frac{2\pi}{\lambda_q} - \frac{2\pi}{\lambda_0}\right) = \frac{d}{dz}\arg G(z)$$

$$\frac{e}{\gamma_0 mc^2}\left(\frac{\lambda_q}{2\pi\rho_0}\right) J = \frac{k-s}{\tilde{\Lambda}^2}\cdot|G(z)| \quad . \tag{4.33}$$

We see that the deviation of the magnet period from the nominal period is determined by the rate of change of the argument of G. The relation determining the magnitude of the static field is somewhat more complex, due to the form of J [see Eq. (4.22)]. As previously noted, for operation at the fundamental wavelength and for weak magnetic fields, J approaches unity, and in this limit:

$$B_0 = \frac{2\pi}{\lambda_q}\left[\left(\frac{\gamma_0 mc^2}{e}\right)^4 \left(\frac{2}{k}\right) |G(z)|\right]^{1/3} \quad . \tag{4.34}$$

For strong fields, numerical methods must be used to compute B_0. The right hand side of (4.32) is, by construction, the first order term in $d\delta/dz$:

$$\frac{d\bar{\delta}}{dz} = \frac{e}{\gamma_0 mc^2}\,\frac{1}{\lambda_q}\int_{z-\lambda_q/2}^{z+\lambda_q/2} dz'\ \mathcal{E}\cdot\beta_{\perp 0} \tag{4.35}$$

where $\beta_{\perp 0}$ is the zeroth order transverse velocity, as given by Eq. (4.9). Numerical integration of (4.35), and comparison of the result with the right hand side of (4.32), yields directly the relation of B_0 and $|G|$.

In practice, it is possible to evaluate the effectiveness of functional cancellation without actually solving for $B_0(z)$, since the equations of motion are formulated in terms of $(\lambda_q J/\rho_0)$, and this quantity is defined by Eq. (4.33) in a form which can be used in the full (non-linear) equations of motion. As with the constant period magnet, an evaluation of emittance excitation for off-axis initial motion requires the numerical integration of the full equations of motion. An example of functional cancellation is included with the data for the constant period magnets in Sec. VI.

V. MOTION IN THE VERTICAL PLANE

Gain expansion can be implemented with zero vertical dispersion, so that the change in energy during the interaction does not contribute directly to the excitation of the vertical betatron coordinates y_β and y_β':

$$y_\beta = y - \eta_y\delta = y \qquad \text{if } \eta_y = 0 \,. \qquad (5.1)$$
$$y_\beta' = y' - \eta_y'\delta = y'$$

The excitation of y_β and y_β' therefore depend only on the manner in which y and y' change during the interaction.

As with the motion in the orbit plane, our interest is actually in the change in the invariant A_y:

$$A_y \equiv \pi\left[\left(\frac{y_\beta}{\beta_y}\right)^2 + \beta_y\left(y_\beta' - \frac{\beta_y'}{2\beta_y}\, y_\beta\right)^2\right]. \qquad (5.2)$$

We will find that A_y and A_y^2 are changed by the interaction even when $\eta_y = 0$, but that the change is small and should not cause significant complications.

Assuming that the horizontal component of the magnetic field is zero in the orbit plane, the evolution of y will be determined by an equation of the form:

$$\frac{d^2\overline{y}}{dz^2} + \mu^2(1-2\delta)\overline{y} = -2\overline{\delta}'\cdot\overline{y}' \qquad (5.3)$$

where $\overline{y}$ is the averaged position

$$\overline{y} = \left(\frac{1}{\lambda_q}\right)^2 \int_{z-\lambda_q/2}^{z+\lambda_q/2} dz' \int_{z'-\lambda_q/2}^{z'+\lambda_q/2} dz''\, y(z) \,, \qquad (5.4)$$

$\overline{\delta}$ the averaged energy, and μ is determined by the amplitude and gradient of the various components of the horizontal magnetic field.

If the energy $\overline{\delta}$ were constant, the invariant A_y is not altered by Eq. (5.3). Even if $\overline{\delta}$ changes, the linearized form of Eq. (5.3):

$$\frac{d^2\overline{y}}{dz^2} + \mu^2\overline{y} = 0 \qquad (5.5)$$

does not alter A_y. But when $\overline{\delta}$ changes, the non-linear terms, ~ $\overline{\delta}\cdot\overline{y}'$ and $\overline{\delta}\cdot\overline{y}$ change A_y. This can perhaps best be seen if $\overline{y}$ and $\overline{\delta}$ are expanded as power series in the electric field:

$$\overline{y}(z) = \sum_{j=0}^{\infty} y_j\mathcal{E}^j$$
$$\overline{\delta}(z) = \sum_{j=0}^{\infty} \delta_j\mathcal{E}^j \,. \qquad (5.6)$$

Then y_0 and y_1 satisfy, respectively:

$$y_0'' + \mu^2(1-2\delta_0)y_0 = 0$$

$$y_1'' + \mu^2(1-2\delta_0)y_1 = -2(y_0'\delta_1' + \mu^2\delta_1 y_0) \quad . \tag{5.7}$$

The zeroth order term y_0 satisfies the unperturbed equation of motion, and does not contribute to emittance growth. The first order term y_1 must satisfy an inhomogeneous differential equation with a driving term proportional to δ_1. As for motion in the orbit plane, the first order term y_1 will, in general, result in a non-zero value for A proportional to the optical power density and the initial value of A.

Rather than develop a general solution to Eq. (5.7), we will consider the effect of the laser interaction in the limit $\mu L \ll 1$ as an example. We are guided in the choice of this limit primarily by expediency; although $\mu L \gtrsim 1$ in most cases of interest, the assumption $\mu L \ll 1$ substantially simplifies the analysis. In this limit, the change in $\overline{y}'$ during the interaction is due to the term $2\overline{y}'\overline{\delta}'$, and we can write:

$$\begin{aligned}\Delta\overline{y}' &\approx L\cdot y'' \\ &\approx -2\overline{y}'\overline{\delta}'L \\ &\approx -2\overline{y}'(\Delta\overline{\delta}) \quad .\end{aligned} \tag{5.8}$$

The change in $\overline{y}$ is simply $\Delta\overline{y} = \overline{y}'L$. If, as in Sec. II, we consider the effect of the laser interaction on an ensemble of electrons each with the same values of A_y, but uniformly distributed around the perimeter of the phase space ellipse, the first moment of the change in the ensemble average of A_y during the interaction is:

$$\begin{aligned}\langle\delta A_y\rangle &= \pi\left\langle\frac{2\overline{y}\Delta\overline{y}}{\beta_y} + 2\beta_y\,\overline{y}'\Delta\overline{y}'\right\rangle \\ &\approx -2A_y\left\langle\Delta\overline{\delta}\right\rangle\end{aligned} \tag{5.9}$$

where we have assumed that $\Delta\delta$ and $\overline{y}'$ are uncorrelated, and $\beta_y' = 0$. The second moment of the change in A_y is:

$$\begin{aligned}\langle\delta A_y^2\rangle &= \left\langle\left(\frac{2\overline{y}\Delta\overline{y}}{\beta_y} + 2\beta_y\,\overline{y}'\Delta\overline{y}'\right)^2\right\rangle \\ &\approx 6A^2\left\langle\Delta\delta^2\right\rangle \quad .\end{aligned} \tag{5.10}$$

As described in the appendix, the distribution of the electrons depends on both δA_y and δA_y^2. From the Fokker-Planck equation,

$$P(A) \approx \exp\left[\int_0^{A_y} dA \, \frac{\langle\delta A\rangle - \frac{1}{2}\frac{d}{dA}\langle\delta A^2\rangle}{\frac{1}{2}\langle\delta A^2\rangle}\right], \tag{5.11}$$

where the changes $\langle\delta A\rangle$ and $\langle\delta A^2\rangle$ include the contributions of both the laser and the storage ring. The significance of the contribution of the laser to $\langle\delta A_y\rangle$ and $\langle\delta A_y{}^2\rangle$ will be determined by the effect of laser operation on the numerator and denominator of the integrand in Eq. (5.11):

Quantum fluctuations and synchrotron damping ordinarily result in values for these quantities equal to:

$$[\langle\delta A\rangle - \frac{1}{2}\frac{d}{dA}\langle\delta A^2\rangle] = -\frac{1}{N_D}(A + \varepsilon_0)$$
$$\langle\delta A^2\rangle = \frac{2\varepsilon_0}{N_D}(A + \varepsilon_0) \tag{5.12}$$

where N_D is the number of turns to damp the vertical coordinate, and ε_0 is the natural vertical emittance.

The contribution of laser operation to the first moment of δA is $\langle\delta A\rangle = -2A\langle\Delta\delta\rangle$. However, the average energy loss $\Delta\delta$ would be replaced in a storage ring by the RF cavity, and by the Robinson-Liouville Theorem,[24] this would result in an equal and opposite contribution to $\langle\delta A\rangle$ in the storage ring. The net change in δA due to the laser for a full orbit around the ring would therefore be zero. However, the second moment of the change in A due to the laser is not cancelled, and we must compare its magnitude with the natural value of $\langle\delta A^2\rangle$. We find that

$$\frac{\langle\delta A^2\rangle_{\text{laser}}}{\langle\delta A^2\rangle_{\text{quant. fluct.}}} = 3\,\frac{A^2\langle\Delta\bar{\delta}^2\rangle}{\varepsilon_0(A + \varepsilon_0)}\,N_D \,. \tag{5.13}$$

If we confine our interest to the tails of the distribution ($A \gg \varepsilon_0$), we find that the condition for the neglect of vertical plane laser excitation is:

$$\langle\Delta\bar{\delta}^2\rangle \ll \frac{1}{N_D}\,\frac{\varepsilon_0}{A} \,. \tag{5.14}$$

Since $\langle\Delta\bar{\delta}^2\rangle$ is, in general, proportional to the optical power density, there is implicit in (5.14) a limit to the laser power output. Since $\langle\Delta\bar{\delta}^2\rangle$ will itself be a function of A, the details will depend on the magnitude of ε_0 and the functional form of $\langle\Delta\bar{\delta}^2\rangle$. If ε_0 is small, as is typical for the vertical plane emittance, and if $\langle\Delta\bar{\delta}^2\rangle$ remained independent of A into the wings of the distribution, it would be reasonable to require $\langle\Delta\bar{\delta}^2\rangle \ll (1/10\,N_D)$ to insure that the probability distribution was not disturbed. Assuming the ratio of the first moment of $\langle\Delta\bar{\delta}\rangle$ to the second moment $\langle\Delta\bar{\delta}^2\rangle$ is of the order of the number of magnet periods,[2] we infer a limit to the radiated energy per electron in this case equal to:

$$\langle\Delta\bar{\delta}\rangle \ll \frac{\#\text{ of magnet periods}}{10\cdot N_D} \,. \qquad (5.15)$$

This is not a particularly confining limit for long magnets.

VI. NUMERICAL RESULTS

The analytical treatment of the preceding sections has several shortcomings as noted in those sections. Considerations have been restricted to the linearized form of the equations of motion, and to the effects of laser operation on the terms in δ, x and x' proportional to the first power of the optical electric field. Those portions of the analysis relating to the excitation cancellation were further generally restricted to the case of on-axis initial motion.

While the results in these sections provide important guidance in the design of the laser magnet, numerical methods must be used to determine the actual motion of the electrons and the effects of laser operation on the emittance. The procedure for the numerical calculation has been outlined in Sec. II. To compute the probability distribution for the transverse invariant A in storage ring FELs, we need to integrate the Fokker-Planck equation (appendix). Computation of the integral requires knowledge of the moments $\langle\delta A\rangle$ and $\langle\delta A^2\rangle$ as functions of A [Eqs. (2.7) and (2.9)]. Given the equations of motion (3.1), $\langle\delta A\rangle$ and $\langle\delta A^2\rangle$ follow from the magnet design, laser operating wavelength and power, and the value of the betatron function $\beta(z)$ at the start of the interaction.

The effects of excitation cancellation on emittance, the variation of $\langle\delta A\rangle$, $\langle\delta A^2\rangle$ and the probability distribution P(A) with A, and the effects of the higher order terms in the optical electric field are best illustrated by consideration of some selected examples.

Table 1 lists the computed values of $\langle A\rangle \equiv \int A\, P(A)dA$ for a number of excitation cancelling and non-excitation cancelling magnets in a storage ring laser operated as an amplifier at a wavelength of one micron. Each of the magnets considered in the table have the same length (20 m), period (5.0 cm), and magnetic field strength (6.05 KG). For the constant period magnets, the operating energy was adjusted to yield net optical slips $qL = 1.6\pi$, 1.8π, 2.0π, 2.2π, and 2.4π radians. The transverse gradient for each of these magnets, nominally about 7 Kgauss/cm, was adjusted to maintain a betatron phase advance $\tilde{\Lambda}L = 4$ radians. The beta function at $z = -L/2$ was assumed to be $\beta(-L/2) = 10$ meters with $\beta' = 0$. The optical power density in the interaction region was assumed to be 10^5 watts/m^2, and it was assumed that the electrons entered the interaction region

Table 1.

qL	$\langle A \rangle = \int A\ P(A)dA$	Magnet Type
1.6π	5.45×10^{-7} cm-radians	Constant Period
1.8π	3.21×10^{-7}	"
2.0π	1.38×10^{-8}	"
2.2π	2.92×10^{-7}	"
2.4π	5.47×10^{-7}	"
2.0π	2.32×10^{-8}	Functional Cancellation

with random optical phase. The damping time for transverse betatron motion was assumed to be 10^5 turns, and the excitation due to quantum fluctuations in the synchrotron radiation was neglected.

We can identify the value of $\langle A \rangle$ in Table 1 with the emittance. The magnet with optical phase slip $qL = 2\pi$ satisfies the excitation cancelling condition [Eq. (3.23)], and has the lowest emittance, $1.38 \cdot 10^{-8}$ cm radians. The emittance with the non-cancelling alternatives, $qL = 1.6\pi$, 1.8π, 2.2π and 2.4π, ranged from $2.92 \cdot 10^{-7}$ cm to $5.45 \cdot 10^{-7}$ cm, larger by a factor of 20 to 40.

A magnet employing functional cancellations, with $f(z) = 1$, was also analyzed. With the same length, period, magnetic field, and betatron phase advance as the constant period magnets in Table 1, and with a nominal optical phase slip $qL = 2\pi$, the emittance was computed to be $\langle A \rangle = 2.32 \cdot 10^{-8}$ cm radians, within a factor of two of the result obtained with the excitation cancelling constant period magnet.

The actual variation of $\langle \delta A \rangle$, $\langle \delta A^2 \rangle$, and $P(A)$ with A for the excitation-cancelling constant period magnet is shown in Figs. 3, 4 and 5. The magnitude of $\langle \delta A \rangle$ decreases with A, while $\langle \delta A^2 \rangle$ increases. The probability distribution $P(A)$ decreases rapidly with A, from $P(A) = 10^8$ at $A = 1 \cdot 10^{-8}$ cm radians, to 10^{-20} at $A = 2 \cdot 10^{-7}$.

The effect of the non-linear terms in $\mathcal{E}$ is illustrated in Table 2 which lists the computed values of $A = \int A\ P(A)dA$, $\langle \delta A \rangle$ and $\langle \delta A^2 \rangle$ for the excitation cancelling constant period magnet ($qL = 2\pi$, $\tilde{\Lambda}L = 4\pi$) at optical power densities of 10^5 and 10^6 watts/cm^2. The moments $\langle \delta A \rangle$ and $\langle \delta A^2 \rangle$ are calculated for an initial value of the transverse invariant equal to $1 \cdot 10^{-8}$ cm radians.

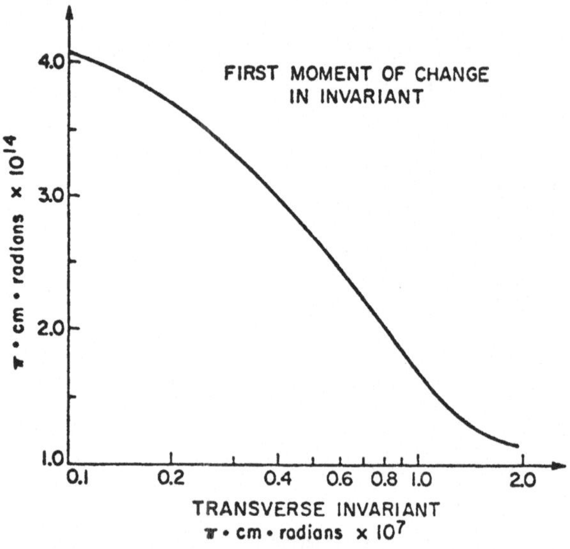

Fig. 3. The figure indicates the functional dependence on A of $\langle \delta A \rangle$, the first moment of the change in the transverse invariant due to laser operation. The laser is assumed to be operating as an amplifier at an optical power density of 10^5 watts/cm^2.

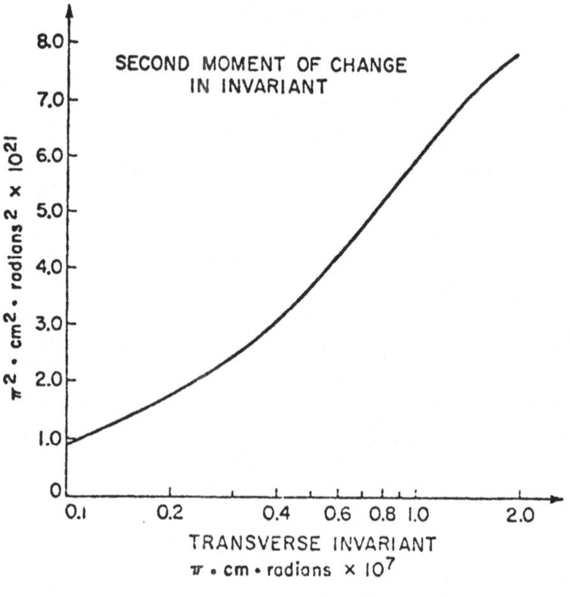

Fig. 4. The figure indicates the functional dependence of the second moment, $\langle \hat{\delta} A^2 \rangle$, of the change in the invariant. The operating conditions are the same as in Fig. 3.

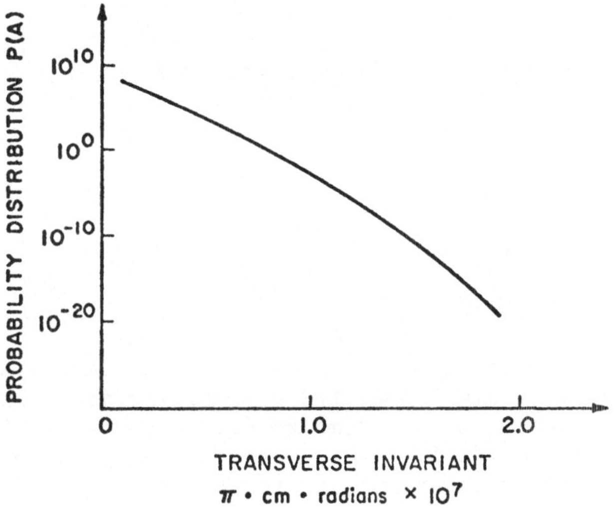

Fig. 5. The figure indicates the form of the probability density function P(A) computed from the data in Figs. 3 and 4.

From the discussion in Sec. III [Eq. (3.26)], the moments $\langle \delta A \rangle$ and $\langle \delta A^2 \rangle$ should scale as $\mathcal{E}^2$ if the first order terms in x and x' are dominant. But note that these quantities scale in Table 2 as $\mathcal{E}^4$, indicating the dominance of the 2nd order terms. Reflecting the increases in $\langle \delta A \rangle$ and $\langle \delta A^2 \rangle$, the emittance increased to $A = 1.15 \cdot 10^{-7}$ cm radians at 10^6 watts/cm^2, an order of magnitude larger than at 10^5 watts/cm^2.

While the second order terms did not result in the loss of the beam in this example, their presence, and the resulting effect on the emittance, indicates that careful attention must be paid to the non-linear terms when evaluating laser operation at high power.

Table 2.

Optical Power Density	$\langle \delta A \rangle$	$\langle \delta A^2 \rangle$	$\langle A \rangle$
10^5 watts/cm^2	4.14×10^{-14}	8.83×10^{-22}	1.38×10^{-8}
10^6 watts/cm^2	4.19×10^{-12}	8.80×10^{-20}	1.48×10^{-7}

Acknowledgments

It is a pleasure to acknowledge discussions with David Deacon, Robert Taber, and Helmut Wiedemann. The discussion of the vertical plane emittance follows lines developed originally by David Deacon.

APPENDIX: THE DISTRIBUTION FUNCTION

If the various moments of the change in the transverse invariant due to laser operation are known, the probability distribution function for the electrons can be estimated from the Fokker-Planck Equation. If $\langle\delta A\rangle$, $\langle\delta A^2\rangle$, $\langle\delta A^3\rangle$, . . . , are the computed moments of the change in the invariant per turn around the ring, and ΔP is the change per turn in the distribution function, then:[25]

$$\Delta P = \frac{\partial}{\partial A}\{\langle\delta A\rangle P(A) - \frac{1}{2}\frac{d}{dA}[\langle\delta A^2\rangle P(A)] + \frac{1}{6}\frac{d^2}{dA^2}[\langle\delta A^3\rangle P(A)] + \ldots\} . \qquad (A.1)$$

If we consider the equilibrium case, $\Delta P = 0$, and Eq. (A.1) can be solved explicitly for P(A). If we consider only the effects of $\langle\delta A\rangle$ and $\langle\delta A^2$, we obtain:

$$P(A) \sim \exp \int_0^A dA \frac{\langle\delta A\rangle - \frac{1}{2}\frac{d}{dA}\langle\delta A^2\rangle}{\frac{1}{2}\langle\delta A^2\rangle} . \qquad (A.2)$$

This is the result which is normally quoted. If we include the third order moment we obtain the formal result:

$$P(A) \sim \exp \int_0^A dA \frac{\langle\delta A\rangle - \frac{1}{2}\frac{d}{dA}\langle\delta A^2\rangle + \frac{1}{6}\frac{d^2}{dA^2}\langle\delta A^3\rangle + \frac{1}{6}\langle\delta A^3\rangle\frac{d^2}{dA^2}P(A)}{\frac{1}{2}\langle\delta A^2\rangle - \frac{1}{3}\frac{d}{dA}\langle\delta A^3\rangle} \qquad (A.3)$$

For this study, we have used both (A.2) and (A.3) to compute P(A). Equation (A.2) is valid if the third and higher order moments are small, but the test of "smallness" is unfortunately not well defined for the present application. The primary value of (A.3) is as a check on the convergence of (A.2). The complicating factor in (A.3) is the presence of P(A) within the integrand on the right hand side. But this does not interfere with its value as a check on (A.2): We can simply require that the value of P(A) calculated using (A.2) replicate itself when plugged into the integrand of (A.3). All the data presented here passed this test.

Natural values of $\langle \delta A \rangle$ and $\langle \delta A^2 \rangle$

Laser operation changes the probability distribution P(A) by modifying the "natural" values of $\langle \delta A \rangle$ and $\langle \delta A^2 \rangle$. The values of these quantities in the absence of the beam are determined by the synchrotron damping and quantum fluctuations. The natural first moment is simply $\langle \delta A \rangle = -A/N_D$, where N_D is the number of turns to damp. The second moment can be computed from the known form of the emittance distribution:

$$P(A) \sim e^{-A/\varepsilon_0}$$

$$\Rightarrow \langle \delta A^2 \rangle = \frac{\varepsilon_0}{N_D} (\varepsilon_0 + A) \ . \qquad (A.4)$$

While the natural damping due to synchrotron radiation was included in the calculation of the probability distribution function P(A) discussed in Sec. VI, the contribution of quantum fluctuations to the second moment was assumed small in the expectation that it would be dominated by the effects of laser operation.

REFERENCES AND FOOTNOTES

1. T. I. Smith, J.M.J. Madey, L. R. Elias, and D.A.G. Deacon, Reducing the Sensitivity of a Free Electron Laser to Electron Energy, J. Appl. Phys. 50:4580 (1979).
2. J.M.J. Madey, Scaling Relations for the Power Output of Gain Expanded Storage Ring Free Electron Lasers, HEPL-853, June 1979.
3. M. Sands, The Physics of Electron Storage Rings — An Introduction, SLAC Report 121* (November 1970), Eqs. (2.42) and (2.45).
4. M. Sands, ibid.
5. J.M.J. Madey and R. C. Taber, Equations of Motion in a Free Electron Laser Magnet with a Transverse Gradient, in: "Physics of Quantum Electrons, V. 7," S. Jacobs, H. Pilloff, M. Sargent, M. Scully, and R. Spitzer, eds., Addison-Wesley, Reading, MA (1980).
6. J.M.J. Madey, Gain and Saturation in Excitation Cancelling Gain-Expanded Free Electron Lasers, HEPL-855, June 1979.
7. J.M.J. Madey, op. cit. (see Ref. 2).
8. J.M.J. Madey, Optimization of Parameters for Excitation Cancelling Gain-Expanded Free Electron Lasers," HEPL-871, June 1980.
9. M. Sands, op. cit., Eq. (2.56).
10. M. Sands, op. cit., Eqs. (2.42) and (2.45).
11. M. Sands, op. cit., Eq. (2.29).

*A preliminary version of these notes was published in Physics with Interacting Storage Rings, B. Touschek, ed. (Academic Press, 1971).

12. M. Sands, op. cit., see Sec. V.
13. A. Bambini and A. Renieri, Free-Electron Laser — Single-Particle Classical-Model, Lett. Nuovo Cim. 21:399 (1979);
G. Dattoli and A. Renieri, Classical multimode theory of the free electron laser, Lett. Nuovo Cim. 24:121 (1979).
14. The members of the ensemble are assumed to be uniformly distributed in the betatron phase.
15. J.M.J. Madey, D. Deacon, and T. I. Smith, Free-electron lasers, boundary deformation, and the Robinson-Liouville theorem, J. Appl. Phys. 50:7875 (1979).
16. Note that the area A enclosed by ensemble in the two dimensional (x,x') phase space can also be changed by interactions which rotate the ensemble out of the (x,x') plane.
17. J.M.J. Madey and R. C. Taber, op. cit.
18. M. Sands, op. cit., Eq. (2.64).
19. D.A.G. Deacon, Theory of the Isochronous Storage Ring Laser, Ph.D. dissertation (Stanford University, 1979), unpublished;
D.A.G. Deacon and J.M.J. Madey, Isochronous Storage-Ring Laser: A Possible Solution to the Electron Heating Problem in Recirculating Free-Electron Lasers, Phys. Rev. Lett. 44:449 (1980).
20. H. Bruck, "Accelerateurs Circulaire de Particules," Press Universitaires de France, Paris (1969).
21. K. Steffen, Selected topics of beam optics relevant to storage ring design, in: "Physics with Intersecting Storage Rings," B. Touschek, ed., Academic Press, New York (1971).
22. We assume similar series expansion for $\overline{\delta}$ and $\Delta\psi$:

$$\overline{\delta} \equiv \sum_{j=0}^{\infty} \delta_j \&^j \;, \quad \Delta\psi \equiv \sum_{j=0}^{\infty} \Delta\psi_j \&^j \;.$$

23. W. B. Colson, Free Electron Laser Theory, Ph.D. dissertation (Stanford University, 1977), unpublished.
24. J.M.J. Madey, D. Deacon, and T. I. Smith, op. cit.
25. F. Reif, "Fundamentals of Statistical and Thermal Physics," McGraw-Hill, New York (1965), p. 577.

AN INTRODUCTION TO THE THEORY OF THE ISOCHRONOUS STORAGE RING LASER

David A. G. Deacon

High Energy Physics Laboratory
Stanford University
Stanford, California 94305

INTRODUCTION

Many configurations have now been proposed to take advantage of the simplicity of the free electron laser interaction mechanism. The primary goals in this endeavor have been to construct an efficient, powerful source of radiation in the infrared through the ultraviolet. In this paper, I discuss a technique for utilizing the beam stored in a storage ring to drive the laser mechanism, in which the efficiency of the system is preserved by making the electron optics nearly isochronous: hence the name isochronous storage ring laser.

The storage ring based system has the initial advantage of retaining the electron energy from pass to pass so that only the energy lost to the laser and the other loss mechanisms needs to be replenished during operation. The problem with the storage ring laser is that any high frequency energy change induced on the electron beam (bunching) is preserved, and can accumulate, interfering with the energy extraction in the laser on subsequent passes. However, since the laser interaction conserves the volume of the electron distribution in its six dimensional phase space,[1] this kind of problem can be avoided in an appropriately designed system.

Renieri[2] has shown that in a system in which the energy modulation is allowed to accumulate, the laser power is limited to a small fraction of the power emitted by the electrons as synchrotron radiation. The isochronous storage ring laser eliminates this buildup of the energy spread, thereby increasing the efficiency and extractable power of the system. The difference between the two types of systems is demonstrated schematically in Fig. 1. The

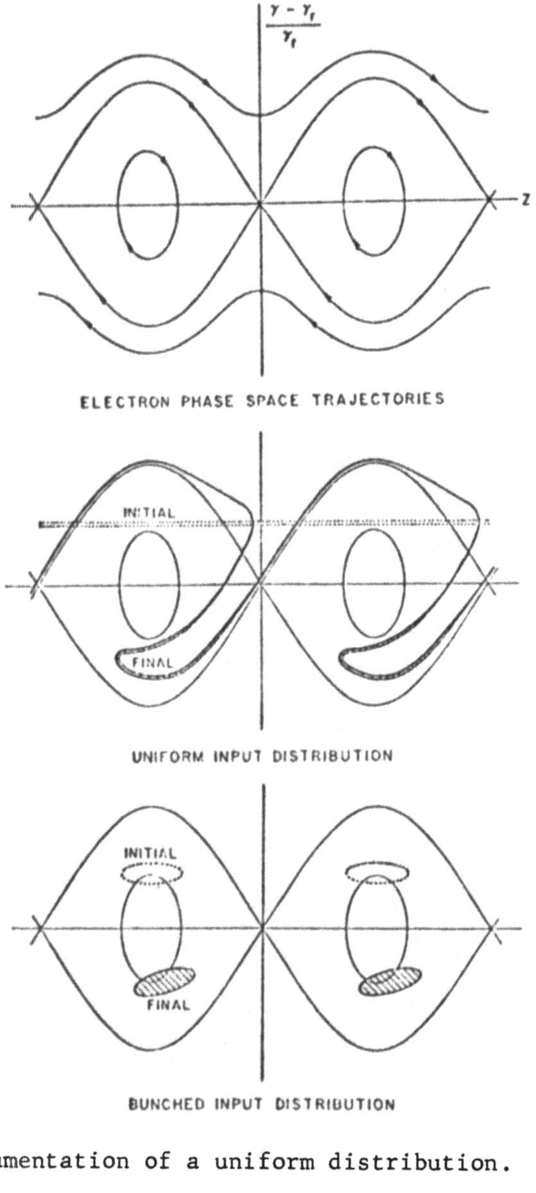

Fig. 1. The filamentation of a uniform distribution.

upper panel shows the orbits described by the electrons in their longitudinal phase space. The electron trajectories are either closed or open, depending on their initial coordinates in the phase space. If a uniform distribution is injected into the laser, it will evolve into the complicated distribution shown in the central panel. Since the phase space transformation produced by the storage ring operates on a wavelength much larger than the wavelength of the laser light, the distribution cannot be untangled. Subsequent passes through the laser then accumulate energy spread. However, if the electron distribution is initially bunched, as shown in the lower panel, the microbunches can be injected into the closed orbit regions where the phase space evolution is simple. An acceleration provided by the rf cavity suffices to return the distribution close to its initial configuration so that the process can be repeated indefinitely without the introduction of additional damping.

This type of system has a number of unique properties. The circulating electrons pick up energy from the rf cavity on each pass, only to give it up again on the average to the laser field. The microbunched electrons act as a frequency transformer with the phase relationship between the rf cavity and the laser radiation maintained by the isochronism of the electron transport system. The emission into the laser is strongly enhanced by the bunching of the electrons. The mean energy exchange now occurs through the first order interaction, so that the extracted energy is proportional to the laser electric field and gain is no longer an appropriate concept. In addition, both the starting and saturation mechanisms are novel: the laser starts oscillation through a progressive creation of laser potential wells and redistribution of the electron charge, and saturates by equilibrating the linear energy transfer to the field with the quadratic energy loss through the mirrors.

It should be recognized at the outset that difficulties will be encountered in making such a system work. The ratio of the frequencies involved is on the order of $kc/\omega_{rf} \approx 10^7$ which implies that the tolerances on the isochronism requirement will be tight. For this reason, it was long considered impossible to take advantage of the electron bunching in a storage ring. I therefore attempt to select parameters in the example I use to demonstrate that an isochronous storage ring laser can be built with known technology.[3]

Although the ideas discussed here are specifically directed at the storage ring system, the same analysis also applies to a linear system in which a number of free electron lasers are strung together with accelerating cavities to replenish the extracted energy (Fig. 2). Indeed, the linear system is easier to realize in some respects since the momentum compaction is zero in the absence of bending fields. However, the quality of electron beams available from linear electron accelerators is orders of magnitude poorer than that available in storage rings. For this reason, a linear device would require a much higher laser intensity. In addition, synchrotron damping is no longer present, so that a special interaction region would be desirable at the start to establish the trapping of the uniform input beam. While my discussion will focus on the storage ring system, the reader may wish to keep in mind that the results also apply to the linear configuration.

I begin by deriving the conditions under which the optical traps exist, and characterizing their dependence on the various deviations from isochronism in the arc of the storage ring. Following a discussion of the unique operational characteristics of the laser, the system performance is illustrated by a numerical example.

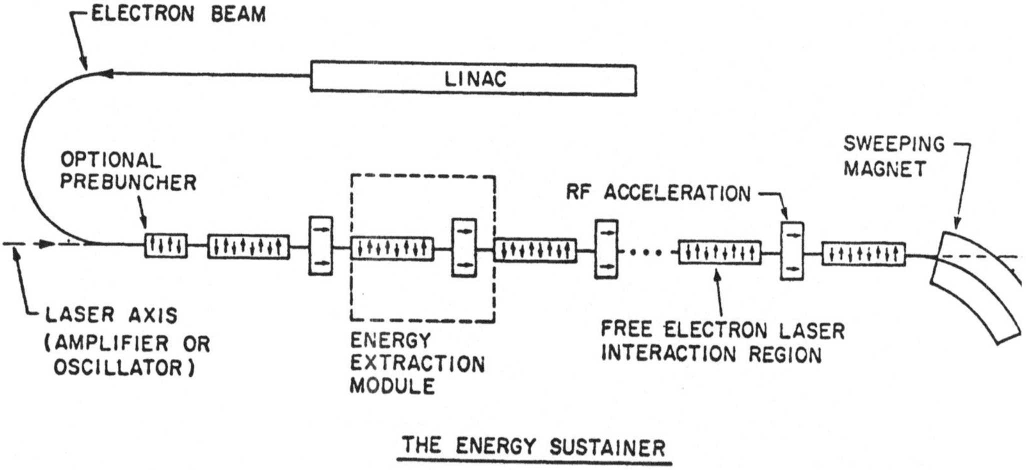

Fig. 2.

EXISTENCE OF THE OPTICAL TRAPS

The equation of motion for the electrons in the laser section of the storage ring laser is Colson's[5] pendulum equation:

$$\ddot{\theta} = -\Omega^2 \sin\theta$$

where θ is the optical phase seen by the electron $\theta \equiv k\delta z$, k is the wavenumber of the light, and δz is the longitudinal displacement of the electron from the resonant particle which experiences no energy change. The frequency Ω^2 is linear in the electric field of the light ε and the magnetic field B, $\Omega^2 = e^2\gamma B/\gamma^2 m^2 c^2$, where γ is the electron energy in units of mc^2. In a circuit of the arc of the storage ring outside the laser, the electron experiences an acceleration depending on the rf voltage V and its time displacement τ from the central particle τ_o :[6]

$$\Delta E = eV[\sin \omega_{rf}(\tau + \tau_o) - \sin \omega_{rf}\tau_o] \tag{2}$$

The second term is negligible at lower energies, and will be neglected in what follows. A longitudinal displacement is also produced:[4]

$$c\Delta\tau = -\frac{L}{4}\left(\frac{a_x^2}{\beta_x} + \frac{a_y^2}{\beta_y}\right) - \alpha L \frac{\delta\gamma}{\gamma} - \alpha_2 L \left(\frac{\delta\gamma}{\gamma}\right)^2 \tag{3}$$

where a_x, a_y, β_x, and β_y are the horizontal and vertical amplitudes and beta functions[6] of the storage ring optics, L is the path length of the trajectory from the exit of one laser to the entrance of the next, and α and α_2 are the first and second order momentum compaction factors

$$\alpha \equiv \frac{1}{L}\int_0^L \frac{\vec{\eta}\cdot\vec{\rho}}{\vec{\rho}^{\,2}}\,ds - \frac{1}{\gamma^2} \tag{4}$$

$$\alpha_2 \equiv \frac{1}{L}\int_0^L \frac{\vec{\eta}'^{\,2}}{2}\,ds + \frac{3}{2\gamma^2} \tag{5}$$

More terms have been retained in this expression than are familiar from Ref. 6 because of the sensitivity of the laser to small phase shifts.

I will discuss the laser phenomena in terms of the normalized variables θ and Y

$$\theta \equiv k\delta z$$
$$Y \equiv \frac{4\pi N}{\varphi}\frac{\delta\gamma}{\gamma} \tag{6}$$

where φ is the phase advance per pass in the laser $\varphi \equiv \Omega\ell/c$, and N is the number of magnet periods. The existence of the traps can be shown through a linearized analysis and a matrix propagator approach. Linearizing about $\theta = 0$ for an electron with time displacement τ_i, the equations (1), (2), and (3) can be combined as

$$\begin{pmatrix}\theta\\ Y\end{pmatrix}_{n+1} = \begin{pmatrix}1 & -A\varphi\\ 0 & 1\end{pmatrix}\left[\begin{pmatrix}\cos\varphi & \sin\varphi\\ -\sin\varphi & \cos\varphi\end{pmatrix}\begin{pmatrix}\theta\\ Y\end{pmatrix}_n + \begin{pmatrix}-D\\ U/\varphi\end{pmatrix}\right] \tag{7}$$

where the coefficients A, D, and U have been defined as

$$A \equiv \frac{kL}{4\pi N} \alpha \tag{8}$$

$$D \equiv \frac{L}{4}\left(\frac{a_x^2}{\beta_x} + \frac{a_y^2}{\beta_y}\right) - \xi \tag{9}$$

$$U \equiv 4\pi N \frac{eV \sin \omega_{rf}\tau_i}{E} \tag{10}$$

The electron coordinates oscillate stably about fixed points if the transformation matrix has eigenvectors with unity magnitude.[4] Optical traps therefore exist in the region defined by the boundaries

$$A\varphi = 2 \tan \frac{\varphi}{2} \tag{11}$$

$$A\varphi = - 2 \cot \frac{\varphi}{2} \tag{12}$$

and illustrated in Fig. 3. As the laser intensity increases, the operating point traces out a line from the origin with slope A . For a given value of momentum compaction, the laser intensity is limited to a maximum value. The optimum operating region lies in the band $|A| \leqslant 1$ where the laser intensity limit is maximized

$$\alpha \leqslant \frac{4\pi N}{kL} \tag{13}$$

The more heavily shaded region in the figure is the stability region for traps which exist near the $\theta = \pi$ fixed point. A stable region evidently exists for any α , no matter how large, and no matter in which direction.

A numerical simulation not limited by the linear approximation has produced the trajectory plot shown as Fig. 4. The coordinates at the laser of several electrons propagated around the system 200 times with $\alpha = 0$ form a set of closed trajectories. The solid curve shows for reference the separatrix of the motion in the laser alone. Figure 5 demonstrates the effect of nonzero α on the trajectories, with $A = -5$ on the left, and $A = 0.7$ on the right. Negative values of A reduce the size of the traps and while small positive values increase their area, the storage ring taken as a whole approaches the unstable transition energy at $A = 1$.

The size of the traps can be calculated in the smooth approximation, which is valid provided that no element of the system induces a phase space displacement large compared to the trap size. In this approximation, the ordering of the elements can be neglected, and the equations of motion integrated to yield the constant of the motion:

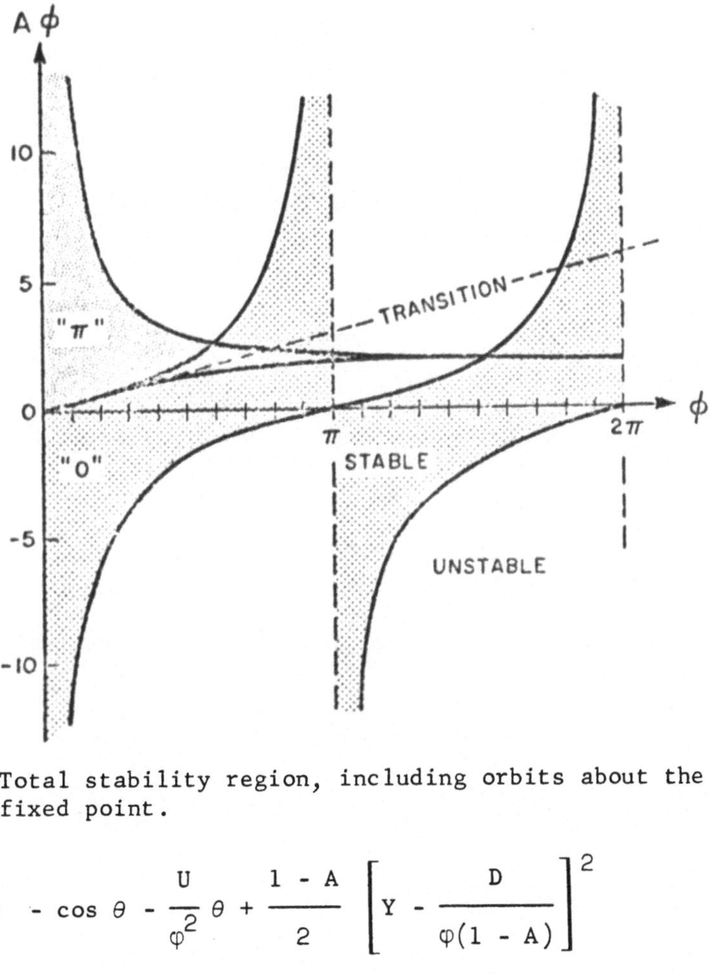

Fig. 3. Total stability region, including orbits about the "π" fixed point.

$$H = -\cos\theta - \frac{U}{\varphi^2}\theta + \frac{1 - A}{2}\left[Y - \frac{D}{\varphi(1 - A)}\right]^2 \tag{14}$$

The first two terms produce the potential (Fig. 6) in which the electrons oscillate. The fixed points occur at the extrema of the potential

$$\sin\theta = \frac{U}{\psi^2} \tag{15}$$

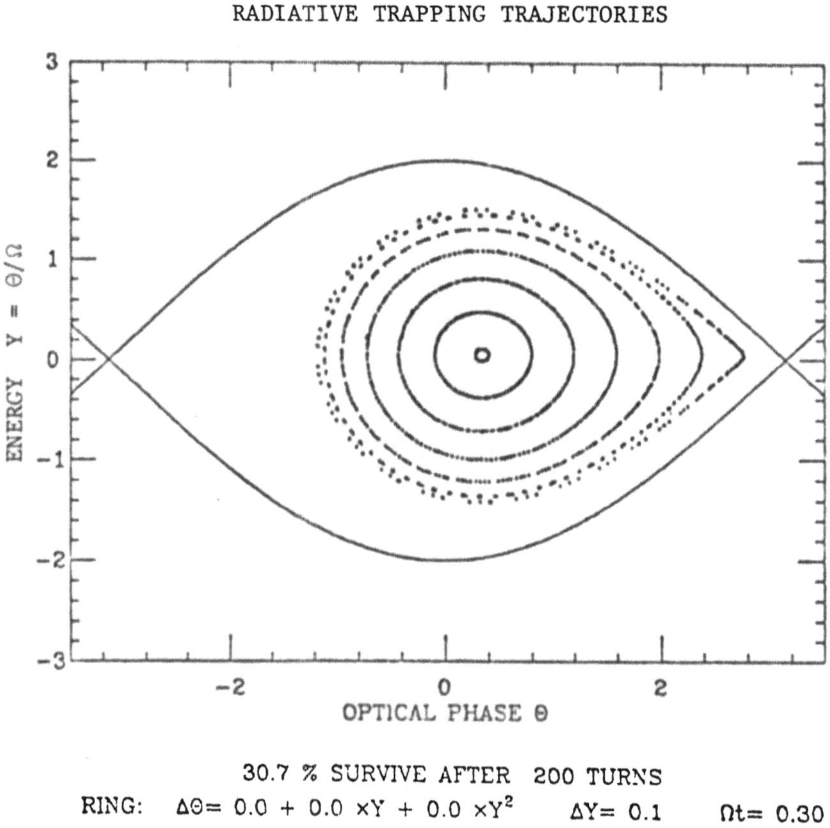

Fig. 4. Electron trajectories in optical trap for isochronous ring.

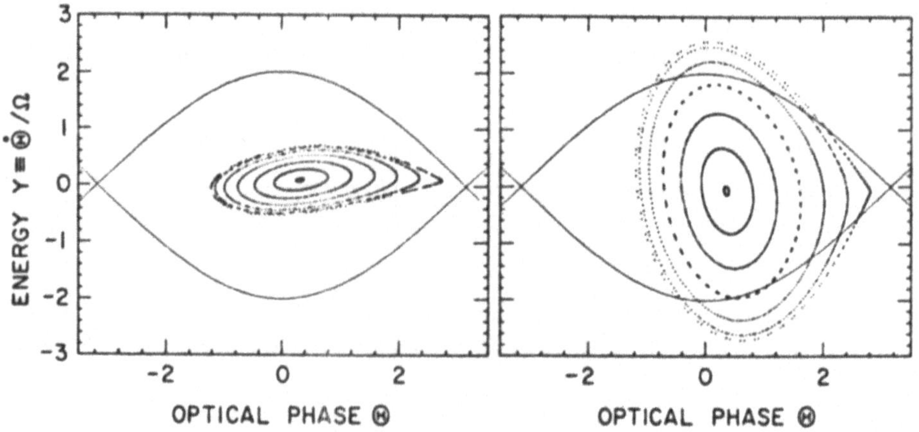

Fig. 5. The effect of momentum compaction A on the electron trajectories in the trap.

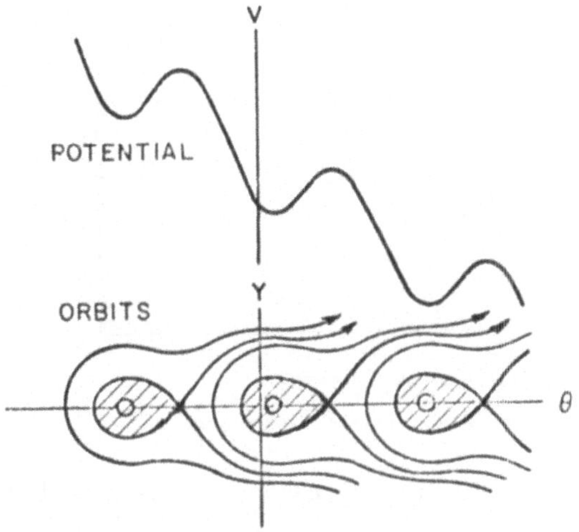

Fig. 6. Sketch of potential, with orbits, for small phase advance.

the electrons oscillate about one of the fixed points, and the electron at the other point traces out the separatrix of the trap. Increasing the voltage U drives the two fixed points together, reducing the trap size until the condition is violated, and the trap disappears entirely. Fig. 7 shows the effect on the trap of Fig. 4 when the voltage is doubled. For a given laser intensity, traps will exist symmetrically distributed along the longitudinal axis about $\tau = 0$ out to a maximum time displacement defined by (16). The energy aperture of a given trap will then vary according to its rf phase

$$\Delta Y = Y_{max} - Y_{min} = \frac{4}{\sqrt{1 - A}} \left(\sqrt{1 - \frac{U^2}{\varphi^4}} - \frac{U}{\varphi^2} \cos^{-1} \frac{U}{\varphi^2} \right)^{\frac{1}{2}} \quad (17)$$

A population of particles with a given transverse amplitude will oscillate about a phase stable point displaced in energy by (14)

$$Y_{center} = \frac{D}{\varphi(1 - A)} \quad (18)$$

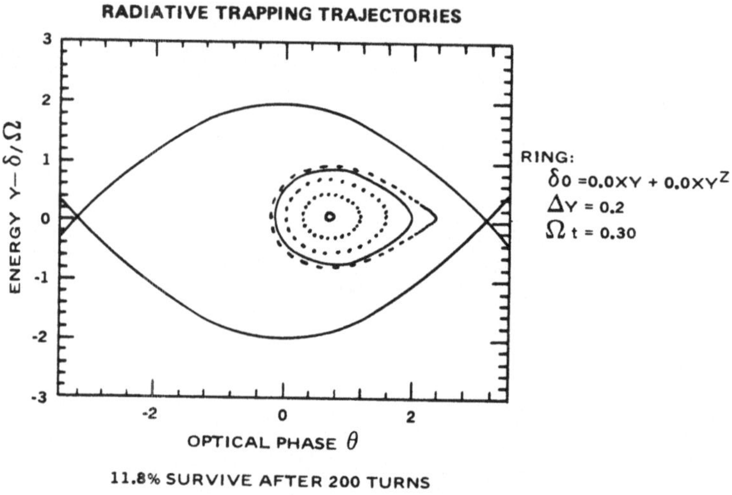

Fig. 7. Trajectories with larger voltage.

It is easy to show by integrating (1) for small φ that a particle with energy displacement Y interacts less strongly with the laser

$$\Delta Y = -\varphi \sin\left(\theta + \frac{Y\varphi}{2}\right) \frac{\sin Y\varphi/2}{Y\varphi/2} \tag{19}$$

The energy exchange is reduced by the $\sin x/x$ factor as shown in Fig. 8. The transverse motion is limited by the requirement that

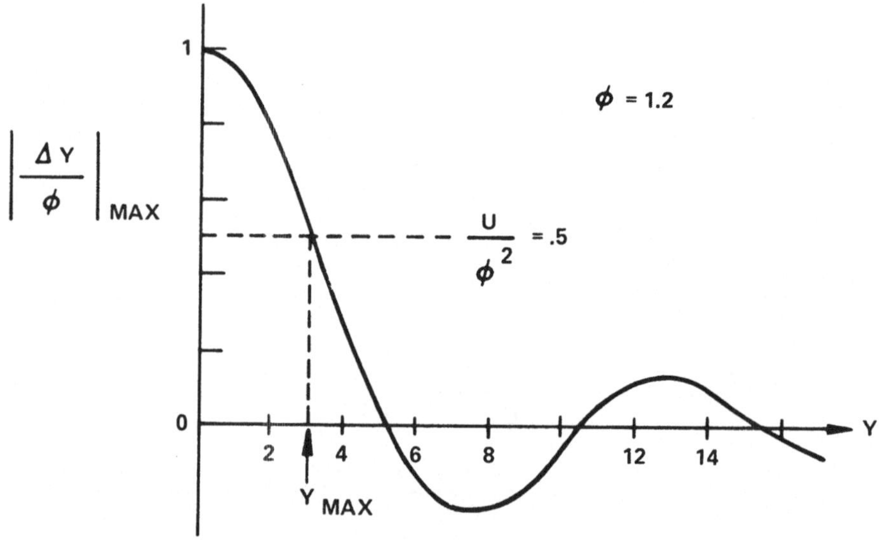

Fig. 8. Laser energy loss reduction vs. the injection energy coordinate.

energy balance exist at the trap center

$$\frac{\sin Y\varphi/2}{Y\varphi/2} > \frac{U}{\varphi^2} \tag{20}$$

This limitation is shown by the dotted lines in Fig. 8 for $U/\varphi^2 = \frac{1}{2}$. Populations with different transverse amplitudes will occupy traps displaced from resonance as shown in Fig. 9. As the displacement approaches the limit (20), the trap size diminishes quadratically. For $U/\varphi^2 = \frac{1}{2}$, the maximum aperture for transverse motion is approximately

$$D < 7.6(1 - A) \tag{21}$$

OPERATIONAL CHARACTERISTICS

The parameters in an isochronous storage ring laser should be chosen so that during operation, the entire equilibrium distribution is contained deep within the optical traps. The energy spread and emittance of the beam must fit inside the longitudinal and transverse apertures (17) and (21) for about six standard deviations to maintain a long trapping lifetime

$$\frac{\sigma_e}{E} < \frac{1}{6} \frac{\varphi}{2\pi N\sqrt{1 - A}} \tag{22}$$

$$\frac{\varepsilon}{\beta} < \frac{1}{6}\, 7.6(1 - A)\, \frac{4\pi}{kL} \tag{23}$$

The emitted power under these circumstances, will be

$$P_{extracted} = \int d\theta \rho(\theta)\, \frac{eV}{T} \sin \omega_{rf}\tau(\theta) \tag{24}$$

where $\rho(\theta)$ is the electron distribution inside the traps. Charge circulating outside the traps make no contribution to the power if the synchronous energy of the storage ring is made the same as the laser resonant energy.

The completely trapped condition can be achieved either by initiating the optical field externally, or by letting it grow up out of the spontaneous radiation.

In the first case, large traps are immediately established, and a fraction of the charge is stored. The charge which falls outside the traps now orbits around them on perturbed trajectories in the rf bucket.

RADIATIVE TRAPPING TRAJECTORIES

RING:	$\Delta\theta = -2.0 + 0.0 \times Y + 0.0 \times Y^2$	$\Delta Y = 0.25$	$\Omega t = 0.30$
RING:	$\Delta\theta = -1.5 + 0.0 \times Y + 0.0 \times Y^2$	$\Delta Y = 0.25$	$\Omega t = 0.30$
RING:	$\Delta\theta = -1.0 + 0.0 \times Y + 0.0 \times Y^2$	$\Delta Y = 0.25$	$\Omega t = 0.30$
RING:	$\Delta\theta = -0.5 + 0.0 \times Y + 0.0 \times Y^2$	$\Delta Y = 0.25$	$\Omega t = 0.30$
RING:	$\Delta\theta = 0.0 + 0.0 \times Y + 0.0 \times Y^3$	$\Delta Y = 0.25$	$\Omega t = 0.30$

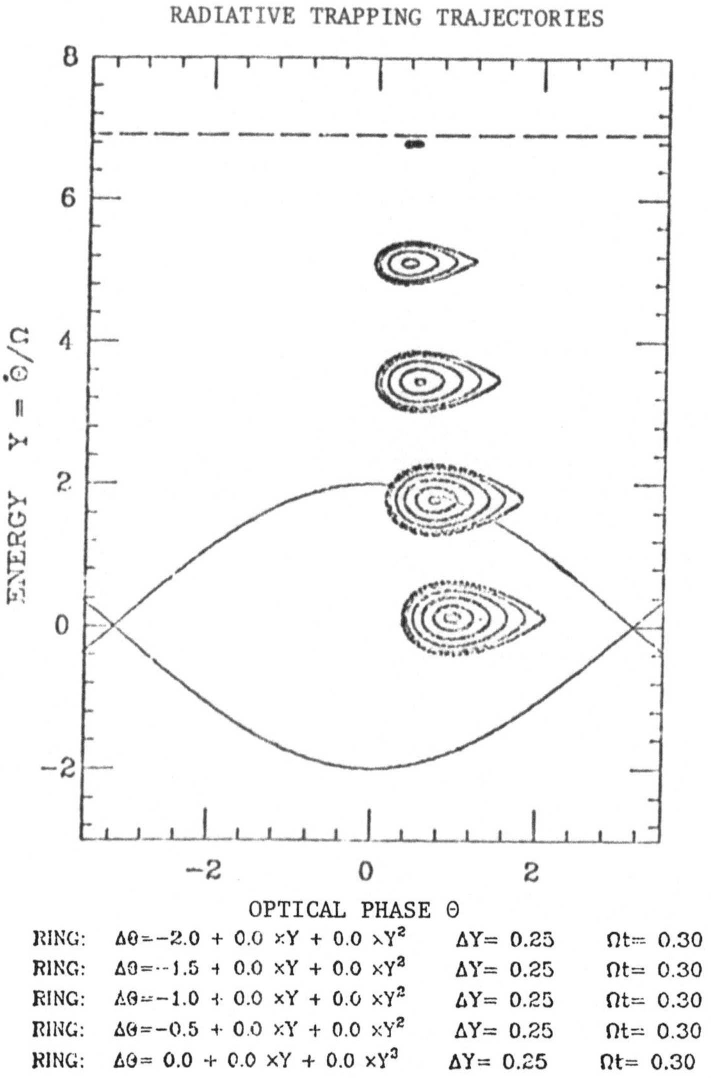

Fig. 9.   Particle trajectories as a function of phase shift D .

Since the presence of the optical traps displaces these orbits to much higher energies, these particles rapidly damp down again to their equilibrium energy distribution, finding their way into the traps within a damping time. At this point the power (24) is still zero because as many traps absorb as emit. However, if the orbit length of the electron is made a rational fraction of the laser round trip distance, the same electrons see the same photons, and the absorbing electrons eliminate their own traps, releasing themselves back into the rf bucket. After an absorption time and a few damping times, all of the charge will have redistributed itself among the emitting traps, and the laser will emit power according to the mean rf phase. The laser output power can then be controlled by shifting the phase of the accelerating cavity.

If the system is to establish itself from the spontaneous radiation, the threshold condition must be satisfied

$$\int d\theta \rho(\theta) \frac{eV}{AT} \sin \omega_{rf}\tau(\theta) \quad > \quad \gamma I$$

where $\rho(\theta)$ is the charge distribution in the traps established by the spontaneous radiation which builds up in the cavity, A is the electron beam area, I is the laser intensity, and γ is the laser cavity loss per round trip. For constant charge density, the extracted power P(I) increases as the energy aperture of the traps (φ), the area in the transverse phase space planes (φ for each), the number of traps (φ^2), and the mean phase of the traps (φ^2), so that the power varies as φ^7 or $P(I) \propto I^{7/4}$. The power grows at an ever increasing rate until the intensity becomes large enough to heat the distribution above its natural energy spread and the charge density begins to drop. If enough charge was present in the bunch at turn-on, the traps will grow to completely enclose the equilibrium distribution: the original electrons will have been either trapped or driven into the walls of the storage ring by the laser heating. The system saturates at an intensity at which the energy extraction which is linear in $\mathcal{E}$ equals the quadratic energy loss.

A NUMERICAL EXAMPLE

For a laser which operates à 1 μ in a magnet of 175 periods and a storage ring of 100 MeV, making A = - 1 implies a value for the momentum compaction of the system of 0.7 x 10^{-5}. If the laser losses are adjusted to produce saturation at 1 MW/cm^2, the laser phase advance per pass becomes φ= 1.2. With a peak cavity voltage above 64 kV, an acceleration of $U/\varphi^2 = \frac{1}{2}$ is achieved less than half way to the maximum phase. The zero current energy aperture is 13.6 σ_e and the transverse aperture for an emittance of $\mathcal{E} = 10^{-8}$ m rad is 7σ_a, so that the trap aperture is large enough to contain the equilibrium distribution for several hours.

If the trapped charge is concentrated in the region $U/\varphi^2 = \frac{1}{2}$, the output power will be

$$P_{extracted} = 32 \text{ kW/amp}$$

while the incoherent synchrotron radiated power is only 100 W/amp for a damping constant of $N_D = 10^6$. The output power is clearly very interesting for stored currents on the order of a fraction of an ampere, and the efficiency will be limited primarily by the dissipation in the rf cavity, the magnet coils, etc. By retaining the electron bunching through the use of isochronous electron optics, the efficiency limitation of Ref. 2 has been eliminated.

REFERENCES

1. J. M. J. Madey, D. A. G. Deacon, and T. I. Smith, J. Appl. Phys. 50:12 (1979).
2. A. Renieri, "The free electron laser: the storage ring operation," CNEN-Frascati Report 77-33, Edizioni Scientifiche, C.P. 65, 00044 Frascati, Rome, Italy (1977).
3. A discussion of the tolerances which must be held to realize the isochronous storage ring laser is contained in Ref. 4, Chapter 5.
4. D. A. G. Deacon, Ph.D. Dissertation, Stanford University (1979). Available from University Microfilms International, 300 N. Zeeb Rd., Ann Arbor, MI 48106.
5. W. B. Colson, Phys. Lett. 64A:190 (1977).
6. M. Sands, "The physics of electron storage rings: an introduction," SLAC Report No. 121, Stanford Linear Accelerator Center, Stanford, CA 94305 (1965).

FEL ACTIVITY IN FRASCATI INFN NATIONAL LABORATORIES

S. Tazzari

I.N.F.N. - Laboratori Nazionali di Frascati
C.P. 13 - 00044 Frascati, Italy

1. INTRODUCTION

The Frascati National Laboratory (LNF) is operating a 1.5 GeV electron storage ring, Adone. The ring, designed and operated until 1978 mainly for high energy physics colliding beam experiments, is now dedicated to nuclear physics[1] and synchrotron radiation[2] research. A wiggler, designed to provide a hard X-ray beam,[3] is installed and operational. Although it has only three full periods, it can be operated in the undulator mode, to provide spontaneous radiation at visible wavelengths.

I am reporting on the work in progress on free electron lasers, which has developed along two lines: the study of the spontaneous radiation emitted by the existing undulator (wiggler) magnet, and the study of a FEL recirculated beam experiment to be performed on the storage ring.

The first measurements on the wiggler "coherent" beam have been performed[4] in the framework of a collaboration between LNF, the University of Naples, and the University of Trento.* They are discussed in Section 2.

---

*LNF: R. Barbini, M. Bassetti, M. E. Biagini, R. Boni, M.T. Capria, A. Cattoni, V. Chimenti, S. Guiducci, A. Luccio, M. Preger, C. Sanelli, M. Serio, S. Tazzari, F. Tazzioli, G. Vignola. Naples: E. Burattini, N. Cavallo, M. Foresti, C. Mencuccini, E. Pancini, P. Patteri, R. Rinzivillo, U. Troya. Trento: G. Dalba, F. Ferrari, P. Fornasini.

The FEL experiment, originally proposed[5] by R. Barbini* and G. Vignola*, has been approved by INFN early in 1980 and funded. It will be carried out in collaboration by LNF and the Universities of Naples and Bari. The outlines of the proposal are discussed in Section 3.

2. STUDY OF THE SPONTANEOUS RADIATION FROM THE WIGGLER MAGNET

2.1 Undulator Theory

We summarize the main features of undulator radiation:

a) For each observation angle θ, the radiation is composed of harmonics centered at

$$\lambda_h = \frac{\lambda_q}{2h\gamma^2}\left(1+K^2+(\gamma\theta)^2\right) \quad (h=1,2,3,\ldots) \tag{1}$$

where $\gamma = E/m_oc^2$ is the electron energy in unit of rest mass, K the undulator parameter and λ_q its wavelength.

b) The width of spectral peaks in the forward direction ($\theta = 0$) is

$$\frac{\Delta\lambda}{\lambda_h} \sim \frac{1}{hN} \tag{2}$$

where N is the number of undulator periods. For $\theta \neq 0$, there also appears an inhomogeneous broadening given by

$$\frac{\Delta\lambda}{\lambda_h} \sim \frac{\gamma^2\,\theta^2}{1+K^2} \tag{3}$$

c) The above line structure extends beyond $\theta_o = 1/\gamma$.

2.2 Spectrum Measurements in the Forward Direction: Preliminary Results

The main parameters of the Adone wiggler magnet[6,7] are listed in Table I.

By operating the magnet at very low current (51 A, corresponding to $B \simeq 200$ Gauss) a K of ~ 1, characteristic of the undulator regime, can be obtained.

*On leave from CNEN, Centro di Frascati.

With storage ring energies in the range from 500 to 800 MeV spontaneous coherent emission over the visible wavelength range is observed.

The experimental set-up to study the spectral characteristics of radiation emitted in the forward direction is schematically shown in Fig. 1.

The background radiation originating in the bending magnets next to the straight section where the wiggler is installed can be measured when the wiggler magnet is not energized, and subtracted out. Since the λ dependence of the background is known, the efficiency of the detector photomultiplier can also be taken care of.

Experimental results at $\gamma = 1361.5$ and $\gamma = 1166.8$ are shown in Fig. 2.

Table I. Wiggler Magnet

No. of poles	5 full + 2 half-poles
Gap height	40 mm
Total length	2100 mm
Pole-to-pole distance ($\lambda_q/2$)	327 mm
Maximum field	1.85 T
Max amperturns per pole	31,500
Max current	4500 A
Max power consumption	189 KW

The peak wavelength and intensity ratios are in agreement with the computation if K = 1.1, entirely consistent with the expected error on the estimate of K, is assumed.

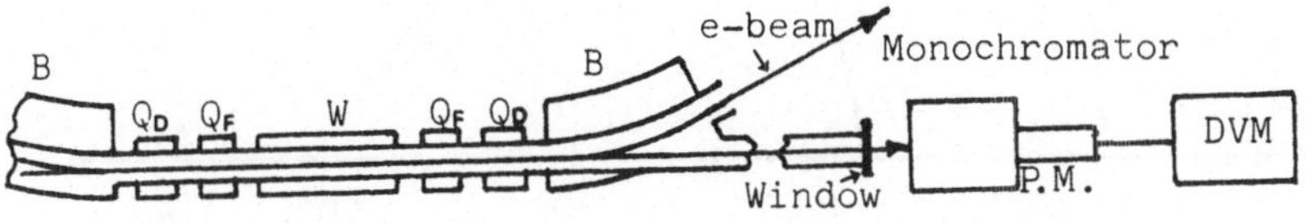

B, Q_D, Q_F are the bending magnets and quadrupoles of the ring; W is the wiggler magnet.

Fig. 1. Schematic layout of experiment.

The shape and width of the measured distributions (∿ 40% f.w.h.h.) are not in agreement with the computation (∿ 30% f.w.h.h.). Possible causes for the discrepancy are being investigated, but it should be recalled that rather large systematic errors may be present in the subtraction procedure.

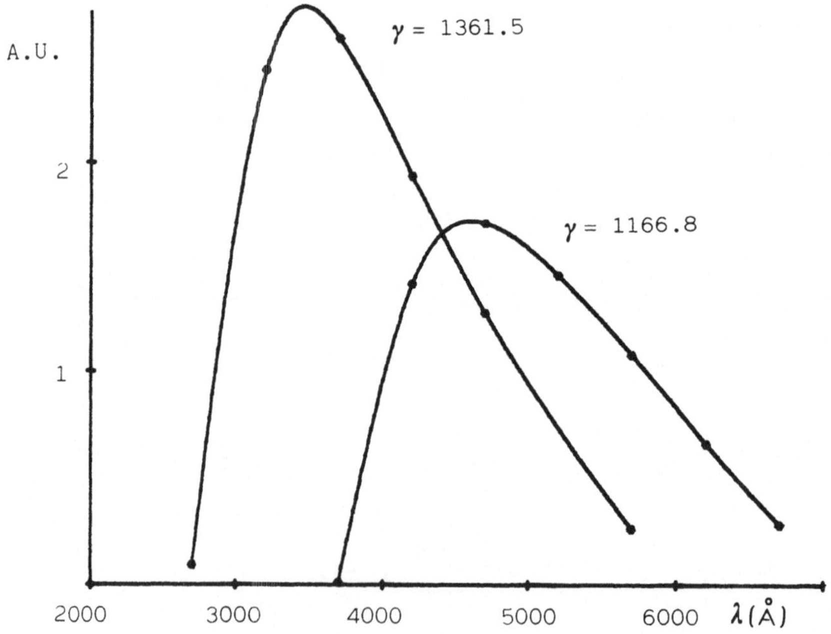

Fig. 2. Measured spectral distribution (first harmonic, $\theta = 0$) versus wavelength after background subtraction. The angular and wavelength resolutions are $\Delta\theta \simeq .1$ mr and $\Delta\lambda \simeq 380$ Å respectively.

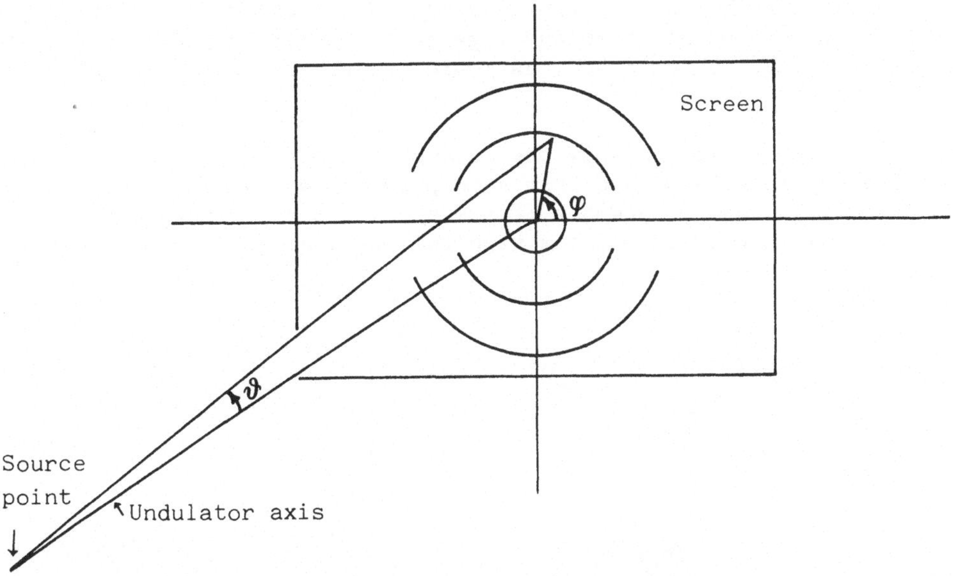

Fig. 3. Sketch of radiation pattern as observed on a screen.

2.3 Angular Distribution Measurements: Preliminary Results

The characteristic colored ring appearance of the undulator radiation as observed on a screen normal to the wiggler axis is schematically shown in Fig. 3. The dependence of peak wavelength λ on θ and on the harmonic number is shown in Fig. 4.

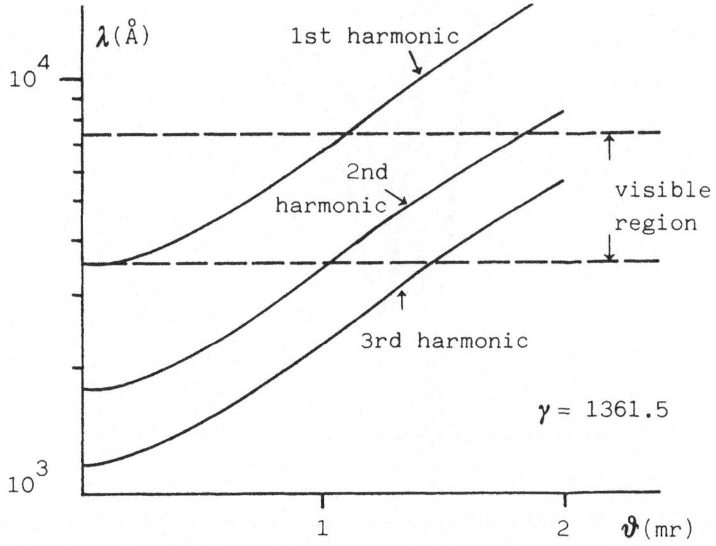

Fig. 4. Peak wavelength vs. θ.

In a number of patterns observed on a screen 14 meters away from the center point of the wiggler, at several machine energies in the range .5 to .8 GeV, harmonics up to the third were clearly visible.

Figures 5 and 6 show the intensity distribution of the central wavelength, as obtained from a computer code[8], as a function of θ and of the harmonic number, for $\psi = 0$ and $\psi = \pi/2$ respectively. The visible portions of the spectrum are indicated by a heavy line.

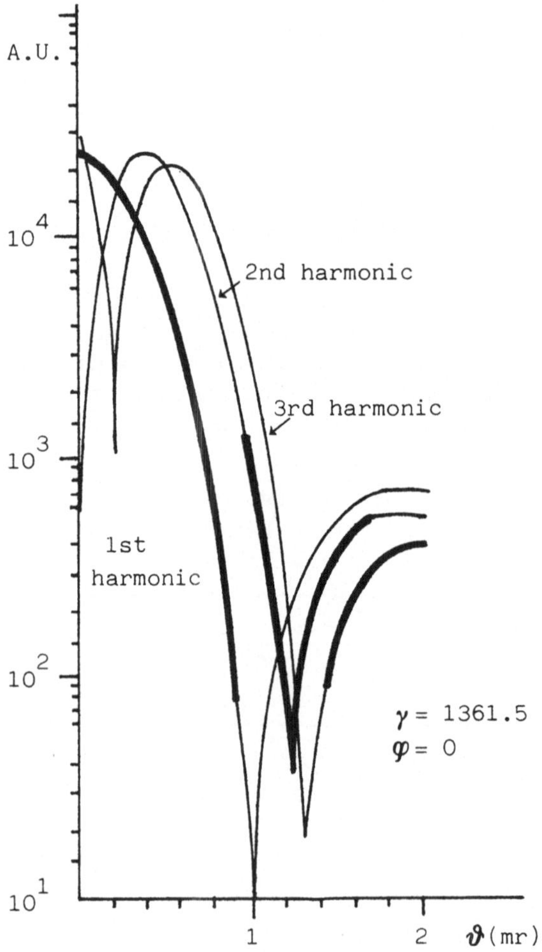

Fig. 5. Intensity distribtuion of peak wavelength vs. θ for $\psi = 0$.

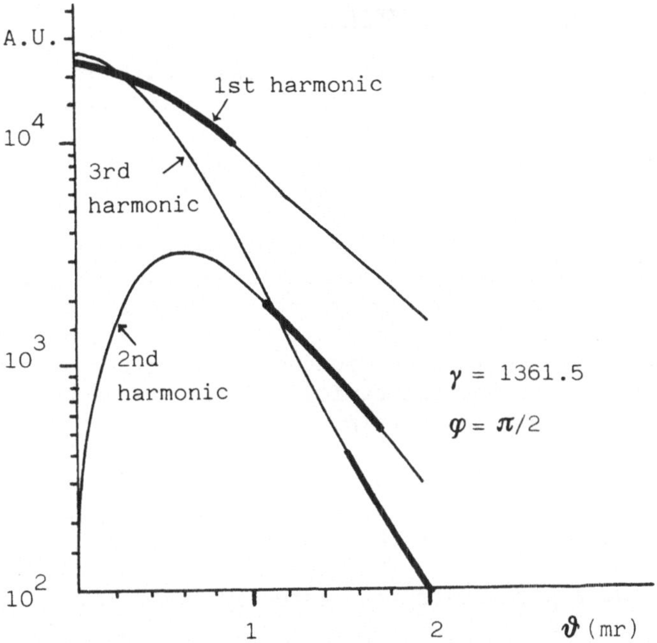

Fig. 6. Intensity distribution of peak wavelength vs. θ for $\psi = \pi/2$.

The computed distributions seem to explain very well the observed patterns; in particular the expected intensity drops at around $\psi = 0$ clearly appear as dark spots at the correct values of θ.

More precise measurements are needed, including a densitometric analysis to check that intensities are in the predicted ratios.

3. THE DESIGN OF A FREE ELECTRON LASER EXPERIMENT ON ADONE (LELA)

3.1 Goals

The main goal of the LELA experiment[5] is to collect information on the following topics:

- amplification of radiation with the aid of an external "seed" laser (Argon laser λ = 5145 Å);
- wavelength and optical gain as functions of electron energy and undulator magnetic field;
- transient behavior of the laser radiation;
- steady-state interaction between laser radiation and stored electrons (i.e., optical gain vs. electron energy, maximum optical output power, optical spectrum).

3.2 FEL and Storage Ring Parameters

It is proposed to install a transverse undulator on one of the Adone straight sections. An external laser beam can be sent along the straight section axis to interact with the electron beam and the undulator field (see Fig. 11), or an optical cavity can be built by adding a mirror at each end of the straight section.

In order to minimize the laser beam losses in the optical cavity (oscillator experiment) and to avoid head-on collisions between photons and electrons, it is convenient to operate Adone with three electron bunches and to adjust the optical cavity length to half the distance between two consecutive bunches: a single photon bunch will then travel inside the optical cavity and will meet one of the electron bunches once every round trip.

The spontaneous radiation wavelength observed on the undulator axis is given by (1) (with $\theta = 0$, $h = 1$), and can be tuned by varying either the electron energy or the undulator magnetic field or both.

The undulator parameters are listed in Table II.

For a pure cosine-like vertical magnetic field

$$B_z = B_o \cos \frac{2\pi}{\lambda_q} y \qquad (4)$$

the parameter K is given by

$$K = \frac{eB_o \lambda_q}{\sqrt{2}\; 2\pi m_o c} \simeq 6.6\; B_o\,(\mathrm{KG})\; \lambda_q(\mathrm{m}). \qquad (5)$$

By defining the R.M.S. magnetic field on axis by

$$\overline{B} = \left[\frac{1}{\lambda_q} \int_o^{\lambda_q} |B_z(y)|^2 \, dy \right]^{1/2} \qquad (6)$$

one can also write

$$K = 9.33\; \overline{B}\,(\mathrm{KG})\; \lambda_q(\mathrm{m}). \qquad (7)$$

In order to avoid all unnecessary technical complications, it is best to operate in the visible wavelength region with magnetic fields that can be achieved by standard magnet technology. The electron energy should be the highest possible in order to achieve the highest possible peak current (see § 3.3). The undulator period should be designed so as to accomodate the maximum number of periods in the fixed length of the Adone straight section (2.5 m). On the other hand, the achievable magnetic field on axis will depend both on the undulator period and on the gap height.[9,10] Finally, electron energy, undulator period and magnetic field are connected by the wavelength equation (1).

Taking into account the above constraints and using the results of magnetic field calculations[11] one ends up with the basic FEL parameters listed in Table II.

The FEL gain has been demonstrated[12] to be inversely proportional to the square of the total spontaneous radiation linewidth, which, in turn, is made up of two contributions, homogeneous and inhomogeneous broadening, adding quadratically. The inhomogeneous broadening is usually required to be negligible with respect to the homogeneous one.

Table II. FEL Parameters

Undulator period	$\lambda_q = 11.6$ cm
Number of periods	$N = 20$
Undulator length	$L_W = 2.32$ m
Homogeneous broadening	$[(\Delta\lambda/\lambda)]_o = \lambda_q/2L_W = 2.5\%$
RMS magnetic field on axis	$\bar{B} = 3153$ G; $K = 3.412$
Electron energy	$E = 610$ MeV
Radiation wavelength	$\lambda = 5145$ Å
Optical cavity length	$L = 17.5$ m

This places upper limits on the e-beam angular divergence and energy spread. It also requires the off energy η function to vanish at the place where the undulator is mounted.

By increasing the number of independent quadrupole families from 2 to 4, Adone can be made into a six-period machine with η vanishing in alternative straights, thus satisfying the requirements. The standard cell and the resulting functions are shown in Fig. 7. The main machine parameters for this new structure are listed in Table III. In the table, it has been assumed that a new 51.4 MHz RF cavity, at present under test, will be installed and operating (see also § 3.3).

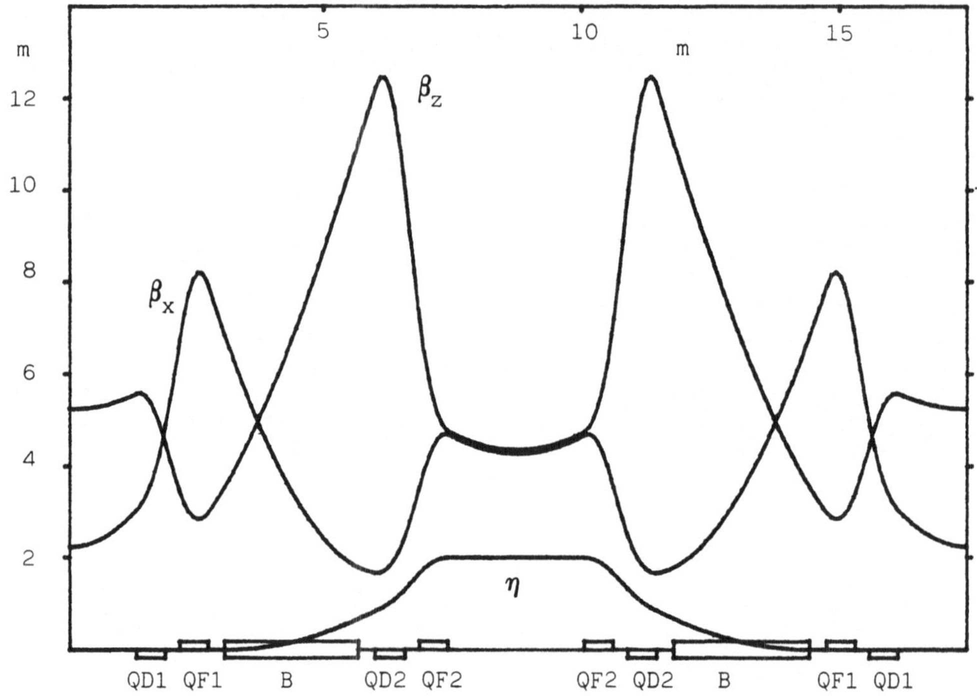

Fig. 7. Optical functions for the standard cell.

3.3 Small Signal Gain

The FEL small signal gain per pass in the homogeneous broadening regime and for a monochromatic e-beam is:[13]

$$g_o = -32\sqrt{2}\,\pi^2\lambda^{3/2}\,\lambda_q^{1/2}\,\frac{K^2}{(1+K^2)^{3/2}}\,\frac{I_p}{I_A}\,\frac{N^3}{\Sigma_L}\,f(x) \qquad (8)$$

where

$I_A = \frac{ec}{r_o} = 17.000$ A, $\qquad \gamma_o$ = operating energy (in unit of m_oc^2),

I_p = peak current/bunch, $\qquad \gamma_R$ = resonance energy = $\left[\frac{\lambda_q}{2\lambda}(1+K^2)\right]^{1/2}$,

$x = 4\pi N\,\frac{\gamma_o - \gamma_R}{\gamma_R}$, $\qquad f(x) = -\frac{1}{x^3}\left\{\cos x - 1 + \frac{1}{2}x\sin x\right\}$.

The function f(x) is plotted in Fig. 8 and is proportional to the derivative of the spontaneous emission lineshape

$$\left(\frac{\sin x/2}{x/2}\right)^2.$$

By assuming it has its maximum value (= -0.0675 for x = 2.6056) and taking for the E.M. beam cross section the value $\overline{\Sigma}_L = L_w\lambda/\sqrt{3}$ (see § 3.5), with the parameters listed in Table II the gain g_o can be written

$$g_o = 6.7 \times 10^{-4}\ I_p(A) \simeq 3 \times 10^{-3}\ \frac{i(mA)}{\sigma_y(cm)} \qquad (9)$$

$I_p(A)$ is the peak current/bunch, i(mA)is the mean current/bunch, and σ_y is the R.M.S. bunch length.

Table III. Machine Parameters

Electron energy	E = 610 MeV
Momentum compaction	$\alpha_c = 1.36 \times 10^{-2}$
Fractional energy spread	$\sigma_p = 2.3 \times 10^{-4}$
Invariant	$\langle H \rangle = 0.38$ m
Radial emittance (off coupling)	$A_x = 0.25$ mm x mrad
Energy loss in bending magnets	$U_o = 2.45$ KeV/turn
Energy loss in undulator	$U_w = 109$ eV/pass
Radial betatron tune	$\nu_x = 5.15$
Vertical betatron tune	$\nu_z = 3.15$
Radial natural chromaticity	$C_x = -1.06$
Vertical natural chromaticity	$C_z = -1.61$
Damping partition numbers	$J_s = 2;\ J_x = J_z = 1$
Damping times	$\tau_i = 184/J_i$ msec
Revolution frequency	$f_o = 2.856$ MHz
RF frequency	$f_{RF} = 51.4$ MHz
Harmonic number	h = 18
Number of bunches	$n_b = 3$
1 RF cavity: RF peak voltage	$V_{RF} = 300$ KV
RF acceptance	$\varepsilon_{RF} = 3.58\%$
2 RF cavities: RF peak voltage	$V_{RF} = 600$ KV
RF acceptance	$\varepsilon_{RF} = 5.06\%$

The optical gain per pass must exceed the cavity losses: the computed diffraction losses are negligible, and it is assumed that mirror absorption and transmissivity can reasonably be kept below 4% total. The gain is therefore required to be at least of the order of a percent. This value can be obtained with mean currents of some tens of mA/bunch and σ_y's of the order of a few cm. At 610 MeV the anomalous bunch lengthening phenomenon must however be properly taken into account.

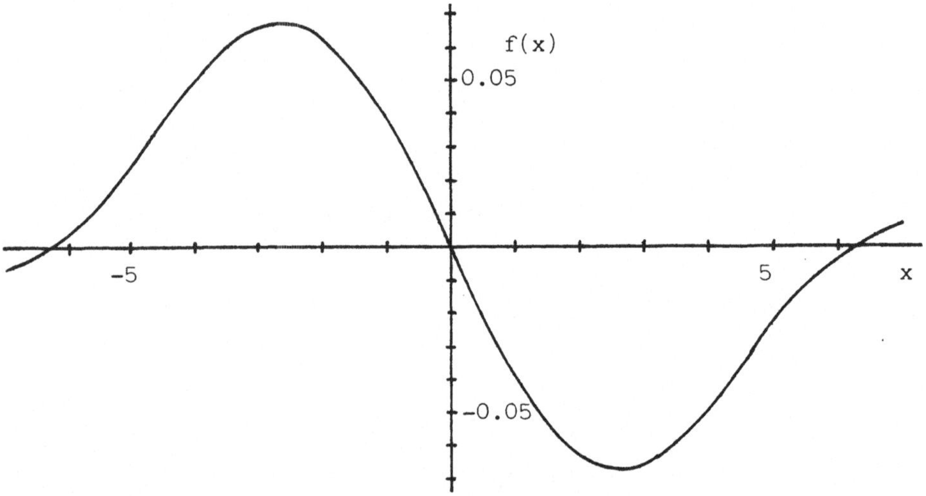

Fig. 8. The gain function.

3.4 Bunch Lengthening

By extrapolating the results on bunch lengthening obtained for Adone to the new structure, using a Chao-Gareyte[14] model, σ_y can be written as

$$\sigma_y \text{ (cm)} \simeq 29 \left[\frac{i\text{(mA)}}{V_{RF}\text{(KF)}}\right]^{.37} \tag{10}$$

By inserting eq. (10) into (9) one obtains:

$$g_o \simeq 10^{-4}\, i_{(mA)}^{.63} \times V_{RF\,(KV)}^{.37} \tag{11}$$

In Fig. 9, the R.M.S. bunch length σ_y is plotted vs. bunch current both with and without anomalous lengthening, and for the two cases V_{RF} = 300 KV and V_{RF} = 600 KV. Figures 10a and 10b show the corresponding gain curves.

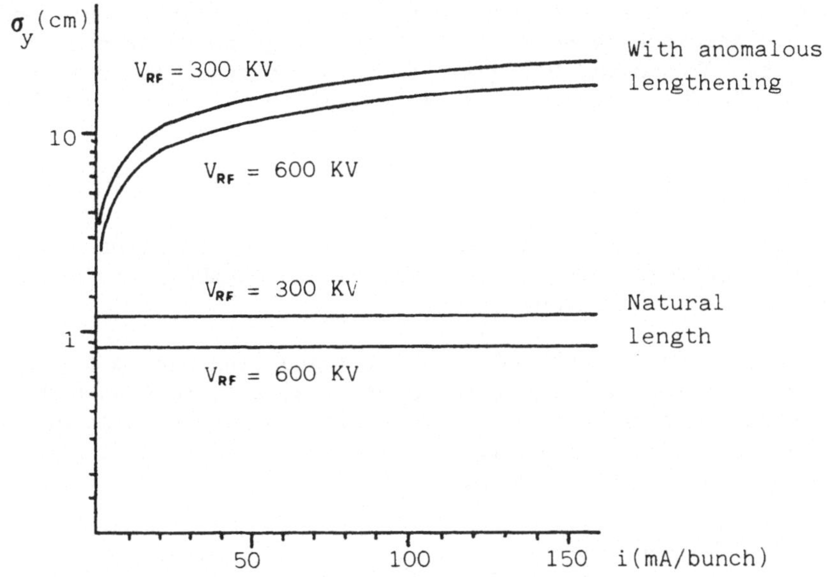

Fig. 9. Natural and anomalous bunch length σ_y of V_{RF} = 300 KV and V_{RF} = 600 KV.

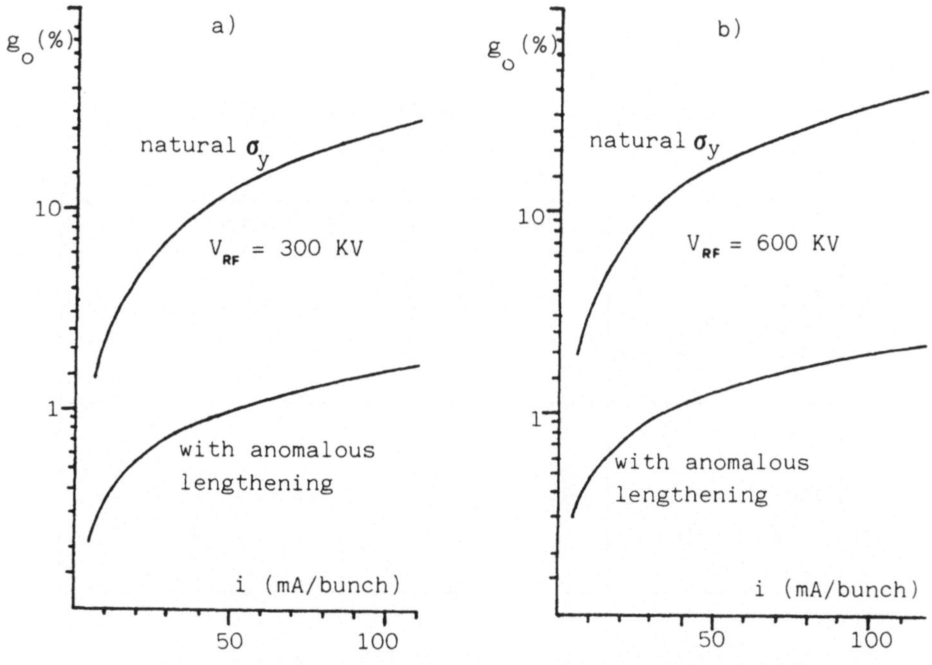

Fig. 10. Optical gain/pass vs. mean current bunch.
a) V_{RF} = 300 KV. b) V_{RF} = 600 KV.

We conclude that the anomalous lengthening significantly affects the gain. Accurate measurements at the energy and currents considered, with the new RF system, will have to be performed. It is at present felt that the Chao-Gareyte type extrapolation gives an upper limit to the anomalous lengthening and, consequently, a lower limit to the gain.

3.4 Electron Beam Lifetimes

In evaluating electron beam lifetimes, the contributions from single and multiple Touschek effects, vacuum chamber aperture and RF acceptance, have to be considered.

Lifetimes are calculated assuming a gaussian e-beam distribution. Actually, in a steady state laser operation, electrons should have a harmonic oscillator energy distribution whose tails are steeper than those of a gaussian. This could lead to lifetimes longer than those computed in the following sections.

3.4.1 Touschek Effect

A computer code developed by F. H. Wang[15] shows that, for i = 50 mA/bunch, the Touschek lifetimes τ_T (including multiple Coulomb scattering) and the beam cross section enlargement ratios ζ are those listed in Table IV (see also ref. 16).

Table IV. Touschek Lifetimes

	V_{RF} = 300 KV		V_{RF} = 600 KV	
	τ_T (hours)	ζ	τ_T (hours)	ζ
Without anomalous lengthening	22.9	2.05	48.2	2.24
With anomalous lengthening	449	1.08	1056	1.10

3.4.2 Vacuum Chamber Aperture

We recall[17] that the total radial spread can be written as

$$\sigma_x^2 = \sigma_{x\beta}^2 + \sigma_{x\varepsilon}, \tag{12}$$

where

$$\sigma_{x\beta}^2 = \sigma_p^2 \langle H\rangle_{mag} \left(\frac{J_s}{J_x}\right) \beta_x \frac{1}{1+\chi^2} \tag{13}$$

is the betatron contribution (χ^2 is the coupling coefficient for betatron oscillations) and $\sigma_{x\varepsilon}^2$ is the energy spread contribution.

Without making any assumption on the transient behavior, it can reasonably be assumed that if a steady state is to be reached, the R.M.S. electron energy spread will reach the equilibrium value[18]

$$\frac{2\Delta\gamma}{\gamma} = \frac{1}{2N} , \tag{14}$$

so that

$$\sigma_{x\varepsilon} \approx \frac{\Delta\gamma}{\gamma} \approx \eta \frac{1}{4N} \tag{15}$$

This will soon become the dominant contribution to the total spread (12) where $\eta \neq 0$.

At equilibrium, the beam lifetime is[16]

$$\tau_{qx} = \tau_x \, e^{(d/\sigma_{x\varepsilon})^2} \left(\frac{\sigma_{x\varepsilon}}{d}\right)^2 \tag{16}$$

where d is the half width of the vacuum chamber. With d = 7 cm, τ_x = 174 msec and η = 2 m, we get:

$$\tau_{qx} \approx 56 \text{ sec.} \tag{17}$$

3.4.3 RF Acceptance

With similar arguments the beam lifetime for energy oscillations, assuming a gaussian energy spread distribution function, can be written:

$$\tau_q = \tau_s \, e^{(\varepsilon_{RF}/\frac{\Delta\gamma}{\gamma})^2} \left(\frac{\Delta\gamma/\gamma}{\varepsilon_{RF}}\right)^2 . \tag{18}$$

With τ_s = 87 msec one has

$$\tau_{q\varepsilon} \simeq 38 \text{ sec} \quad \text{(1 RF cavity)}, \tag{19}$$

$$\tau_{q\varepsilon} \simeq 19 \text{ hours} \quad \text{(2 RF cavities)}. \tag{20}$$

3.5 Optical Parameters

According to Eq. (8), the FEL gain is inversely proportional to the optical mode cross section $\bar{\Sigma}_L$ in the interaction region, provided the electron beam is fully contained with the laser beam ($\Sigma_e < \bar{\Sigma}_L$).

The laser beam cross section $\bar{\Sigma}_L$ is given by

$$\bar{\Sigma}_L = \frac{1}{L_w} \int_{-L_w/2}^{L_w/2} \Sigma(y)dy = \frac{2\pi w_o^2}{L_w} \int_0^{L_w/2} \left[1 + \left(\frac{\lambda y}{\pi w_o^2}\right)^2\right] dy \tag{21}$$

where L_w is the length of the interaction region and w_o is the beam waist for a TEM_{oo} gaussian mode.

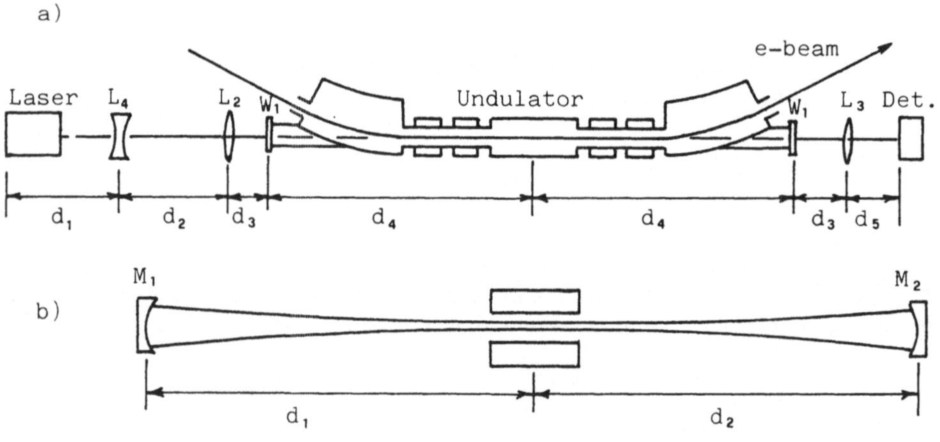

Fig. 11. a) Schematic layout of the amplification experiment.
b) Schematic layout of the oscillation experiment.

The optimum beam waist can be found by minimizing the mode cross section

$$\frac{d}{dw_o} = \overline{\Sigma}_L = 0 , \tag{22}$$

which implies

$$w_o = \sqrt{\frac{L_w \lambda}{2\pi \sqrt{3}}} . \tag{23}$$

By substituting Eq. (23) into Eq. (21), we have:

$$\overline{\Sigma}_L = \frac{L_w \lambda}{\sqrt{3}} . \tag{24}$$

For $\lambda = 5145$ Å (our experiment) the optimum beam waist becomes $w_o \approx 0.35$ mm.

3.6 The Amplification Experiment

A possible layout of the amplification experiment is sketched in Fig. 11a. Here L_i are spherical lenses with focal lengths F_i and W_i are quartz windows. The "seed" laser to be used will be the Spectra-Physics SP 164-09 model operating in single line mode at 5145 Å. In order to have $w_o = 0.35$ mm at the undulator mid-point[17] distances and focal lengths can be chosen as follows:

$d_1 = 2.60$ m; $d_2 = 2.69$ m; $d_3 = 0.40$ m; $d_4 = 6.0$ m; $d_5 = 0.91$ m

$F_1 = -2.5$ m; $F_2 = 2.5$ m; $F_3 = 0.8$ m.

3.7 The Oscillator Experiment

The optical cavity parameters must also be chosen so as to produce a waist $w_o = 0.35$ mm at the center of the interaction region for a wavelength $\lambda = 5151$ Å.

In Fig. 11b, M_i are concave mirrors with curvature radii R_i.

Table V shows the computed optical cavity parameters. W_i are the beam sizes on the mirrors and the corresponding Fresnel numbers, $N = a^2/d\lambda$, are calculated assuming the mirror size, a, is equal to the largest w_i.

Fig. 12 shows the e-beam and laser beam profiles along the interaction region both in the radial and in the vertical plane.

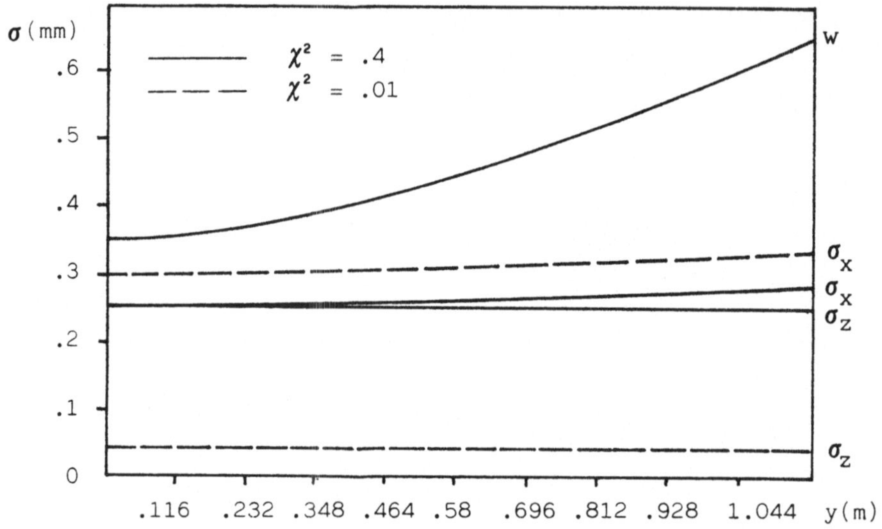

Fig. 12. Electron beam and laser beam profiles, along the interaction region.

Table V. Optical Cavity Parameters

d_1(m)	d_2(m)	w_o(mm)	w_1(mm)	w_2(mm)	R_1(m)	R_2(m)	$\mathcal{N}$	$\left(1-\frac{d_1+d_2}{R_1}\right)\left(1-\frac{d_1+d_2}{R_2}\right)$
8.75	8.75	0.35	4.14	4.14	8.81	8.81	2.2	0.973

The e-beam dimensions are plotted for two values of the betatron oscillation coupling factor $\chi^2 \approx 0.42$ (circular e-beam) and $\chi^2 \approx 0.01$, the latter corresponding to the minimum coupling so far obtained in Adone and gives a rather flat beam.

The detailed design of the optical cavity under vacuum is in progress.

3.8 Optical Klystron

The design of the proposed undulator makes it rather easy to change the coil interconnections in such a way as to produce an optical klystron type magnetic field configuration.[19]

Another possibility is to have two undulators in two adjacent straight sections, using the ring bending magnet in between as a dispersive drift space.

An optical klystron set-up, even with its inherently lower saturation output power, could become very interesting if the gain of the conventional FEL is found to be marginal.

Calculations for our case are in progress and some of the preliminary results have been presented during this school[20] by I. Boscolo.

Acknowledgements

The help from R. Barbini and G. Vignola in critically reading and in the editing of this paper, is gratefully acknowledged.

REFERENCES

1. R. Caloi et al.: A new monochromatic and polarized photon beam at Frascati. Proc. Intern. Conf. on Nuclear Physics with Electromagnetic Interactions, Mainz, 1979; LNF-79/30(P).
2. P.U.L.S. (CNR/GNSM-INFN): Rapporto di attivita 1977. LNF.
3. M. Bassetti, M. E. Biagnini, R. Boni, E. Burattini, M. T. Capria, A. Cattoni, N. Cavallo, V. Chimenti, G. Dalba, F. Ferrari, M. Foresti, P. Fornasini, S. Guiducci, A. Luccio, C. Mencuccini, E. Pancini, P. Patteri, M. Preger, R. Rinzivillo, C. Sanelli, M. Serio, S. Tazzari, F. Tazzioli, U. Troya: The Adone wiggler facility (to be published).
4. R. Barbini, E. Burattini, C. Cattoni, M. T. Capria, G. Vignola: Prime misure sulla radiazione coerente dal wiggler. Adone Int. Memo RM-22 (1/9/80).
5. R. Barbini, G. Vignola: LELA: A free electron laser experiment in Adone. Frascati Report LNF-80/12(R), March 1980.
6. M. Bassetti, A. Cattoni, A. Luccio, M. Preger and S. Tazzari: A. transverse wiggler magnet for Adone. Frascati Rep. LNF-77/26 ('77).
7. M. Bassetti, S. Tazzari in "Wiggler meeting," Frascati, June 29-30, 1978, ed. by A. Luccio, A. Reale and S. Stipcich (LNF, 1978).

8. R. Barbini and G. Vignola: The LELA undulator as a source of synchrotron radiation. Adone Internal Memor G-32 (1979).
R. Barbini, M. T. Capria, G. Vignola: Angular distribution of the Adone wiggler synchrotron radiation. Adone Internal Memo G-35 (1979).
9. Groupe "Supraconducteurs", CEN-Saclay Report DPh/Pe-STRIP SUPRA 78-35 (1978).
10. A. Cattoni: Progetto di massima di un ondulatore per il FEL; Adone Internal Memo MA-45 (1979).
A. Cattoni, C. Sanelli: Calcolo di un ondulatore per Adone: minimo λq. Adone Int. Memo MA-46 (1979).
B. Dulach: Ondulatore: calcolo delle deformazioni meccaniche. Adone Int. Memo M-7 (1979).
11. R. Barbini, M. E. Biagini and G. Vignola: Sul campo magnetico dell'on dulatore per il FEL. Adone Int. Memo MA-44 (1979).
12. L. R. Elias, W. M. Fairbank, J. M. J. Madely, H. A. Schwettman and T. I. Smith, Phys. Rev. Letters 36, 717 (1976).
D. A. G. Deacon, L. R. Elias, J. M. J. Madey, G. J. Ramian, H. A. Schewettman and T. I. Smith, Phys. Rev. Letter 38, 892 (1977).
W. B. Colson, Phys. Letters 59A, 187 (1976); 64A, 190 (1977).
13. C. Pellegrini, IEEE Trans. on Nuclear Sci., NS-26, 3791 (1979).
C. Pellegrini in "Report on FEL Workshop", Riva del Garda, June 4-6, 1979, ed. by G. Scoles.
14. A. W. Chao and J. Gareyte: Scaling law for bunch lengthening in SPEAR II, Report SPEAR 197/PEP 224 (1976).
S. Tazzari: Scaling dell'allungamento anomalo, Adone Int. Memo T-93 (1978).
15. F. H. Wang: Touschek lifetime at Adone and beam size. Adone Int. Memo T-113 (1979).
16. H. Bruck, Accelerateurs Circulaires de Particules (Presses Univ. de Feance, 1966).
17. M. Sands: The physics of electron storage rings. An introduction Report SLAC-121, UC-28(ACC) (1970).
18. G. Dattoli, A. Renieri: Nuovo Cimento B, 59, 1 (1980).
19. N. A. Vinokuronov, A. N. Shrinsky, Proc. of the 6th National Conf. on Charged Particle Accelerators. Dubna, 1978, vol. 2.
A. S. Artamov, N. A. Vinokurov, P. D. Voblyi, E. S. Gluskin, G. A. Kornyukhin, V. A. Kochubei, G. N. Kulipanov, V. N. Litvinenk N. A. Mezentsev and A. N. Skrinsky: The first experiments with an optical kylstron installed on the VEPP-3 storage ring. Novosibirs Preprint, 1980.
20. I. Boscolo and V. Stagno: The coherent emission from a bunched electron beam in a wiggler. These Proceedings.

HIGH-GAIN MILLIMETER AND SUBMILLIMETER FREE ELECTRON LASERS

V. L. Granatstein, P. Sprangle, and R. K. Parker

Naval Research Laboratory

Washington, D.C. 20375

INTRODUCTION

The relativistic electron beams produced by pulse-line accelerators (induction Linacs[1], IRED accelerators[2], and radial line accelerators[3]) can have suffiicent intensity to shift the laser interaction to the regime where collective electron oscillations participate in the wave amplification process. Free electron lasers with collective interactions can have sufficient gain so that practical amplifiers can be developed in addition to oscillators. Also there is potential for high efficiency in converting electron kinetic energy to photon energy. The electron energy characteristic of pulse-line accelerators (roughly 1-10 MeV) is relatively modest and lasers employing these acclerators are expected to operate in the wavelength range extending from millimeters to the near infrared. Electron beam pulse duration 10 ns - 10 μs is much longer than with r.f. accelerators, so that linewidth of the radiation may be compatible with good coherence even with operation in the millimeter wave regime.

BASIC PROCESSES IN COLLECTIVE INTERACTION LASERS: STIMULATED RAMAN SCATTERING

Collective free electron lasers amplify coherent radiation in a medium consisting of a dense, cold stream of relativistic electrons which are wiggled transversely either by a periodic wiggler magnet or by a strong electromagnetic wave. In the rest frame of the electron beam, the wiggler field appears to be a pump electromagnetic wave at frequency ω_o' and wave number k_o' (rest frame quantities denoted by primes). The incident pump wave scatters into a counter-propagating scattered electromagnetic wave (ω_s', k_s') and an electron density wave at the frequency, ω_e'. Both the backscattered e.m. wave and the plasma wave grow exponentially.

The rest frame geometry is shown in Figure 1. The incident pump e.m. wave has a transverse electric field $E_o' e_y$ which excites a transverse oscillation of the electrons with velocity $v_o' = -e_y e E_o' (1 - v_o'^2/c^2)^{1/2}/m\omega_o'$ where e is the magnitude of electron charge and m is the non-relativistic electron mass. In the presence of an incipient scattered wave with magnetic field $B_s' e_y$, an axial force $e\, v_o' B_s' e_z$ is exerted on the electrons. The coupling between the incident e.m. wave and the scattered e.m. wave thus produces a radiation pressure force (pondermotive force) which leads to a low frequency density modulation of the electrons. The complete expression for the pondermotive force is $F = -e(v_o' \times B_s' + v_s' \times B_o')$. The frequency and wave number of the electron density modulation satisfies the following conservation laws:

$$\omega_e' = \omega_o' - \omega_s' \text{ and } k_e' = k_o' + k_s' \tag{1}$$

where k_e', k_o' and k_s' are positive real quantities denoting wave number magnitudes.

The growth of the density modulation gives increasing coherence to the scattering process, resulting in a growing scattered wave which in turn increases the denstiy modulation still further. Thus, there is a feedback mechanism in this process which may result in an instability and exponential growth of both the scattered wave and the density modulation.

If the electron distribution is sufficiently dense and cold, it has a natural frequency of collective oscillation at the plasma frequency, ω_p. In that case the stimulated scattering process will be greatly strengthened by synergism with the growth of plasma waves when $\omega_e' = \omega_p$. The frequency of the scattered wave is then displaced from the incident frequency by a characteristic frequency of the medium (viz., $\omega_s' = \omega_o' - \omega_p$) and the process is called Stimulated Raman Scattering.

A number of theoretical analyses of stimulated Raman scattering of e.m. radiation from relativistic electron beams have appeared in the literature.[4-14] The paper by Sprangel et al[14] treats analytically the dynamics of the free electron laser in the non-linear regime. Particle simulation codes have also been used to study the nonlinear problem.[12] Recently both particle simulation studies[13] and analytical work[14] have shown substantial efficiency enhancement by appropriate contouring of wiggler period and amplitude. Several review articles covering eleective free electron lasers have also appeared in the literature.[15-19] We note especially the article by Sprangle et al[17] from which we have adapted many of the expressions for growth rates and efficiency which follow.

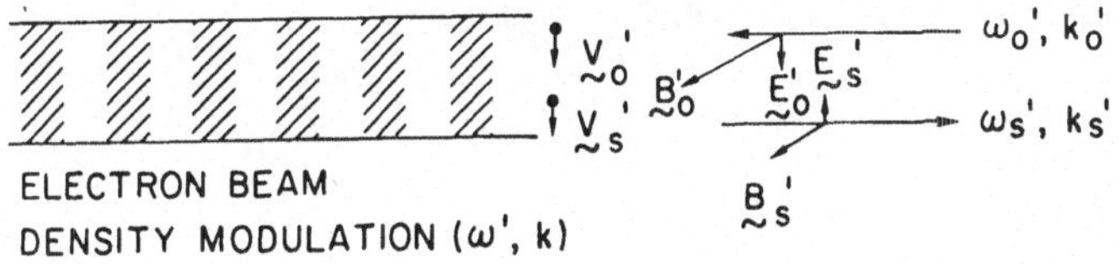

Figure 1. Simulated scattering in the electron beam rest frame.

THE DOPPLER-SHIFTED LASER FREQUENCY

Eq. (1) gives the relationship between the wave frequencies in the rest frame of the electron beam. In the remainder of this paper, for the convenience of the reader all expressions will be given in the laboratory frame where the electrons are streaming with velocity v_z.

In the beam frame, the stimulated scattering is a Stokes process with the frequency of the scattered output wave being smaller than the incident pump wave frequency. However, in transforming from the beam frame to the laboratory frame, the pump frequency is down-shifted while the backscattered wave frequency is up-shifted. This double doppler shift results in a scattered output wave in the laboratory frame at much shorter wavelength than the wavelength of the wiggler pump.

For a scattered e.m. wave propagating at the speed of light in the same direction as the electron beam, we have the following expression for output frequency

$$\omega_s = \gamma_z (1 + \beta_z) (\gamma_z \beta_z \kappa_o c - \omega_p) \tag{2}$$

where

$$\kappa_o = 2\pi/\ell \tag{3a}$$

when the pump wave is a magneto-static wiggler of period ℓ,

and

$$\kappa_o = \omega_o (1 + v_z/v_{ph})/v_z \tag{3b}$$

when the pump is an electromagnetic wave propagating counter to the streaming electrons with phase velocity v_{ph}, and frequency ω_o. In equations (2) and (3), $\beta_z = v_z/c$, $\gamma_z = (1 - \beta_z{}^2)^{-1/2}$, and the frame invariant plasma frequency

$$\omega_p = (e^2 n/ {}_o m \gamma)^{1/2} \tag{4}$$

where n is the electron density, $\gamma = (1 - v_z{}^2/c^2 - v\ {}^2/c^2)$, v is the transverse electron velocity, and we are employing rationalized m.k.s. units.

For the case of a highly relativistic beam, the laser output frequency of Eq. (2) may be as large as

$$\mathrm{MAX}\ (\omega_s) = 2\gamma_z{}^2 (2\pi/\ell)c \tag{5a}$$

for a magneto-static wiggler pump. For an electromagnetic pump wave with $v_{ph} = c$,

$$MAX\ (\omega_s) = 4\gamma_z^2\ \omega_o \tag{5b}$$

The relationship between the frequencies and wave numbers of the pump wave, the scattered output wave, and the idler wave at ω_p are shown in the Stokes diagrams, Fig. 2. Fig. 2a shows the beam rest frame relationships, while Fig. 2b shows the laboratory frame relationships. Fig. 2 depicts the case where the pump wave is a zero-frequency magneto-static ripple in the laboratory frame, and the scattered wave propagates at the speed of light.

GROWTH RATE OF THE OUTPUT SCATTERED WAVE

As discussed above, the interaction of a cold dense electron beam with a wiggler field can result in an instability in which the scattered output wave grows exponentially. In that case, an electromagnetic wave at ω_s will be amplified as it propagates along the z axis colinear with the electron beam; the wave amplitude will increase as $e^{\Gamma z}$, where Γ is the amplitude growth rate.

The expressions presented below for the growth rate were derived[7,17] assuming that the pump wave is RHCP in the case of an electromagnetic wiggler and helical in the case of a magnetostatic wiggler. It was also assumed that for an interaction of length L, $\Gamma L >> 1$ so that end effects are not important in the wave amplification process.

The wiggler field excites a transverse electron velocity with magnitude given by

$$v = \frac{e\ B_w}{m\,\gamma\,\kappa_o} \tag{6}$$

where for a magnetostatic wiggler with transverse magnetic field $b_r = b_r(\hat{e}_x \cos 2\pi\ z/\ell + \hat{e}_y \sin 2\pi\ z/\ell)$,

$$B_w = b_r \tag{7a}$$

while for an electromagnetic wiggler with electric field $E_o = E_o\ (\hat{e}_x \cos\ (k_o z + \omega_o t) + \hat{e}_y \sin\ (k_o z + \omega_o t))$,

$$B_w = E_o\ (1 + v_z/v_{ph})/v_z \tag{7b}$$

It should be noted that the relationship between the streaming electron energy, γ_z, and the total electron energy γ depends on the magnitude of transverse velocity, $v_\perp$ as

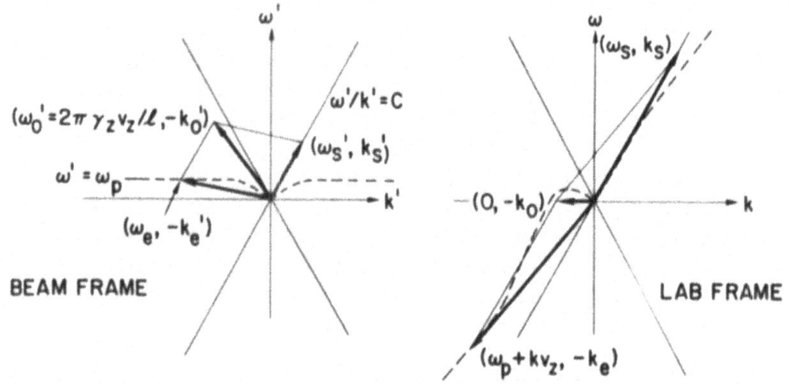

Figure 2. Stokes diagrams for Free Electron Laser based on stimulated Raman scattering for case where pump wave is a magnetostatic wiggler. Dispersion curves for space change waves.

$$\gamma_z = \gamma/(1 + \gamma^2\beta^2)^{1/2} \tag{8}$$

where

$$\beta_\perp = v /c.$$

Thus, when a magnetostatic wiggler is used to excite a strong $v_\perp$, the streaming energy and the doppler shift decrease; a magnetostatic wiggler does not increase the total energy.

When the electron beam is propagated along a strong uniform magnetic guide field of strength $\overline{B}$, the transverse velocity excited by the wiggler may be enhanced[5] by resonance effects near the cyclotron frequency $\overline{\Omega} = eB/m\gamma$. In that case, the right hand side of Eq. (6) should be multiplied by the enhancement factor $v_z\kappa_0(v_z\kappa_0 - \overline{\Omega})$; of course, in a physical system, the quantity $v_z\kappa_0 - \overline{\Omega}$ cannot be made arbitrarily small[5] and its magnitude has been calculated[5] for the case where the pump wave is a normal mode of the system as $v_z\kappa_0 - \overline{\Omega} = 1/2(\overline{\Omega}^2 + 4\omega_p{}^2)^{1/2} - 1/2\overline{\Omega}$.

In reference 17, a general analysis of the free electron laser mechanism is performed utilizing a right-handed circularly-polarized, spatially-periodic, magnetic wiggler. The analysis is fully relativistic and performed in the laboratory frame of reference. The electron beam distribution function, associated with the relativistic Vlasov equation, is expanded to first order in the excited fields about the exact equilibrium distribution in the wiggler field. The first order distribution function is then used to obtain the driving currents for both the radiation as well as the space charge fields. The wave equations for the fields are used to obtain the spatial evolution (including the transient regime) of the radiation and space charge waves.

Several distinct interaction modes are analyzed in Ref. 17; the growth rates as well as the saturation efficiencies for the various interaction regimes are calculated. The various interaction regimes are basically characterized by the magnitude of the magnetic wiggler field and electron beam density. The present paper deals primarily with high-gain free electron lasers; hence, only the high-gain, collective regime (Raman regime) and high-gain noncollective regime (strong wiggler regime) will be considered. The dispersion relation, relating the radiation wave number k_s and the frequency ω_s, can be put into the form where F is the filling factor, i.e., the ratio of the electron beam area to the area of the electromagnetic beam being amplified with the limitation $F < 1$, and $\xi = \omega_p/c\kappa_o$ is the electron beam strength parameter. The dispersion relation in Eq. (9) describes the high-gain free electron laser in both the Raman and strong wiggler regimes. The spatial growth rate of the radiation field is found by taking k_s to be complex and solving Eq. (9), using the usual techniques, for the imaginary part of k_s. In general, the expressions presented in the following portion of the paper assume $\gamma_z >> 1$.

Raman, High-Gain, Collective Regime

The amplitude growth rate in the stimulated Raman scattering regime is given by

$$\Gamma_R = 0.5\ (\gamma_z\xi)^{1/2}\ \beta\ \ F^{1/2}\ \kappa_o \tag{10a}$$

Equation (10a) is valid when space charge forces dominate the pondermotive forces requiring

$$\beta\ <<\ \beta_p \tag{10b}$$

where

$$\beta_p = 4\xi^{1/2}\ (\gamma_z{}^3F)^{-1/2}\ . \tag{10c}$$

Strong Wiggler Regime

When the wiggler field is very strong, the pondermotive forces may modify the plasma wave dispersion. In the case the wave amplification process is altered. The pondermotive wave dominate the collective space charge wave. The expression for amplitude growth becomes

$$\Gamma_A = \frac{\sqrt{3}}{2}\ \Gamma_R(\beta\ /\beta_p)^{-1/3} \tag{11a}$$

validity of Eq. (11a) requires

$$\beta\ >>\ \beta_p \tag{11b}$$

which has been called the "strong pump regime."

Growth rate is plotted as a function of $\beta_\perp$ in Fig. 3; the case of a magnetostatic wiggler has been assumed and free electron laser parameters have been taken as $\gamma = 5$, $\ell = 3$ cm, $F = 1$, and electron density $n = 10^{11}$ cm^{-3} in Fig. 3. For these parameters, it may be seen from the figure that the division between the Raman scattering regime and the strong pump regime occurs at $\beta\ = \beta_p = 0.1$; this value of $\beta_\perp$ corresponds to a wiggler magnetic field of 1.8 kG. The normal

Raman scattering regime has been the regime of operation in intense beam experiment conducted to date. It may be shown that for $\xi > (\gamma^2 - 1)/(4\gamma)^2$ the strong scattering regime disappears completely.

Last, it should be noted in Fig. 3 that amplitude exponentiating lengths are on the order of 10 cm so that practical amplifiers or reasonable length are possible.

LASER EFFICIENCY

For a wiggler with axially unfirom perido and amplitude, efficiency of converting electron kinetic energy into output wave energy is given by[17]

$$\eta_R = \xi/\gamma_z \tag{12}$$

when

$$\beta \ll \beta_p.$$

and by

$$\eta_A = \nu_R \, (\beta / \beta_p)^{2/3} \tag{13}$$

when

$$\beta_\perp \gg \beta_p.$$

The expression for efficiency in the Raman regime may be expressed in terms of the laser output frequency ω_s and shown to have an upper bound given by

$$\eta_R < 2\gamma\omega_p/\omega_s$$

where we have used $\omega_s = 2\gamma_z{}^2 \, c\kappa_o$ and $\gamma_z < \gamma$. This upper bound efficiency is plotted as a function of ω_s in Fig. 4 for values of accelerator voltage and current density appropriate to intense relativistic electron beam accelerators. (In present intense beam accelerators current densities of 10^4 A/cm^2 are routinely achieved while $J = 10^5$ A/cm^2 is a reasonable estimate of what could be attained with careful design.)

Note that good efficiency, on the order of 10%, is predicted in Fig. 4 for wavelengths longer than about 100 μm. It should also be stressed that recent theoretical studies have indicated the possibility of greatly increasing efficiency above the values predicted by equations (12) and (13) by such techniques as decreasing the wiggler period (while keeping lb_r constant) in the region where the extraction of energy from the electrons is beginning to saturate.[14]

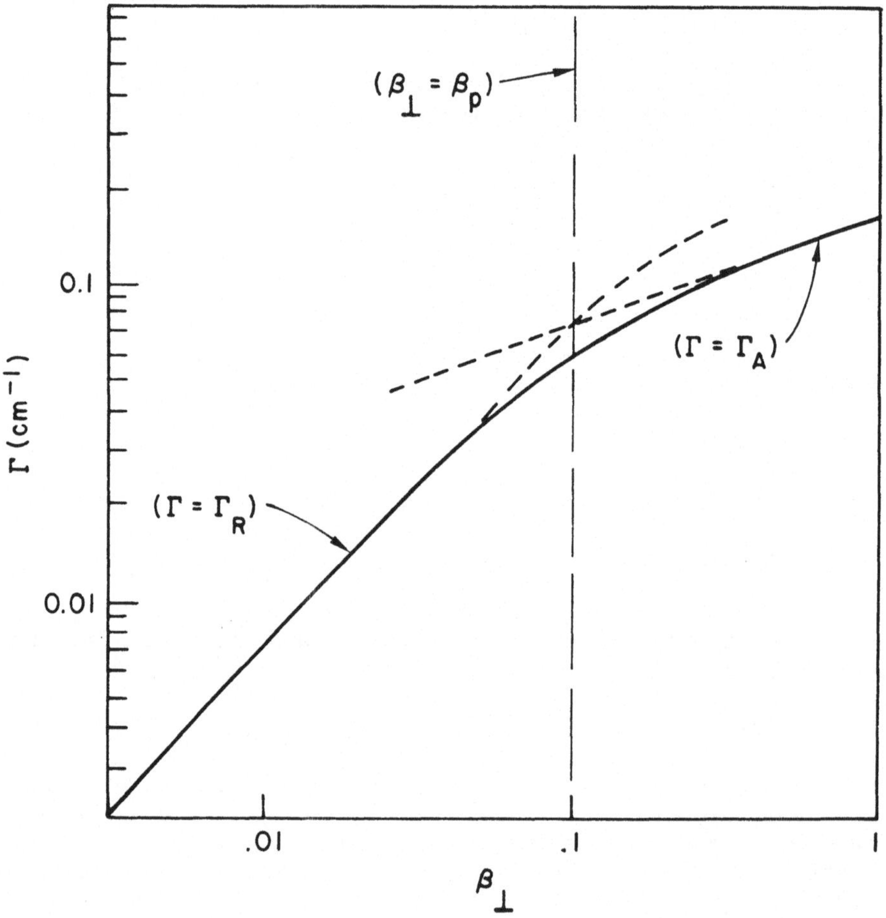

Figure 3. Wave amplitude growth rate as a function of transverse electron velocity.

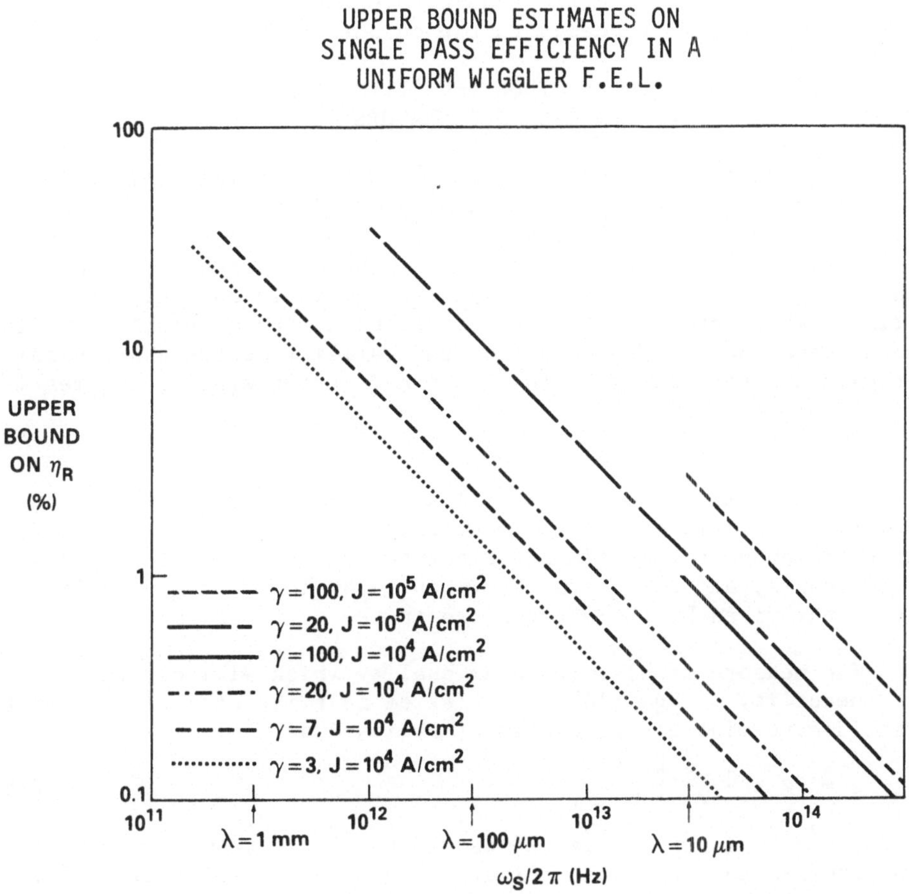

Figure 4. Upper bound estimates of single-pass, unenhanced, Raman REL efficiency as a function of output frequency. Several examples of electron ucrrent density and accelerator voltage are given.

Transfer of energy from the electrons to the output e.m. wave saturates when the electrons become trapped in the minima of the total longitudinal wave potential (i.e., the pondermotive plus space-charge waves), and the electrons are no longer free to maintain a phase relationship conducive to energy loss. This process in a uniform wiggler leads to the expressions for saturated efficiency of energy transfer given by equations (12) and (13). However, when the wiggler period is reduced the phase velocity of the total longitudinal wave is also reduced and the trapped electrons are forced to slow down and give up more energy to the output radiation.

LIMITATIONS IMPOSED BY ELECTRON VELOCITY SPREAD, AND THE REQUIREMENT FOR HIGH QUALITY ACCELERATORS

In order that the preceding expressions for gain and efficiency be valid the electron beam must have a sufficiently small axial velocity spread, Δv, at a given time and spatial position. If Δv is excessive, bunching of the electrons becomes smeared out and stimulated scattering tends to take place from individual electrons (stimulated Compton scattering) rather than from a density wave in the electron beam. The condition for validity of the expressions for gain and efficiency in the Raman and strong wiggler regimes is

$$\Delta v/c \ll \eta/\gamma_z^2 \qquad (14)$$

where η is given by equations (12) and (13). If Eq. (14) is not satisfied, stimulated Compton scattering holds, and both the gain and efficiency of the stimulated scattering process fall rapidly as Δv increases, and in general are much below the values of gain and efficiency given by the expressions in the sections above.

The measure of electron beam quality which will determine if the inequality in Eq. (14) is satisfied is emittance, ε, defined for a solid streaming electron beam of radius R as

$$\varepsilon = \pi R <\theta^2>^{1/2} \qquad (15)$$

where $<\theta^2>^{1/2}$ is the r.m.s. spread in the angle between the electron trajectories and the axis. Clearly a spread in orientation of the electron trajectories implies a spread in the velocity components. For small values of $<\theta^2>^{1/2}$, the spread in axial velocity is related to emittance as

$$\Delta v/c = \varepsilon^2 \beta_z / 2\pi^2 R^2 \qquad (16)$$

Then Eq. (14) and Eq. (16) imply that for R = 3 mm, η = 10% and γ = 4, the emittance must satisfy $\varepsilon\beta_z\gamma_z \ll 130\ \pi$ mrad cm, this is a realizable but not a trivial requirement for an intense electron beam.

Aside from variations of Δv at a given time and spatial position, there can also be systematic variations of Δv in time and space which also will degrade free electron laser performance. Variation with time arises from fractional fluctuations in accelerator voltage, $\Delta V/V$. The inequality in Eq. (14) implies that

$$\Delta V/V << \eta \tag{17}$$

in order that stimulated Raman scattering occur. Marshall et al[18] have shown that if $\Delta V/V > \eta$, the free electron laser radiation will be absorbed during a portion of the accelerator current pulse.

A systematic spacewise variation in electron energy occurs because of space charge effects in an intense electron beam. For a solid beam on axis the potential variation across the beam radius is

$$\Delta V = I/4\pi\ \varepsilon_0 v_z \tag{18}$$

where I is the beam current and ε_0 is the permittivity of free space. For a thin annular beam of thickness t and radius a,

$$\Delta V = (I/4\pi\ \varepsilon_0 v_z)\ (t/a) \tag{19}$$

In order for stimulated Raman scattering to occur ΔV due to electrostatic effects must satisfy the inequality in Eq. (17); this will place a limitation on beam current. For example, if η = 10% and V = 3 MV, then from Eq. (17) and Eq. (18), one has the requirement for a solid beam that I << 10 kA; of course, it is clear from the form of Eq. (19) that larger total currents are permitted for annular electron beams.

Last, there will also be a potential variation across the beam radius due to the transverse gradient of the wiggler magnetic field[11,30] which is necessary to satisfy $\nabla \cdot b_r = \quad x\ b_r = 0$. For a solid beam on axis, the fractional potential variation across the beam radius due to the transverse gradient in b_r is

$$\Delta V/V = (b_r\ Re/2mc)^2 \tag{20}$$

where it is assumed that $2\pi R/l < 1$. The shear due to the transverse dependence of the wiggler field opposes the shear due to space charge effects.

G. Experiments

Data on laboratory studies of stimulated scattering from intense relativistic electron beams which resulted in millimeter and submillimeter radiation are summarized in Table I; the experiments listed were carried out by the Naval Research Laboratory,[20,23] Columbia University[21-23] or the Lebedev Institute.[24] Experimental studies are also in progress at M.I.T. where a periodic electrostatic wiggler is being employed,[25] the Ecole Polytechnique,[26] TRW Inc.,[27] and the Hebrew University of Jerusalem.[28]

The intense-electron-beam studies listed in Table I are characterized by electron energy in the range 0.7 - 2 MeV and corresponding output wavelengths that range from 3 mm down to 400 μm. The current is large (4.5 - 30 kA) making possible the participation of collective electron beam modes in the stimulated scattering process and resulting in large gain; amplified spontaneous emission was demonstrated in relatively short interaction lengths (30 - 60 cm) and in one experiment[20] the single pass amplitude gain $G_L > 2$. As yet, there is no continuous power capability in the intense beam studies, the experiments being characterized by a single electron pulse of 10 - 50 nsec in duration.

Two types of pump waves were investigated: a magnetostatic wiggler (examples 2 and 3 in Table I) and a powerful electromagnetic wave (examples 1 and 4 in Table I). In the experiments which used an electromagnetic pump wave, it was generated by a portion of the same electron beam in which the stimulated backscattering occured. The experimental configuration corresponding to example 1 in Table I is sketched in Fig. 5. The pump wave with wavelength 2 - 3 cm was generated in a region of the beam far downstream from the cathode near the output end of the experiment. "Conventional" processes were used to generate the pump wave (i.e., the electron cyclotron maser instability in example 1, and the usual BWO instability in a tube with periodically rippled walls in example 4). The pump wave was then made to propagate upstream; when it encountered the cold streaming electrons near the cathode, stimulated scattering occurred, resulting in backscattered millimeter and submillimeter radiation. In addition to the wiggler field, the intense beam experiments typically had an externally imposed uniform axial magnetic field that was large enough so that magnetic resonances in the output power were observed.[25]

Theory predicts that intense-beam free-electron lasers with output in the millimeter and submillimeter will have saturated efficiencies on the order of several to several tens of percent (when operated as traveling wave amplifiers). The efficiency measured for the free electron laser oscillators listed in Table I has been lower by one to two orders of magnitude. A more detailed discussion of one of the experiments may shed light on this apparent discrepancy. Figure 6 is a schematic of the free electron laser oscillator corresponding to example 3 in Table I. The free electron laser oscillator was created by passing a 1.2 MV electron beam of 25 kA through a spatially periodic, linearly polarized, magnetostatic field which had a ripple amplitude B_W = 400G in the radial direction and a period of $l = 8$ mm. As shown in Figure 6, an annular electron beam was expanded radially by passing it through an adiabatic reduction in the confining magnetic field. Beam expansion not only reduced the transverse electron energy spread but also placed the beam within the resonant volume of the Fabry-Perot cavity formed by the two mirros. However, the short beam duration (30 nsec) and the long cavity length (1.5 m) limited the feedback radiation to a maximum of three interactive passes. The double-doppler shifted radiation was diffraction coupled from an aperture in the output mirror. The peak output power was measured to be 1 MW at 400 μm, corresponding to an efficiency of 0.02%.

TABLE I

	Experimental Configuration	Wiggler Fields	Electron Beam	Output Radiation	Reference
(1)	Amplified spontaneous emission	Electromagnetic wave $l = 2$ cm $E_0 = 4 \times 10^4$ V/cm $L = 0.3$ m	2 MeV, 30 kA $a = 1.8$ cm, $t = 2$ mm $n = 3 \times 10^{12}$ cm^{-3} $\Delta V/V$ (electrostatic) $= 4\%$	$\lambda_s = 400 \mu m$ 1 MW $G_L = \int_0^L e^{\Gamma z} dz > 2$	20
(2)	Amplified spontaneous emission	Magnetostatic wiggler $l = 0.6$ cm, $b_r = 0.5$ kG $L = 0.36$ m	0.86 MeV, 5 kA $a = 1$ cm, $t = 1.5$ mm $n = 10^{12}$ cm^{-3} $\Delta V/V$ (electrostatic) $= 2\%$	$\lambda_s = 1.5$ mm 8 mW	21,22
(3)	Oscillator with optical cavity Mirror trans. = 2%	Magnetostatic wiggler $l = 0.8$ cm, $b_r = 0.4$ kG $L = 0.4$ m	1.2 MeV, 25 kA $a = 2.2$ cm, $t = 1$ mm $n = 4 \times 10^{12}$ cm^{-3} $\Delta V/V$ (electrostatic) $= 3\%$	$\lambda_s = 400 \mu m$ $\Delta\lambda_s/\lambda_s = 2\%$ 1 MW	23
(4)	Amplified spontaneous emission	Electromagnetic wave $l = 3.2$ cm $E \approx 10^5$ V/cm $L = 0.6$ m	0.7 MeV, 4.5 kA $a = 1$ cm, $t = 1.5$ mm $n = 1.4 \times 10^{12}$ $\Delta V/V$ (electrostatic) $= 4\%$	$\lambda_s = 3.2$ mm 20 W	24

This study was instructive and did result not only in substantial output power, but also in a narrowing of the linewidth of the output radiation to $\Delta\lambda_s/\lambda_s \sim 2\%$ compared with $\Delta\lambda_s/\lambda_s > 10\%$ in the case of the amplified spontaneous emission. However, variations in accelerator voltage during the "constant" portion of the accelerator waveform ($\Delta V/V \approx 5 - 10\%$) may well have caused periodic reversal in the energy transfer from beam to wave. Moreover, computer studies of the electron trajectories within the interaction volume have shown that the annular electron beam is composed of an outer layer with large velocity spread and an inner layer with small velocity spread.[29] Calculations using the linear theory of Raman scattering and the results of the electron trajectory analysis indicate that the growth in the scattered wave amplitude was limited to approximately one exponential increase during a single pass through the 40 cm ripple length. Since feedback was limited to only three passes by the short beam duration, it is highly unlikely that the process became saturated.

An improved experiment in which the stimulated Raman scattering process could reach saturation would require an electron beam of improved quality as well as a much stronger wiggler magnet. We calculate that with 1.7 kA of 2 MeV electrons in a solid beam with R = 0.3 cm, a wiggler field with b_r = 2.5 kG and period 2 cm would result in a Raman amplitude growth rate $\Gamma_R = 0.17$ cm^{-1} or 10 e-folds of the wave amplitude in a 60 cm interaction length. The predicted saturation efficiency is η_R = 7% corresponding to an output power of 200 MW at a wavelength of 480 μm.

Finally, one should consider for future experimentation the possibility of two-stage free electron lasers. Free-electron laser operation at short wavelengths has the disadvantage of requiring very high electron kinetic energy. Accelerator voltage requirements could be reduced for intense beam systems by using a two-stage approach. To illustrate this technique, a near-infrared radiation source is considered. In a two-stage laser, two consecutive and distinct scattering interactions take place within a single electron beam. The output radiation from the first stage, in which the pump is a circularly polarized static magnetic field, is reflected back on the beam and used as an electromagnetic pump wave in the second stage. The final wavelength of the output radiation, from the second stage, is $\lambda \approx 1/8\gamma^4$ instead of $1/2\gamma^2$ as would be the case in a single stage device. Hence, in a two-stage laser, far shorter output wavelengths can be realized for the same electron kinetic energy. For example, a 3 MeV electron beam with a total current of 7 kA is passed in the first stage through a wiggler with a period of 2 cm. This interaction would produce radiative power pulse of 3.6 GW at a wavelength of 340 μm. Reflection of this intermediate frequency back on the beam would result in a second scattering, upshifting the output radiation to a wavelength of 10 μm. With efficiency enhancement, the output power could be 200 MW for an efficiency of $\sim$ 1%. Sequential acceleration of the electron beam during the second scattering interaction would have a similar effect on efficiency enhancement as decreasing the period in a wiggler magnet. This two-stage free electron laser concept is depicted in Fig. 6.

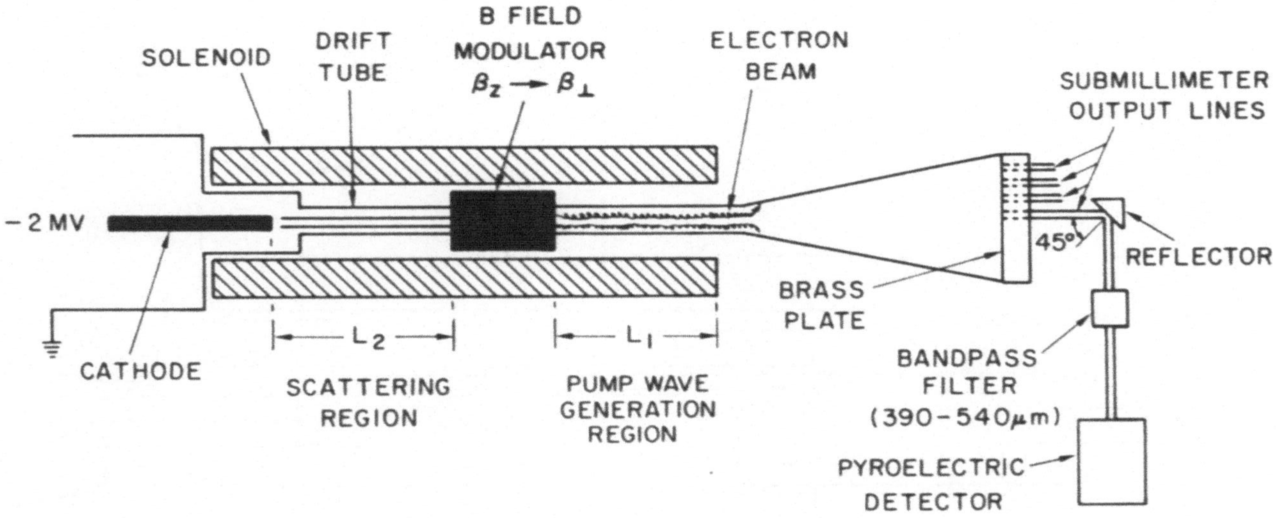

Figure 5. Experimental arrangement in study of stimulated scattering in the high-gain collective regime with electromagnetic pump wave (example 1 in Table 1).

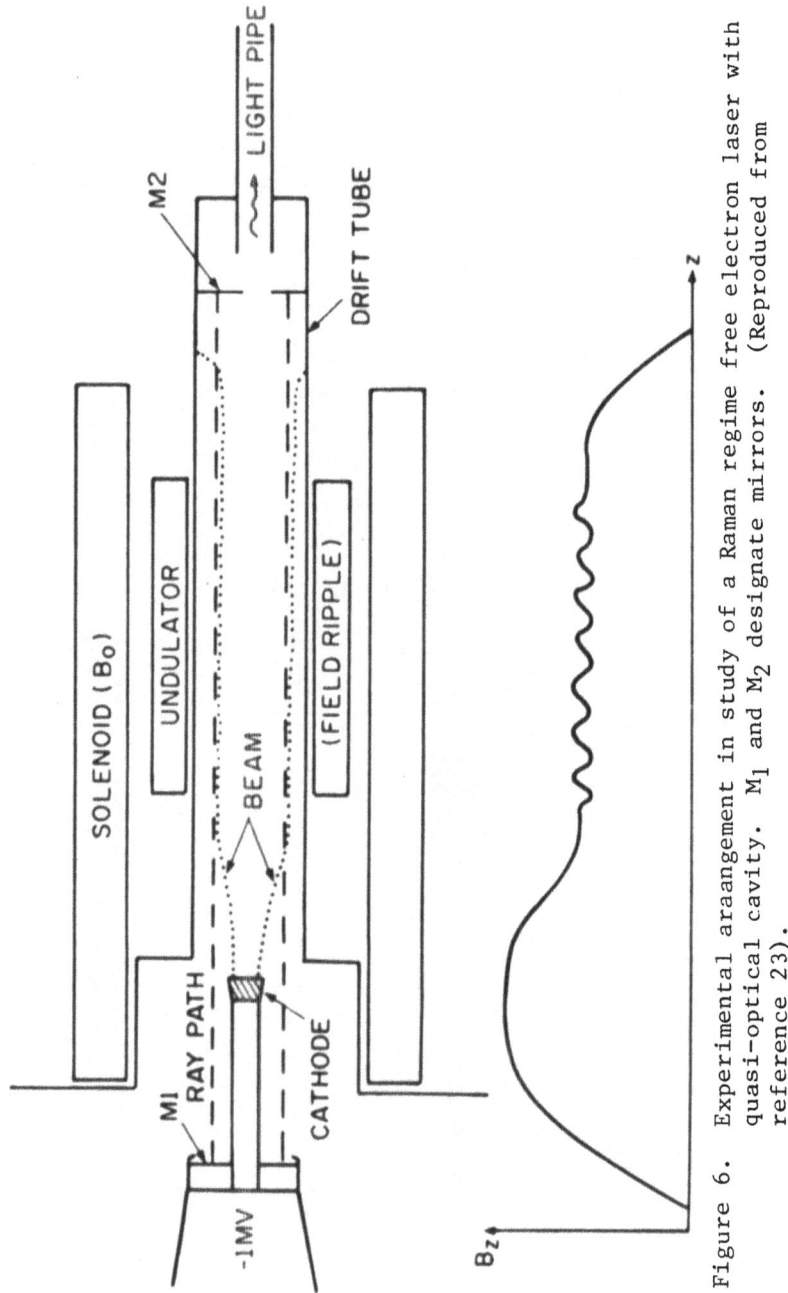

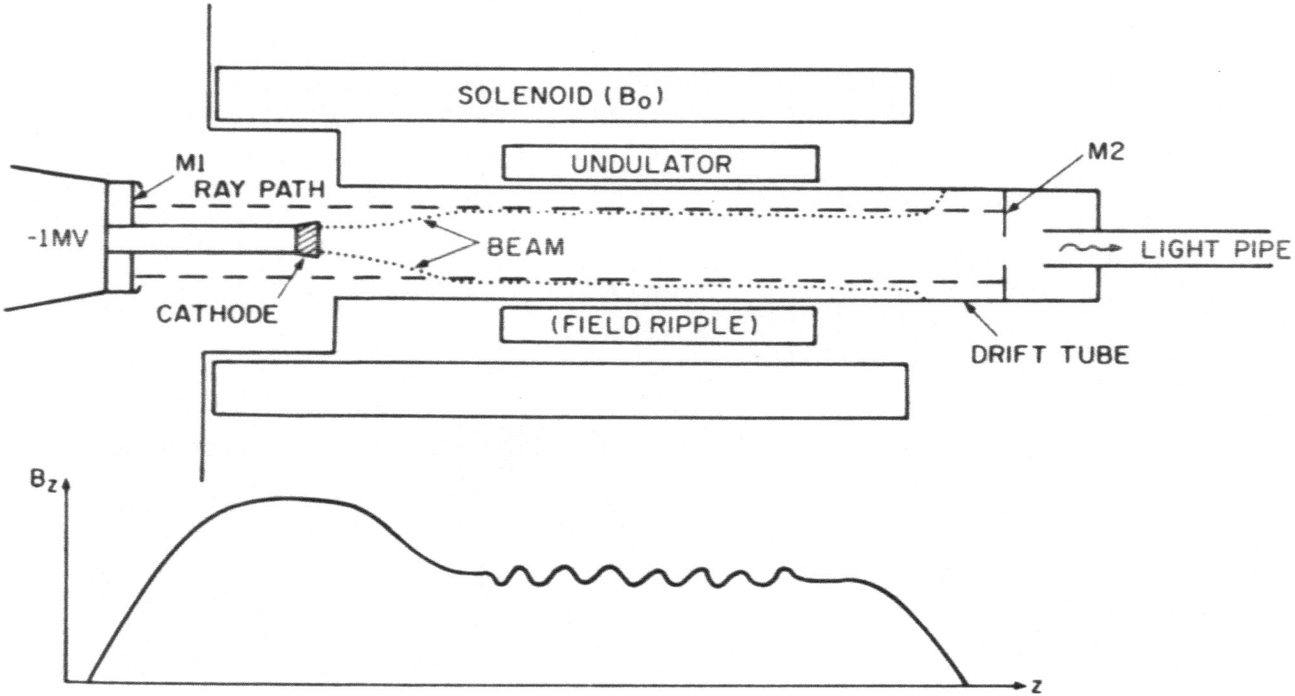

Figure 6. Experimental araangement in study of a Raman regime free electron laser with quasi-optical cavity. M_1 and M_2 designate mirrors. (Reproduced from reference 23).

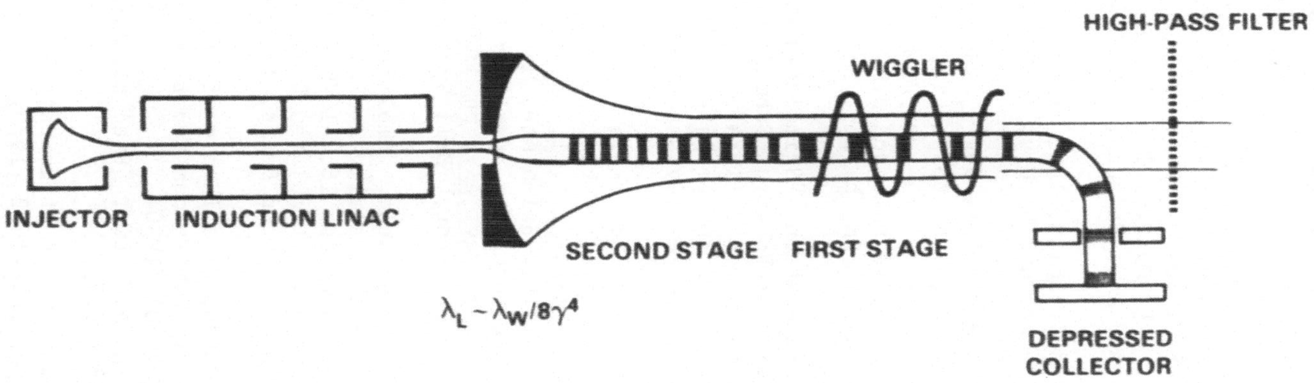

Figure 7. Conceptual operation of a two-stage Raman free electron laser with output in the near infared. An electron accelerator producing a beam with normalized emittance $\varepsilon\beta\gamma \lesssim$ 10 mrad. cm. would be requred for operation in the Raman regime at $\lambda_s \sim 10$ um; this representa s substantial improvement over the present state of the art in induction linacs and other high current accelerators. Energy recovery by employing a collector which is depressed in potential is depicted as a means of improving overall laser efficiency.

REFERENCES

1. e.g., N.C. Christofilos, R.E. Hester, W.A.S. Lamb, D.D. Reagan, W.A. Sherwood, and R.E. Wright, Rev. Sci. Instrum. 35, 886 (1964); R. Avery, G. Behrsing, W.W. Chupp, A. Faltens, E.C. Hartwig, H.P. Hernandez, C. MacDonald, J.R. Meneghetti, R.G. Nemetz, W. Popenuck, W. Salsig, and D. Vanecek, IEEE Trans. Nucl. Sci. NS18, 479 (1971); and R.E. Hester, D.G. Bubp, J.C. Clark, A.W. Chesterman, E.G. Cook, W.L. Dexter, T.J. Fessenden, L.I. Reginato, T.T. Yokota and A.A. Faltens, IEEE Trans. Nucl. Sci. NS26, 4180 (1979).
2. e.g., B. Bernstein, and I. Smith, IEEE Trans. Nucl. Sci., NS20, 294 (1973); and R.K. Parker and M. Ury, IEEE Trans. Nucl. Sci., NS22, 983 (1975).
3. e.g., A.I. Pavlovskii and V.S. Bosamykin, Sov. At. Energy 37, 942 (1974); A.I. Pavlovskii, V.S. Bosamykin, G.D. Kuleshov, A.I. Gerasimov, V.A. Tananakin, and A.P. Klementev, Sov. Phys. Dokl. 20, 441 (1975); D. Eccleshall and J.K. Temperley, J. Appl. Phys. 49, 3649 (1978); and I. Smith, Rev. Sci. Instrum. 50, 714 (1979).
4. P. Sprangle and V.L. Granatstein, "Stimulated Cyclotron Resonance Scattering and Production of Powerful Submillimeter Radiation," Appl. Phys. Lett. 25, pp. 377-379 (1974).
5. P. Sprangle, V.L. Granatstein and L. Baker, "Stimulated Collective Scattering from a Magnetized Relativistic Electron Beam," Phys. Rev. A12, pp. 1697-1701 (1975).
6. V.I. Miroshnichenko, "Stimulated Coherent Scattering of an Electromagnetic Wave by a Relativistic Electron Beam in a Magnetic Field," Sov. Tech. Phys. Lett. 1, pp. 453-454 (1975).
7. N.M. Kroll and W.A. McMullin, "Stimulated Emission from Relativistic Electons Passing through a Spatially Periodic Transverse Magnetic Field," Phys. Rev. A17, pp. 300-308 (1978).
8. A. Hasegawa, "Free Electron Laser," Bell System Tech. J. 57, pp. 3069-3089 (1978).
9. I.A. Bernstein and J.L. Hirshfield, "Amplification on a Relativistic Electron Beam in a Spatially Periodic Transverse Magnetic Field," Phys. Rev. A20, pp. 1661-1670 (1979).
10. P. Sprangle and A.T. Drobot, "Stimulated Backscattering from Relativistic Unmagnetized Electron Beams, J. Appl. Phys. 50, pp. 2652-2661 (1979).
11. P. Sprangle and R.A. Smith, "The Theory of Free Electron Lasers," Naval Research Laboratory Memorandum Report 4033, 1979; also Phys. Rev. A21, pp. 293-301 (1980).
12. T. Kwan, J.M. Dawson and A.T. Lin, "Free Electron Laser," Phys. Fluids 20, pp. 581-588 (1977).
13. A.T. Lin and J.M. Dawson, "High Efficiency Free Electron Laser," Phys. Rev. Lett. 42, pp. 1670-1673 (1979).

14. P. Sprangle, C.M. Tang and W.M. Manheimer, "Non-Linear Theory of Free Electron Lasers and Efficiency Enhancement," Phys. Rev. A21, pp. 302-318 (1980); also "Non-Linear Formulation and Efficiency Enhancement of Free Electron Lasers," Phys. Rev. Lett. 43, pp. 1932-1936 (1979).
15. V.L. Granatstein and P. Sprangle, "Mechanism for Coherent Scattering of Electromagnetic Waves from Relativistic Electron Beams," IEEE Trans. MIT25, pp. 545-550 (1977).
16. A. Gover and A. Yariv, "Collective and Single-Electron Interactions of Electron Beams with Electromagnetic Waves and Free-Electron Lasers," Appl. Phys. 16, pp. 121-133 (1978).
17 P. Sprangle, R.A. Smith and V.L. Granatstein, "Free Electron Lasers and Stimulated Scattering from Relativistic Electron Beams," Infrared and Millimeter Waves, Vol. 1, ed. K.J. Button, Academic Press, New York, 1979, pp. 279-327.
18. T.C. Marshall, S.P. Schlesinger, and D.B. McDermot, "The Free Electron Laser: A High Power Sub-Millimeter Radiation Source," in Advances in Electronics and Electron Physics, Vol. 53, ed. L. Marton, Academic Press, New York, 1980 (to be published).
19. V.L. Bratman, N.S. Ginzburg and M.I. Petelin, "Ubitrons and Scattrons," in Relativistic High Frequency Electronics, ed. A.V. Gaponov, Academy of Sciences of the U.S.S.R., Institute of Applied Physics, Gorki'i, 1979, pp. 217-248 (in Russian).
20. V.L. Granatstein, S.P. Schlesinger, M. Herndon, R.K. Parker, and J.A. Pasour, "Production of Megawatt Submillimeter Pulses by Stimulated Magneto-Raman Scattering," Appl. Phys. Lett. 30, pp. 384-386 (1977).
21. T.C. Marshall, S. Talmadge and P. Efthimion, "High-Power Millimeter Radiation from an Intense Relativistic Electron-Beam Device," Appl. Phys. Lett. 31, pp. 320-322 (1977).
22. R.M. Gilgenbach, T.C. Marshall, and S.P. Schlesinger, "Spectral Properties of Stimulated Raman Radiation from an Intense Relativistic Electron Beam," Phys. Fluids 22, pp. 971-977 (1979).
23. D.B. McDermott, T.C. Marshall, S. P. Schlesinger, R.K. Parker, and V.L. Granatstein, "High-Power Free-Electron Laser Based on Stimulated Raman Backscattering," Phys. Rev. Lett. 41, 1368 (1978).
24. P.G. Zhukov, V.S. Ivanov, M.S. Rabinovich, M.D. Raizer, and A.A. Ruchadze, "Stimulated Compton Scattering from Relativistic Electron Beam," Proc. of the Third Internat'l Topical Conf. on High Power Electron and Ion Beam Research and Technology, Novosibirsk, 3-6 July 1979 (preprint).
25. R.E. Shefer, K.K. Jacobs, and G. Bekefi, "Quasistatic Pump for Free Electron Lasers," Bull, Am. Phys., Soc. 24, 1067 (1979).
26. H. Boehmer, J.M. Buzzi, H.J. Doucet, B. Etlicher, H. Lamain, and C. Rouille, "Resonance Effect on Relativistic Electron Beam Propagation for Collective Free Electron Laser," Bull. Am. Phys. Soc. 24, 1066 (1979).

27. H. Boehmer, J. Munch, and M.Z. Caponi, "Free Electron Laser Experiment with a Spatially Varying Pump Amplitude," Bull. Am. Phys. Soc. 24, 1066 (1979).
28. Private discussions with J. Hirshfield.
29. R.H. Jackson, R.K. Parker, and V.L. Granatstein, "Beam Quality Studies for Intense Beam Free Electron Lasers," Digest of Fourth Internat'l Conf. on Infrared and Millimeter Waves and Their Applications, Miami, 10-15 December 1979, IEEE CAT. No. 79 CH 1384-7 MTT, pp. 96-97.
30. P. Sprangle and C.M. Tang, "Three-Dimensional, Non-Linear Theory of the Free Electron Laser," Naval Research Laboratory Memorandum Report 4280 (1980).

FREE ELECTRON LASERS BASED UPON STIMULATED RAMAN BACKSCATTERING: A SURVEY

T.C. Marshall

Plasma Laboratory
Columbia University
New York, N.Y. 10027

I. INTRODUCTION

In this chapter I discuss a type of Free Electron Laser (FEL) which is based upon stimulated Raman (SR) Backscattering of a magnetostatic pump wave from a cold, dense, relativistic electron beam. Radiation from such a laser[1] is of megawatt level and tuneable over the millimeter to far infrared spectrum (~2mm - 100μ), using an electron beam of "modest" energy (0.5 - 2 MEV). One incentive to develop such a laser is that the production of intense far infrared coherent sources is difficult; conventional lasers are troubled by lack of energy storage, highly selective level excitation, and low efficiency--the latter usually following from a downconversion process which is driven by a powerful CO_2 laser pump. Nevertheless, several far infrared lasers--at widely spaced intervals in the spectrum--have achieved noteworthy power output (~1 MW); as a particular example, consider the CH_3F system, which radiates at 496μ, and which is Raman-like.[2] The Raman process is a nonlinear optical process requiring multi-photon absorption.

The SR version of the FEL is a system in which a plasma (longitudinal) wave mode and an electromagnetic scattered mode feed on the energy of the pump (frequency ω_0); it utilizes a convective instability which can amplify an electromagnetic signal in a travelling-wave amplifier fashion. As such, a device can operate in three ways: (1) as an amplifier for noise [superradiant or superluminescent model]; (2) as an amplifier for a coherent signal; (3) as an oscillator [e.g., a cavity laser with quasi-optical feedback]. Theory is most highly developed for case (2), but experimentally (1) and (3) have been explored to date.

We now briefly review the results of a simplified theory of the interaction.[3] Solution for the real part of the dispersion relation shows that--for excitation of the space charge wave idler--the back-scattering frequency will be

$$\omega_s = \frac{\omega_o - \omega_p/\gamma}{(1 - \beta)} \approx 2\gamma^2\omega_o \tag{1}$$

where $\beta = V_b/c$, $\gamma = (1-\beta^2)^{-\frac{1}{2}}$, $\omega_p{}^2 = 4\pi n_b e^2/\gamma m$, and $\omega_o = 2\pi\beta c/\ell$, ℓ=period of the magnetostatic rippled field (superimposed as a rule upon a guiding magnetic field). (See Fig. 1) The backscattered wave will grow exponentially along the rippled zone (the "wiggler" or "undulator") in the small signal approximation, weak pump regime with growth rate

$$\mathrm{Im}\omega \sim (\omega_p/\gamma\omega_o)^{\frac{1}{2}}\,\omega_\perp \quad \text{or} \quad \Gamma = \text{gain/cm} = \omega_\perp\left(\frac{\omega_p\ell}{4\pi\gamma c^3}\right)^{\frac{1}{2}}, \tag{2}$$

where $\omega_\perp = eB_\perp/mc$ is related to the pump amplitude. [We note here that there is another collective mode on the beam--the cyclotron mode--for which Eq. (1) obtains with ω_p replaced by $\Omega_c = eB_{zo}/mc$. The cyclotron mode is a transverse wave; however, in situations where there is an inhomogeneity of the pump amplitude transverse to B_{zo}, there develops a space charge oscillation at Ω_c which is driven by the ponderomotive action of the pump and the scattered Raman wave, as in (1).] In Fig. 2 we show[3] how the gain of the space charge idler mode varies with the beat frequency between the pump and

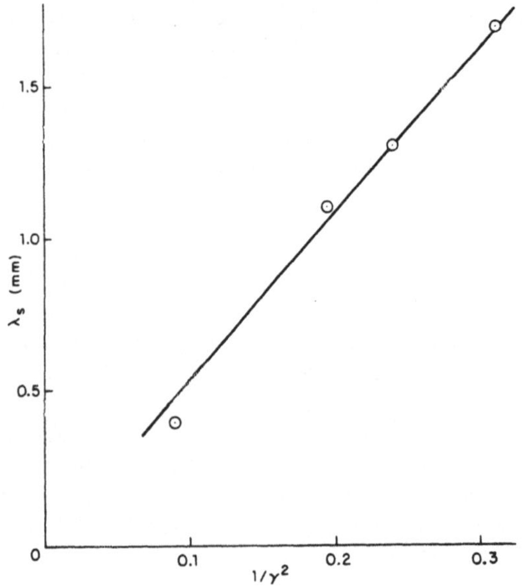

Fig. 1. Experimental data showing $\lambda_s \sim \ell/2\gamma^2$ where ℓ = 8 mm.

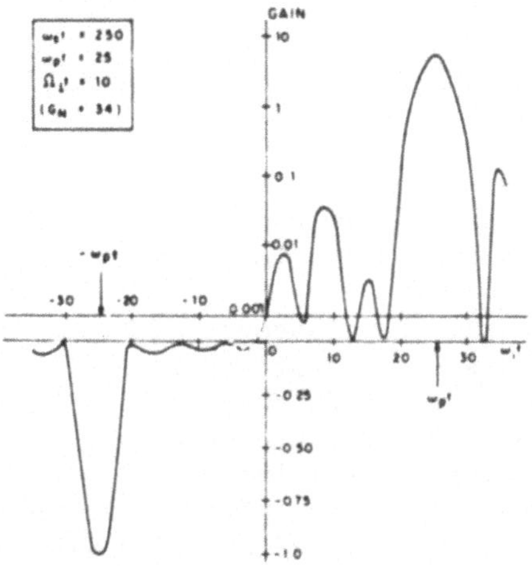

Fig. 2. Gain[3] of the Columbia-NRL laser experiment versus $\omega_i t$, where $\omega_i = \frac{\omega_s}{2\gamma} - \gamma\omega_o$.

signal waves. When this beat is at ω_p, the gain is very large, and we have the stimulated Raman emission (Stokes). If the beat is at $-\omega_p$, we have stimulated Raman absorption (anti-Stokes). Off-three-wave resonance, there is some gain due to the finite-length two-wave beat--which incidentally accounts for the lasing action of the first reported FEL.[4] In practice, however, the net gain for the laser system off-resonance is negative, owing to excitation of the higher-order sidebands: the strong intercavity scattered wave can act as a pump on any other wave. We note the three-wave resonance has a finite bandwidth given by

$$\Delta\omega_s \sim \mathrm{Im}\omega \tag{3}$$

which is characteristic of parametric processes.

The efficiency of the device will be bounded by pump-depletion, but nonlinear saturation of the idler (when the beam fluctuation amplitude $\delta\eta$ becomes ~ beam density, n) will intervene as a rule, in which case the efficiency is

$$\eta \sim \omega_p/2\omega_o\gamma \sim 5\% \tag{4}$$

in our case. We shall also distinguish between the efficiency of the laser and the amplifier; when one limits the efficiency to energy conversion to only a single mode, then

$$\eta \sim \left(\frac{\omega_p}{\gamma\omega_o}\right)^{\frac{1}{2}} \frac{\omega_\perp}{\gamma\omega_o} \sim 1\% \qquad (5)$$

in our case. Thus the laser has less efficiency than the amplifier; but this is usually the case in electronics, and laser oscillators have other desireable properties (e.g., frequency-stability).

More recent theories have uncovered the following interesting effects. Davidson[5] has found that the span of unstable wavenumbers is very broad for high amplitude pump and for high γ, high density electron beams. Thus one might expect high gain FEL Raman amplifiers to be troubled by parasitic oscillation, although they should be broad-band amplifiers. Another difficulty has surfaced when two-dimensional effects are considered numerically:[6] stimulated Raman forward-scattering appears at sufficiently high pump amplitude and sufficiently long wiggler as a low frequency, absolute instability; it can grow to high amplitude and disrupt the backscattering process. Thus we fully expect that implementation of FEL's as devices will not be immune to problems which have occurred in other high power systems.

We now comment upon the effects of finite beam energy spread upon gain. Excitation of the collective beam mode will occur only if the scattered wavelength exceeds the Debye distance on the beam. This leads to a requirement on the parallel beam energy spread $(\Delta\gamma/\gamma)_\parallel$

$$(\Delta\gamma/\gamma)_\parallel < \frac{1}{2}\frac{1}{\gamma}\left(\frac{\omega_p}{\omega_o}\right) . \qquad (6)$$

Should (6) not obtain, the gain of the system is reduced (approximately an order of magnitude, in cases to be discussed here) and becomes

$$\mathrm{Im}\omega \approx \frac{\omega_p}{\omega_s} \frac{\omega_\perp^2}{(k_o+k_s)^2} \frac{1}{V_{th}^2} , \qquad (7)$$

where $k_o=2\pi/\ell$, $k_s=\omega_s/c$ and V_{th} is the spread of electron energies $V_{th}=c(\Delta\gamma/\gamma)_\parallel \gamma^{-2}$, which is the stimulated Compton result.[7] Recently, McDermott has calculated the influence of beam energy spread on two- as well as three-wave lasers, and finds that the gain is always diminished; however if $(\Delta\gamma/\gamma)_\parallel \ll 1/N$, where N is the number of ripple periods of the wiggler, the thermal effect will cause no important change in gain. This suggests that the optimum system may be a comparatively short section of wiggler with high gain, in a laser cavity.

II. APPARATUS

In this section we briefly summarize some of the salient features of the electron beam and its power source. Here the noteworthy element is the high pulsed power-- $\simeq 10^{10}$w -- in the beam: Even at low efficiency, impressive power can be obtained from an FEL operating on this system.

Power is delivered to a simple diode from a capacitive energy-store (a Marx-Bank) via a pulse-forming transmission line. In our unit, the Marx stores 10kJ at 100kV, which is erected to 2MV. The energy is transmitted and stored in a dielectric ($\epsilon \approx 80$) coaxial line (20Ω) which is matched to the load; at the end of the line, a pressurized switch connects to the diode, which is in parallel with a variable disk resistor (the latter permits a wide choice of diode geometries). This line also minimizes the effect of diode closure (change of diode impedence due to release of plasma), which is appreciable at 0.1μsec. Shown in Fig. 3 is a layout of the machine (Physics International Pulserad 220) and diode, together with the performance waveforms. This installation permits adjustment to diode voltage flatness < 2%.

The electron beam is emitted in foilless-diode geometry. The gross properties of this beam can be observed via signals from a

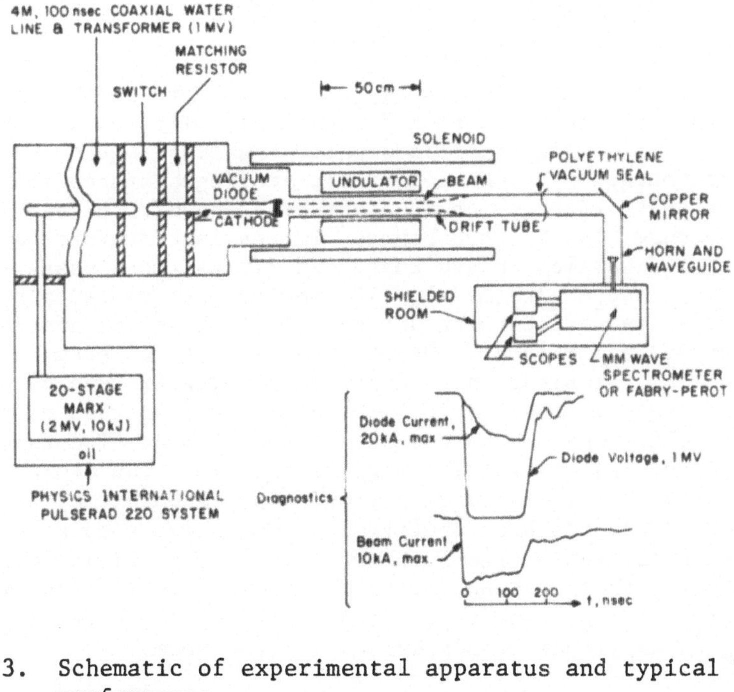

Fig. 3. Schematic of experimental apparatus and typical beam performance.

Faraday cup, calorimeter, witness plate patterns, microwave noise, and x-rays. However, detailed studies of beam emissivity and/or energy distribution have lagged.

The random spread ($\langle\theta^2\rangle^{1/2}$) of electron trajectories in the beam defines an emittance $\varepsilon=\pi R\sqrt{\langle\theta\rangle^2}$ where R is the beam half-width; this translates into an effective parallel energy spread $(\Delta\gamma/\gamma)_{\parallel} \sim \frac{1}{2}(\gamma\varepsilon/\pi R)^2$. It is not unreasonable to find $\sqrt{\langle\theta^2\rangle} < 1/10$ and therefore $\varepsilon \sim 10\pi$ mrad-cm for intense beams, in which case $(\Delta\gamma/\gamma)_{\parallel} \sim 4\%$ and SRS is possible down to $\lambda_s \sim 100\mu$, i.e., $\lambda_s > \lambda_{Debye} \sim (\Delta\gamma/\gamma)_{\parallel} c/\omega_p\gamma$. Diode geometry plays an important role here.

In an unneutralized beam, rotation is necessary to balance the radial forces,[8] and beam rotation has been observed.[9] In addition to beam emittance effects, the beam rotation and space charge will spread the beam parallel energy; these effects increase linearly in a laminar fashion across the finite beam thickness δ (note $\delta \simeq \lambda_s$), but are in opposite tendencies:

$$\left(\frac{\Delta\gamma}{\gamma}\right)_{\parallel} \approx \frac{\omega_p^2\delta^2}{4c^2}\left(1 - \omega_p^2/\omega_c^2\right) . \tag{8}$$

If we choose $\omega_p \simeq \omega_c$ on the average inside the beam, then the net contribution from this effect will be $\sim 10^{-2}\omega_p^2\delta^2/c^2$ which, even for dense beams, is only ~0.1%. (Note for beam equilibrium, $\omega_p/\omega_c \leqq \gamma/\sqrt{2}$).

Another source of beam energy spread is more troublesome: the shear in particle quiver motion ($\Delta\tilde{\beta}_\perp = \Delta\tilde{V}_\perp/c$) owing to the radial dependence of the wiggler field amplitude acting together with the beam space charge. The characteristic scaling distance for this radial field is $\sim \ell/2\pi$ which is of order of beam thickness in the case of intense, low γ, thin beams. Increasing ℓ unfortunately means increasing the wavelength and therefore diffraction losses. If we take $(\Delta\gamma/\gamma)_\perp = \gamma^2\beta_\perp\Delta\beta_\perp/2 \equiv (\Delta\gamma/\gamma)_{\parallel} \sim 1\%$ then we must restrict $B_\perp < 1000G$ and accept a tradeoff between diffraction loss, gain, and beam energy-spread in the laser. Deleterious effects of beam heating due to the initial transient upon entering the wiggler have been predicted;[10] experimental observations--using lateral witness papers[11] show: (1) electrons are thrown out of the beam continuously by the field ripples; (2) the initial wiggler transient can de-focus the beam radially, and (3) strong cyclotroning components of the motion are set-up. This has been confirmed by numerical simulations of the particle orbits, which clearly show the effect of finite orbit size and beam space charge in enhancing the spread of electron axial velocities. Thus we reserve judgement upon proposed FEL's combining high wiggler amplitude with a dense electron beam.

Among the new experiments planned at Columbia in the next two years is a detailed study of beam energy spread, to be measured in a Thomson-backscattering experiment. A powerful 10.6μ CO_2 laser signal can be scattered from the electron beam, and its backscattered wave ($\lambda_s \sim \frac{1}{2}\mu$) will show a Doppler-broadened profile which yields the spread in parallel electron velocities.

Looking downstream, we require many devices capable of detecting millimeter and submillimeter signals. The latter are conveyed via "light pipe" (over-moded waveguide) to a shielded room, wherein are located Schottky-barrier crystal and/or pyro-electric detectors. Analysis of scattered wavelength is recorded using either a grating spectrometer[12] (0.5-5.0 mm) and/or a Fabry-Perot interferometer. In the latter, the "mirrors" are actually plane reflecting mesh. These devices, having resolving power ~50-100, also permit observation of spectral width.

III. A--Superradiant Experiments

Motivated in part by theoretical work by Sprangle et al.[13] and Manheimer and Ott,[14] Efthimion and Schlesinger[15] undertook to study radiation produced by the interaction of an electron beam with a rippled magnetic field. Interest in this subject dates back to Motz,[16] and of course, Phillips,[17] who actually developed a device (the "Ubitron") which used a tenuous, non-relativistic beam. To interpret their experiments, Efthimion and Schlesinger investigated the spectrum of radiation, which results from the intersection of the doppler-shifted negative-energy space charge and cyclotron waves with the wave-guide modes of the drift tube containing the beam (only the negative energy--or stokes--modes are unstable; the positive energy, or anti-stokes, modes are stable). This radiation appeared in the 10-30 Ghz spectral region; analysis is complicated by the multiplicity of waveguide modes and the presence of beam noise.

Operation at short undulator period simplified the study, and permitted generation of multi-megawatt radiation in the millimeter spectral region.[18] Previously, Schlesinger and Granatstein[19] had also discovered intense submillimeter radiation produced by stimulated backscattering from a strong microwave pump field (the latter produced by a cyclotron maser process). Thus a new interpretation gained acceptance: the role of the rippled field was to produce a strong (~100MW/cm^2) pump wave in the electron frame; the ponderomotive force of the beat wave produced by the pump and the growing scattered wave bunched the space-charge of the beam--whereupon the pump wave scattered from the periodic space charge fluctuations to reinforce the scattered wave, etc. Noise present on the beam is amplified, since the Stimulated Raman Backscattering process is a convective instability.

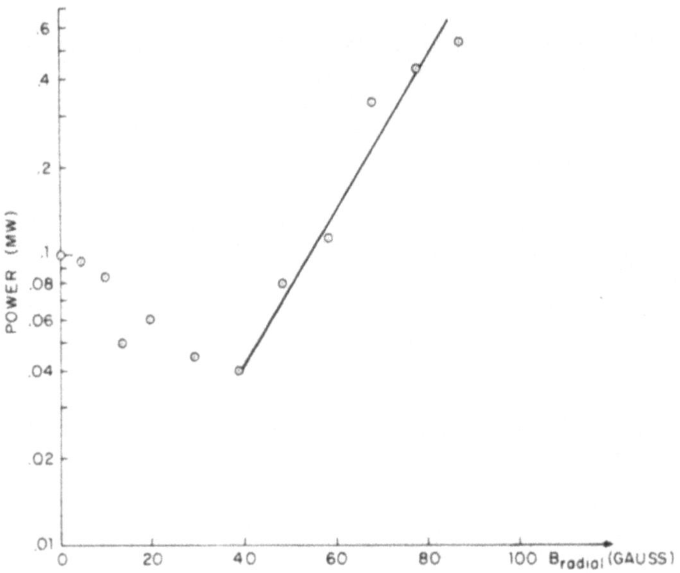

Fig. 4. Dependence [21] of superradiant stimulated Raman Backscattering upon pump amplitude.

We now turn to the first detailed study of this process, done by Gilgenbach et al.[20,21] who used a short-pulse (15nsec), moderate-density electron beam ($\gamma \approx 2$, $\omega_p/2\pi \approx 2$ Ghz). First, we examine the dependence of power upon pump-amplitude (the strength of the rippled field was controlled by designing pulsed wigglers). Fig. 4 shows that--after a threshold is achieved--the signal increases exponentially, with coefficient which is linear in pump (wiggler) amplitude. The "threshold" is characteristic of the three-wave parametric process, and is more related to the wave-coupling- rather than the dissipative-process. Using a spectrometer, two modes were identified: that due to scattering from the space charge mode and that due to scattering from the cyclotron mode (Fig. 5). The latter arises from the finite radial geometry in the experiment, and--in spite of strong cyclotron damping [21]--accounts for the lion's share of radiation in view of the reduced beam density: The growth rate of the cyclotron mode [14] varies as ω_p rather than $\omega_p^{1/2}$, and its damping rate varies as ω_p^2. By moving the intersection of the beam line with the waveguide modes near the light-line, single "resonances" were produced, and by variation of γ and ω_c, full identification of the mode-coupling processes was made. By increasing the pump amplitude, the power of the cyclotron mode was caused to saturate; investigation of the dependence of scattered power upon undulator length revealed the convective nature of the instability.[22] Saturation may have been caused by beam-dynamical effects.

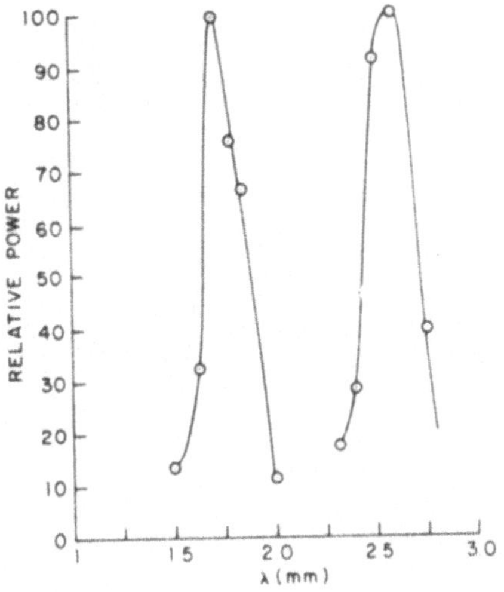

Fig. 5. Superradiant power spectrum[20] (arbitrary scale): ℓ = 8mm, $\gamma \sim 2$; the peak on left is scattering from the plasma (ω_p) idler and the peak on the right is scattering from the cyclotron idler (Ω_c).

Whenever, in the electron frame of motion, the cyclotron frequency of the electron, eB_{zo}/mc, becomes equal to the pump wave frequency $\gamma\omega_o$ or the scattered frequency $\omega_s=\gamma\omega_o-\omega_p$, the efficiency of 3-wave scattering is enhanced.[23] To demonstrate this, we constructed a helical wiggler (ℓ=18mm) positioned over a small low-current (γ=2, I=1kA, dia.=½cm) pencil electron beam. Two resonances in scattered power were found [Fig. 6] when the electron gyration was in the same sense as the helical field lines. When $\underline{B}_{zo}$ is reversed (dotted curve), the helical lines are opposed to the electron gyration and no magnetoresonant effect is obtained. One should note that when the pump wave is magnetoresonant, rapid conversion of $V_{\parallel}$ to $V_{\perp}$ motion occurs, and beam electrons can be lost to the walls of the drift-tube.

There are many limitations in superradiant operation: one of these is that power output is dependent upon noise present at the input of the beam into the wiggler. Another has to do with the bandwidth of the radiation. Theory shows that the spectrum of unstable modes broadens as the gain (pump amplitude) is increased (eq. 3)--this was observed experimentally (Fig. 7). In the Raman regime, the line broadening mechanism should be homogeneous.

III. B--Laser Experiments

In order to solve some problems listed above, it is desireable to encourage the system to oscillate in a quasi-optical Fabry-Perot

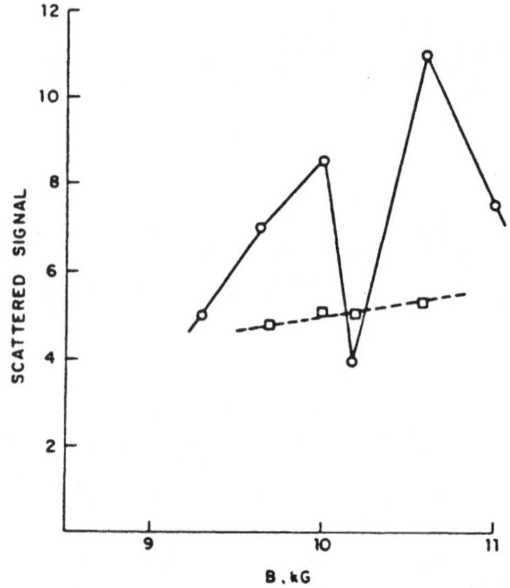

Fig. 6. Magnetoresonant superradiant signal at $\lambda_s \approx 4$mm, $\gamma_{\parallel} \sim 1.8$. The right-hand peak corresponds to $\Omega_c = \gamma\omega_o$ and the left-hand peak corresponds to $\Omega_c = \omega_s = \gamma\omega_o - \omega_p$ in the electron frame. Data courtesy of D.S. Birkett.

resonator. Pump amplitude can be kept low, while the power is high since feedback painlessly extends the effective length of the wiggler. The oscillator also involves features such as frequency-stability and

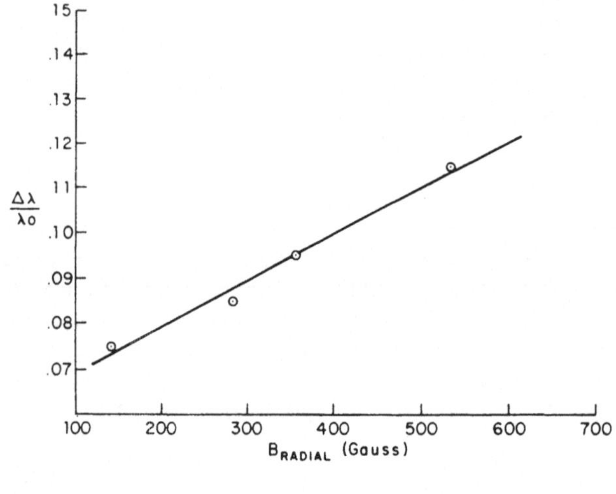

Fig. 7. Dependence[20] of superradiant power spectral width upon pump amplitude.

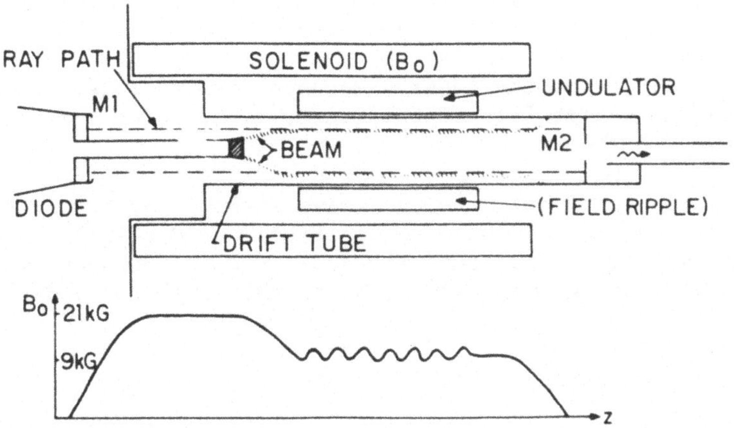

Fig. 8. Layout of the Columbia-NRL laser experiment.

narrow linewidth, very desireable properties of a source of tuneable radiation in a spectral zone where more conventional sources are lacking.

The apparatus of the first "collective" FEL,[1] tested at NRL by a group from Columbia and NRL, is shown in Fig. 8. The wiggler length (60 periods, 50 cm) was chosen to be on the order of the Rayleigh range $\pi a^2/\lambda_s$, where a is the thickness of the optical beam waist defined by the annular space between the cathode and the anode (~8mm). Radial expansion of the beam via magnetic profiling not only places the beam in the ray-path of the mirrors, but also cools the electron beam via conservation of the adiabatic invariant $\gamma T_\perp/B_{zo}$, and permits the wiggler amplitude to increase gradually along the electron trajectory. When the flat mirrors were precisely aligned (<0.1°) lasing was observed.

Analysis of the laser signal[1,3] showed several things:

a) The wavelength was 0.4 mm, corresponding to $\gamma=3$ as measured from the Diode voltage.

b) The spectral width narrowed from several percent to ≈2% as lasing set in.

c) The wavelength of the signal held constant despite sizeable fluctuation in diode voltage; this is due to the fact that only signal having the same frequency as the ambient cavity radiation is regenerated.

d) The power was estimated at ~1 MW and showed no evidence of saturation.

e) As the diode voltage falls, the laser power falls abruptly due to stimulated absorption (see Figs. 2 and 10).

f) Analysis of the laser waveshape permits determination of the amplification factor ΓL: in this case, $\Gamma L \approx 2$, which compares satisfactorily with theory (eq. 2).

In this experiment, the pump wave (ℓ=8mm) was far from magnetoresonance, and the guiding magnetic field had negligible effect on the laser gain (eq. 2). The high density electron beam ($\omega_p/2\pi \sim 8$ Ghz) provided damping against excitation of the cyclotron modes.

IV. LASER EXPERIMENTS--INTERPRETATION

In conventional lasers, we deal with systems where there is a narrow, stable linewidth and excitation of just one, or a few, cavity modes. In our FEL, the bandwidth for positive gain is enormous by atomic scale, and many cavity modes are excited. However--the laser will regenerate radiation which lies within only the narrow range necessary to maintain three-wave resonance, and therefore the linewidth of the laser will be narrower than the superradiant system, which is very sensitive to fluctuations in beam energy ($\Delta\lambda/\lambda \sim 2\Delta\gamma/\gamma$).

Amplification of the cavity radiation can be interrupted if the energy of the electron beam drifts off-resonance (see Fig. 2); in this event, the beat wave between the pump and the signal is not a normal mode of the system. Should the electron energy wander too far--by an amount $\Delta\gamma \sim \omega_p/\omega_o$, then the anti-stokes mode is excited and the cavity radiation is rapidly quenched by stimulated absorption--observed on the trailing edge of the laser signal, Fig. 10. It is also easy to see that the strong cavity radiation can itself act as a pump on other electromagnetic waves. While these "sidebands" are "parasitic", at least one may be useful: that which causes further upconversion of scattered frequency by a multiple of $8\gamma^4$ times the pump frequency.[22] Parasitic modes will cause the gain curve of Fig. 2 to be slightly negative for all energies off the stokes three-wave resonance; otherwise there might be residual net gain off-resonance due to the much weaker two-wave beat process.

One way to stabilize the laser frequency, improve the coherence, and eliminate the mirrors (one of which is mounted on the cathode) is to incorporate a distributed feedback element in the laser cavity (Fig. 9). Here, carefully machined quarter-wave deep groves on a grounded inner conductor interact with radiation having λ_s=2d. Radiation is scattered forward and backward by this "grating", while the cavity length is effectively a multiple of $\lambda_s/2$ so that only one

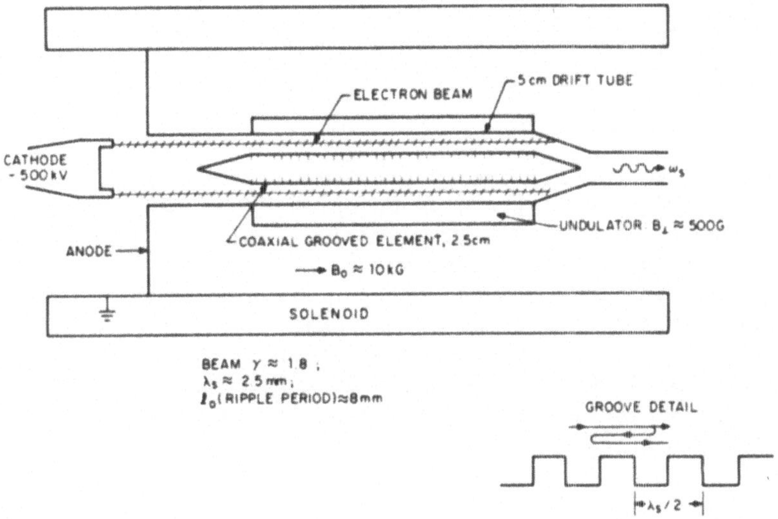

Fig. 9. Layout of an FEL using distributed Feedback (experiment by (P. Divorkis and D.S. Birkett).

axial mode is excited. Distributive feedback experiments on CH_3F lasers have proved successful,[24] and in the summer of 1979, Birkett and Divorkis at Columbia were successful in incorporating distributed feedback on our FEL and observed line-narrowing. Clearly, the wavelength of the FEL could be changed only by replacing the feedback element.

The efficiency of the first Columbia-NRL laser was thought[1] to be about an order of magnitude shy of the theoretical value (eq. 5); however, recently Jackson et al.[25] have studied the energy distribution of the beam electrons using numerical methods, and have concluded that only the inner core of the beam was cold enough to sustain stimulated Raman emission--in which case the experimental efficiency result is more nearly in line with theory (note the laser gain is very insensitive to beam density: $n^{1/4}$).

To obtain high efficiency[26] one must consider an FEL amplifier (one or more stages following the laser oscillator) driven by a strong pump, the period of which varies along the wiggler so that the frequency of the system is a constant. As we have seen, it is very likely a high gain FEL will have very broad bandwidth, and may suffer from self-oscillation and instability; but this can be cured by adequate isolation. The high gain, high efficiency system would find best application as a single pass amplifier stage following a low-gain laser oscillator stage located in the same electron beam line.

V. RECENT EXPERIMENTS

A. Amplifier Experiments

In order to compare theory more closely with experiment, the Columbia group (principally D. McDermott) is collaborating with H. Fetterman of Lincoln Labs to construct and perform an experiment in which a strong coherent test wave (10-50 kW) is injected into the input of the FEL; this serves to replace noise as the input, and permits quantitative measurement of the Gain, saturation effects, and coupling loss. The coherent source for this experiment is a CH_3F laser which is pumped by a 50 MW, 9.6μ pulse from a CO_2 TEA laser. The CH_3F system oscillates at 496μ and--if the C^{12} isotope is replaced by C^{13}--at 1200μ.

B. Recent Laser Experiments

The new Pulserad 220 facility at Columbia permits operation of a collective FEL--very similar to the original design[1]--but with a long (150 nsec-about 10 bounces of cavity radiation), well-regulated ($\Delta V/V<2\%$) diode voltage pulse. Both superradiant (~100kW) and laser (~ MW) operation at $\frac{1}{2}<\lambda_s<1$mm is under investigation by D.S. Birkett at the Columbia laboratory. Lasing is being studied in a long (~150 cm) rather low Q cavity (average mirror reflectivity ~85%) from which power is coupled by diffraction. Typical pulse shapes take the form of a staircase structure with 10-20 nsec steps (the cavity bounce time) as shown in two examples in Fig. 10. In Fig. 10(B) the diode voltage is rather well-regulated, and Fig. 10(A) shows an example where the beam energy varies in time. Superradiant studies showed that without field profiling, radiation would be emitted from the system for only ~20-50 nsec, whereas the profiled electron beam lases over the entire accelerator pulse. This is a consequence of diode-closure effects, which are sensitive to beam geometry, and which spoil the beam emissivity.

VI. DIRECTION OF FUTURE WORK

The high-gain laser we have discussed is approximately the size of conventional lasers (i.e., 1-2 m), however it is attached to a complex electron beam-generating source. Research on future FEL beam sources and theory involve the following:

1) Attention to repetitive pulsing or CW operation: here, the development of the compact Microtron source, to power a tuneable FEL at Bell Laboratories[27] is noteworthy. Since the average beam current available is quite limited (≈100 mA), use of a pencil-beam

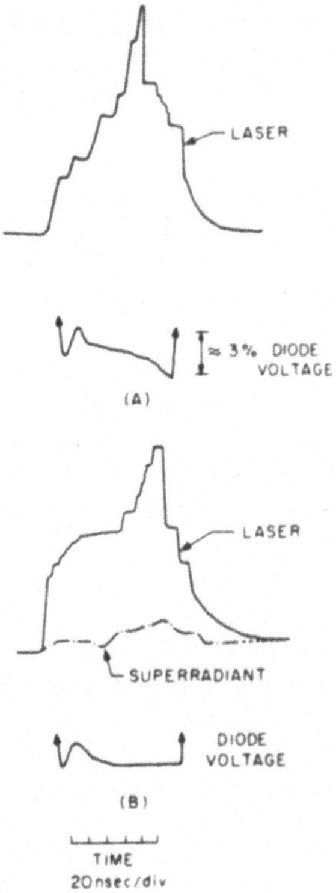

Fig. 10. Laser signals from a long-pulse FEL, $\gamma \sim 2.5$. Diode voltage waveform is shown below on an expanded scale ($V_D \sim 900kV$).

is a useful way to enhance the current density. Operation in the range $\lambda_s \sim 100\text{-}400\mu$ using Energy ~20-9 MeV is foreseen. A wiggler having ℓ=20cm, length=10m will permit 40%-70% gain per pass. CW power of several watts is expected, and pulse power should be several hundred kilowatts.

2) If a dielectric is introduced into the drift tube, then the scattered frequency becomes

$$\omega_s = \frac{\omega_o - \omega_p/\gamma}{(1 - \beta n')} \qquad (9)$$

where n' is the effective refractive index of the dielectric-loaded waveguide drift tube. If the factor in the denominator of eq. (9) above becomes ($\beta n'$-1), this is the Cerenkov-Raman effect.[28] Thus beam energy can be lowered if a dielectric is inserted into the region interior or exterior to the beam.

3) With regard to beam energy spread, in addition to studies already described, development of a hot-cathode source is desireable, as well as theoretical insight into preparation of cold (Brillouin flow) beams using appropriate diode geometry.

4) Simulation of electron beam and FEL operation in 2D, realistic beam geometries[6] will help isolate problems inherent in operation of the FEL amplifier in the efficient, strong-pump regime. Here, design of an appropriate wiggler becomes essential.

5) Whereas the SRS version of the FEL operates with exponential gain, two-wave gain also becomes large if the beam is dense, the pump is strong, and the length of the wiggler is chosen appropriately.[29] Exploration of high gain, two wave versions is therefore essential, incorporating beam thermal spread into analysis.

6) Development of novel beam recycling systems[30] to improve efficiency and pulsed induction Linacs (NRL) is proceeding. Also, whereas theory suggests that FEL efficiency is potentially high, experimental work is required to demonstrate this in practical application.

A detailed review of the high-power sub-millimeter FEL will appear shortly.[31]

REFERENCES

1. McDermott, D.B., Marshall, T.C., Schlesinger, S.P., Parker, R.K., and Granatstein, V.L. (1978). Phys. Rev. Lett. 41, 1368.
2. DeTemple, T.A., (1979). In "Infrared and Submillimeter Waves" (K.J. Button, Ed.) vol. 1, p. 129. Academic Press, N.Y.
3. McDermott, D.B., and Marshall, T.C. (1980). In Physics of Quantum Electronics, ed. Jacobs, Pilloff, Sargent, Scully, and Spitzer (Addison-Wesley, Reading, MA.), Vol. 7, p. 509.
4. Deacon, D.A.G., Elias, L.R., Madey, J.M.J., Ramian, G.J., Schwettman, H.A., and Smith, T.I. (1977). Phys. Rev. Lett. 38, 892.
5. Davidson, R.C. and Uhm, H.S., (1979). Bull. Am. Phys. Soc. 24, 1066.
6. Cary, J.R. and Kwan, T.J., (1979). Bull. Am. Phys. Soc. 24, 1066.
7. Hasegawa, A. (1978). Bell System Tech. J. 57, 3069.
8. Davidson, R.C. (1974). "Theory of Nonneutral Plasmas." W.A. Benjamin, Inc., Reading, MA., p. 36.
9. Talmadge, S, Marshall, T.C., and Schlesinger, S.P. (1977). Phys. Fluids 20, 974.
10. Hasegawa, A., Mima, K., Sprangle, P., Szu, H.H., and Granatstein, V.L., (1976). Appl. Phys. Lett. 29, 542.

11. Gilgenbach, R.M., McDermott, D.B., and Marshall, T.C., (1978). Rev. Sci. Instr. 49, 1098.
12. Pasour, J.A., and Schlesinger, S.P. (1977). Rev. Sci. Instr. 48, 1355.
13. Sprangle, P., Granatstein, V.L., and Baker, L. (1975). Phys. Rev. A. 12, 1697
14. Mannheimer, W.M., and Ott, E. (1974). Phys.Fluids 17, 1413.
15. Efthimion, P.C., and Schlesinger, S.P. (1977). Phys. Rev. A, 16, 633.
16. Motz, H. (1951). J. Appl. Phys. 22, 527.
17. Phillips, R.N. (1960). IRE Trans. Elec. Devices ED-7, 231.
18. Marshall, T.C., Talmadge, S., and Efthimion, P. (1977). Appl. Phys. Lett. 31, 320.
19. Granatstein, V.L., Schlesinger, S.P., Herndon, M., Parker, R.K., and Pasour, J.A. (1977). Appl. Phys. Lett. 30, 384.
20. Gilgenbach, R.M., Marshall, T.C., and Schlesinger, S.P. (1979). Phys. Fluids 22, 971.
21. Gilgenbach, R.M., Marshall, T.C., and Schlesinger, S.P. (1979). Phys. Fluids 22, 1219.
22. McDermott, D.B., Marshall, T.C., and Schlesinger, S.P. (1978). Comments Plasma Phys. and Contr. Fusion 3, 165.
23. Chen,Y.G., Leheny , R.F., and Marshall, T.C., (1965). Phys. Rev. Lett. 15, 184.
24. Kneubühl, F.K., reported in Lasers '79 Int'l Conf., Orlando, Fla., December 1979, p. 812.
25. Jackson, R., Parker, R., Efthimion, P., Granatstein, V., Sprangle, P., and Smith R. (1979). Bull. Am. Phys. Soc. 24, 1077.
26. Sprangle, P., Physics of Quantum Electronics, S. F. Jacobs, H.S. Pilloff, M. Sargent, M. O. Scully, R. Spitzer, eds. Addision-Wesley, Reading, MA. (1980), Vol. 7, p. 207.
27. Shaw, E.D., and Patel, C.K.N., Physics of Quantum Electronics, S.F. Jacobs, H.S. Pilloff, M. Sargent, M.O. Scully, R. Spitzer, eds. Addison-Wesley, Reading, MA. (1980), vol. 7, p. 665.
28. Walsh, J.E., Physics of Quantum Electronics, S.F. Jacobs, H.S. Pilliff, M. Sargent, M.O. Scully, R. Spitzer, eds. Addison-Wesley, Reading, MA (1980), Vol. 7, p. 255 and 301.
29. Johnston, S. (1979). Bull. Am. Phys. Soc. 24, 1066.
30. Brau, C.A., and Cooper, R.K., Physics of Quantum Electronics, S.F. Jacobs, H.S. Pilloff, M. Sargent, M.O. Scully, R. Spitzer, eds. Addision-Wesley, Reading, MA. (1980), Vol. 7, p. 647.
31. Marshall, T.C., Schlesinger, S.P., and McDermott, D.B. in Advances in Electronics and Electron Physics, C. Marton, ed., vol. 53, Academic Press, N.Y. (1981).

ACKNOWLEDGEMENTS

Research has been supported by AFOSR and ONR. The author expresses his appreciation to his colleagues at Columbia University.

A FREE ELECTRON LASER EXPERIMENT

H. Boehmer, M. Zales Caponi and J. Munch

TRW DSSG
One Space Park
Redondo Beach, CA 90278

INTRODUCTION

A Free Electron Laser (FEL) device is a source of high power, coherent electromagnetic radiation that can be tuned to operate from cm to visible wavelengths. The radiation is generated by scattering an external electromagnetic wave or "pump" from an intense relativistic electron beam. The coherence of the radiation is due in part to the axial self-bunching of the beam in the presence of the pump. The FEL operates in different regimes depending on the magnitude of the macroscopic system parameters. For extremely intense electron beams collective effects may play a dominant role in the dynamics of the system [collective regime]. This is to be contrasted for example with the case of extremely energetic, but low intensity electron beams where single particle effect dominate the behavior [Compton regime]. Other macroscopic parameters that determine the regime of operation include the electron beam energy spread (ΔE), the pump amplitude (I_p) and wavelength (λ_w), and the length (L) of the system.

We have built an experiment that allows the characterization of the FEL in different regimes. In addition, we have developed a theoretical model that describes the FEL interaction in a self consistent manner for <u>any</u> set of values of the macroscopic parameters.[1,2] Thus by driving the system at those particular modes of oscillation that are most unstable, theoretical predictions can be experimentally verified and the FEL gain and efficiency can be optimized.

Recent experiments at Stanford[3,4] and Columbia/NRL[5,6] have shown the feasibility of generating infrared,[3,4] (10.6 μ amplifier and 3.4 μ oscillator) and millimeter[5,6] radiation by the free electron laser mechanism. However, each experiment was operated in a different FEL regime, not driven into saturation and at low efficiencies ($\eta \leq 10^{-3}$). The experiment reported in this paper can operate in both the Compton regime (as the Stanford experiment) and the collective regime (as the Columbia/NRL experiment). It operates in the microwave region and unlike the UBITRON and the Columbia/NRL experiment, it utilizes a helical wiggler as the electromagnetic pump and a solid electron beam of high repetition rate and long pulse length which facilitates diagnostics.

General Theory

The details of the physical mechanism associated with the FEL interaction are a function of the system parameters that in turn determine the FEL regime of operation. However, the basic mechanism is the same in all regimes. In the presence of the pump, the streaming (say in the z direction) beam of electrons oscillates with a transverse velocity $v_{0\perp}$ that is proportional to and perpendicular to the pump field. A longitudinal (along the z axis) low frequency Lorentz force is induced by the coupling of the oscillatory velocity $v_{0\perp}$ and the initial high frequency radiation. The induced longitudinal force, also called ponderomotive or radiation pressure force, produces an axial bunching of the beam or a density perturbation. For a sufficiently intense electron beam, collective effects become important and the self-space charge potential set up by a density perturbation influences nearby density perturbations and vice versa creating a periodic oscillating density perturbation or density wave. In any case, the density perturbation couples with the oscillatory velocity $v_{0\perp}$ and induces a current that in turn generates high frequency electromagnetic radiation (stimulated backscattering) that reinforces the initial radiation. The process is then unstable and results in the stimulated emission of high frequency radiation. The intensity of the emission increases with the pump amplitude and it is advantageous to use an external ripple magnetic field (zero frequency pump) instead of an external electromagnetic wave. For example, a 1 kG ripple magnetic field is equivalent to a 300 MW/cm^2, cw external source.

To calculate the natural modes of this system we have developed a self-consistent formulation[2,7]. Within the assumptions of a one spatial dimension variation and small signal (the energy in the radiation fields is smaller than the energy in the pump field) the model is valid for a multimode system and any set of microscopic parameters. Thus, the results are not limited to confined regimes of operation. The details of this analysis have been summarized in Ref. 2 and Ref. 7. Essentially, it is based in the linearization of Vlasov Maxwell's equations and includes space charge effects and the presence of a uniform external guide field. A set of differential equations is then obtained that describes the time-space evolution of the electrostatic potential ϕ and the right and left circularly polarized vector potential A_+ and A_-.

By appropiately Fourier transforming in time and Laplace transforming in space for a given set of periodic boundary conditions (e.g., absorbing boundaries) the differential equations are transformed in an infinite set of coupled algebraic equations. The set is reduced to one equation for the normal modes $[k = k(\omega)]$ or dispersion relation by additional simplifying assumptions as follows. The pump field (wiggler) is taken to be circularly polarized and the initial perpendicular momentum of the electron beam is assumed to be only that created by the pump. In addition, for small gain lengths (GL < 1), the field spectrum is taken to be approximately constant for modes k near the most unstable modes (negligible otherwise).

The calculated dispersion relation has a complicated form in terms of energy integrals of the distribution function and the external macroscopic parameters ($B\omega$, λ_w and the length L of the system). This relation can be numerically solved to find the value of the FEL gain and wavelength for a given frequency and electron distribution function. Note that due to the electron medium the relation $k_r = \omega_r/c$ is no longer valid. In what follows we describe some of the properties of the obtained dispersion relation and illustrate some of the characteristics by: 1) showing some of the results obtained from its numerical solution; 2) writing down the simplified form obtained for the case of an infinite length, cold beam situation.

The complete dispersion relation can be schematically represented as:

$$EL \times EM = CT \tag{1}$$

where EL is the well known electrostatic dispersion relation, that for a cold electron beam is simplified to

$$EL = 1 - \frac{\omega_p^2}{\gamma^3(\omega - kV_b)^2} \tag{2}$$

where ω_p is the plasma frequency, $\omega_p^2 = 4\pi\, q^2 n/m$. EM is the electromagnetic dispertion relation in the presence of an electron medium interacting with a wiggler field (this includes the ponderomotive effects) and CT is a coupling term. Depending on the size of the macroscopic parameters EL $\lessgtr$ 1. This determines the characteristics of different FEL regimes. For example if EL $\simeq$ 1 and CT << 1,

Equation (1) is satisfied by EM $\simeq$ 0. In this case the "collective effects" that give raise to "plasma waves" are negligible $[(\omega_p/\gamma^{3/2}) << (\omega - kV_b)]$ and the system can be described with a single particle picture (strong pump regime if GL > 1 or Compton regime if GL < 1).

For EL $\simeq$ 0, the collective effects become dominant and Equation (1) describes the Raman regime. In addition, depending on the length of the system and the electron beam velocity relative to phase velocity of the electrostatic wave and/or the ponderomotive potential $(\omega/(k_\omega + k_r)$, where k_r = radiation wavevector), the characteristics of the different regimes change. Figures 1 and 2 for example, show the variation of the maximum gain with density for different values of wiggler magnetic field amplitude and wavelength, electron beam energy and system length. The behavior is extremely sensitive to the choice of macroscopic parameters. It is interesting to notice the oscillatory behavior in the "RFL" regime (Raman, finite length).

The dependence of the output frequency with beam energy and density at maximum gain is illustrated in Figure 3. The macroscopic parameters used in this figure are quite different than those of Figures 1 and 2 and correspond to experimental parameters for a microwave system now in operation at TRW and whose characteristics will be described in next section. The points are experimental values.

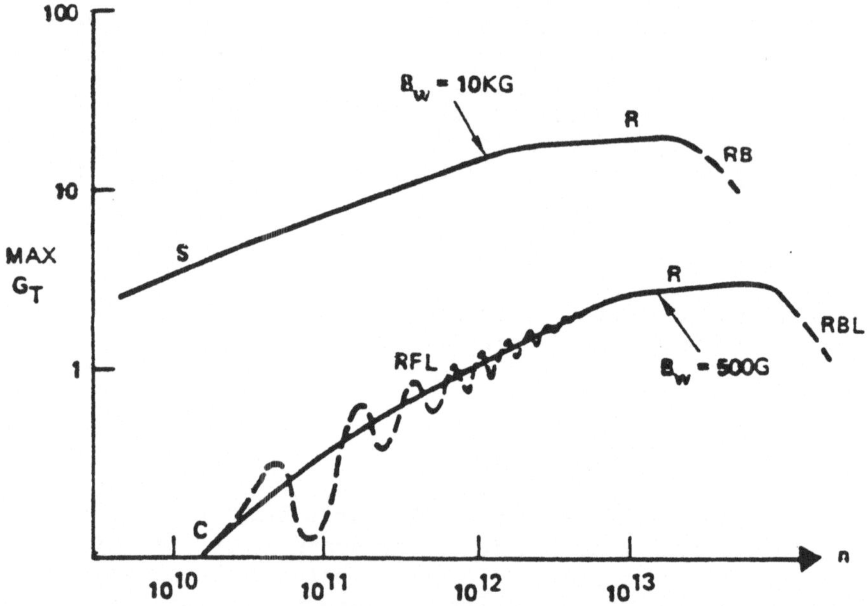

Figure 1. Total Maximum Gain vs Density for Different Pump Amplitudes with a Cold Electron Beam Equilibrium γ = 100, L = 10 m, λ_0 = 2 cm.

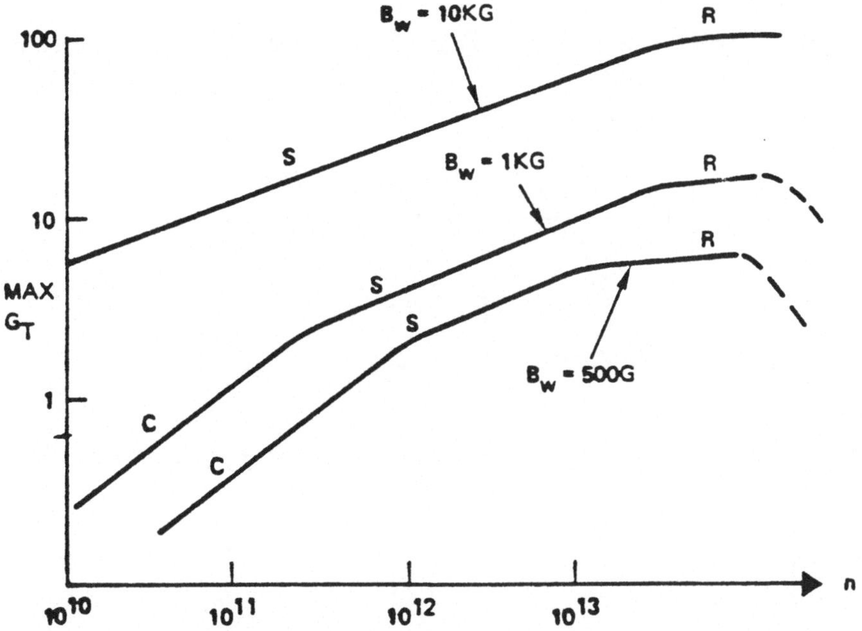

Figure 2: Total Maximum Gain vs Density for Different Pump Amplitudes with a Cold Electron Beam Equilibrium $\gamma = 10$, $L = 1m$, $\lambda_0 = 2$ cm.

The frequency of radiation for maximum gain and a cold beam is "approximately (but not <u>exactly</u>) given by the formula:

$$\omega = \frac{2\pi c}{\lambda_w} \beta_z \gamma^2 \left(1 - \frac{\omega_p \lambda_w}{2\pi\gamma^{3/2}\beta c}\right)\left[1 + \beta_z \left(1 - \frac{\omega_{cef}^2 \lambda_w^2}{2\pi\ c^2}\right)\right]\left(1 + \frac{b_w^2}{k_w^2}\right) \quad (3)$$

where: $\beta_z = V_b/c$,

$$\omega_{cef}^2 = \frac{\omega_{co}}{\beta^2\ \gamma^2[1 - (\omega_p \lambda_w)/2\pi\gamma^{3/2}\ \beta_z c)]^2}$$

ω_{co} is the waveguide cut-off frequency. The cut-off frquency correction is of impontance only for either mm or microwave radiation and $b_w = qB_w/mc$ with B_w = wiggler field amplitude.

In addition for a high gain system (GL > 1) a cold beam

$\left(\frac{\Delta p}{p\gamma^2} < \frac{\omega}{k_r V_b} - 1 \right)$ and no guiding magnetic field the dispersion relation can be exactly written as:

$$D(\omega,k) = \left[1 + \frac{\omega_p^2}{\gamma^3(\omega - kV_b)^2} \left(1 + \frac{b_w^2}{k_w^2 c^2 \alpha} \right) \right] \tag{4}$$

$$+ \frac{b_w^2}{2k_w^2 c^2 \alpha} \frac{\omega_p^2 \, \varepsilon^{em}(k)}{\gamma^3(\omega - kV_b)^2} \left[\frac{1}{\varepsilon^{em}(k_+)} + \frac{1}{\varepsilon^{em}(k_-)} \right]$$

where $\varepsilon^{em}(k) = \frac{\omega^2}{k^2c^2} - 1 - \frac{\omega_p^2}{\gamma k^2 c^2}$ and $k_\pm = k \pm k_w$ and $\alpha = \frac{1 + \omega_p^2}{\gamma k_w^2 c^2}$.

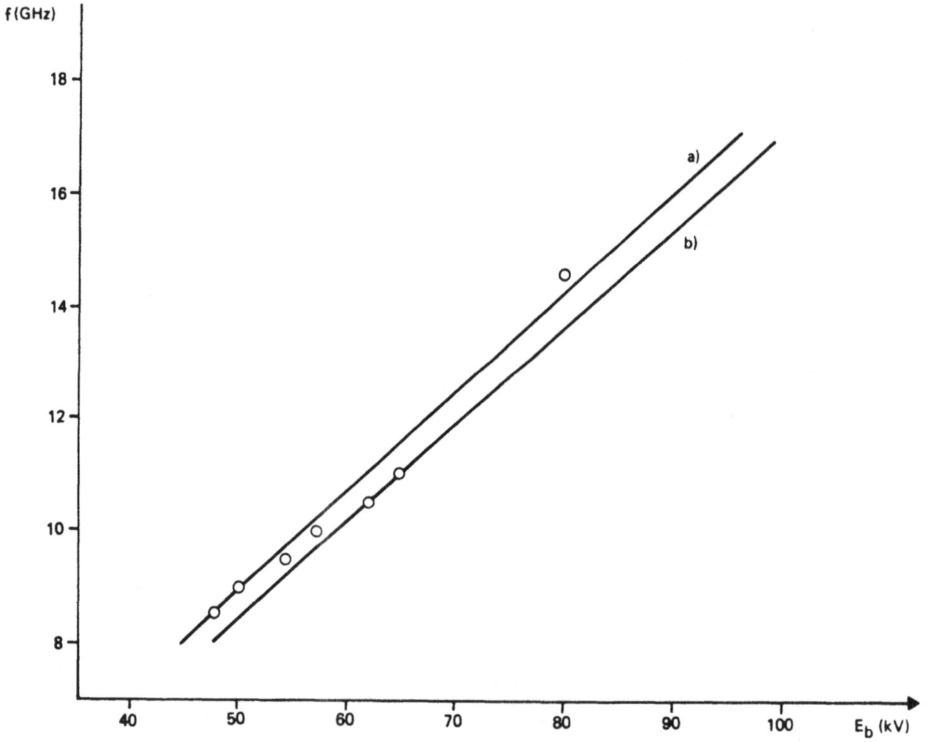

Figure 3: Frequency versus beam voltage for FEL operation in a circular waveguide of 7.29 GHz cut-off frequency. a) Assuming negligible small beam density n_b and wiggler field B_{wi} b) $n_b = 3 \times 10^{10} cm^{-3}$; $B_w \triangleq 64$ G; $(\Delta p/p) = .001$. The circles are experimental points.

Equation (4) includes the self-consistent equilibrium, space charge effects and both the forward and backward wave. Although these terms are usually neglected, they play an important role for low γ experiments, shifting the frequency and varying the gain value. Note that as $k_w \to 0$ Equation (4) becomes $\varepsilon^{el}\ \varepsilon^{em} \to 0$ and therefore separable into electrostatic and electromagnetic relations with only stable modes as solutions. That is, $G \to 0$ indicates that there is no longer any coupling.

The predictions obtained from the linear theory are utilized in the gain optimization of the TRW low γ experiment. In addition, in order to calculate the system efficiency a nonlinear theory is presently being developed that includes the possibility of a wide band spectrum. In general, the analysis is numerical due to the complexity of the problem. For a single mode system Ref. 8 describes the analysis in detail. It is based in following the equations of motion for the particles together with Maxwell's equations. For a multimode system, the equations presented in Ref. 8 can be easily generalized. In any case, an order of magnitude for the efficiency of standard (constant b_w and k_w) free electron laser can be obtained by assuming for example that the dominant saturation mechanism is trapping. In that case, the efficiency is given by

$$\eta_T = \frac{\gamma(V_b) - \gamma\left[\omega/(k_w + k_r)\right]}{\left[\gamma(V_b) - 1\right]}$$

If the system is multimode, the particles might remain "untrapped" (or trap and untrap as shown in some particle simulations)[7]. In that case, an estimate of the efficiency is given by assuming that the radiation at a particular wavelength stops when the electron beam slows down sufficiently that it no longer radiates with the bandwidth of the initial gain curve. We call this the frequency downshift efficiency and is given by

$$\eta_D \sim \frac{\Delta\omega_{1/2}}{\omega_{max}} \simeq \frac{2\ \mathrm{Gain}}{k_w}$$

where $\Delta\delta_{1/2}$ is the bandwidth of the initial radiation spectrum. These efficiencies and their predicted values for the TRW experiment will be further discussed in next section.

Experimental Device

The experimental arrangement is illustrated in Fig. (4). The electron source is cathode and anode assembly taken from a commercial high power klystron. The heated cathode is 5.7 cm in diameter and produces up to 20 A at 100 kV. The electron beam is pulsed at up to 60 pps and the pulsewidth is 1.6 μs. A converging magnetic field, designed to produce Brillouin flow, guides the electrons through a hole in the anode into the interaction region. The resulting instantaneous axial energy spread of the beam, $\Delta E_{\parallel}/E_{\parallel}$, is expected to be of the order of 10^{-4}.

The interaction region consists of a circular waveguide, 1 m long and 2.54 cm in diameter, along the axis of which the electron beam propagates. On the outside of the waveguide is wound a double helical coil which produces the perpendicular magnetic field perturbations. Two different waveguide-helix assemblies are used, one with λ_w = 2.5 cm, the other with λ_w = 1.9 cm. Tapered end sections of the wiggler, in which the helix current is changed form zero to its full value over 5 wiggler periods, were found to be extremely important for the undisturbed propagation of the electron beam. The wiggler field amplitude can be as high as 500 gauss.

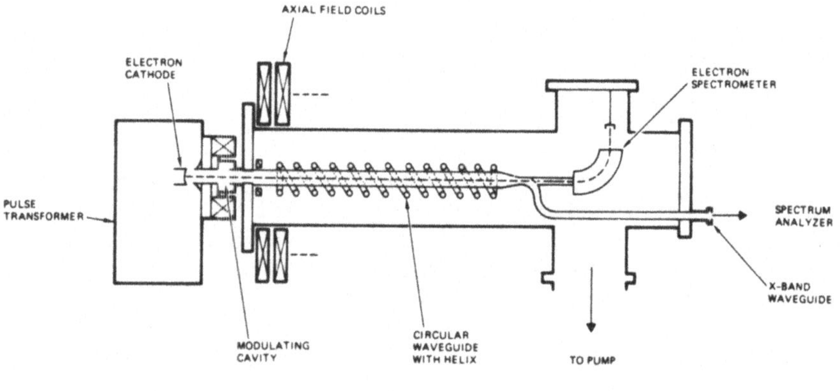

Figure 4: Schematic of the 9-16 GHz FEL Experiment.

The whole assembly is contained in a vacuum vessel which is surrounded by 24 large pancake coils. These coils produce an axial solenoidal magnetic field up to 6 kG which confines the electron beam to < 5 mm diameter.

Downstream from the interaction region, waveguides extract the electromagnetic radiation while the electron beam is analyzed with a magnetic spectrometer. The emitted microwaves, which range from

8 to 16 GHz, depending on the beam voltage and helix, are detected with a hetrodyne receiver. This permits the detailed analysis of the emission frequency spectra. Using a box-car averager, the spectra can be time resolved within the duration of the electron beam.

Experimental Results

Typical spectra of the microwave radiation generated by the interaction of the rippled magnetic field and the electron beam are shown in Fig. 5. It should be noted that each frequency f_o appears at $f = f_{LO} \pm f_{IF}$ where f_{LO} is the local oscillator and f_{IF} the intermediate frequency of the reciever. The parameters of Fig. 5 are: beam voltage 90 kV, beam current 12 A, beam density $1.2 \times 10^{10} cm^{-3}$, and external magnetic field 2.1 kG. By increasing the helix current to a certain value I_o, a narrow band mode is excited. At $0.9\ I_o$, no microwave power was observed while at $1.2\ I_o$ several modes are present with downshifted frequencies. Both the onset and frequency shift with the pump field is predicted by theory. A threshold pump field is necessary to obtain a positive growth rate and the frequency shift is evident from the last term of Eq. 3. The helical magnetic pump field was of the order of 100 gauss.

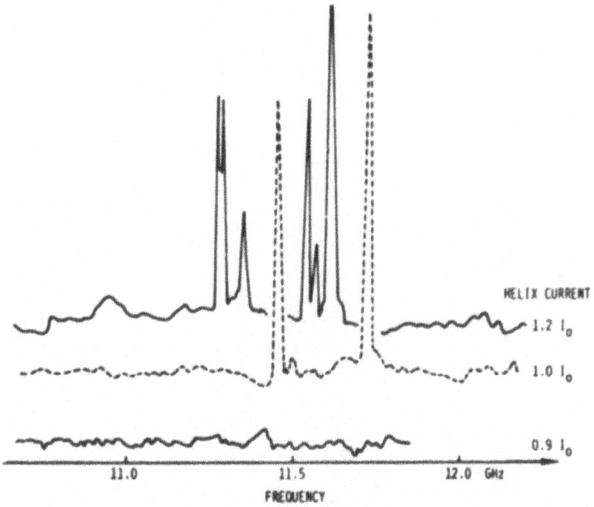

Figure 5. Frequency spectrum of electromagnetic power output as a function of helical magnetic field.

The high operating density in excess of $10^{10} cm^{-3}$ lets the interaction take place in the collective regime. This regime has a significantly larger growth rate compared to the single particle interaction. Correspondingly, the superradiant excitation of microwaves was observed for beam densities larger than $5 \times 10^9 cm^{-3}$ only.

The system was also operated as an oscillator. To this end, an aperture was inserted into the beam input side of the circular waveguide allowing the electron beam to pass, but reflecting microwaves. At the opposite end of the system, a small fraction of the microwaves was coupled out with a 20 db coupler while the bulk of the microwave power was reflected back into the system with a movable short, Fig. 6 shows microwave spectra obtained with the same system parameters as Fig. 5. The lower part of Fig. 6 shows a multimode spectrum obtained at an arbitrary reflector position. By moving the reflector, i.e. by changing the cavity length, one particular mode can be selected for amplification (upper part of Fig. 6) and the amplitude of that mode is enhanced, in this example by 20 db.

Several factors were found to contribute to the complexity of the observed emission spectra: 1) the beam pulse voltage is not constant in time, different frequencies were observed to be excited at different times; 2) normally, the beam mode interacts with the

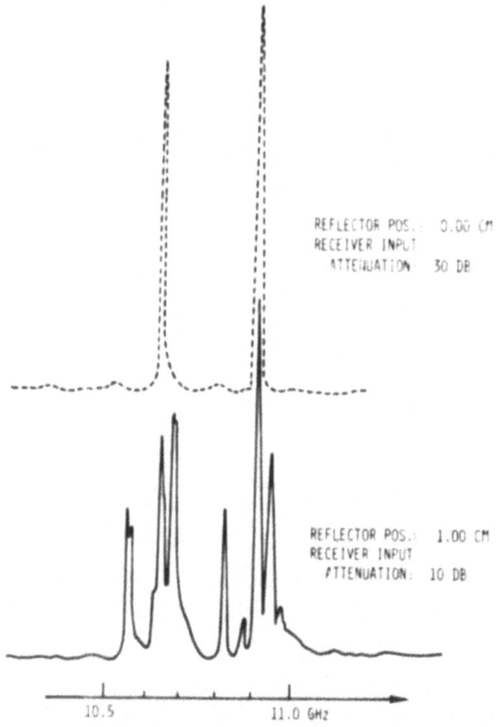

Figure 6. Frequency spectra of the oscillator emission as a function of the position of the movable short.

E.M. waveguide mode at a well defined point in ω and k; for some choices of k_w and V_b, the beam mode can be almost tangential to the waveguide mode, so that the interaction frequency is not well defined; 3) if the longitudinal magnetic field is increased to a value at which the electron cyclotron frequency is above the waveguide cut-off frequency, the gyrotron mode is easily excited.

Electron Beam Prebunching

A new methode of amplifier operation of a FEL was investigated. Rather than injecting an EM wave of proper wavelength into the system, like in the normal amplifier operation, the beam is velocity modulated in a microwave cavity or waveguide. As in a klystron, the propagating beam transforms the velocity into density modulations. The wiggler pump then scatters off these enhanced density fluctuations. Figure 7 shows the microwave signals received with prebunching. Without wiggler field, the system acts as a klystron, the amplitude is low. Turning on the magnetic wiggler increases the output amplitude by more than 20 db due to the FEL interaction.

The use of beam prebunching facilitates the measurement of the FEL dispersion relation and gain curves. If the system is operated at a gain value at which it is not superradiant, then the maximum output power is obtained if the modulation frequency coincides with the point of intersection of the EM waveguide mode and the beam mode. By varying the beam voltage, this point of intersection can be changed and with a constant modulation frequency, a gain curve can be obtained. An experimentally obtained example is shown in Fig. 8. In order to keep the interaction linear, the parameters were deliberately choosen to result in a small gain. The experimental gain curve was found to be considerably wider than the theoretical prediction. The exact reason for this discrepancy is unknown. Integration over time interval during which the beam voltage and therefore the output frequency varies, could simulate a large bandwidth. The optimum beam voltage obtained from gain cruves like Fig. 8 for various modulation frequencies, are included in Fig. 3. The group of points between 48 kV and 70 kV were obtained with constant cathode temperature, so that the beam density should increase with beam voltage. This might explain the scatter of the data points between the theoretical curves for $n_b \approx 0$ and $n_b = 3 \times 10^{10} cm^{-3}$.

Efficiency Enhancement

Although it has not yet been demonstrated experimentally, it is generally assumed that saturation of the FEL interaction is due to beam particles being trapped in the potential well of the ponderomotive wave. As described in the theory section, an estimate of the FEL efficiency in the different operating regimes can be calculated by assuming that the electron beam characteristics (ΔE, n)

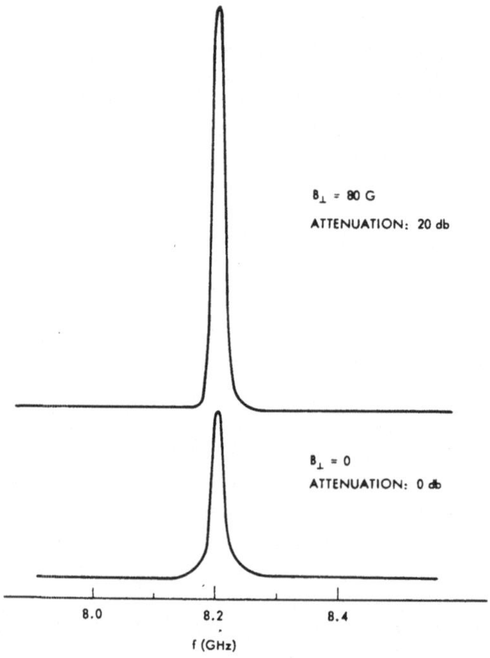

Figure 7. Microwave Emission of Prebunched Beam Without and with Wiggler Field $B_\perp$.

remain constant as it slows down and all the mean energy lost by the electron beam is transformed into radiation at a given wavelength. Saturation due to trapping can occur when the beam velocity is of the order of the electrostatic potential phase velocity V_{ph} and the electrons are trapped.

Alternatively, the interaction can be limited by frequency downshift. Figure 8 shows the experimental gain curve as a function of beam velocity for a given operating frequency. A beam energy loss, due to transfer of energy to the wave, will shift the point of interaction off the maximum gain resulting in system saturation.

Theoretical values for both the trapping efficiency η_T and the downshift efficiency η_D are shown in Table I for typical parameters of the experiment.

In the experiment described above, the maximum efficiency could not yet be determined because the output amplitude did not reach saturation. A new microwave power source is being prepared which should enable the system to reach saturation by beam modulation at a sufficiently high amplitude.

TABLE I

V_b	n_b	B_w	η_T	η_D
50 kV	$3 \times 10^{10} cm^{-3}$	80 G	0.5%	0.6%
100 kV	$10^{11} cm^{-3}$	450 G	2.1%	6%

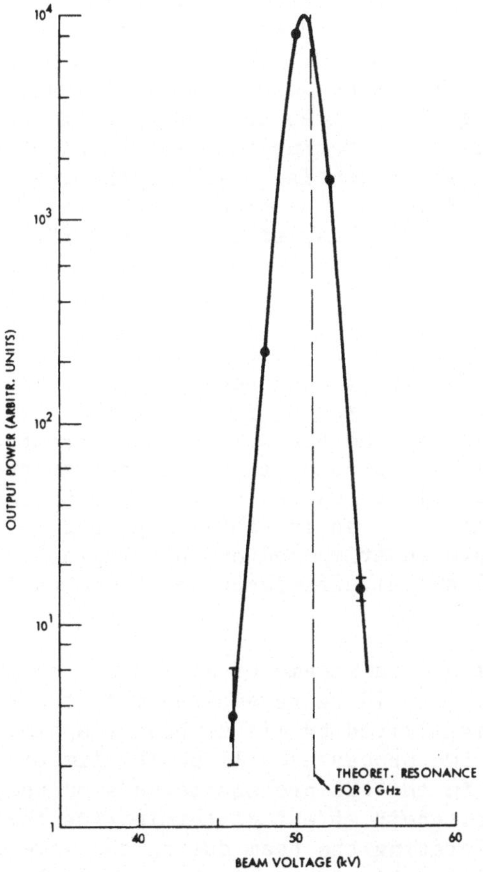

Figure 8. Microwave power as a function of beam energy for a beam prebunched at 9 GHz.

Recently,[8] it has been suggested that the FEL efficiency can be dramatically enhanced by adiabatically tapering the wiggler field amplitude and/or wavelength. In this manner, the energy and shape of the ponderomotive potential well ("bucket") formed by the interaction of the wiggler field with the electrons and the amplified laser signal can be spatially varied. The electrons that are initially trapped in the bucket tend to remain trapped if the motion is sufficiently adiabatic. As the bucket energy decreases, the trapped electrons mean energy is reduced. The extracted electron energy goes into radiation providing the amplification of the input laser signal. The relevant details of this concept and corresponding theoretical approach have been discussed in Reference 8.

Instead of the variable wiggler method of efficiency enhancement, it is also feasible to keep the wiggler constant and to increase the beam energy at the point of interaction. Two different methods are possible: one involves the trapped electrons, the other relies on keeping the distribution untrapped. In the first scheme the beam particles are trapped in the same fashion as in the variable wiggler case. An accelerating field is then applied to both groups of electrons. In adiabatically accelerating the trapped electrons, the accelerator "pushes against the potential hill" of the wave. In this way, the increment of energy given to the trapped beam is immediately transformed into wave amplitude. This method may be difficult to implement since the acceleration has to be done slowly to prevent beam untrapping, while in most accelerators acceleration occurs at discrete points with a step function voltage. Furthermore, the re-acceleration efficiency is reduced since energy also flows into the electrons that are not trapped. The efficiency enhancement by acceleration of trapped beam electrons has been considered by Rowe[9] for the case of the traveling wave tube. The case of the FEL was recently treated by Lin[10] using particle simulation. He found that half of the acceleration energy goes into runaway electrons and half into enhanced radiation. The interaction was limited by an instability producing a long wavelength ponderomotive wave superimposed on the original short wavelength wave. The total system efficiency was found to increase by a factor of two.

In considering the second scheme of efficiency enhancement by beam re-acceleration, it should be remembered that the efficiency of the basic FEL process is limited by either beam trapping or gain downshift which of the two processes will be the limiting factor in extracting beam energy in the FEL process depends on the system parameters. In either case, their effect of terminating the interactions can be avoided by accelerating the beam during the wave emission process. Since the acceleration does not have to be done adiabatically as in the trapped particle case, a stepwise acceleration is permissible thereby making this scheme easily applicable in an actual FEL system. If the temperature does not increase, the system amplitude can be expected to increase continously with the normal "linear," small signal gain. Moreover, all electrons participate in the radiation process suggesting higher ultimate efficiencies may be feasible. An increase in beam temperature, for example by the velocity modulation due to the ponderomotive force, would be undesirable since it would move the system from the high gain Raman into the low gawn warm beam regime. Although this would not limit the process, it would necessitate a larger interaction region.

In order to study the efficiency enhancement by beam reacceleration, the present experimental system will be modified by inserting one or more acceleration gaps into the waveguide structure. By choosing a particular premodulation power, the system can be made to saturate at a given point along the axis of the systems. By choosing the point of saturation downstream from the acceleration gaps, the efficiency enhancement by prevention of trapping can be studied. If saturation and therefore trapping occurs before the gaps, efficiency enhancement by acceleration of the trapped particles can be studied.

REFERENCES

1. H. Boehmer, J. Munch, M. Zales Caponi, IEEE Transact. in Nucl. Sc. V 26, 3830 (1979).
2. M. Zales Caponi, J. Munch, H. Boehmer, Physics of Quantum Electronics, V VII, Ed. by S. Jacobs et.al, Addison Wesley Pub. Co., pg. 523-553, (1980).
3. D.A.G. Deacon, L.R. Elias, J.M.J. Madey, G.J. Ramian, H.A. Schwettman and T.I. Smith, Phys. Rev. Lett. 38, 892, (1977).
4. L.R. Elias, W.M. Fairbank, J.M.J. Madey, H.A. Schwettman and T.I. Smith, Phys. Rev. Lett. 36, 717, (1976).
5. D.B. McDermott, T.C. Marshall, S.P. Schlesinger, R.K. Parker and V.L. Granatstein, Phys. Rev. Lett. 41, 1368 (1978).
6. R.M. Gilgenbach, T.C. Marshall and S.P. Schlesinger, Phys. Fluids 22, 971 (1979).
7. P.C. Liewer, T. Lin, J. Dawson, M. Zales Caponi, to be published in Phys. Fl.
8. Phys. of Quantum Electronics, S. Jacobs Ed., Add. Wesley Publ., (1980) (cf. P. Sprangle, C.M. Tang, W.M. Manheimer, Pg. 207-253 or N.M. Kroll, P.L. Morton, M.N. Rosenbluth, pg. 89-145).
9. J.E. Rowe, "Nonlinear Electron-Wave Interaction Phenomena," Academic Press (1965).
10. A.T. Lin, "Enhancement of Electromagnetic Radiation from Free Electron Lasers by Applying a DC Electric Field," Report #PPG-456, January 1980, Center of Plasma Physics and Fusion Energy, UCLA.

THE OPTICAL KLYSTRON FOR COHERENT X-RAY GENERATION*

F. De Martini

Istituto di Fisica "G. Marconi"

Universita di Roma, 00185 Italy

and

J. A. Edighoffer

Electrical Engineering Department

Stanford University, Stanford, California 94305

INTRODUCTION

This paper presents the theory of the modified Optical Klystron designed for the generation of coherent electromagnetic radiation in the far UV or X-band of the spectrum.[1] The Optical Klystron (OK) which has been first proposed by Vinokurov and Skrinsky (1977)[2,3] in connection with the development of the free-electron laser (FEL)[4] is composed of two undulators and of a dispersive magnetic system placed in the optical cavity of the laser.

In this configuration, a large enhancement of the gain is provided by the nonlinear interaction of the stimulated field with the bunched electron beam. Such bunching process developed after a periodic velocity modulation is induced on the beam in one of the undulators called "the buncher"[5].

The device we consider in the present work is essentially different from the original one: the optical cavity is absent and no free-electron laser action is involved.

*Work supported in by Consiglio Nazionale delle Ricerche and Centro Ricerche FIAT, S.p.A.

Still, in one of the undulators, the buncher, the electrons, interact with an external laser field undergoing a klystron-type velocity modulation. This again results in bunching the particle beam, i.e., in the generation of an array of electron-density fringes travelling at relativistic velocity. These ones are made to coherently backscatter, and frequency upconvert, an incoming electromagnetic field that can be associated with a laser beam or with a magneto-static (or electrostatic) wiggler structure.

The involved scattering process is a linear one, to first approximation, and it does not suffer from the limitations affecting the behavior of the free-electron laser at short wavelengths[6]. It is expected that the frequency dependence of the FEL gain may prevent in the future the operation of this device at wavelengths belonging to the UV or X-ray region of the spectrum[7].

The process of bunching and scattering are considered separately in the paper. An analytical solution is given, for both processes, for an idealized electron beam, i.e., in absence of energy spread ($\delta\gamma/\gamma = 0$) and for zero emittance. The theory of the bunching process is paralleled by a Monte Carlo computer simulation analysis which has been carried out for an electron beam in realistic conditions.

One numerical example is given at the end of the paper.

BUNCHING IN THE RELATIVISTIC KLYSTRON

The coherent phasing of electrons in a beam can be obtained by several particle-field interaction processes. In most types of accelerator machines, an effect of bunching is produced by the requirement that the particles stay in phase with the acceleration fields themselves within a limited phase angle. Another cause of bunching is the interaction of the electron beams with external laser fields or with the field in the cavity of a free-electron laser. Inverse sinchrotron radiation or inverse Cerenkov effects[8] are to be classified among such types of bunching processes.

In the present chapter, we shall analyze the bunching of a relativistic electron beam after a nearly collinear interaction with an external laser beam (with wavelength λ_b) has taken place in the buncher. This one is assumed to be composed of a magnetostatic transverse plane undulator of period λ_w and length l_w.

We assume that the reaction of the particles on the field is zero, i.e., we neglect gain processes in the buncher. Another simplifying assumption is that, after leaving the buncher, the particles drift for a length $z_b \gg l_w$ in vacuum, i.e., in absence of magnetic structures enhancing the dispersive effect on the beam[3]. These are the conditions relative to the generation of klystron-type bunching in the relativistic regime[9].

If we suppose that the charge density and the energy of the electron beam correspond to the "ballistic" regime for the particles[1], we can consider the evolution of a single particle after leaving the buncher ($z = 0$) at time t_0 (cfr. Fig. 1). The kinetic energy of the particle is $T = (\gamma - 1)m_0c^2$, where $\gamma \equiv (1 - \beta^2)^{-1/2}$, $\beta \equiv v/c$. Let us assume that a harmonic energy modulation is imposed on the beam in the buncher: $T = T_0 + \Delta T \cos\omega_b t_0$.

From the equations above we get: $\Delta T = m_0c^2 \Delta\gamma = (\gamma^3 \beta_0 m_0c^2)\Delta\beta$ where $\Delta\gamma$ and $\Delta\beta$ are the amplitude modulations of the corresponding quantities about the values γ_c and β_0 relative to the unperturbed velocity v_0 of the particles. If t is the time at which the electron reaches the drift coordinate, we have $t = t_0 + z \,|v_0 \times (1 + \Delta\beta/\beta_0 \cos \omega_b t_0)|^{-1}$. Assume an idealized uniform beam. The charge current at $z = 0$ may be written as $I_0 = e\, dn/dt_0$ where n is the number of particles leaving the buncher at time t_0. The bunched current as function of time at z can be found by writing $I(z/t) = e\, dn/dt = I_0\, dt_0/dt$. To evaluate te $I(z/t)$, we differentiate the above equation $t = f(t_0)$ with respect to t_0.

If we define a dimensionless quantity proportional to z, $\bar{z} = z\,(\omega_b/c)\,(\Delta T/m_0c^2)\,(\gamma\beta_0)^{-3}$(bunching parameter) we obtain: $dt \simeq dt_0 (1 + \bar{z} \sin \omega_b t_0)$. The current is then given by $I \simeq I_0 (1 + \bar{z} \sin\omega_b t_0)^{-1}$. This is not a convenient formula for which to determine the time dependence of $I(z/t)$ for it expresses I in terms of t_0 rather than t. Nevertheless by Fourier transformation we can express $I\,(z/t)$ in terms of the running variable $\tau = t - z/v_0$):

$$I\,(z/t) = I_0 \{1 - 2 \sum_{n=1}^{\infty} J_n\,(n\bar{z}) \cos \,|n\,(\omega_b\tau + \pi/2)|\} \qquad (1)$$

$J_n\,(n\bar{z})$ is the Bessel function of order n.

Equation (1) shows that $(I - I_0)$ is a superposition of harmonic waves each of which reaches its absolute maximum at values of the argument equal to $(n\xi)$ where $\xi = 1 + 0.8086\, n^{-2/3} \sim 1$ for large n [10]. If we integrate (1), we obtain the space-time evolution of the electron density. This one can be expressed again in terms of harmonic waves all travelling at the same speed v_0 in the laboratory frame. The scattering of a coherent electromagnetic field by one of these density waves is then a coherent scattering process that conserves the polarization state of the incoming field and gives rise to a new coherent electromagnetic field with doppler-converted frequency.[1]

It is interesting to assess the validity of the conditions under which the "ballistic" regime is valid for particles of the beam. Let us treat the beam as a relativistic electron plasma and solve in the small-signal approximation the Poisson and the continuity equations in the electron rest frame (e.r.f.)[11]. Consider the longitudinal space charge modes moving along z. In these conditions, the current density and the velocity are modulated by the mutual interference along z of one slow and one fast plasma modes[12]. The velocity modulation of the beam can be expressed in the form $v = v_0 + \Delta v \cos(\omega_p z/ v_0) \cos \omega_b t_0$, where $\omega_p = (4\pi e\rho_0/\gamma\, m_0)^{1/2}$ and ρ_0 are respectively the invariant plasma frequency and the electron density for the unbunched beam. If we differentiate $t = t_0 + \int_0^z dz/v$ with respect to t_0, we obtain an expression of the derivative dt/dt_0 which is formally identical to the one obtained for the "ballistic" regime. However, a new expression of the bunching parameter is obtained:

$$\overline{z} = (\Delta\beta\gamma\omega_b/\beta_0\, \omega_p)\ \text{sine}\ (\omega_p z/\gamma\, v_0).$$

Making use of previous formulas, we can easily verify that this expression coincides with the one corresponding the "ballistic" regime if we approximate the sin function with its argument. This one must be evaluated at the drift distance z_b at which the coherent doppler scattering is taking place. This argument establishes the regime at which the optical klystron is working. If $(\omega_p z/ \gamma_0) << \pi/2$, the ballistic regime is verified and the electrons can be considered as "free" particles in the interaction. As the above quantity approaches $\pi/2$, space charge effects become increasingly important and the beam goes in a "collective" or plasma regime. As the argument of the sin function is proportional to $\gamma^{-3/2}$ we conclude that the "ballistic" regime is generally realized for very relativistic beams ($\gamma \gtrsim 50$) with reasonably small peak currents. (I_p < 1 to 10 Amps.)

In the present work, we assume that the ballistic regime is realized and that expression (1) is valid for an idealized optical klystron. In that case, the coherent e.m. scattering from n^{th} harmonic density wave is proportional to $(J_n)^2$. It is noteworthy that the maximum value of J_n, where scattering is larger, is a relatively slowly decreasing function of n. This value may be expressed by Nicholson's formula:

$$J(n\xi) = 0.6533\ n^{-1/3} \quad \text{for large n } (n \gtrsim 5)$$

so the power radiated into each harmonic decreases as $n^{-2/3}$ (ref. 10).

The idealized optical klystron is for an idealized electron beam at zero energy spread ($\delta\gamma/\gamma$) and emittance (ε), an idealized bunching laser of zero linewidth and divergence and an idealized wiggler. To determine the reduction in harmonic content using realistic parameters for the laser, electron beam and wiggler, a Monte Carlo computer simulation of the bunching process was done. The computer simulation involves integrating the pendulum equation derived by Colson[13]. The energy and phase of each particle was updated at each time step, the bunching wiggler length l_w being divided into 100 times. The electron beam was modeled as having a gaussian distribution in energy, transverse position and divergence angles, but a uniform longitudinal spread. The laser was modeled as a single frequency, constant amplitude laser with gaussian waists in the transverse dimensions, including the divergence. The wiggler was modeled as a sinusoidal magnetic field only in one transverse dimension with $\nabla \cdot B \neq 0$.

The results of the simulation are shown in Fig. 2 and 3. In the first figure, the idealized result shows a very slow fall of the harmonic number n in very good agreement with the Bessel function formulation of Eq. 1. In this case, the 25th harmonic is only down a factor of 5 from the fundamental. In Fig. 3, an emittance of $3 \cdot 10^{-9}$ meter radians was added. The 5th harmonic is down to about 5% of the fundamental. In that last case, the harmonics higher than 10 are lost. The simulation has shown that emittance of the electron beam is the most crucial parameter as it causes the most severe spreading of the particles out of bunches for realizable parameters.

The computer simulation agrees with a calculation of the current reduction caused by the effect of the emittance and energy spreads. It is calculated by integrating over all emission angles θ and energies γ of the corresponding gaussian distributions the phase terms determined by the longitudinal displacement of the electrons which participate in the bunching process. Calling ψ_b the tilting angle with respect to β of the planes of constant energy modulation in the buncher, the reduction terms relative to the angular and energy spreads are given by, respectively:

$$r_\theta = \frac{1}{\sqrt{2\pi}\,\sigma_\theta} \int_{-\infty}^{+\infty} \exp\,(-\theta^2/2\sigma_\theta^2) \times \cos(2\pi f_n(\theta)/\lambda_b) \times d\theta$$

$$r_\gamma = \frac{1}{\sqrt{2\pi}\,\sigma_\gamma} \int_{-\infty}^{+\infty} \exp\,(-(\gamma - \gamma_0)^2/2\sigma_\gamma^{\,2}) \times \cos(2\pi\, g_n(\gamma)/\lambda_b) \times d\gamma \qquad (2)$$

where $f_n(\theta) \simeq n l_b \cdot (\psi_b^{\theta} - \theta^2/2)$, $g_n(\gamma) = n l_b (\gamma - \gamma_0)/\beta\gamma_0^{3}$ and l_b is the drift space, i.e., the coordinate z of the first maximum of J_n where we assume the coherent scattering takes place. Using these reduction factors, the electron current relative to the nth harmonic is given, after Eq. 1, by: $I_n = I_0 J_n(n\xi)\, r_\theta \cdot r_\gamma$. This formula allows a fairly good estimate of the coherently scattered field even from a high order harmonic. The next section will be devoted to the detailed analysis of the coherent Thomson scattering from a bunched electron burst interacting with a coherent electromagnetic field. This one may be associated either with a radiation plane wave with a magnetostatic (or electrostatic) wiggler structure owing to the Weizsacker-Williams approximation[14,15].

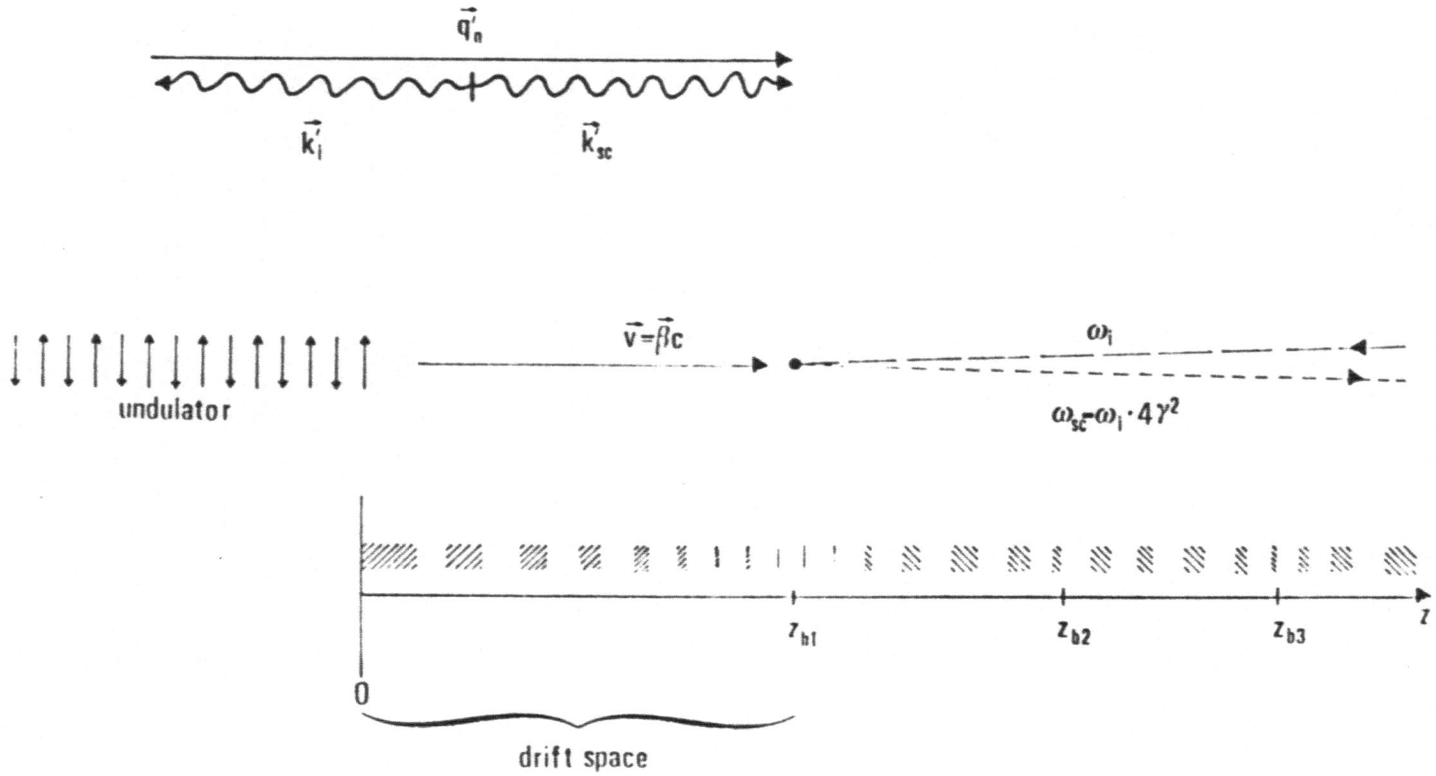

1. Kinematics of the backscattering process.

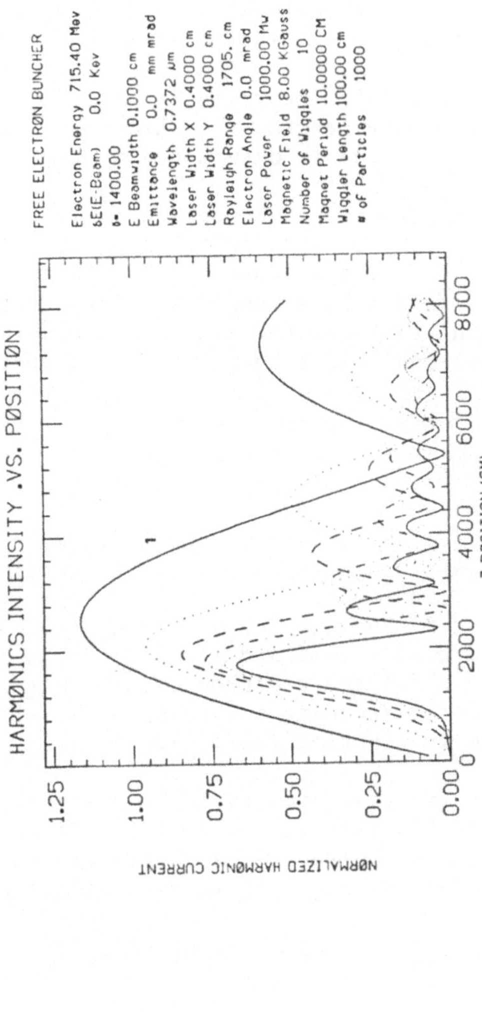

Figure 2. Monte Carlo simulation of the bunching process (ideal case). The intensities of the harmonics 1 to 6 are shown.

COHERENT SCATTERING

Our analysis takes into consideration in (e.r.f.) the basic elastic (or inelastic) scattering phenomena that are common in low energy plasma physics as well as in solid-state spectroscopy. In particular we consider here a process of elastic scattering of a coherent electromagnetic wave with an inhomogeneously broadened distribution of electrons whose spatial density along z is represented by a static grating given by Eq. (1). This process is similar to the X-ray diffraction from a crystal structure. Let us assume first that the interaction takes place between plane electromagnetic and electron density waves. Refer to Fig. 1 for the kinematics of the interaction. If the wavelength of the incoming radiation λ'_i, in (e.r.f.), is much larger than the Compton wavelength λ'_C (= h/m_0c = 0.0243 Å), the prevailing scattering process is determined by the Thomson cross-section with no frequency shift, i.e., $\lambda'_i = \lambda'_{sc}$. Assume that the coherent scattering interaction takes place in (e.r.f.) with a nth harmonic density wave limited in a parallel-piped volume of size $L_x \cdot L_y \cdot L'_z = V'$ (ref. to Fig. 4). Suppose that the planes of constant phase of the density wave make an angle ψ_b with respect to $\overline{z}'$. This may be due to bunching the beam through the interaction of the electron beam with an external laser beam having a direction of propagation making an angle ψ'_b with respect to $\hat{\beta}$ (in Lab.f.) within the buncher. The transformation formula is tan $\psi'_b = \gamma \tan \psi_b$. As we have said, the electron density wave fringes are at rest in (e.r.f.). However, in our scattering interaction, the end surfaces of the scattering volume are moving along z' with speed $(-\beta_0 c)$. An electromagnetic sqaure-wave pulse of length $_i$, with a carrier frequency $f'_0 = c/\lambda_{0i}$, is propagating along the direction (-z'). Suppose that at the initial time t = 0, the center of mass of the incoming e.m. pulse is at $z' = z'_0$ and the coordinate of the end surfaces of the electron burst are at z' = 0 and $z' = L'_z$.[16] In addition, let us assume that in (e.r.f.) the electron burst length is much larger than the electromagnetic incoming pulse: $L'_z \gg {}'_i$. This condition is largely satisfied in practice (Ref. 1).

The expression of the scattered field by a single particle is given in Ref. 14 page 683 (see also Ref. 17). If we integrate over all particles of the pulse, we can write the far-field in the direction determined by the angle ψ'_s in the following form[18,19]

$$E'_{sc} = A.\ E'_i\ .\exp(i(k'_b\cos\psi'_b.L'_z/2 + k'_b\sin\psi'_b.L_y/2)).$$
$$.\int_{-\infty}^{+\infty} G(h').\mathrm{sinc}(\sigma_x h').\mathrm{sinc}(\sigma_y(h' - \bar{h}'_y)).\mathrm{sinc}(\sigma_z\ (h' - \bar{h}'_z)).$$
$$.\exp\ (i\ 2\pi\ h'\ \overline{z})\ .\ dh' \qquad (3)$$

where $A = V'\ (r\rho'_{max}\ \cos\psi'_s/r')$, $r_0 = (e^2/m_0c^2) = 2.82 \times 10^{-13}$cm is the classical electron radius, ρ'_{max} is the amplitude of the electron density wave, R' is the polar coordinate of the observer in the direction determined by ψ_s', $k_b' = 2\pi/\lambda_b'$, $\alpha_x = 1$, $\alpha_y = \sin\psi_s'$, $\alpha_z = 1 + \cos\psi_s'$, $h_0' \equiv f_0'/c$, $h' \equiv f'/c$, $G(h') = \sigma'$ sinc $|(h' - h_0')\sigma'|$. exp $(-i\ 2\pi h' z_0')$ is the Fourier transform of the incoming electromagnetic square pulse to be scattered, $\bar{h}_y' = k_b' \sin\psi_b'/2\pi\ \alpha_y$, $\bar{h}_z' = k_b' \cos\psi_b'/2\pi\ \alpha_z$, $\sigma_x = -\alpha_x L_x$, $\sigma_y = -\alpha_y L_y$, $\sigma_z = -\alpha_z L_z'$ and $\bar{z} = (\sigma_x + \sigma_y + \sigma_z)/2 + (R' + z_0') - ct'(1 - \beta\alpha_z)$. Other quantities are defined in Fig. 4. We can see that Eq. 3 expresses the scattered field as a triple convolution of functions which represent the shape of the particle and radiation pulses. The integral in (3) is easily calculated if we think that the electron pulse length in (e.r.f.) L_z' is, in actual cases, by far larger than L_x, L_y and σ'. In that case we can approximate the function $\sin(\sigma_z \cdot (h' - \bar{h}_z'))$ with a δ function peaked at $h' = \bar{h}_z'$. We avoid here the details of the calculations. The most relevant results are that the scattered pulse is a square pulse of duration equal to the duration of the electron pulse in (lab.f.) and that the Doppler-transformed wavelength il (lab.f.) is $\lambda_{sc} = \lambda_i \cdot \gamma^2 \cdot (I + \beta) \cdot \alpha_z = \lambda_i\ 4\gamma^2$, in case of an incoming radiation pulse with $k_i' =$ (=z'), as in Fig. 4.

The ratio between the e.m. power scattered in a solid angle $\Delta\Omega$ and the incoming power in (lab.f.) is given by the expression

$$\eta = \frac{\Delta P_{sc}}{\Delta\Omega} / P_{inc}) = |(r_0 \cdot N_n \cdot \gamma \cdot \sigma_{inc})^2/(L_x L_y\ \sigma_{sc}'^2)|\ x$$

$$\cdot\ |\cos\psi_s'\ (\sin\psi_s' + (\cos\psi_s' + \beta)\gamma^2|\ /(\cos^2\psi_s + \gamma^2 \sin\psi_s) .$$

$$\cdot\ \mathrm{sinc}^2\,\Gamma_p \cdot \mathrm{sinc}^2\,\Gamma_x \cdot \mathrm{sinc}^2\,\Gamma_y . \qquad (4)$$

where $\Gamma_p = \sigma_p'(\bar{h}_z' - h_0')$, $\Gamma_x = \sigma_x'\ \bar{h}'$, $\Gamma_y = \sigma_y'\ (\bar{h}_z - \bar{h}_y)$ (cfr. ref. 18), N_n is the total number of particles belonging to the n harmonic.

The momentum convservation (phase matching) in the coherent interaction is determined by the maxima of the function sinc Γ_p. The absolute maximum of this function, corresponding to 1st order scattering establishes the maximum scattered intensity at an angle $\psi_s' = 2.\ \psi_b'$, i.e., for specular reflection from the electron-density fringes (cfr. Fig. 4). The order of harmonics involved in 1st order scattering is $n = \lambda_b/\lambda_{sc}$, i.e., the generation of a short wavelength requires a short wavelength of the bunching laser and a high content of harmonics. The functions sinc Γ_x and sinc Γ_y determine the far-field diffraction pattern of the scattered radiation. The emission solid angle corresponding to the main diffraction lobe is given by the expression[17]: $\Delta\Omega = \lambda_{sc}^2/L_x L_y$. For $\lambda_{sc} = 10$ Å, a transverse section $L_x L_y = (0.3\ \mathrm{mm})^2$, the emission solid angle is 10^{-9} sterad, i.e., the X-radiation is emitted with a divergence = 1.8×10^{-5} rad. The linewidth of the scattered radiation can be very small as it is determined essentially

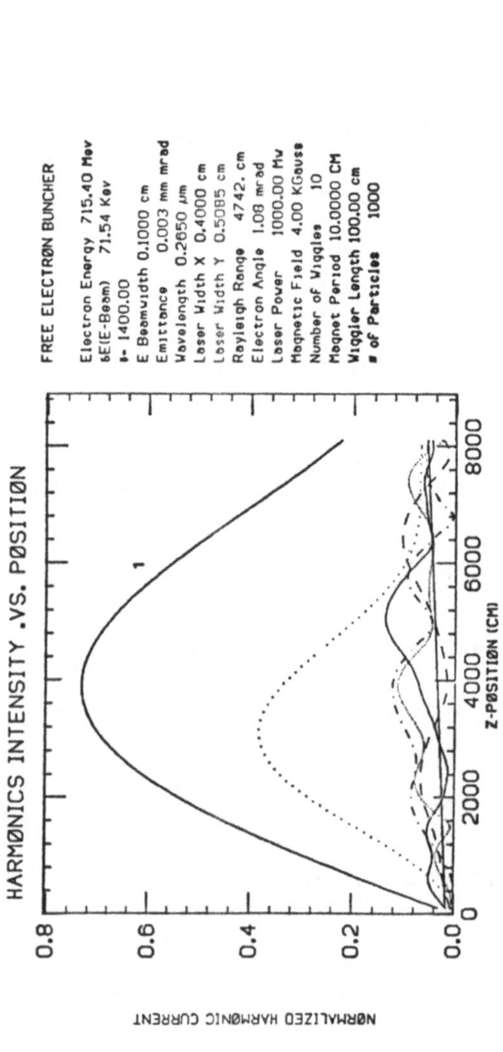

Figure 3. Monte Carlo simulation of the bunching process (emittance of the electron beam $3 \cdot 10^{-9}$ m rad). The various harmonics are identified graphically by correspondence with Fig. 2.

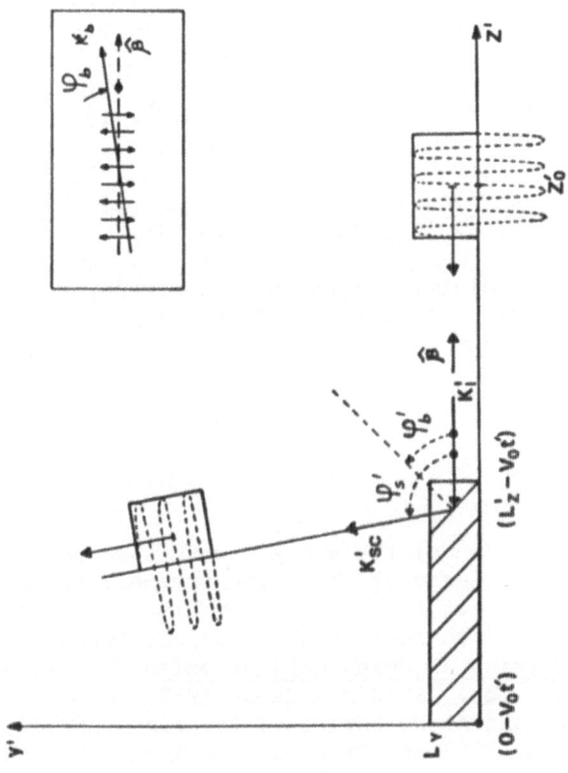

Figure 4. Kinematics of the scattering process in (e.r.f.) involving e.m. and electron pulses of finite lengths. The inset refers to the kinematics of the bunching process in (lab.f.).

by the harmonic content of the electron and electromagnetic pulses participating in the scattering process. This appears evident from our analysis involving scattering from harmonic density waves travelling at the average electron velocity $v_0 = \beta_0 c$.[20]

Let us give a numerical example based on the parameters of the NSLS booster synchrotron now operating at the Brookhaven Laboratories.[21] Assume that bunching is provided by the 5th harmonic of the Nd-Yag laser, $\lambda_b = 2120$ Å. Assume also that $\gamma = 1400$, $L_z = 1$ cm, $L_x L_y = 9.10^{-2}mm^2$ and that the incoming e.m. pulse is associated with a wiggler with $\lambda_w = 3$ cm and 1 - 2 m, and B = 5 KGauss. A soft X-ray coherent radiation with wavelength $\lambda_{sc} = 150$ Å is emitted at $\psi_s = 0$. The duration of the emitted pulse is $\Delta\tau = 30$ psec and the peak power is $P_{sc} = 100$ KW.

In conclusion we believe that the proposal presented in this work may represent a sensible solution to the long searched problem of generation of coherent far-UV or X-ray radiation. The applications of this device in spectroscopy, holography, nuclear physics as well as in medicine and biology look fairly attractive.

We acknowledge the valuable contributions of Professor R. H. Pantell of Stanford University and Dr. S. Baccaro of the University of Roma.

REFERENCES

1. A preliminary version of the present work has been reported by F. De Martini, "An X-ray relativistic free-electron frequency unconverter," Proc. of the Workshop on Free-Electron Generators of Coherent Radiation, Telluride, Colorado, August 13-17, 1979, published in <u>Physics of Quantum Electronics</u>, Vol 7, Addison Wesley Pub. Co., Inc., Reading, Mass. 1979.
2. N. A. Vinokurov and A. N. Skrinsky, preprints INP 77.59 and 77.67 Novosibirsk 1977 and Proc. of the National Conf. on Charged Particle Accelerators, Dubna, 1978, Vol 2, p. 233.
3. R. Coisson, review paper on optical Klystron. To be published.
4. D. A. G. Deacon, R. L. Elias, J. M. J. Madey, J. J. Ramian, H. A. Schwettmann and T. I. Smith, Phys. Rev. Lett. <u>38</u>, 892 (1972).
5. C. Shif and A. Yariv in <u>Physics of Quantum Electronics</u>, Vol. 7, (cfr. Ref. 1).
6. In the present paper we limit ourselves to consider the linear doppler regime. Parametric gain effects in the interaction can be nevertheless important, and they will give rise to an enhancement of the unconversion efficiency which is calculated here. The nonlinear parametric coupling in our process will be considered by the present authors in a forthcoming paper.

7. C. Pellegrini in 1979 Particle Accelerators Conference Proceedings, San Francisco, March 1979, IEEE Trans. on Nuclear Sciences NS 26, 3, 3791 (1979).
8. J. A. Edighoffer and R. M. Pantell, J. Appl. Phys. 50, 6120, 1979 M. A. Piestrup, G. B. Rothbart, R. N. Fleming and R. H. Pantell, J. Appl. Phys. 46, 132 (1975). See the list of references reported at Ref. 5 in the work by F. De Martini, in Physics of Quantum Electronics, Vol 7, Addison Wesley Publ. Co., Reading, Mass. 1979 (cfr. Ref. 1).
9. Klystron-bunching in nonrelativistic regime is analyzed in the context of microwave tubes by J. C. Slater, Microwave Electronics, Van Nostrand, Princeton (1980).
10. E. Jahnke and F. Emde, Tables of Functions, Dover, New York (1945).
11. Primed symbols will correspond hereafter to the definition in electron-rest-frame (e.r.f.) of the corresponding quantities expressed in laboratory frame (lab.f.).
12. The space-charge approach has been adopted in the paper of Ref. 1.
13. W. B. Colson, Phys. Lett. 64A , 190, 1977.
14. J. D. Jackson, Classical Electrodynamics, Wiley, N.Y. (1975).
15. E. Fermi, Zeit Phys. 29, (1924), E. J. Williams, Proc. Roy. Soc. A139, 163 (1933).
16. These time and space coordinates are not to be confused with the ones used in connection with the bunching process. For instance, in our present case, the space coordinates are defined about the first maximum of the intensity profile of the nth harmonics.
17. J. A. Stratton, Electromagnetic Theory, McGraw Hill Co., N.Y. 1952, Chapter VIII.
18. We adopt the Fourier transform formulation of P. M. Woodward, Probability and Information Theory, Pergamon Press, N.Y. (1963).
19. In the present analysis we neglect diffraction effects in x' direction.
20. R. Coisson and F. De Martini, submitted for publication.
21. J. Galayda et al., in 1979 Particle Accelerator Conference Proc., San Francisco, Ca., Cfr. Ref. 7. In the numerical example, the effects of the emittance and of the energy spread are taken care of in the calculation of the harmonic content of the beam through the reduction integrals given by Eq. (2).

COHERENT EMISSION FROM A BUNCHED ELECTRON BEAM IN A WIGGLER

Ilario Boscolo
Comitato Nazionale Energia Nucleare, Centro di Frascati
00044 Frascati, Rome, Italy and Università di Lecce
Italy

Vincenzo Stagno
Università di Bari, Italy

1. DESCRIPTION OF THE SYSTEM

The physical mechanism we are going to describe for the coherent emission by an electron beam has three separate phases. In the first, the beam with an initial uniform charge density is energy modulated; in the second this modulation is transformed into a density modulation; in the last phase the coherent bunched beam interacts with the static magnetic field of a wiggler (for the spontaneaous coherent emission) or with a potential pattern build up by an external electromagnetic wave travelling with the EB and the static periodic magnetic field (for the stimulated coherent emission).

The schematic of the device through which the EB passes is shown in Fig. 1. It has a buncher where the combination of an external wave with the static field of a wiggler yields the necessary periodic force for the energy modulation of the EB. The attached drift region accomplishes the transformation of that modulation into the density fluctuation.

The last section of the device is the radiator where the coherent emission occurs.

Concerning the radiator, we have to say that two configurations must be investigated: one with an input wave and the other without (Fig. 1). In the first case the electron bunches emit spontaneously and coherently owing to the sinusoidal motion in the periodic mag-

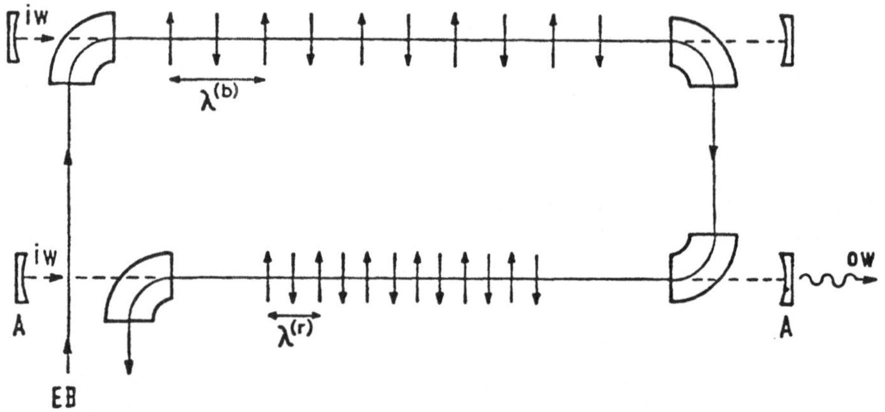

Fig. 1 Schematic of the Converter: EB, electron beam; iw, input wave; ow, output wave. The mirrors AA are added only in the version with stimulated emission. When the period of the radiator $\lambda^{(r)}$ is equal to the period of the buncher, $\lambda^{(b)}$, the device becomes a TOK

netic field and the coherence of the bunches. In the second case an input electromagnetic wave runs with the bunches and, if they are correctly phased relative to the electric field, a resonant energy exchange between the wave and the EB can be achieved.

Since the electron bunching is unharmonic[1], an up-conversion of the input buncher frequency wave can be obtained by matching the radiator period with higher harmonics[1,2].

When the device operates on the fundamental harmonic, the input wave either of the buncher or of the radiator can be provided by the radiation emitted by the same EB. In this configuration the device operates like a klystron. Since the transverse motion of the electrons and the transverse electric field are involved in the emission it is called Transverse Optical Klystron (TOK).

However, contrary to the microwave klystron, in our case we can have coherent emission without the interaction between a wave and the bunches; that is, we can have spontaneous coherent emission. The device in this configuration will be called SOK, Spontaneous Optical Klystron.

For the sake of completeness, it has to be said that when the device operates at higher harmonic it will be called a Converter.

2. BEAM DYNAMICS WITHIN THE BUNCHER

The static magnetic field of the wiggler in the small region along the axis is approximately[3]

$$\vec{B}^{(b)} = B_o^{(b)} \left\{ \frac{1}{2j} \exp\left[jk^{(b)} z \right] + cc. \right\} \hat{x} \tag{1}$$

being $k^{(b)} = 2\pi/\lambda^{(b)}$ the wavenumber of the buncher undulator. All quantities of the buncher, drift space and radiator will be superscripted with the indices b, d and r respectively.

We take the input electromagnetic wave with the electric field polarized along the y-direction

$$\vec{E}^{(i)} = E_o^{(i)} \left\{ \frac{1}{2} \exp\left[j\left(k^{(i)} z - \omega^{(i)} t \right) + \varphi \right] + cc. \right\} \hat{y} \tag{2}$$

where $\omega^{(i)} = k^{(i)} \beta_i c$, $\beta_i c$ being the phase velocity of the wave travelling in the z-direction.

The relativistic EB moves along an approximately sinusoidal trajectory[4] (Fig. 2)

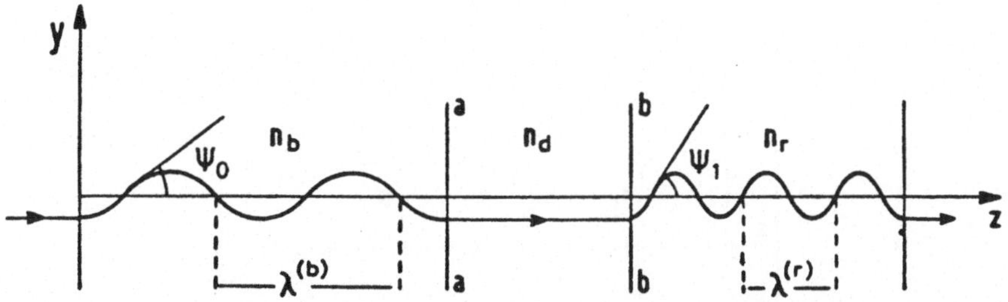

Fig. 2. Diagram of the electron trajectory in the laboratory frame: aa and bb are the interface between the buncher and the drift space, and between the drift space and the radiator respectively; n_b, n_d, and n_r are the relative refractive indices; ψ_0, ψ_1 are the maximum deflextion angles with respect to the z-axis within the buncher and the radiator respectively.

$$\vec{r}(z) = \frac{b}{k^{(b)}} \sin\left[k^{(b)} z\right] \hat{y} + z \hat{z} \tag{3}$$

Here the parameter b is related to the particle momentum and the maximum deflection angle ψ_o with respect to the z-axis by

$$b = \frac{e B_o^{(b)}}{k^{(b)} p} = \sin \psi_o \tag{4}$$

The Electron Beam average velocity along the z-axis is

$$v_b = \beta_b c \simeq \beta c \left[1 - \frac{b^2}{4}\right] \tag{5}$$

whose corresponding Lorentz factor will be γ_b.

The dynamic of electrons is studied in the reference frame which moves with the average velocity of the EB, where the electrons are non relativistic. It will be called $(EBS)_b$ when referred to the buncher and $(EBS)_d$ when referred to the drift space. All the quantities in these frames will be labeled by further indices b and d respectively.

The electric component of the transformed magnetic field (1) and the transformed wave (2) are

$$\vec{E}_b^{(b)} = E_{bo}^{(b)} \left\{ \frac{e^{j\left(k_b^{(b)} z_b + \omega_b^{(b)} t_b\right)}}{2} + cc. \right\} \hat{y} \tag{6}$$

$$\vec{E}_b^{(i)} = E_{bo}^{(i)} \left\{ \frac{e^{j\left[\left(k_b^{(i)} z_b - \omega_b^{(i)} t_b\right) + \varphi\right]}}{2} + cc. \right\} \hat{y} \tag{7}$$

with obvious meaning of symbols.

In order to find the tightness of the bunching, the phase space evolution of the electron distribution will be studied. The one-dimensional collisionless Vlasov equation is used

$$\frac{\partial \rho_b}{\partial t_b} + \dot{z}_b \frac{\partial \rho_b}{\partial z_b} + \dot{p}_{bz} \frac{\partial \rho_b}{\partial p_{bz}} = 0 \tag{8}$$

The three-dimensional equation reduces to the simple (8) with some considerations on the single electron Hamiltonian (see Ref. 5 and 6)

$$H = \left(\vec{P}_b - e\,\vec{A}_b\right) \Big/ \left(2m_o\right) \tag{9}$$

By taking into account the physical case in which the synchronism condition is fulfilled,

$$\omega_b^{(i)} = \omega_b^{(b)} = \omega_b \tag{10}$$

from the motion equation

$$\dot{p}_{bz} = - \partial H/\partial z_b \qquad \dot{z}_b = \partial H/\partial p_{bz} \tag{11}$$

we have that the only important term in the Hamiltonian is the term which has a purely spatial dependence.

It has the form

$$V_b^{(b)} = V_{bo}^{(b)} \left\{ \frac{e^{j\left(k_{b+} z_b + \varphi\right)}}{2} + cc. \right\} \tag{12}$$

with

$$\begin{cases} k_{b+} = k_b^{(i)} + k_b^{(b)} \\ V_{bo}^{(b)} = \dfrac{e^2 \left(1 - \beta_i \beta_b\right) E_{bo}^{(i)} \; E_{bo}^{(b)}}{2m_o c^2 \; \beta_b \left(\beta_i - \beta_b\right) \; k_b^{(b)} \; k_b^{(i)}} \end{cases} \tag{13}$$

The potential (12) gives rise to the modulation of electron cloud and to the pendulum motion of the single electron with an approximate frequency

$$\Omega_{b+} \approx \left\{ V_{bo}^{(b)} \Big/ m_o c^2 \right\}^{1/2} k_{b+} \, c \tag{14}$$

The initial condition we assume for the evolution equation is the usual uniform density with a gaussian energy spread in the LAB system which corresponds in the phase spase to

$$\rho(z,\varepsilon,0) = \frac{\rho_o}{\sqrt{2\pi}\,\sigma_\varepsilon\,\varepsilon_o} \exp\left[- \frac{\left(\varepsilon - \varepsilon_o\right)^2}{2\sigma_\varepsilon^2 \, \varepsilon_o^2} \right] \tag{15}$$

where σ_ε is the fractional energy spread.

In order to simplify the expression of the eq. (8) and to work with the significant parameters we make the following variables transformation

$$\begin{cases} \zeta = k_{b+} \, z_b \\ \tau = \Omega_{b+} \, t_b \\ q = \left(m_o \, V_{bo}^{(b)} \right)^{-1/2} p_{bz} \\ \sigma = \left\{ V_{bo}^{(b)} \Big/ m_o c^2 \right\}^{-1} \sigma_\varepsilon^2 \end{cases} \tag{16}$$

Equation (8) becomes

$$\frac{\partial \rho}{\partial \tau} + q \frac{\partial \rho}{\partial \zeta} + \sin \zeta \, \frac{\partial \rho}{\partial q} = 0 \tag{17}$$

The initial condition (15) in the $(EBS)_b$ with dimensionless parameters becomes (see Appendix)

$$\rho_b(\zeta,q,0) = \frac{\rho_{bo}}{\sqrt{2\pi\sigma}} \exp\left\{ - \frac{q^2}{2\sigma} \right\} \tag{18}$$

If the buncher is short ($\tau < 1$), we can look for a solution of the evolution equation of the Taylor kind

$$\rho_b(\zeta,q,\tau) = \rho_b(\zeta,q,0) \sum_{0r}^{\infty} D_r(\zeta,q)\, \tau^r \qquad (19)$$

The complete calculation of the terms D_r is made in references 1 and 7.

With the integration of (19) over all the dimensionless momenta q at the end of the buncher, the electrons distribution along ζ-axis is obtained directly as a Fourier series

$$\rho_b(\zeta,\tau_b) = \rho_{bo} \sum_{0m}^{\infty} A_m(\tau_b) \cos(m\zeta) \qquad (20)$$

The bunching coefficients A_m take the form

$$A_m(\tau_b) = (2\pi\sigma)^{1/2} \sum_{0r}^{\infty} \left[\sum_{0\ell} (2\ell-1)!!\, a_\ell^r\, \sigma^\ell\right] \tau_b^{2r} \qquad (21)$$

where a_ℓ^r is a coefficient dependent on the initial energy spread and the potential depth.

In Figure 3 the momentum and density modulation at the end of the buncher numerically calculated with a Runge-Kutta method[8] from eq. (17) are plotted. We have taken for the parameter σ the value 0.2 only for definiteness (it corresponds to possible practical values of electron beam, laser beam and wiggler parameters).

In Table 1 the maximum value of the first harmonics is reported

TABLE 1

m	1	2	3	4	5	6	7	8	9	10
$\|A_m\|$	.96	.55	.30	.13	.05	.02	.006	.002	$5\cdot10^{-4}$	10^{-4}

3. BEAM DYNAMICS WITHIN THE DRIFT SPACE

When the length of the buncher is as short as the density modulation is unappreciable, in order to obtain a sufficient modulation a drift space must be set after the buncher. This technique is particularly suited with high energy beam. As an example we report below in Tab. 2 the bunching coefficients in both cases (ε = 600 MeV, L = 2 m) with and without a drift region.

The evolution of the distribution function $\rho_d(z_d, p_{dz}, t_d)$ is

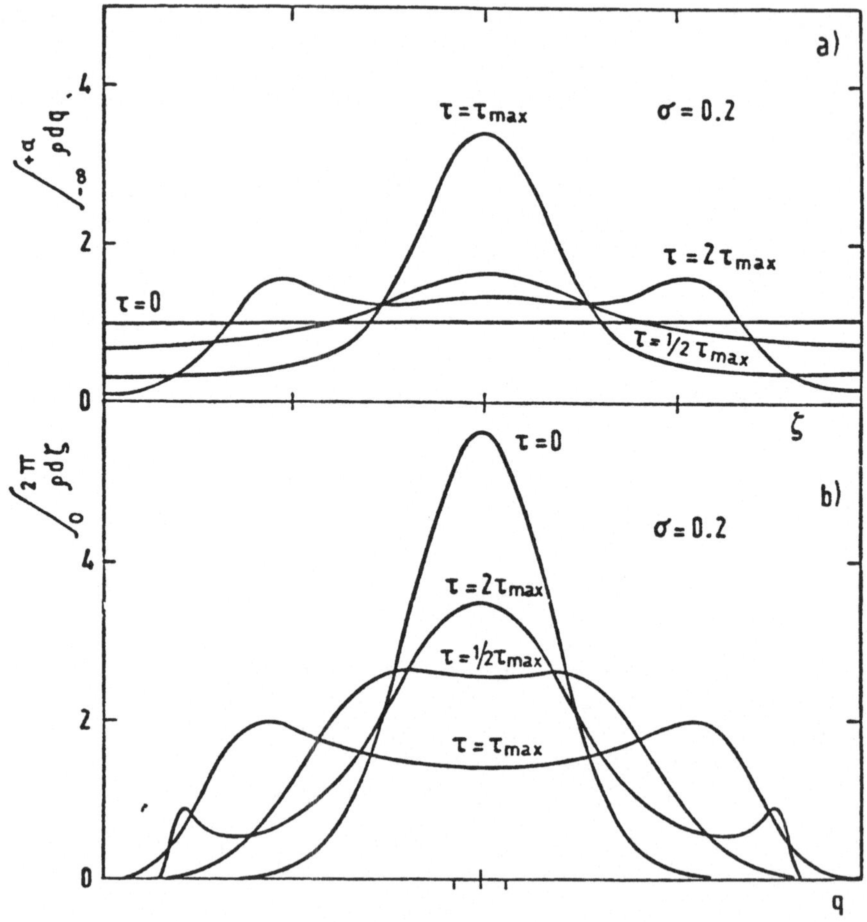

Fig. 3. Time evolution of the numerically calculated (a) density distribution and (b) momentum distribution function for different interaction times within the buncher. τ_{max} is the time at which the electron beam density for $\zeta = \pi$ takes the maximum value.

now studied in the phase space associated with the $(EBS)_d$ by the Vlasov eq. (8) with the last term dropped out

$$\frac{\partial \rho_d}{\partial t_d} + \dot{z}_d \frac{\partial \rho_d}{\partial z_d} = 0 \tag{22}$$

TABLE 2

m	1	2	3	4	5	6	7	8	9	10
$\|A_m(\tau_b)\|$	.05	.002	10^{-4}	$3\cdot10^{-6}$	10^{-7}	$3\cdot10^{-8}$	10^{-8}	$7\cdot10^{-10}$	10^{-10}	$2\cdot10^{-11}$
$\|A_m(\tau_b,\tau_d)\|$	.70	.30	.10	.03	.01	.003	$6\cdot10^{-4}$	10^{-4}	$2\cdot10^{-5}$	$4\cdot10^{-6}$

The drift space must be dispersive otherwise it would be too long for a practicable device. Therefore, it will be made by one or more sectors of bending magnets. The momentum compaction α_c defined as[9]

$$\alpha_c = \frac{p}{L}\frac{\Delta L}{\Delta p} \tag{23}$$

will characterize the dispersive properties of the system. It will relate the electrons momentum spread Δp with the lengthening ΔL of the electron path length L.

When the electrons move with constant speed the relation between space and time is given by

$$\frac{\Delta t}{t} = \frac{\Delta L}{L} - \frac{1}{\gamma^2}\frac{\Delta p}{p} \tag{24}$$

Combining (23) and (24) we get

$$\frac{\Delta t}{t} = \left(\alpha_c - \frac{1}{\gamma^2}\right)\frac{\Delta p}{p} \tag{25}$$

However, for a free space $\alpha_c = 0$, therefore

$$\frac{\Delta t}{t} = -\frac{1}{\gamma^2}\frac{\Delta p}{p} \tag{26}$$

If $\alpha_c \neq 0$ we can define an effective γ_{eff} for the particle given by

$$-\frac{1}{\gamma_{eff}^2} = \alpha_c - \frac{1}{\gamma^2} \tag{27}$$

With this assumption Eq. (25) can be rewritten

$$\frac{\Delta t}{t} = -\frac{1}{\gamma_{eff}^2}\frac{\Delta p}{p} \qquad (28)$$

We can conclude that the dispersive drift space may be considered as a free drift space for particles having the effective mass in the LAB frame

$$m_{eff} = \gamma_{eff}\, m_o = \gamma \frac{m_o}{\left(1-\alpha_c\gamma^2\right)^{1/2}} \qquad (29)$$

This mass transforms in the $(EBS)_d$ to

$$m_{oeff} = \frac{m_o}{\left(1-\alpha_c\gamma^2\right)^{1/2}} \qquad (30)$$

Then in Equation (22) the velocity $\dot{z}_d$ will be

$$\dot{z}_d = p_{dz}/m_{oeff} \qquad (31)$$

The initial condition for equation (22) is given by the distribution at the output of the buncher. Since the average EB velocity (relative to the LAB system) in the buncher and in the drift region are different, we have to perform a conformal transformation in order to pass from $(EBS)_b$ to $(EBS)_d$.

The detailed calculations are reported in ref. 7.

The final result is

$$\rho_d\left(z_d,p_{dz},t_d\right) = J \cdot \rho_d\left(z_d,p_{dz},0\right) \cdot \sum_0^\infty {}_r D_r\left(z_d - \frac{p_{dz}}{m_{oeff}}t_d, p_{dz}\right) t_{bl}^r \qquad (32)$$

where J is the Jacobian that relates the new and the old phase-spaces.

Proceeding at this point with the integration over all the

momenta at the end of the drift space and returning to the initial variables, the (32) reads

$$\rho_d\left(z_d, t_{b1}, t_{d2}\right) = \rho_{do} \sum_{o}^{\infty} m\, A_m\left(t_{b1}, t_{d2}\right) \cdot \cos\left(mk_{d+} z\right) \tag{33}$$

where $\rho_{do} = (\rho_{bo}/\gamma_o)$, γ_o being the Lorentz factor between the $(EBS)_b$ and $(EBS)_d$. The Fourier coefficients of the electron density perturbation, A_m, are dependent on the times t_{b1} and t_{d2} of interaction within the buncher and the drift region. Their full involved expression is given in ref. 7.

In the LAB the electron distribution function has the form

$$\rho(z,t) = \rho_o \sum_{o}^{\infty} m \left\{ \frac{A_m\left(t_{b1}, t_{d2}\right)}{2} \exp\left[jm\left(k_+^{(d)} z - \omega_+ t\right)\right] + cc. \right\} \tag{34}$$

The momentum and density modulation diagrams at the output of the drift space are shown in Fig. 4

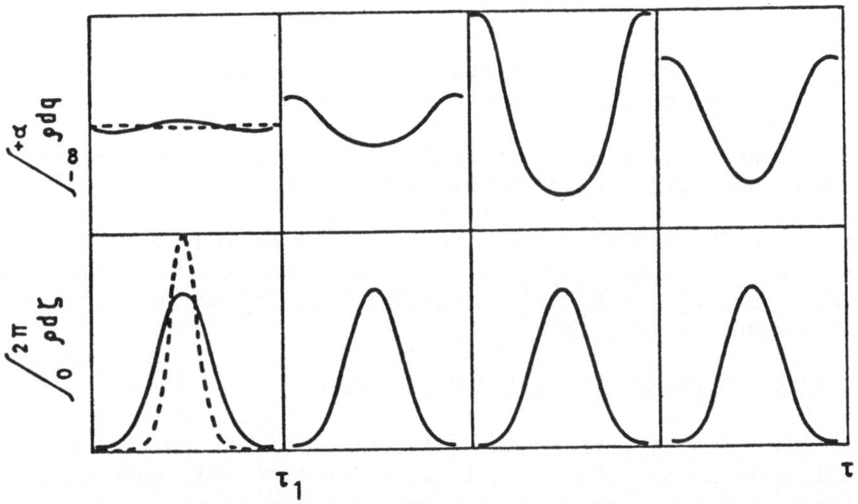

Fig. 4. Evolution with time of the electron distribution function with a buncher and a drift space. For $\tau < \tau_1$ the beam passes trough the buncher (the dashed lines represent the initial conditions for the beam). In the drift region $(\tau > \tau_1)$ a negative mass effect is present because the electron's energy has been chosen over the transition.

In Figure 5 the evolution within the drift space of some har-

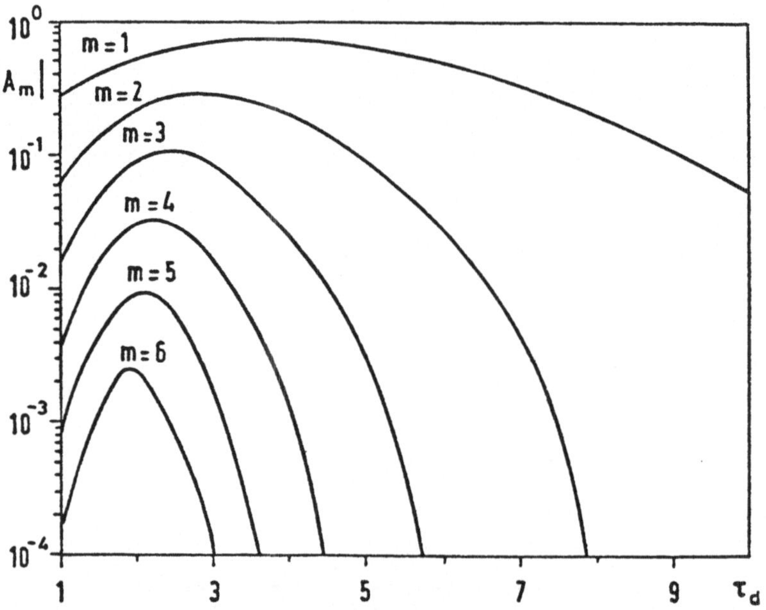

Fig. 5. Time evolution of the bunching coefficients $|A_m|$ in the drift space for $\sigma = 0.05$ and $\tau_1 = 0.314$

monics is reported. We must notice that the "plateau" in the diagrams shortens rapidly with the harmonic number.

It is worth remarking that the frequency of the constructed wave is the same as the input wave $\omega_+ = \omega^{(i)}$ and the wavelength is quasi the same, $k_+^{(r)} \simeq \beta_i k^{(i)}$. In addition its speed is $v_r = \beta_r c$ and its wavenumber is $k_+^{(r)} = \omega_+/v_r$.

4. SPONTANEOUS RADIATION BY A BUNCHED BEAM

We want firstly perform the calculation of the emitted intensity and power in the SOK case, in which we have only spontaneous emission by a bunched beam; in the next section we will deal with the stimulated emission by the bunched beam.

Since the electrons travel at approximately speed c, we can say that the EB remains substantially unchanged over reasonable radiator wiggler length.

The emitted energy per unit solid angle per unit frequency bandwidth (this calculation is made in C.G.S. units unlike all others in the article) is[10]

$$\frac{d^2 I}{d\omega\, d\Omega} = \frac{\omega^2}{4\pi^2 c^3} \left| \int_{-\infty}^{+\infty} dt \int d^3 \vec{r} \left[\hat{n} \times (\hat{n} \times \vec{J})\right] \cdot \exp\left[j\omega\left(t - \hat{n} \cdot \frac{\vec{r}}{c}\right)\right] \right|^2 \tag{35}$$

where the geometrical quantities $\hat{n}$ and $\vec{r}$ are defined in Fig. 6 and

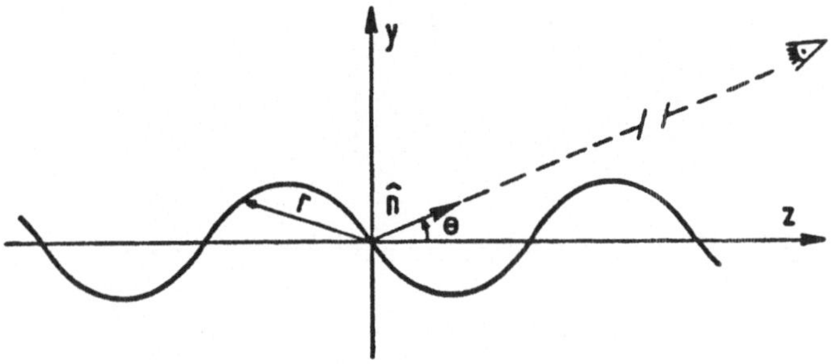

Fig. 6. Sinusoidal trajectory of the electrons in the radiator, showing the geometrical parameters used in the text.

$\vec{J}$ is the beam current density

$$\vec{J}(\vec{r}, t) = e\, c\, \rho(\vec{r}, t)\, \vec{\beta}(\vec{r}) \tag{36}$$

which recalling (34) and assuming for all electrons the same velocity dependent only on z (neglecting the energy dependence and the angular divergence) gives

$$\vec{J}(z,t) = ec\vec{\beta}(z)\ \rho_o \sum_{o} \left\{ \frac{A_m}{2} \exp\left[jm\left(k_+^{(r)}\ z - \omega_+ t\right)\right] + cc \right\} \tag{37}$$

This allows an analytical solution of the integral (35) without losing any physical content.

Combining equations (3), (35), (36) and (37) (the complete calculations are performed in ref. 1), for the emitted radiation intensity we get

$$\frac{d^2I}{d\omega\, d\Omega} = \frac{e^2\omega^2}{4\pi^2 c} \rho_o^2 \left| \sum_{o}^{\infty}{}_m \left\{ \frac{A_m}{2} I_m + cc \right\} \right|^2 \tag{38}$$

Here

$$\begin{cases} I_m = \int_{-\infty}^{+\infty} dt \int_{-L/2}^{+L/2} dz \left[\hat{n} \times (\hat{n} \times \vec{\beta}) \right] \exp\left[j\, \Psi_m \right] \\ \Psi_m = \omega\left(t - \hat{n} \cdot \frac{\vec{r}}{c}\right) + m\left(k_+^{(r)} z - \omega_+ t\right) \end{cases}$$

Considering only the radiation emitted in the plane of the particle paths at the angle θ, for an infinite EB length the calculation of the integral (35) leads to

$$\frac{d^2I}{d\omega\, d\Omega} = \frac{e^2}{c\left(k^{(r)}\right)^2} \left(\frac{\omega}{\omega_+}\right)^2 \rho_o^2 \left\{ \left[\sum_{-\infty\, \ell,m}^{+\infty} A_m\, \delta\left(\frac{\omega}{\omega_+} - m\right) F_\ell^y(\theta) \frac{\sin N_r \pi \left[(\alpha_m - \ell)\right]}{(\alpha_m - \ell)} \right]^2 + \right. \tag{39}$$

$$\left. + \left[\sum_{-\infty\, \ell,m}^{+\infty} A_m\, \delta\left(\frac{\omega}{\omega_+} - m\right) F_\ell^z(\theta) \frac{\sin N_r \pi \left[(\alpha_m - \ell)\right]}{(\alpha_m - \ell)} \right]^2 \right\}$$

Here $k^{(r)}$ is the wavenumber of the radiator, N_r the number of periods, $\alpha_m = (mk_+^{(r)} - k \cos\theta)/k^{(r)}$, ω and k are referred to the output radiation and F_ℓ^y, F_ℓ^z are complicated function of the observation angle θ.

In the forward direction ($\theta = 0$), with N_+ bunches interacting with the wiggler field ($N_+ \gg N_r$), at the condition of the maximum emission ($\alpha_m = 1$, that is the condition of coherent emission from the successive bunches), the intensity is given by

$$\left.\frac{d^2 I}{d\omega\, d\Omega}\right|_{\theta=0} = \left.\frac{d^2 I_e}{d\omega\, d\Omega}\right|_{\theta=0} \cdot \frac{\beta^2}{\beta_r^2} \left(\rho_0 \frac{\lambda_+}{2}\right)^2 \sum_{1}^{\infty}{}_m A_m^2 \frac{\sin^2\left(N_+ \pi \nu_m\right)}{\nu_m^2} \tag{40}$$

where

$$\left.\frac{d^2 I_e}{d\omega\, d\Omega}\right|_{\theta=0} = \frac{N_r^2 e^2 b^2}{4c} \left(\frac{\beta_r}{1 - \beta_r}\right)^2 \tag{41}$$

is the single electron spontaneous radiation intensity in the forward direction in a plane polarized wiggler at $\theta = 0$ and $\nu_m = (\omega/\omega_+) - m$.

Equations (40) gives for the coherent power emitted on the m-th harmonic

$$\frac{dP_m}{d\Omega} = \left(\frac{dP_s}{d\Omega}\right) \cdot \frac{\beta^2}{\beta_r} \rho_0 \frac{\lambda}{4} N_r A_m^2 \tag{42}$$

where the spontaneous power $(dP_s/d\Omega)$ is

$$\frac{dP_s}{d\Omega} = \frac{e^2 b^2}{4} \rho_0 N_r \left(\frac{\beta_r}{1 - \beta_r}\right)^2 k_r \beta_r c \tag{43}$$

The relation (42) have a good physical meaning; in fact:

(a) the power emitted by electron bunches coherently interacting with a wiggler field is enhanced with respect to the spontaneous incoherent power by a factor $\sim \rho_0 (\lambda/4) N_r A_m^2$. This represents the number of particles packed with the right phase in a quarter of the emitted wavelength and in addition the bunch to bunch coherence.

(b) combining the equations (42) and (43) we can see that the coherent spontaneous power is proportional to the square of the beam current.

In addition from equation (39) we can also deduce the expected fractional linewidth (FWHM)

$$\frac{\Delta\omega}{\omega} \simeq \frac{1}{m\, N_+} \tag{44}$$

The bunching coefficients A_m are in general dependent on the adimensional parameter σ and on the adimensional time τ_b of interaction within the buncher.

A very rough evaluation of the first harmonic coefficient A_1 for short interaction time τ_b within the buncher field can be obtained from the (19) provided that the condition $\tau \ll 1$ is fulfilled.

In this case the maximum for A_1 in the drift region is given by

$$|A_1| \approx 10^{-4} \frac{\lambda^{(i)}\, E_o^{(i)}\, B_o^{(b)}\, L_b}{\sigma_\varepsilon} \tag{45}$$

The coherent emitted power (42) has a significant value with respect to the incoherent one only if

$$\rho_o \frac{\lambda^{(i)}}{4} N_r\, A_1^2 > 1 \tag{46}$$

Combining (45) and (46) we get an upper bound for the initial electron beam energy spread

$$\sigma_\varepsilon < 5.6 \cdot 10^{-5}\, \rho_o^{1/2}\, N_r^{1/2}\, \lambda^{(i)^{3/2}}\, E_o^{(i)}\, B_o^{(b)}\, L_b \tag{47}$$

In Table 3 are reported some numerically calculated harmonic coefficients A_m and the corresponding spontaneous coherent emitted power for two interesting practical cases. In the first the whole bunching is accomplished within the buncher wiggler (a drift space can be avoided with low energy beam, typically $\varepsilon < 50$ MeV) and refer to an input wavelength $\lambda = 10\ \mu$. In the second ($\varepsilon = 600$ MeV, Storage Ring Adone case) the device has the three sections and the input wavelength is $\lambda = 0.5\ \mu$.

Before to conclude this argument we note that the real limit in the above calculations (Eq. (36) and following), seems to be the approximation of the same velocity for all the particles.

TABLE 3

a) Electron beam: $\varepsilon = 30$ MeV; $\sigma_\varepsilon = 10^{-3}$; $I = 10$ mA

Input wave: $\lambda^{(i)} = 10\ \mu$; $E_0^{(i)} = 4$ MW/m

Wigglers: $\lambda^{(b)} = \lambda^{(r)} = 3.2$ cm; $N_b = N_r = 60$; $B_0^{(b)} = B_0^{(r)} = 5$ kG

m	1	2	3	4	5	6	7	8	9	10
$\lvert A_m \rvert$	.93	.70	.48	.32	.19	.10	.056	.026	.012	.0056
$P_m^{(co)}$ (Watts)	.43	.25	.12	.05	.02	$5\cdot10^{-3}$	$2\cdot10^{-3}$	$.3\cdot10^{-3}$	$70\cdot10^{-6}$	$16\cdot10^{-6}$

b) Electron beam: $\varepsilon = 600$ MeV; $\sigma_\varepsilon = 2.3\cdot10^{-4}$; $I = 0.1$ A

Input wave: $\lambda^{(i)} = 0.5\ \mu$; $E_0^{(i)} = 2.6$ MV/m

Wigglers: $\lambda^{(b)} = \lambda^{(r)} = 11.6$ cm; $N_b = N_r = 20$; $B_0^{(b)} = B_0^{(r)} = 4.5$ kG

m	1	2	3	4	5	6	7	8	9	10
A_m	.70	.30	.10	.03	.01	.003	$6\cdot10^{-4}$	10^{-4}	$2\cdot10^{-5}$	$4\cdot10^{-6}$
$P_m^{(co)}$ (Watts)	13	2.4	.3	.02	$3\cdot10^{-4}$	$3\cdot10^{-3}$	$10\cdot10^{-6}$	10^{-6}	$10\cdot10^{-9}$	$.5\cdot10^{-9}$

An evaluation of the coherence properties can be made with a single particle model. In this case each electron emits a wavelength

$$\lambda \simeq \lambda_r \Big/ \left(2\,\gamma_r^2\right) \tag{48}$$

owing to the momentum spread there is a linewidth

$$\frac{\Delta\lambda}{\lambda} \simeq 2\,\frac{\Delta\lambda_r}{\lambda_r} > 2\sigma_\varepsilon \tag{49}$$

The single bunch's radiation is then a superposition of monochromatic components distributed over a wavelength range given by

(49) or in the other words, a random succession of finite wave trains whose length is the well known coherence length[12]

$$\ell_{co} \approx \frac{\overline{\lambda}^2}{\Delta\lambda} \simeq \frac{\lambda}{2}\frac{\gamma_r}{\Delta\gamma_r} < \frac{\lambda}{2\sigma_\varepsilon} \tag{50}$$

This implies a restriction on the path difference between quasi-monochromatic rays emitted by the single bunches that is, the difference of optical path must be small compared with to coherence length (50) of the light.

If now we consider a wiggler with N_r periods the beam particles which cooperate for the coherent emission are, by the Lorentz transformations, those contained in an effective length

$$\ell \simeq N_r \lambda \tag{51}$$

Since this length is the maximum optical path difference between emitted waves, it must not to be greater than the coherence length (50). Then we deduce an upper limit for the periods number of the radiator wiggler

$$N_r < \frac{1}{2}\frac{\gamma_r}{\Delta\gamma_r} \tag{52}$$

5. STIMULATED EMISSION BY A BUNCHED BEAM

When an input wave travels with the electron beam in the radiator the same situation of the buncher is created. Therefore we will have a potential pattern

$$V^{(r)} = V_o^{(r)} \left\{ \frac{e^{j\left[k_+ z - \omega_+ t + \varphi\right]}}{2} + cc \right\} \tag{53}$$

running with a velocity $v_+ = (\omega_+/k_+) = \beta_r c$ when the frequency of the wave is resonant with the frequency of the radiator periodic magnetic field (synchronism condition). The peak value $V_o^{(r)}$ is found to be

$$V_o^{(r)} = \frac{e^2}{2\gamma_r \; m_o c} \; \frac{E_o^{(r)} \; B_o^{(r)}}{k^{(r)} \; k^{(i)}} \tag{54}$$

If now the initial phase of the electronic wave is $\varphi = \pi/2$ relative to the phase of the potential, the EB bunches are, at the beginning, acted upon the crest value of the radiation pressure force (also called ponderomotive force)

$$\vec{F}^{(r)} = -\frac{\partial V^{(r)}}{\partial z}\hat{z} = k_+ V_0^{(r)} \left\{ \frac{e^{j\left[k_+ z - \omega_+ t + \pi/2\right]}}{2j} + cc \right\} \hat{z} \tag{55}$$

Since the electron and the force wave have about the same velocity, the relative phase is conserved and the electron bunches are decelerated for reasonable radiator lengths. Therefore, they are stimulated to give up energy to the radiation field. The rate of energy interchange is

$$\frac{dW}{dt} = \frac{1}{e} \int \vec{F}^{(r)} \cdot \vec{J}\, d^3 \vec{r} \tag{56}$$

and from (37) and (55)

$$\frac{dW}{dt} = \int_{-L/2}^{+L/2} F_z\, \rho(z,t)\, \beta_z(z)\, c\, dz \tag{57}$$

In Equation (57) we ahve assumed the electron beam section less than the radiation beam section, otherwise it ought to multiply the right side by the filling factor.

Neglecting both the energy spread and the oscillatory motion along the z-direction when the wiggler is plane, and furthermore the decrease of the electron average velocity, we may assume in first approximation

$$\beta_z(r)c \cong \beta_r c \tag{58}$$

Equation (55) with (41) becomes

$$\frac{dW}{dt} = \int_{-L/2}^{+L/2} k_+ V_o^{(r)} \left\{ \frac{1}{2j} \exp\left[j\,(k_+ z - \omega_+ t + \pi/2) \right] + cc \right\} \cdot$$

$$\cdot\ \beta_r c\ \rho_o \sum_{o_m}^{\infty} \left\{ \frac{A_m}{2} \exp\left[jm\left(k_+^{(r)} z - \omega_+ t\right) \right] + cc \right\} dz \tag{59}$$

Since $k_+ = m^\star k_+^{(r)}$ (see Fig. 2), $m^\star$ being the harmonic number of the radiator wiggler period, we have

$$\frac{dW}{dt} = m^\star k_+^{(r)} V_o^{(r)} \beta_r c\ \rho_o$$

$$\cdot \sum_{om}^{\infty} \int_{-L/2}^{+L/2} \left\{ \frac{A_m}{4j} \left[\exp j\left[(m + m^\star) \left(k_+^{(r)} z - \omega_+ t\right) + \pi/2 \right] + \right. \right. \tag{60}$$

$$\left. \left. + \exp j\left[(m^\star - m) \left(k_+^{(r)} z - \omega_+ t\right) + \pi/2 \right] \right] + cc \right\} dz$$

Making this integral over an integer number of $\lambda_+^{(r)}$, all the oscillating terms give zero contribute to the average power loss except the $m = m^\star$ term, then

$$P_m^{(st)} = m\ k_+^{(r)} V_o^{(r)} \beta_r c\ \rho_o A_m \frac{L}{2} \tag{61}$$

With usual device parameters Eq. (61) transforms

$$P_m^{(st)} = \frac{1}{2\pi} \frac{e}{m_o c} E_o^{(r)} B_o^{(r)} \lambda\ \gamma_r \beta_r L A_m I \tag{62}$$

Since the average energy lost by a beam of length L in traversing the radiator is

$$\Delta W_m = \int_o^{L/\beta_r c} P_m^{(st)} dt \tag{63}$$

finally we obtain

$$\Delta W_m = \frac{1}{2\pi} \frac{e}{m_o c^2} E_o^{(r)} B_o^{(r)} \lambda \gamma_r L^2 A_m I \tag{64}$$

In Table 4 the stimulated peak power $P_m^{(st)}$ on the m-th harmonic for the case b) of Tab. 3 is listed. The remarkable difference with the spontaneous coherent power of Tab. 3 must be noticed.

TABLE 4

m	1	2	3	4	5	6	7	8	9	10
$P_m^{(st)}$(Watt)	$3{\cdot}10^3$	$1.3{\cdot}10^3$	440	130	40	13	2.6	.44	.09	.02

The only case of ultrarelativistic electrons is reported because outerwise the assumptions are too drastic.

It must emphasize that Equations (62) and (64) give only a crude estimate for low energy, ignoring the change of electrons energy during the interaction and the phase change due to the momentum spread.

By energy conservation arguments, all energy lost by the beam transforms into radiation energy. Gain (G) in a section of the electron beam of volume V is taken to be the fractional increase of the radiation energy in that volume and is given by

$$G_m = \frac{W_m}{\varepsilon_o E_o^2 V} = \frac{1}{2\pi \varepsilon_o} \frac{e}{m_o c^2} \frac{B_o^{(r)}}{E_o^{(r)}} \gamma_r \lambda L A_m \left(\frac{I}{S}\right) \tag{65}$$

where S is the beam section. This equation gives a gain 2-3 orders of magnitude greater than the analogous in the free electron laser scheme[13]. This gain lowers the threshold current for generation.

In addition the current is further depressed by the consistent radiation production by the spontaneous coherent emission. In fact in our device the radiation intensity is proportional to

$$\left(\vec{E}_c + \vec{E}_i\right)^2 = E_c^2 + E_i^2 + 2E_c E_i \cos \Psi \tag{66}$$

Here the angle Ψ is the relative phase between the coherent and input fields, E_c^2 and E_i^2 are the coherent and input term respectively. The interference term $2 \cdot \vec{E}_c \cdot \vec{E}_i$ is the term of the stimulated emission. It can be much higher that the coherent one if $E_i \gg E_c$ and $\cos \Psi = 1$.

6. CONCLUSIONS

The mechanism for the radiation production in the converter and in the TOK allows us to say that these generators are in known class of hormodotrons[14] and klystrons. The emitted power is higher than a FEL because the initial electrons phases with respect to the radiation pressure are rearranged so as to collect a major number of electrons in a radiation "bin" with correct phase for the emission.

The enhancement of the gain allows to depress the requirement on the beam quality and/or to use shorter wiggler. In order to avoid two separate wigglers (when the device has not be inserted in a circular accelerator), the reflex TOK of Fig. 7 can be used, in

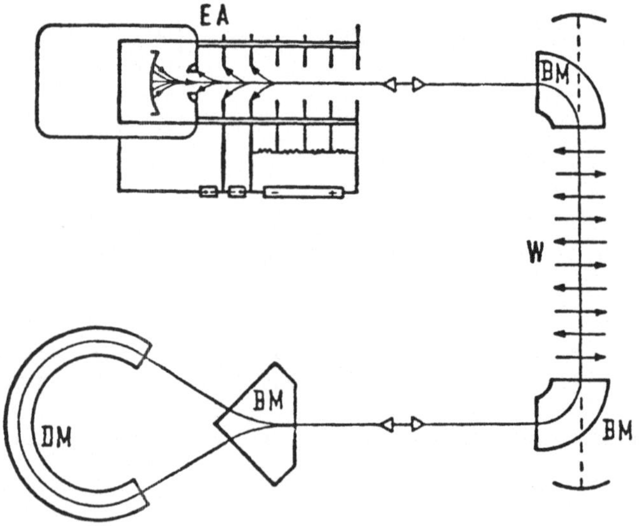

Fig. 7. Layout of the reflex TOK with a ring path for the radiation.

complete analogy with the reflex klystron. When the wigglers are plane polarized, since the cross section of the magnet can be large, the ring TOK of the Fig. 8 can be alternatively used. In this configuration the inverse Compton scattering is avoided.

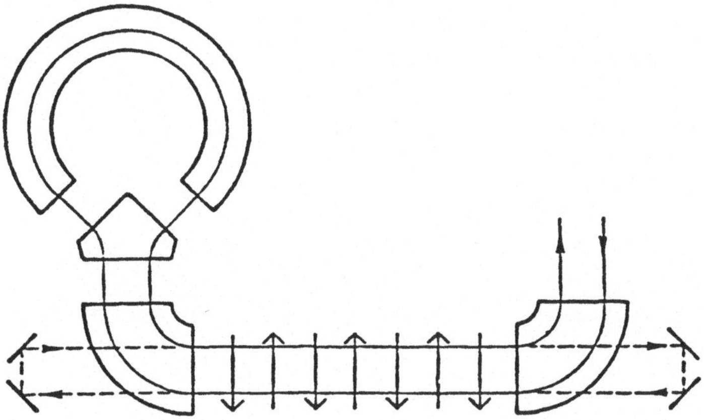

Fig. 8. Layout of the reflex TOK with a ring path for the radiation

Since the bunching is unharmonic, as it is shown by the calculated coefficients of the harmonics, it looks likely to obtain an efficient up-conversion matching the radiator wavelength with a higher harmonic. Thus, generators of submillimeter wavelength can be thought with a power of several watts starting with available microwave generators (10-1 mm).

The improvements reached with the TOK make it better suited than the FEL either for the insertion in the present storage ring where the straight sections are not enough long or for an efficient UV production.

APPENDIX

Transformation of the initial electrons distributions function from the LAB to the $(EBS)_b$ system.

We start with the gaussian energy distribution function in the LAB system

$$\rho(z,\varepsilon) = \frac{\rho_o}{\sqrt{2\pi}\,\sigma_\varepsilon\,\varepsilon_o} \exp\left\{-\frac{\left(\varepsilon-\varepsilon_o\right)^2}{2\,\sigma_\varepsilon^2\,\varepsilon_o^2}\right\} \tag{A-1}$$

Since in our case $p_\perp = 0$ will be also

$$\rho(z,\varepsilon)\, d\varepsilon = \rho(z,p_z)\, dp_z \tag{A-2}$$

and

$$\varepsilon^2 = p_z^2 c^2 + m_o^2 c^4 \tag{A-3}$$

thus

$$\frac{d\varepsilon}{dp_z} = \frac{p_z c^2}{\varepsilon} \simeq \beta_b c \tag{A-4}$$

Combining Equation (A-2) and (A-4) we get

$$\rho(z,p_z) \simeq \beta_b c\, \rho(z,\varepsilon) \tag{A-5}$$

The passage from the LAB to the $(EBS)_b$ is carried out with the Jacobian associated to the transformation

$$\rho_b(z_b,p_{bz}) = J\, \rho(z,p_z) \tag{A-6}$$

The Lorentz transformations are

$$\begin{aligned} z &= \gamma_b(z_b + \beta_b c\, t_b) \\ p_z &= \gamma_b(p_{bz} + (\beta_b/c)\, \varepsilon_b) \\ \varepsilon &= \gamma_b(\varepsilon_b + \beta_b c\, p_{bz}) \end{aligned} \tag{A-7}$$

and the associated Jacobian is

$$J = \begin{vmatrix} \dfrac{\partial z}{\partial z_b} & \dfrac{\partial z}{\partial p_{bz}} \\ \dfrac{\partial p_z}{\partial z_b} & \dfrac{\partial p_z}{\partial p_{bz}} \end{vmatrix} = \left(1 + \beta_b c \ \frac{p_{bz}}{\varepsilon_b}\right) \tag{A-8}$$

Since the particles in the $(EBS)_b$ are not relativistic we get

$$J \simeq 1 \tag{A-9}$$

and in the same approximation for $(\varepsilon - \varepsilon_o)/\varepsilon_o$ with eq. (A-7) we obtain

$$\frac{\varepsilon - \varepsilon_o}{\varepsilon_o} \simeq \frac{p_{bz}}{m_o c} \tag{A-10}$$

Remembering also (A-5), the final expression for the momentum distribution function in the $(EBS)_b$ is

$$\rho_b(z_b, p_{bz}) = \frac{\rho_o}{\gamma_b \sqrt{2\pi}\, \sigma_\varepsilon \, m_o c} \exp\left\{- \frac{p_{bz}^2}{2\, m_o^2 c^2 \, \sigma_\varepsilon^2}\right\} \tag{A-11}$$

With the dimensionless parameters (16) and dimensionless average electron density

$$\rho_{bo} = \frac{\rho_o}{\gamma_b} \left(m_o \ V_{bo}^{(b)}\right)^{-1/2} \tag{A-12}$$

we rewrite (A-11) classically

$$\rho_b(\zeta, q, o) = \frac{\rho_{bo}}{\sqrt{2\pi\sigma}} \exp\left\{- \frac{q^2}{2\sigma}\right\} \tag{A-13}$$

REFERENCES

1. V. Stagno, G. Brautti, T. Clauser and I. Boscolo, Nuovo Cimento, 56B, 219-236 (1980)
2. F. De Martini, Phys. of Quantum Electronics (Addison-Wesley, Reading, Mass, 1980), v. 7
3. J.P. Blewett and R. Chesman, J. Appl. Phys. 48, 2692 (1977)
4. G. Brautti, T. Clauser, A. Rainò and V. Stagno, Nucl. Instrum. and Meth. 153, 357 (1977)
5. H.A. Abawi, F.A. Hopf and P. Meystre, Phys. Rev. A 16, 666 (1977)
6. A. Bambini, A. Renieri and S. Stenholm, Phys. Rev. A 19, 2013 (1979)
7. I. Boscolo and V. Stagno, Nuovo Cimento, 58B, 267 (1980)
8. G.E. Forsyte and W.R. Wlasov, Finite Difference Method for Partial Differential Equations, Wiley, New York 1960
9. See for example, H. Bruck, Accélérateurs Circulaires de Particules, Press Universitaire de France (1966)
10. J.D. Jackson, Classical Electrodynamics, Wiley, New York 1960
11. B.M. Kincaid, J. Appl. Phys. 48, 2684 (1977)
12. M. Born, E. Wolf, Principles of Optics, Pergamon Press 1959
13. W.H. Louisell, J.F. Lam, D.A. Copeland and W.B. Colson, Phys. Rev. A 19, 288 (1979); J.M.J. Madey, J. Appl. Phys. 42, 1906 (1971); F.A. Hopf, P. Meystre, G.T. Moore and M.O. Scully, Phys. of Quantum Electronics (Addison-Wesley, Reading, Mass., 1977) V. 5
14. M.D. Sirkis and P.D. Coleman, J. Appl. Phys. 28, 944 (1957)
15. W.B. Colson, Free Electron Laser Theory, Ph. D. Thesis, Stanford University (1977)
16. I. Boscolo, G. Brautti, T. Clauser and V. Stagno, Appl. Phys. 47, 19 (1979)

FUNDAMENTALS OF ACCELERATORS FOR SINGLE-PASS FREE ELECTRON LASERS*

Alfredo Luccio

Brookhaven National Laboratory

Upton, New York 11973, U.S.A.

ABSTRACT

The characteristics of electron accelerators suitable for single pass free electron lasers are discussed. In particular, their beam qualities: energy, energy spread, current, emittance and bunch time structure are considered, in general and on a few examples. The matching of an electron beam to a tapered wiggler is finally dealt with in some detail.

I. INTRODUCTION

In "classical" free electron lasers (FEL) a beam of relativistic electrons enters an undulator magnet together with a laser beam of wavelength

$$\lambda = \frac{\lambda_w}{2\gamma_r^2} (1 + \kappa^2) \tag{1}$$

where λ_w is the undulator period, γ_r the "resonant" energy of the electrons in rest mass units (mc^2), and

$$\kappa = \frac{e}{2\pi\ mc} B_w \lambda_w = \frac{1}{2\pi} B^* \lambda_w \tag{2}$$

* Work supported by the U.S. Department of Energy

is an undulator parameter, with B_w the undulator r.m.s. magnetic field, and

$$B^* = \frac{B_w}{mc} \tag{3}$$

the "reduced" field.

In the undulator field, which in the electron rest frame appears as an incoming electromagnetic wave of wavelength λ, stimulated scattering takes place, with the result that the incoming laser beam intensity is amplified at the expenses of the energy of the electrons. Moreover, if we build an optical cavity around the undulator, continuous amplification of the radiation produced spontaneously by earlier electrons may result in a free electron laser oscillator.

For the theory of FEL, we will refer mainly to the paper by C. Pellegrini[1], where a classical (i.e. not quantum) presentation is given. We recall here only that the wavelength of the spontaneous radiation from the undulator depends on the observation angle θ with the undulator axis according to

$$\lambda = \frac{\lambda_w}{2\gamma^2} (1 + \kappa^2 + \theta^2\gamma^2) \tag{4}$$

and that the spectral distribution is naturally broad by

$$\frac{\Delta\lambda}{\lambda} = \frac{1}{2N_w} \tag{5}$$

where N_w is the number of periods in the undulator. For this, see e.g. the paper by A. Hofmann.[2]

Free electron lasers can be built to operate on several wavelengths, by adjusting electron energy and undulator parameters, according to Eq(1). For each wavelength range, different accelerators have been used or proposed, like microtrons, electron linacs, induction accelerators, dc-machines, or storage rings, whose characteristics are widely different. Some of these accelerators can be used for single pass free electron laser operation, by sending an electron beam pulse or a train of pulses through the undulator and dumping the beam thereafter. Others, and notably the storage ring, can be used by recirculating the electron beam through the accelerator.

The difference between the two cases is that in a single pass the operation of the accelerator is not affected by the FEL itself, while in the latter the electron beam qualities, like energy, energy spread and possibly emittance, are modified by FEL action and have to be accounted for in the steady state operation of the accelerator.

In this paper we will analyze some of the aspects of single pass FEL's, and particularly we will discuss electron beam qualities suitable for ordinary and variable period (tapered) wigglers, that promise to expand greatly laser gain.

II. ELECTRON BEAM REQUIREMENTS FOR FEL OPERATION

In addition to the wavelength λ, another quantity of importance is the gain per pass, defined as the relative enhancement of the radiation intensity, per electron bunch pass through the undulator. The gain per pass can be written as follows, for monochromatic electrons, or more generally in the homogeneous broadening regime

$$G_o = \left< \frac{\Delta\gamma}{\gamma} \right> \frac{\gamma I_p}{\Sigma} \frac{mc}{eW} \tag{6}$$

where $<\Delta\gamma/\gamma>$ is the energy lost, on the average, by the electrons, I_p the peak electron beam current, Σ the cross section of the e.m. wave, assumed equal to the electron beam cross section, and W the e.m. energy density.

For a constant period undulator, it is obtained from Eq. (6)

$$G_o = - 32\sqrt{2}\,\pi^2 \lambda_w^{3/2} \lambda^{1/2} \frac{\kappa^2}{(1+\kappa^2)^{3/2}} \frac{I_p}{I_A} \frac{N_w^3 f(x)}{\Sigma} \tag{7}$$

with $I_A = 1.7\ 10^4$ ampere (Alfven current). The function f(x) appearing in Eq. (7), called the gain function, is defined as

$$f(x) = \frac{1}{x^3} (\cos x - 1 + \frac{1}{2} x \sin x) \tag{8}$$

with

$$x = 4\pi N_w \frac{\gamma - \gamma_r}{\gamma_r} \tag{9}$$

The gain function is shown in Fig. 1.

The electron beam entering the undulator must satisfy some requirements to allow for proper operation of a FEL, with a desired wavelength and enough gain.

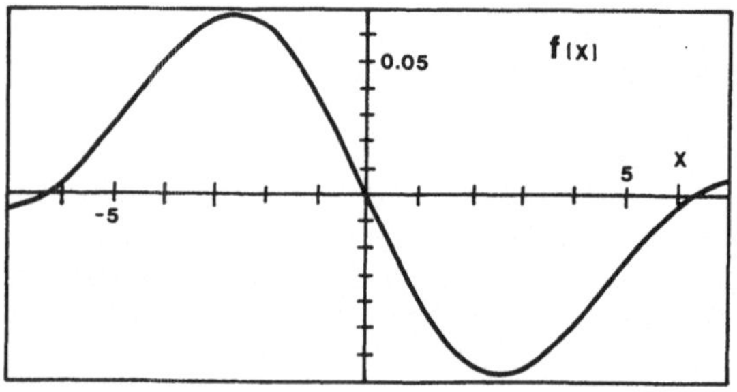

Fig. 1. The gain function.

Electron energy.

Depending on the wavelength chosen, from Eq. (1) and from the values of the parameter κ obtainable in practical undulators[3], we find that the energy E of the electrons can vary from a few MeV (FEL in the sub-millimeter to infrared range) to several hundred MeV (infrared to ultraviolet, and possibly soft X-rays[4]). As the theory shows, the electron energy must be actually slightly higher than the resonant energy that matches Eq. (1) for amplification to occur, since the gain per pass contains the function f(x) of Fig. 1, which is negative (and G_o positive) for $\gamma > \gamma_r$.

Electron energy spread.

If the energy spread of the electron beam is comparable with the width of the gain function, clearly the gain per pass, obtained by folding the monochromatic gain per pass G_o with the energy distribution of the electrons, vanishes. Fig. 1 shows that a limit for the energy spread can be set to

$$\frac{\delta E}{E} = \frac{\delta\gamma}{\gamma} \lesssim \frac{1}{2N_w} \tag{10}$$

Eq. (10) says that in order to insure a non-zero gain, the energy spread of the electrons must not exceed the homogeneous broadening of the spontaneous radiation.

Electron beam emittance.

Eq. (7) shows that the gain is inversely proportional to the photon beam cross section Σ, when the electron beam cross section is matched to it. To obtain a high gain, the electron beam c.s. should not exceed the photon beam c.s. Similar considerations hold for the electron and photon beam angular divergence in the interaction region.

The minimum cross section of the radiation beam is limited by diffraction in the undulator gap to

$$\Sigma = \lambda_w L_w \tag{11}$$

where $L_w = N_w \lambda_w$ is the undulator length. From Eq. (4) it is seen that a spread in the observation angle entails a line broadening. To keep this within the limit defined by Eq. (5), we will consider an angular spread of the radiation not greater than (around $\theta=0$)

$$\delta\theta = \frac{1}{\gamma\sqrt{2N_w}} \tag{12}$$

From Eqs. (11) and (12), for the four dimensional emittance of the electron beam we will therefore set the limit

$$\varepsilon_x \varepsilon_y < \lambda L_w \frac{1}{\gamma^2 2N_w} \simeq \lambda^2 \tag{13}$$

Bunch length.

The laser linewidth is in general much narrower than the limit set by Eq. (5), due to the electron beam time structure and depending on the optical cavity (in the oscillator case).

From a Fourier analysis of the light pulses, which have the same length of the electron beam pulses, approximatively, one sees that the laser output from a FEL consists of an envelope of spectral lines (modes), as it is shown in Figure 2. The width of the envelope is inversely proportional to the bunch transit time τ_e through the undulator

$$\delta\omega_e = \frac{1}{\tau_e} = \frac{2\pi c}{\sigma_e} \tag{14}$$

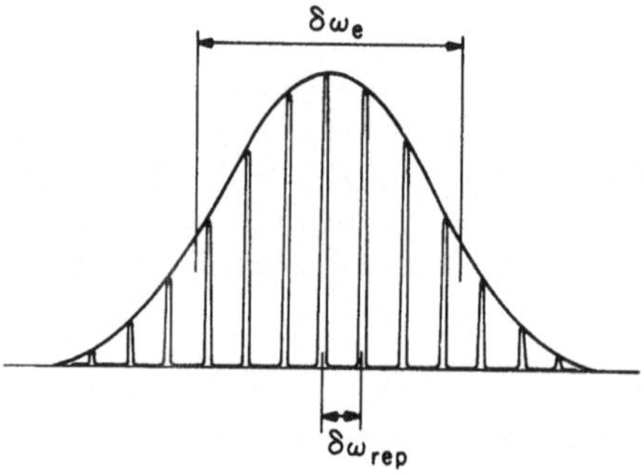

Fig. 2. Structure of a laser line.

σ_e being the (spatial) electron bunch length. The separation between adjacent modes is inversely proportional to the time separation between beam bunches τ_{rep}, and finally the mode width is inversely proportional to the coherence time of the pulse train τ_c.

In an optical cavity, since τ_{rep} must be matched to the photon transit time between mirrors, it is

$$\delta\omega_{rep} = \frac{c}{2L_c} \tag{15}$$

where L_c is the length of the cavity and c the speed of light.

$\delta\omega_c$ is limited in practice by mirror microphonics.[5]

By linewidth, one generally means the expression (14) for the envelope. Write then

$$\frac{\delta\omega}{\omega} = \frac{\lambda}{\sigma_e} \tag{16}$$

This tells us that long bunches are necessary, or a continuous beam, for good resolution. In some cases, debunching of the beam extracted from an accelerator for single pass FEL operation may be necessary.

It is noteworth that electron beam energy spread and emittance limitations can be treated in a similar way, when discussing the requirements of the accelerators.

Let us limit ourselves to the case of a helical undulator and an electron beam with cylindrical symmetry. As it was shown by J. Blewett and R. Chasman[6], in a helical undulator the trajectories of the electrons are helices with wavelength equal to the undulator period λ_w, on which oscillations with much longer wavelength, namely

$$\lambda_f = \sqrt{2}\,\frac{\gamma}{\kappa}\,\lambda_w \qquad (17)$$

are superimposed. Only for electrons injected on the gyration radius

$$r_g = \frac{\lambda_w}{2\pi}\,\frac{\kappa}{\gamma} \qquad (18)$$

(practically on the undulator axis), with a transverse velocity component, in units of c, equal to $\beta_\perp = \kappa/\gamma$, and along the helical path, the amplitude of the long wavelength oscillation is zero.

Define as α the angle between the particle trajectory and the helix (slant angle). We will call central trajectory the one defined by the initial conditions at injection into the undulator

$$r_o = r_g \simeq 0,\ \beta_{\perp o} = \kappa/\gamma,\ \alpha_o = 0 \qquad (19)$$

A particle injected at a finite radius r_0 and with a finite angle, different from $\beta_{\perp o}$, will experience, because of the long wavelength oscillations, a different magnetic field than a central electron (see Figure 3). This change in the effective B_w along the trajectory is equivalent to a change in the undulator parameter κ of Eq. (2), and consequently to a change in the resonant γ_r.

If we define now a transverse beam emittance

$$\varepsilon_\perp = \pi\, r\, \delta\beta_\perp \qquad (20)$$

and differentiate Eq. (1) with respect to κ, we obtain a new expression for the total equivalent energy spread, that can be compared

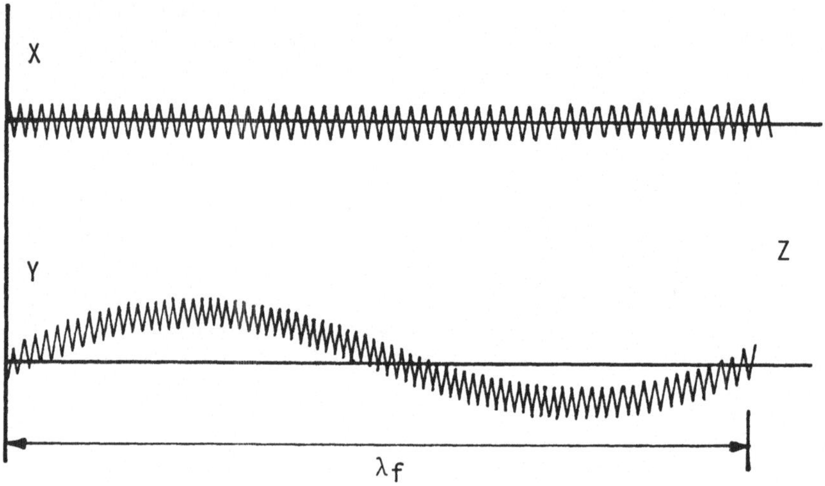

Fig. 3. Trajectories in a helical wiggler. (Adapted from J. Blewett and R. Chasman[6]).

with Eq. (10)[7]

$$\left(\frac{\delta\gamma}{\gamma}\right)^2_{tot} = \left(\frac{\delta\gamma}{\gamma}\right)^2 + \left(\frac{\delta\gamma}{\gamma}\right)^2_{eff} \tag{21}$$

where it is

$$\left(\frac{\delta\gamma}{\gamma}\right)_{eff} = \frac{1}{2}\left(\frac{2\pi}{\lambda_w}\right)^2 \frac{\kappa^2}{1+\kappa^2}\left[r^2 + \bar{\lambda}_f^2 \frac{\varepsilon^2}{\pi^2 r^2}\right] \tag{22}$$

III. EXAMPLES OF EXISTING OR PROPOSED ACCELERATORS FOR SINGLE PASS FEL

Among many existing or proposed accelerator systems for single pass free electron lasers, let us select a few as examples.

Example i): Microtron.

These following data, supplied by A. Renieri and A. Vignati, refer to a conventional microtron at the Frascati-CNEN National Laboratories, Italy.[8] The microtron is being presently upgraded to be used for a single pass FEL in the λ = 10 to 600 μm range.

The data are presented in Table 1.

Table 1. Frascati microtron.[8]

Energy		E	$= 20$ MeV
Energy spread		$\frac{\delta E}{E}$	$= 0.1$ %
Emittance	horizontal	ε_x	$= 6\text{x}10^{-5}$ m-rad
	vertical	ε_y	$= 1.5\text{x}10^{-5}$ m-rad
Peak current		I_p	$= 400$ mA
Beam time structure	macropulse	τ_c	$= 12$ μsec
	rep. time	τ_{rep}	$= 300$ psec
	micropulse	τ_e	$= 25$ psec

Example ii): Superconducting Linac.

This is the one used by J. Madey and Coll. at Stanford University, USA.[9] It has been used for an amplification experiment at $\lambda = 10.6$ μm, and for an oscillator experiment at $\lambda = 3.4$ μm, in conjunction with an helical superconducting undulator. Its main parameters are given in Table 2.

Table 2. Stanford linac.[9]

Energy	E	$= 43.5$ MeV
Energy spread	$\frac{\delta E}{E}$	$= 0.05$ %
Emittance	ε	$= 6\text{x}10^{-8}$ m-rad
Peak current	I_p	$= 2.6$ A
Micropulse	τ_e	$= 4.3$ psec
Repetition time	τ_{rep}	$= 84.6$ nsec
Coherence time	τ_c	$= 0.1$ sec (?)

Example iii): Desirable Linac.

Tells W.E. Stein[10], reporting the results of a 1979 Linac Workshop held at Los Alamos Laboratories, USA, that a "desirable" linac for free electron lasers should have the characteristics listed in Table 3.

Table 3. Desirable Linac.[10]

Energy	E	= 20 to 100 MeV
Energy spread	$\frac{\delta E}{E}$	= ± 1 %
Emittance	ε	≃ 10^{-6} m-rad
Peak current	I_p	= 25 to 250 A
Micropulse	τ_e	= 10 psec
Coherence time	τ_c	= 10 to 50 μsec, long, or CW

Example iv): Electrostatic Generator.

This has been proposed by L.R. Elias for the Quantum Institute, University of California at Santa Barbara,[11] and will be used to produce radiation with λ = 360 μm, either in single pulse or in energy recover mode. Its parameters are in Table 4.

Table 4. QIUCSB DC accelerator.[11]

Energy	E	= 3 to 10 MeV
Emittance	ε	= $1.22\pi\ 10^{-4}$ m-rad
Peak current	I_p	= 2 to 100 A
Pulse length	τ_e	= fraction of μsec
(With recovery)		$\rightarrow \infty$

Example v): Induction Accelerator.

Described e.g. by J.E. Leiss[12], it is characterized by very high current, one-pulse operation. The data of Table 5, supplied by A. Faltens[13], refer to a structure studied for the Livermore Laboratory, University of California.

Table 5. Livermore induction accelerator.[13]

Energy	E	$= 5$ MeV
Energy spread	$\frac{\delta E}{E}$	$= 0.01$ %
Emittance	ε	$= 10^{-3}$ to 10^{-4} m-r
Current	I	$= 500$ A
Pulse length	τ_e	$= 60$ nsec

IV. LONGITUDINALLY VARYING PERIOD UNDULATOR TO INCREASE FEL GAIN

For the case of a single pass FEL, where the quality of the electron beam after the passage through the undulator is of no concern, it has been proposed to expand the efficiency of the energy exchange between electrons and radiation field by a spatial modulation of the undulator parameters.

We will analyze here in some detail one of the proposed schemes, namely the use of an undulator with continuous longitudinally varying period, as proposed by N. Kroll, P. Morton and M. Rosenbluth.[14] We will review the general considerations on the electron beam discussed in the previous sections, as applied to the present case.

Other proposals, like the radial varying field flat wiggler by J. Madey and R. Taber[15], or the Optical Klystron by the Novosibirsk Group[16] will not be discussed here.

The principles of operation of a single pass FEL with longitudinally varying period undulator are the following:

When studying the electron motion in "phase space", ϕ, γ, it appears that under certain conditions an optical "bucket", similar to the RF bucket of the Courant-Snyder theory of synchrotron motion may develop. This bucket is centered on some values, ϕ_r, γ_r' that we will call resonant phase and resonant energy, has widths $\Delta\phi_B$ and $\Delta\gamma_B$, and traps some of the electrons, which will be confined in it thereafter on phase-space oscillatory trajectories.

The variable period makes the bucket move, as it will be shown next, toward smaller and smaller γ values. Provided that this motion is adiabatic enough, the bucket drags the trapped electrons toward lower and lower energy states. The energy lost by the electrons is radiated into laser light, and the e.m. power thus generated can be much more, on the average, than it is obtained by standard undulators.

It is true that the electrons that are not trapped gain energy. However, this gain is not enough to counterbalance the energy lost by the trapped ones.

The theory[7,8] shows that the bucket center coordinates are

$$\gamma_r = \left[\frac{\lambda_w(1+\kappa^2)}{2\lambda}\right]^{1/2} \tag{23}$$

(see also Eq. (1)), and

$$\sin\phi_r = \frac{1}{2\kappa E^*}\frac{d}{dz}(\gamma_r^2) \tag{24}$$

where E^* is the normalized electric field

$$E^* = \frac{eE}{mc^2} \tag{25}$$

and z is the longitudinal coordinate.

The bucket can be represented by a trapping potential well

$$V(\phi) = \cos\phi + \phi\sin\phi_r \tag{26}$$

shown in Fig. 4, of width $\Delta\phi_B$ and of depth ΔV.

It is

$$\Delta\phi_B = 3\pi - \phi_r - \phi^* \tag{27}$$

where ϕ^* is a solution of the equation

$$V(\phi^*) = V(3\pi - \phi_r) \tag{28}$$

and

$$\Delta V = V(\phi^*) - V(\phi_r) \tag{29}$$

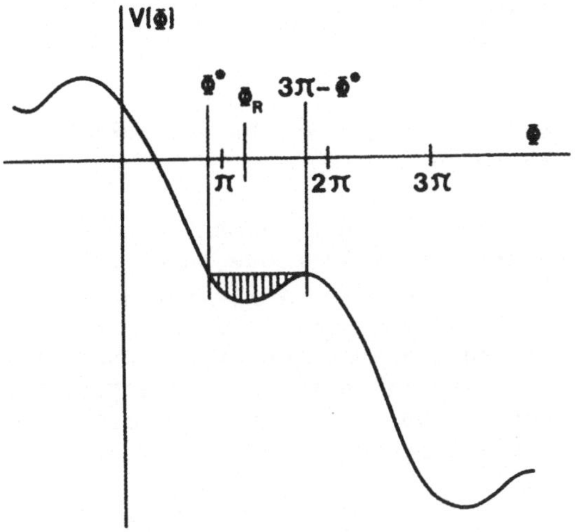

Fig. 4. Variable period undulator. Potential well. (Data of Table 6).

The bucket height is

$$\Delta\gamma_B = \left[\frac{\lambda_w}{2\pi}\,\kappa\, E^*\,\Delta V\right]^{1/2} \tag{30}$$

Eq. (24) shows that for the bucket to form at all, the e.m. intensity must be large enough. From Eq. (23) it appears that the center of mass of the bucket γ_r goes down if the period λ_w decreases with z. Eq. (30) shows that the bucket height depends on the electric field intensity.

Since for a monochromatic beam the number of electrons captured in the bucket is proportional to $\Delta\phi_B$, we can write for the efficiency of energy delivery to the radiation field

$$\eta = \frac{\Delta\phi_B}{2\pi} < \frac{\Delta\gamma}{\gamma} >_{(L_w)} \tag{31}$$

where, from Eq. (23)

$$< \frac{\Delta\gamma}{\gamma} > (L_w) = \left[\frac{\lambda_w(L_w)}{\lambda_w(0)} \; \frac{1+\kappa^2(L_w)}{1+\kappa^2(0)}\right]^{1/2} - 1 \tag{32}$$

The subscripts "0" and "L_w" mean: at the entrance and at the exit of the undulator, respectively.

The gain per pass for this arrangement can be calculated by inserting $< \Delta\gamma/\gamma >_{L_w}$ from Eq. (32) above into (6). The value (32) can be much higher than the value $1/2N_w$ valid for a standard undulator.

Let us now return to the question of the characteristics of an electron beam to match a variable period undulator. Many of the considerations of Sect. 2 apply. However, the total energy spread given by Eq. (21) must not exceed, in the present case, the relative height of the bucket.

Examine the behaviour of a "real" beam, with finite radius and a spread in the angle $\beta_\downarrow$ and in the angle α.

A finite beam size is equivalent to an energy spread, from Eq. (22)

$$\left.\frac{\delta\gamma}{\gamma}\right|_r = \frac{1}{2} \frac{\kappa^2}{1+\kappa^2} \left(\frac{2\pi r}{\lambda_w}\right)^2 \tag{33}$$

and hence a limiting value for capture in the bucket is

$$r_{capt} = \frac{\lambda_w}{2\pi} \left[2 \, \frac{1+\kappa^2}{\kappa^2} \, \frac{\Delta\gamma_B}{\gamma}\right]^{1/2} \tag{34}$$

An angular divergence of the beam around the "center" value κ/γ is also equivalent to an energy spread

$$\left.\frac{\delta\gamma}{\gamma}\right|_\beta = \frac{1}{2} \frac{\kappa^2}{1+\kappa^2} \left(\frac{2\pi}{\lambda_w}\right)^2 \lambda\!\!\!^-{}_f^2 \left(\beta_\downarrow \pm \frac{\kappa}{\gamma}\right)^2 \tag{35}$$

and a limiting value for $\beta_\downarrow$ is also

$$\beta_{\downarrow,capt} = \frac{\kappa}{\gamma} \pm \frac{r_{capt}}{\lambda\!\!\!^-{}_f} \tag{36}$$

Analogously, for the angle α that the vector $\vec{\beta}_\downarrow$ makes with the tangent to the ideal paraxial trajectory, a limit is

$$\alpha_{capt} = \pm \frac{2\pi}{\sqrt{2}} \frac{r_{capt}}{\lambda_w} \tag{37}$$

Now, if we recall the expression (20) for the transverse emittance, and we define a parameter ξ, in analogy with κ

$$\xi = \frac{e}{2\pi mc^2} E \lambda_w = \frac{1}{2\pi} E^* \lambda_w \tag{38}$$

we find that a limiting value for the emittance (squared) is

$$\varepsilon^2 < 2 \Delta V \frac{\xi}{\kappa} \lambda^2 \tag{39}$$

to be compared with the value appearing in (13).

V. A NUMERICAL EXAMPLE

The validity of the model has been tested on a linearly tapered wiggler by computer simulation. A description of the model and of the results can be found in detail in another paper to this School.[7]

The capture mechanism is illustrated in Fig. 5, from the quoted paper (similar plots appear elsewhere in the Literature S.A. Mani[17]). Figures 6 and 7 show the computed r_{capt} and β_{capt}, which appear in good agreement with the values from Eqs. (34) and (36), also shown for comparison.

Some of the parameters used in the simulation are listed in Table 6.

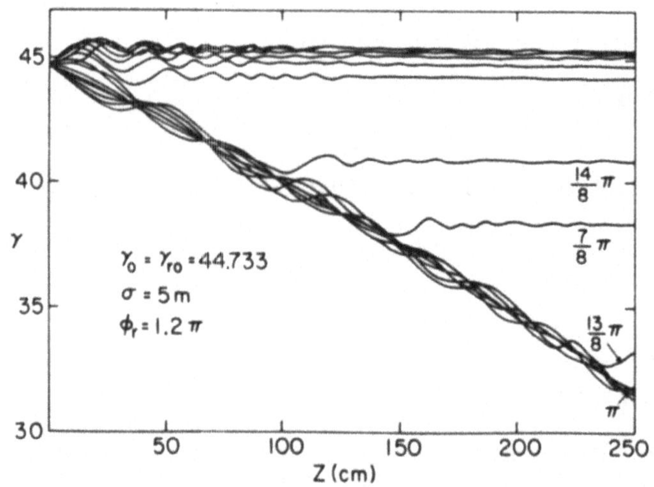

Figure 5. Normalized energy vs. distance in a tapered undulator Monochromatic electron bundle injected on helix.

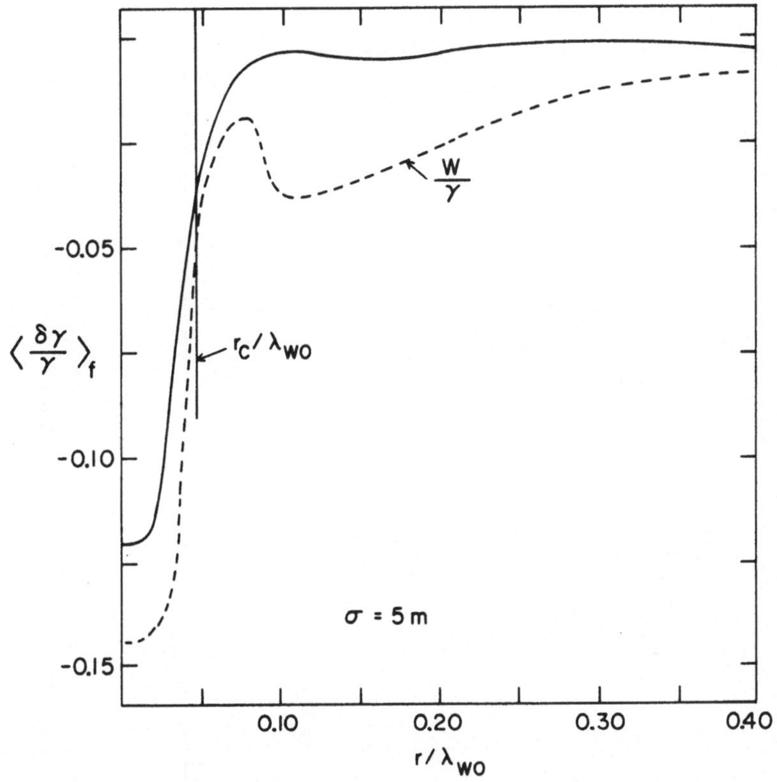

Fig. 6. Average energy loss and relative statistical standard deviation as a function of initial radius. r_{capt} is also shown for comparison.

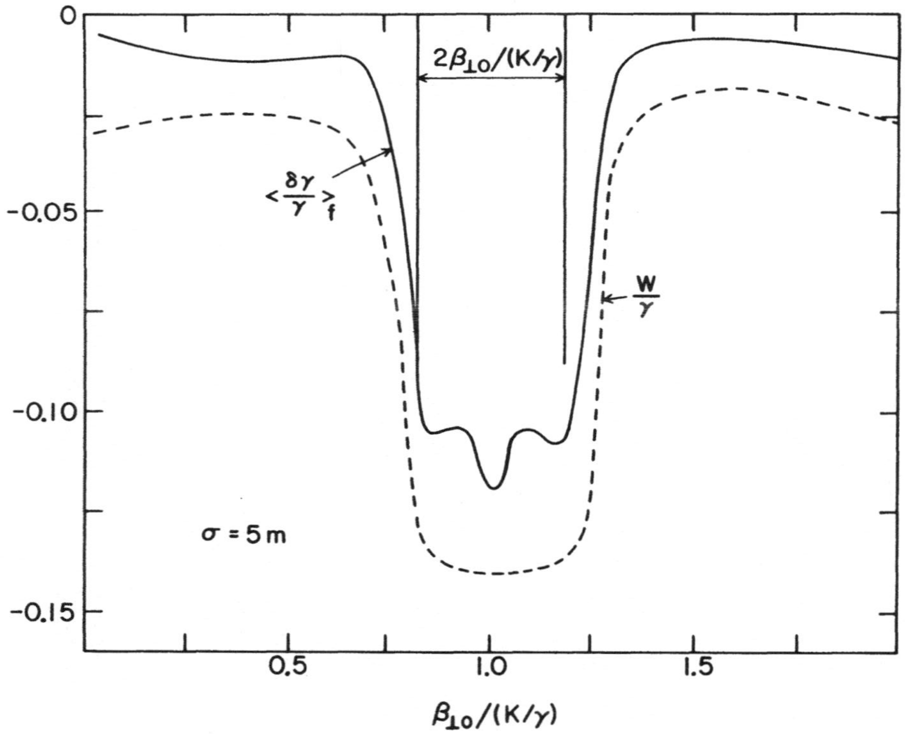

Fig. 7. Average energy loss and standard deviation as a function of the initial transverse velocity. Tapered wiggler.[7]

Table 6. Single pass FEL parameters†. Example.

Electron resonant energy	γ_r	= 45
Undulator period	$\lambda_w(0)$	= 2 cm
Undulator length	L_w	= 2.5 m
Undulator parameter	κ	= 1 (kept const.)
Radiation wavelength	λ	= 10 μm
Radiation parameter	$\xi(0)$	= 1
Potential well width	$\Delta\phi_B$	= π
P. W. center	ϕ_r	= 1.2 π
P. W. depth	ΔV	= 0.5
Bucket relative height	$\Delta\gamma_B/\gamma$	= 0.016
Capture radius, relative	r_c/λ_w	= 0.04
Capture angles, relative	$\beta_{\downarrow,capt}/(\kappa/\gamma)$	= 1 ± 0.180
	$\alpha_{capt}/(\kappa/\gamma)$	= ± 0.180

†Exact values in the paper quoted.[7]

VI. REFERENCES

1. C. Pellegrini, IEEE Trans. Nucl. Sci., NS-26:3791 (1979).
2. A. Hofmann, Nucl. Inst. and Methods, 152:17 (1978)..
3. A. Luccio, Paper II-D to this School.
4. C. Pellegrini, Proc. Japan-USA Seminar on Synchrotron Radiation Facilities, Honolulu, Hawaii, November 5-9, 1979, Report BNL-27317.
5. J. M. J. Madey and J. N. Eckstein, Proc. Workshop on free-electron lasers, G. Scoles Ed., Riva del Garda, Italy, June 4-6, 1979, p. 8.
6. J. Blewett and R. Chasman, J. of Appl. Phys., 48:2692 (1977).
7. A. Luccio and C. Pellegrini, Report BNL-27598, March 1980, also: Contributed paper to this school.
8. A. Renieri and A. Vignati, private communication.
9. D. A. G. Deacon, L. R. Elias, J. M. J. Madey, G. J. Ramian, H. A. Schwettman and T. I. Smith, Phys. Rev. Lett., 38:892 (1977).
10. W. E. Stein, Proc. Linear Accelerator Conf., Montauk, NY (1980), Brookhaven Lab. Report, J. Blewett Ed., p. 57.
11. L. R. Elias, Univ. of California at Santa Barbara, Report QIFEL 004/80 (1980).
12. J. E. Leiss, IEEE Trans. Nucl. Sci., NS-26:3870 (1979).
13. A. Faltens, private communication.
14. N. M. Kroll, P. L. Morton and M. N. Rosenbluth, in "Free-electron Generators of Coherent Radiation," Addison-Wesley Publ. Co., Reading, Mass. (1980), p. 89.
15. J. M. J. Madey and R. C. Taber, ibidem, p. 741.

ELECTROSTATIC ACCELERATOR FREE ELECTRON LASERS

Luis. R. Elias

Quantum Institute
University of California
Santa Barbara, CA 93106

INTRODUCTION

The amplification of short wavelength coherent electromagnetic radiation by relativistic electrons moving through a spatially periodic transverse magnetic field was first demonstrated at Stanford University [1]. These experiments were carried out using the bunched electron beam emerging from a radio frequency linear accelerator. Although the electron beam quality was ideally suited to study the most important operating characteristics of the free electron laser, the small amount of available average electron beam current coupled with only a small laser extraction efficiency contributed to limit both the amount of average laser power produced (P=0.5 watts) and the overall operating efficiency of the device (e<0.1%).

Since the Stanford experiments a considerable amount of work has been done to study various schemes directed toward the development of efficient high power free electron lasers. In some of the schemes high single pass laser extraction efficiency is pursued using for example variable parameter wigglers [2], constant period wigglers consisting of only a few magnet periods [3] and constant period gain-expanded wigglers [4]. In other schemes the electron beam is recirculated several times through the laser interaction region [4,5] to increase total overall efficiency while retaining the characteristically small single pass efficiency of a constant period wiggler.

The present paper addresses the problem of increasing the power and efficiency of free electron lasers from a point of view which is fundamentally different from the schemes mentioned above. The schemes discussed here are based on the utilization of the continuous electron beams generated by electrostatic accelerators. The basic idea is to recover the energy and charge of the electron beam after it has interacted with the free electron laser. This scheme was first suggested by Madey [6] in 1970 and later pursued by Elias [7] in 1978 to develop the two-stage FEL concept.

As will be discussed in detail in the next section two major changes occur as a result of recovering the energy and charge of the electron beam produced by electrostatic accelerators. First, the average amount of electron beam current that can be extracted from the high voltage terminal, at constant voltage, can be increased from typical values of a few hundred microamperes to several amperes of average beam current. Second, even with a low single pass FEL energy extraction efficiency the overall efficiency of the device can be potentially very high because the energy losses occurring during the electron beam recovery stage can be made substantially smaller than the amount of laser energy produced. It will thus be shown here that as a result of electron beam recovery a considerable amount of average laser power ($\bar{P}$>10 kW) can be generated with high overall laser efficiency (e>50%) using electrostatic accelerators. In addition to high current operation, electrostatic accelerators are well suited to provide the excellent quality electron beams demanded by free electron lasers. It is also worth noting that the operation of a free electron laser with electron beam recovery reduces substantially the amount of ionizing radiation normally produced when the spent electron beam is suddenly stopped. This feature alone may be of primary importance to those considering using free electron lasers for commercial or laboratory applications.

Three schemes will be discussed here: a) short pulse operation with no energy recovery, b) CW single-stage operation with energy recovery and c) CW two-stage operation with energy recovery. Also, a review is made of the electron beam quality required by the FEL.

ELECTROSTATIC ACCELERATOR FEL WITH NO ENERGY RECOVERY

The technology of high-voltage electrostatic accelerators is now well established. Since the 1950s these machines have been operated quite reliably to produce very high quality continuous beams of electrons or ions in the medium voltage range from 1 MV to 25MV. The maximum DC beam current (I_B) that can be extracted during conventional operation from these devices is entirely determined by the maximum amount of charging current (I_C) required to maintain the HV terminal charged at constant electric potential. Extracting more beam current than the charging current (I_B>I_C) results in a situation whereby the electric potential of the high-voltage terminal and hence the electron's kinetic energy will decrease steadily with time. During normal operation (no energy recovery) these devices are capable of generating on a steady state basis from a few tens of milliamps of beam current at low voltage to a few hundred microamperes at high voltage. A schematic diagram of a single stage free electron laser using an electrostatic accelerator without electron beam recovery is shown in Figure 1.

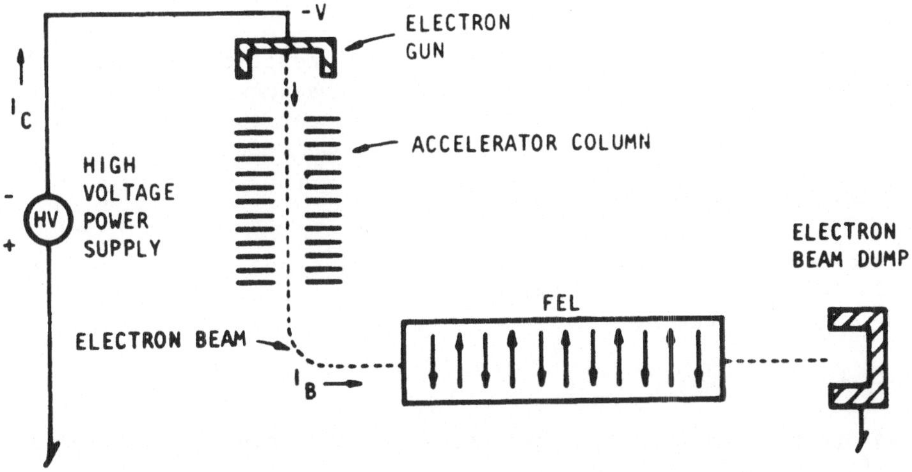

Figure 1. An electrostatic accelerator FEL operating with no beam recovery.

The high-voltage terminal is charged to a potential -V by means of an electrically charged moving belt or pelletron chain. An electron gun located in the HV terminal produces a relatively low voltage electron beam which is subsequently injected into and then accelerated to its final energy by the constant electric field of the accelerating column shown. After interacting with the FEL, the electron beam is stopped at the electron beam dump. There, most of the beam's energy is converted to heat and ionizing radiation. Hence, the overall efficiency of the laser is low since only a small amount of energy is actually converted into laser radiation. It was noted earlier that it is possible to increase the single pass efficiency of the laser by means of variable parameter wigglers. However, the amount of average laser power obtained in this configuration is still small due to the limited amount of average current available from the accelerator in this mode of operation.

Assuming that I_C is the charging current reaching the high voltage terminal and I_B is the electron beam current extracted from the accelerator, then the rate of change of voltage with time can be readily calculated as follows

$$\frac{dV}{dt} = - \frac{[I_B - I_C]}{C} \tag{1}$$

where C is the electrical capacitance to ground of the high-voltage terminal. Typically C=200 picofarad. For constant wavelength operations, the free electron laser operating in the single particle regime requires an electron beam whose energy spread is smaller than the energy width of the gain curve. This requirement imposes a maximum acceptable drop in the HV electrostatic potential of

$$\left[\frac{\Delta V}{V}\right]_{MAX} \simeq \frac{1}{2N} \qquad (2)$$

where N is the number of FEL wiggler periods. Equation (1) and (2) can be combined to yield a value for the maximum electron pulse length that can be used with a free electron laser operating in this mode:

$$[\Delta t]_{MAX} = \frac{CV}{2N(I_B - I_C)} \qquad (3)$$

Before another electron pulse can be initiated, the accelerator HV terminal value must be recharged to its initial potential. The charging rate is given by

$$\frac{dV}{dt} = \frac{I_C}{C} \qquad (4)$$

The total recharging time can thus be calculated combining equations (2) and (4) to obtain

$$[\Delta t]_{CH} = CV/2NI_C$$

It follows that the maximum pulse repetition rate that can be obtained in this mode of operation is:

$$PRR = \frac{1}{[\Delta t]_{MAX} + [\Delta t]_{CH}} \qquad (5)$$

Table 1 below summarizes typical operating characteristic of an electrostatic accelerator FEL when no electron beam energy recovery techniques are used. The results shown were obtained using the above equations with C = 200 picofarad and N = 250.

Clearly a free electron laser operating in the above described configuration can be useful in many laboratory and commercial applications where laser power and efficiency is not of primary consideration and where sufficient protection exists against ionizing

Table 1. Performance of an electrostatic accelerator free electron laser using no electron beam recovery techniques.

V	I_B	I_C	P	$\overline{P}$	$[\Delta t]_{MAX}$	PRR
5MV	2A	500μA	20kW	5W	10^{-6}sec	250Hz
5MV	100A	500μA	1MW	5W	20×10^{-9}sec	250Hz

radiation produced at the electron beam dump. However, if higher power and overall efficiency is required, then the electron beam energy and charge must be recovered. An appropriate technique to achieve this is discussed in the next section.

ELECTROSTATIC ACCELERATOR FEL WITH ENERGY RECOVERY

As noted in the introduction, the power and efficiency of a free electron laser can be substantially improved if the energy of the spent electron beam is recovered. Using electrostatic accelerators this is done in a straight forward way as shown in Figure 2.

After interacting with the FEL the spent electron beam's kinetic energy is reduced from a few megavolts to a few kilovolts by the electrostatic decelerating column shown in the figure. Subsequently, the relatively low kinetic energy beam enters the electron charge collector where the electrons are separated according to energy and captured by the collector surfaces with minimum pro- duction of heat or ionizing radiation. The technique of recovering electron beam energy by means of "depressed collectors" is used frequently with many modern microwave tubes as discussed by Hechtel [8].

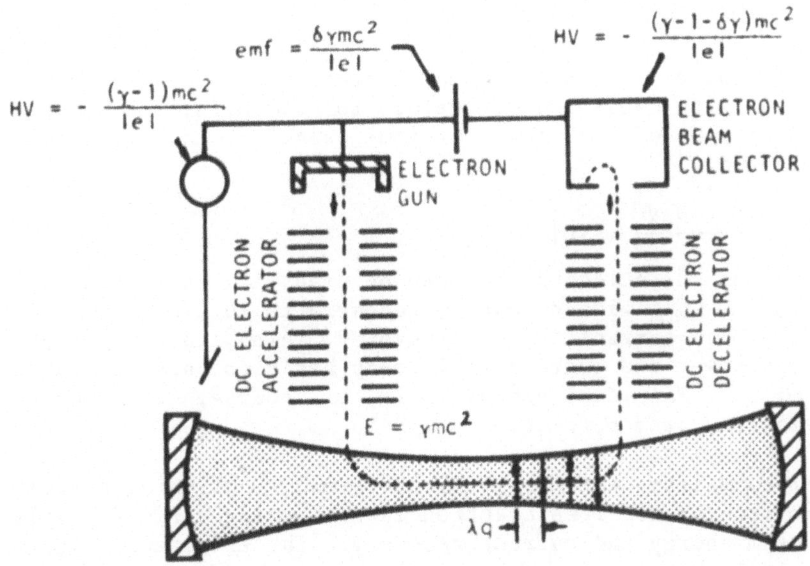

Figure 2. An electrostatic accelerator FEL operating with electron beam energy recovery.

The battery shown between the cathode and collector replaces the energy lost by the electron beam to FEL radiation. I_R

represents the amount of electron beam current recovered. Note that I_R is a conduction current while I_B is a true beam current. Equation (1) must be modified to include the recovered current I_R

$$\frac{dV}{dt} = -\frac{1}{C}[I_B - I_C - I_R] \tag{6}$$

It follows from the above equation that a steady state regime can be obtained ($\frac{dV}{dt} = 0$) when

$$I_B - I_R = I_C \tag{7}$$

That is to say, the potential of the high-voltage terminal will not change if the amount of electron current (I_B-I_R) lost in the system is equal to the charging current I_C. For this scheme to work it is thus important to recover as much of the electron beam current as it is possible. However, even if all of the beam current is not collected it is still possible to operate the FEL with reasonably large values of power and efficiency. For example, assume that I_R=I_B(1-α) where is the fraction of beam current lost. The maximum electron pulse length that can be used with a FEL is obtained by modifying equation (3) to read:

$$[\Delta t]^R_{MAX} = \frac{CV}{2N(I_B - I_R - I_C)} = \frac{CV}{2N(\alpha I_B - I_C)} \tag{8}$$

For example, if 10% of the initial beam current cannot be recovered then using (6) with α=0.1 and the values for C,V,N,I_B, and I_C used in the example discussed in the previous section the following result is obtained:

$$[\Delta t]^R_{MAX} \simeq \frac{[\Delta t]_{MAX}}{\alpha} = 10[\Delta t]_{MAX}$$

Hence, the maximum pulse length that can be used with the FEL has increased from $[\Delta t]_{MAX}$ with no energy recovery to 10 times $[\Delta t]_{MAX}$ when 10% of the beam current is not recovered. Also, in this example the average power and overall efficiency has also been increased by a factor of 10. The ideal situation is, of course, to recover all of the electron beam current.

Table II summarizes the possible performance of single-stage electrostatic accelerator free electron lasers having various levels of electron beam energy and current recovery. The efficiency figure is defines as follows:

$$e = \frac{\text{AVERAGE POWER}}{\text{PEAK POWER}} \times 100\%.$$

Table II. Performance of electrostatic accelerator free electron lasers with various degrees of energy recovery. (V=5MV, C=200 picofarads, I_B=2A, I_C=500 μA, N=250)

α	P(peak)	$\overline{P}$	$[\Delta t]_{MAX}$	Efficiency
1	20kW	5W	10^{-6}sec	.025%
0.1	20kW	50W	10^{-5}sec	.025%
0	20kW	20kW	∞	100%

The results shown in Table II indicate that with electron beam energy recovery ($\alpha < 1$) it is possible to operate FELs at high power and high overall efficiency using electrostatic accelerators even if the charging current I_C is small. Also, since during the electron beam collection process the electrons have only small kinetic energies, the amount of ionizing reaction produced is small. A more detailed discussion of the electron collection process can be found elsewhere in this book under the title "The UCSB FEL Experimental Program".

Note that in the calculation of overall efficiency, power supply losses have not been included. If these losses are taken into account, then in some cases the overall laser efficiencies are expected to be as high as 50% if all the charge and energy in the electron beam is recovered.

TWO-STAGE FREE ELECTRON LASERS USING ELECTROSTATIC ACCELERATORS

The two-stage FEL concept [3] was developed in 1978 at Stanford University as a means of generating tunable coherent radiation at short wavelengths using low energy electron beams, such as the ones available from electrostatic accelerators. If the techniques of electron beam energy recovery discussed previously are also used with two-stage FELs then a considerable amount of laser power can be produced in the 1000Å to 50 μm wavelength range using conventional low voltage electrostatic accelerator. The simplest configuration of a two-stage FEL is illustrated in Figure 3.

As shown in the figure a continuous beam of monochromatic electrons of energy $E=\gamma mc^2$ emerges from the electrostatic accelerator column shown on the left side of the figure. The beam interacts with the FEL wiggler to excite a long wavelength laser TEMoo mode which resonates between the two spherical mirrors. The wavelength of this mode is given approximately by the relation

$$\lambda p = \frac{\lambda o}{2\gamma^2}$$

where λ_o is the period of the magnetic wiggler, λmc^2 is the energy of the incoming electrons. The resonator mirrors are constructed of highly reflective materials at the operating wavelength λ_p to allow the intensity of the optical mode to grow to values in the range 10^8-10^9Mwatts/cm^2. At this high level of optical power density

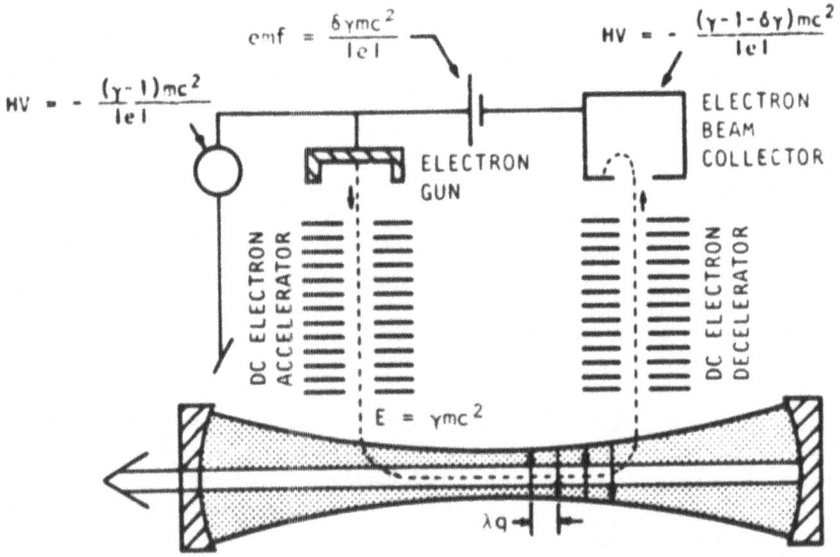

Figure 3. A simple two-stage FEL scheme using electrostatic accelerators with electron beam energy recovery.

the same electron beam can interact again with the intense optical mode to produce coherent radiation at a much shorter wavelength

$$\lambda = \frac{\lambda p}{4\gamma^2} = \frac{\lambda o}{8\gamma^4}$$

The short wavelength optical mode (second-stage FEL) is shown in white as a TEMoo gaussian mode propagating along the axis of the resonator.

A second two-stage FEL scheme is illustrated in Figure 4. Here, separate electron beams are used to excite independently the first and second FEL stages. The major advantages of this scheme are: a) the wavelength of the second stage can be tuned without perturbing the operation of the first stage, b) the small signal gain of the second stage laser can be optimized by choosing correctly the ratio of pump wavelength λ_p to second-stage wavelength λ and c) the FEL interaction length of the second stage can be adequately controlled using independent electron beam optic components. Table IV summarizes the operating characteristics of a two-stage FEL operating at two different wavelengths.

ELECTRON BEAM REQUIREMENTS

A. Beam Quality Requirements. In a free electron laser the axial velocity β_z of the electron beam determines whether or not the electrons radiate coherently. The maximum spread of axial velocities that can be accepted by a constant period FEL wiggler can be calculated from the energy width of the FEL gain curve at fixed wavelength. The maximum velocity spread that can be accepted by a FEL is given by

$$[\delta\beta_z]_{MAX} = \frac{1}{2N\gamma^2} \quad (9)$$

where N is the number of magnetic periods in the wiggler and γmc^2 is the relativistic energy of the electron beam. If β is the total speed of an electron in the beam and $\beta_\perp$ is its total transverse speed then in a FEL:

$$\beta^2 = \beta_z^2 + \beta_\perp^2 = \beta_z^2 + \frac{K^2}{\gamma^2} + \beta_{\perp o}^2 \quad (10)$$

where $\frac{K}{\gamma} = \frac{|e|B\lambda_o}{2\pi mc\gamma}$ is the transverse speed (MKS units) acquired by the electron from the magnetic wiggler. B is the rms value of the magnetic field on axis and λ_o is the periodicity of the magnetic wiggler structure. $\beta_{\perp o}$ is the transverse drift velocity of the electrons with respect to the axis of the wiggler. $\beta_{\perp o}$ is finite if the electron is injected into the magnetic structure at the wrong angle. Changes in β_z can thus originate from variations in B and $\beta_{\perp o}$. Equation (10) can be used to estimate separately the contribution to $[\delta\beta_z]$ from variations in B and $\beta_{\perp o}$. If β is held fixed then a) for fixed $\beta_{\perp o}$

$$[\delta\beta_z]_B = -\frac{K^2}{2\gamma^2\beta_z}\frac{\delta B}{B} \quad (11)$$

and 2) for fixed B

$$[\delta\beta_z]_{B_{\perp o}} = -\frac{\beta_{\perp o}\,\delta\beta_{\perp o}}{\beta_z} - \frac{\beta_{\perp o}^2}{\beta_z} \quad (12)$$

where it has been assumed that $\delta\beta_{\perp o}\ \beta_{\perp o}$. In a constant period magnet the magnetic field at a distance x from the axis is given approximately by: $B(x,z)=B_o\cosh(\frac{2\pi x}{\lambda o})\cos(\frac{2\pi z}{\lambda o})$. For small x, $\delta B/B \simeq \frac{1}{2}(\frac{2\pi x}{\lambda o})^2$ is the fractional change in magnetic field near the axis of the wiggler. Using this relation and equations (9) and (10) it is possible to estimate the maximum electron beam radius R that can be used with a given FEL wiggler.

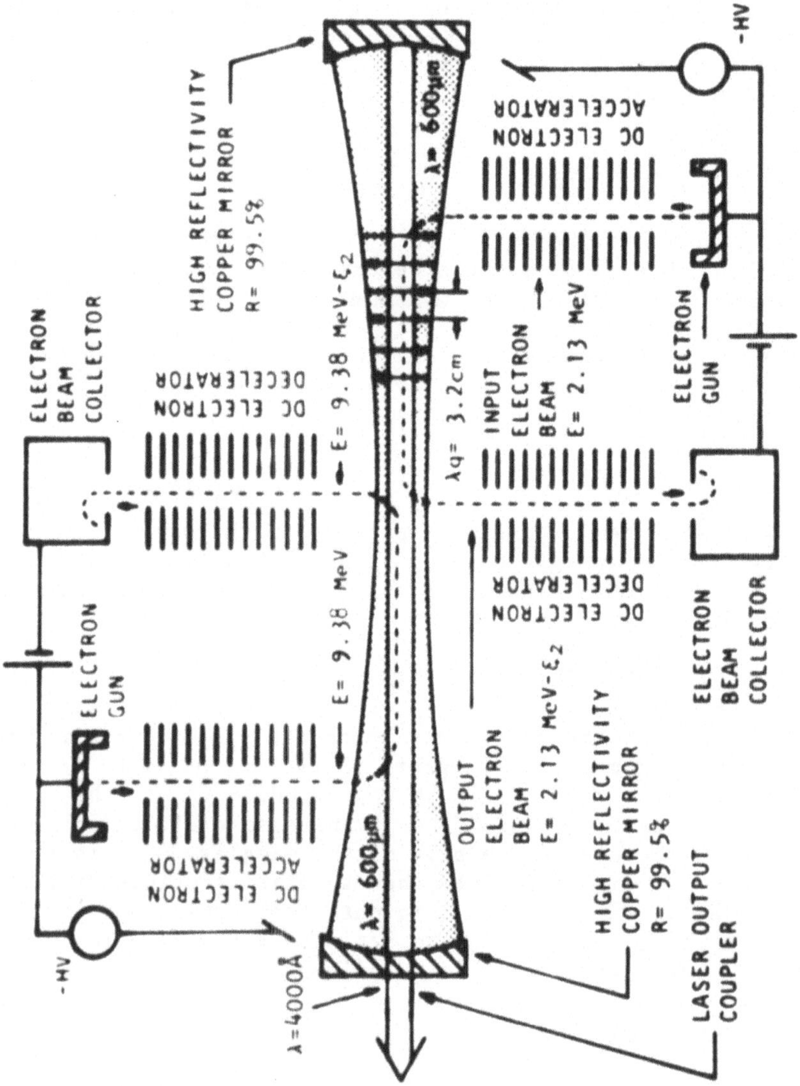

Figure 4. A two-stage FEL using two independent electron beams.

$$R = \frac{\lambda 0}{2\pi K}\sqrt{\frac{\beta_z}{N}} \tag{13}$$

From equations (12) and (9) a maximum acceptable electron beam divergence angle can be derived:

$$\theta = \frac{1}{\gamma\sqrt{2N\beta_z}} \tag{14}$$

And finally from relations (13) and (14) a maximum acceptable transverse electron beam emittance can be obtained:

$$\varepsilon \leq \theta R = \frac{\lambda 0}{2\pi\sqrt{2N\gamma K}} \tag{15}$$

The transverse effects introduced through equation (11) can in principle be minimized provided low wiggler magnetic fields are used. This is, of course, accomplished at the expense of reducing the available small signal optical gain. Also, reducing the transverse dimensions of the electron beam will result in a smaller value of $[\delta\beta_z]$ in equation (11). Equation (12) describes the contribution to $\delta\beta_z$ resulting from the spread in transverse electron velocities in the beam. The sources of such transverse electron velocities can be traced back to the electron gun cathode (thermal effects) and/or to the electron beam accelerator. Their contributions to $[\delta\beta_z]$ can be described as follows:

$$\frac{[\delta\beta_z]_{T_{\parallel}}}{[\delta\beta_z]_{MAX}} = \frac{\frac{1}{\gamma^3}\sqrt{\frac{KT}{mc^2}}}{\frac{1}{2N}} = \left(\frac{2N}{\gamma}\right)\sqrt{\frac{KT}{mc^2}} \tag{16}$$

$$\frac{[\delta\beta_z]_{T_{\perp}}}{[\delta\beta_z]_{MAX}} = \frac{\frac{1}{\gamma^4}\sqrt{\frac{KT}{mc^2}}}{\frac{1}{2N\gamma^2}} = \left(\frac{2N}{\gamma^2}\right)\left(\frac{KT}{mc^2}\right) \tag{17}$$

$$\frac{[\delta\beta_z]_{EMITTANCE}}{[\delta\beta_z]_{MAX}} = \frac{3 \times 10^{-5}\,\frac{J}{\gamma^2\beta^2}}{\frac{1}{2N\gamma^2}} = \frac{6 \times 10^{-5} N\,J}{\beta^2} \tag{18}$$

where T is the cathode temperature, $[\delta\beta_z]_{T_{\parallel}}$ and $[\delta\beta_z]_{T_{\perp}}$ are the contributions to $\delta\beta_z$ from variations in longitudinal and transverse cathode thermal velocities respectively. $[\delta\beta_z]_{EMITTANCE}$ is an empirical relation between transverse emittance e(mm-mred) and electron beam current I(KA) derived by Lawson and Penner [9]:

$$e = \frac{0.3\sqrt{I}}{\gamma\beta}$$

Table 3 lists the electron beam quality requirements necessary to operate single-stage and two-stage FELs.

J_{MIN} is the minimum current density required to operate the FEL with a 10% small signal gain/pass. J_{MAX} has been calculated from the Lawson-Penner conditions and:

$$\frac{[\delta\beta_z]_{EMITTANCE}}{[\delta\beta_z]_{MAX}} = 1$$

$$J_{MAX} = \frac{10^5}{6N} \left(\frac{Amp}{cm^2}\right)$$

Table 3. Electron Beam Quality Calculations.

	SINGLE STAGE FEL		TWO-STAGE FEL	
λ	100μm	4μm	16μm	4000Å
γ	7	50	7	20
R(mm)	2	2	0.4	1.3
N	200	200	600	8000
KT(eV)	0.2	0.2	0.2	0.2
$J_{MIN} \left(\frac{Amp}{cm^2}\right)$ (10% gain)	3	10	100	5000
$\frac{[\delta\beta_z]\, T_\parallel}{[\delta\beta_z]_{MAX}}$	0.02	0.003	0.06	0.28
$\frac{[\delta\beta_z] T_\perp}{[\delta\beta_z]_{MAX}}$	10^{-5}	10^{-6}	3×10^{-5}	10^{-4}
J_{MAX} (Amp/cm^2) (from Lawson-Penner eq.)	100	100	33	3

J_{MAX} depends only on the number of wiggler periods. The calculations listed in Figure 5 indicate that the major source of $\delta\beta_z$ originates from the emittance relations Lawson and Penner. It can be seen from Table 5 that to operate two-stage FELs at short wavelength (=4000A) the transverse emittance of electron beams has to improve by a factor of 100 with respect to the value calculated from the Lawson-Penner relation. Single-stage FELs, on the other hand, can operate with the emittance calculated from the Lawson-Penner relation.

B. Beam Current Requirements and Laser Power Output. The optical small signal particle region can be written in KMS units as follows

$$G = \frac{5.24\ \lambda^{3/2}\ \lambda_0^{5/2}\ B^2\ I\ N^3}{(1+K^2)^{3/2}\ r^2}$$

where:

λ = signal wavelength
λ_o = magnet period
B = RMS magnetic field on axis
I = electron beam current
N = number of magnet periods
r = optical beam radius
$K = q\lambda_0 B/2\pi mc$

The above gain equation has been normalized to give the correct gain value for the Stanford FEL. It is assumed here that the electron beam radius R is smaller or equal to the optical beam radius. At saturation (i.e. when the small signal gain is reduced by a factor of 2) the amount of power that can be extracted from the electron beam as laser radiation is

$$\overline{P} = \frac{IV}{2N}$$

The electron beam requirements and the typical expected performance of a single-stage free electron laser has been incorporated into Table 4. Similarly Table 5 summarizes the operating characteristics of a two-stage FEL based on the scheme illustrated in Figure 4.

Table 4. Performance of a single-stage FEL and required electron beam characteristics.

Wavelength (μm)	360
Magnet period (cm)	3
# magnet periods	100
Magnetic field (Γ)	0.06
Small signal gain (Amp^{-1})	0.60
Average laser power (kW/Amp)	15
Overall efficiency (%)	~50
Elect. beam energy (MeV)	3
Maximum transverse emittance (mm-mrad)	π122
Maximum $\Delta\delta/\gamma$	5×10^{-3}

Table 5. Performance of a two-stage FEL and required electron beam characteristics.

Wavelength (μm)	0.4	16
Pump wavelength (μm)	600	4000
Pumpwave intensity (MW/cm^2)	250	60
Interaction length (m)	2.4	1.2
Small signal gain (Amp^{-1})	5×10^{-3}	8×10^{-3}
Power output (kW/Amp)	0.5	2
Overall efficiency (%)	0.3	1.5
Elect. beam energy (MeV)	9.38	3.55
Maximum transverse emittance (mm-mrad)	$\pi 0.8$	$\pi 10$
Maximum $\Delta\gamma/\gamma$	6×10^{-5}	0.8×10^{-3}

CONCLUSIONS

The operations of single-stage and two-stage free electron lasers using the electron beams produced by electrostatic accelerators has been discussed. The techniques of electron beam energy recovery reviewed in this chapter can be used to produce intense beams of coherent electromagnetic radiation in the far infrared region with high levels of efficiency with present electrostatic accelerator technologies. At shorter wavelength high power laser radiation can also be produced, but only at reduced overall efficiency. It may be possible be produced, but only at reduced overall efficiency. It may be possible to extend the operating wavelength region of single-stage electrostatic accelerator free electron lasers into the visible region with the advent of new HV technology.

REFERENCES

1. L.R. Elias, W.M. Fairbank, J.M.J. Madey, H.A. Schwettman, T.I. Smith, Phys. Rev. Lett. 36, 717(1976). D.A.G. Deacon, L.R. Elias, J.M.J. Madey, H.A. Schwettman, T.I. Smith, Phys. Rev. Lett. 38, 892(1977).
2. N.M. Kroll, P.Morton, M.M. Rosenbluth. SRI Technical Report JSR-79-01, SRI International(1980).
3. L.R. Elias, University of California at Santa Barbara, Quantum Institute Report QIFEL-003/80(1980).
4. J.M.F. Madey and R. Taber. Physics of Quantum Electronics, Vol. 7, Chap. IV, 774(Addison-Wesley Publishing Co.(1980).

5. A. Renieri, Report 77.33, CNEN-Centro di Frascati, Edizione Scientifiche C.1P. 65, 00044, Frascati, Rome, Italy (1977). L.R. Elias, J.M.M. Madey, T.I. Smith, Stanford High Energy Physics Laboratories, Report HEPL-824 (1978). To be published in Applied Physics.
6. J.M.J. Madey, Ph.D. thesis, Stanford University (1970), p. 150, unpublished.
7. L.R. Elias, Phys. Rev. Lett. 42, 977 (1979).
8. J.R. Hechtel, IEE Trans. Electron Devices, ED-24, 9 (1977).
9. V.K. Neil, "Emittance and Transport of Electron Beams in a Free Electron Laser," JASON Technical Report JSR-79-10. SRI International (1979).

PERIODIC MAGNETS FOR FREE ELECTRON LASERS

M. W. Poole

Science Research Council
Daresbury Laboratory
Warrington, England

INTRODUCTION

Periodic magnets for free electron lasers may have either circular or linear field polarization, but in either case there is still little experience in the design, construction and operation of such special magnetic structures. Elias and Madey[1] have discussed the design and testing of the Stanford FEL magnet, and a number of other groups are now designing magnets for the new FEL projects. There is also experience at some laboratories in the operation of the same type of magnet for undulators, and this topic has generated much of the design work on periodic magnets[2-6].

The design principles for FEL magnetics will be presented with particular emphasis on plane electromagnets[7]. Reference will be made to permanent magnet systems that obey many of the same design criteria, and helical magnets will also receive consideration. The limitations of these different magnet types will be pointed out, and their relative merits discussed.

BASIC REQUIREMENTS

The gain of a FEL can be expressed in the following form, assuming the use of a confocal optical cavity:

$$G \quad \alpha \quad \frac{\lambda^{0.5}}{\lambda o^{2.5}} \quad \frac{K^2}{(1+K^2/2)^{1.5}} \quad I \; L^2 \quad \text{, and}$$

λ = output wavelength
λo = magnet period
K = deflection parameter (α λoB)

I = peak electron current
L = magnet length
B = magnetic field

For fixed λ, I and L

$$G \quad \alpha \quad \frac{1}{\lambda o^{2.5}} \quad \frac{K^2}{(1 + K^2/2)^{1.5}}$$

and it can be seen that in principle $\lambda 0$ should be minimized, whilst there also exists an optimum value $K = 2$ for any choice of $\lambda 0$. In fact as $\lambda 0$ decreases the required magnetic field must increase to maintain K, so that there will be a limit to the optimization of the gain. Furthermore it will be demonstrated that the maximum achievable K reduces with $\lambda 0$ due to the existence of technological limits, and that there is also some effect on the linewidth of the output radiation, with a consequent reduction in gain. For some choices of FEL parameters it is possible that these effects lead to an optimum gain at a value of K slightly less than 2.

It should be noted that any FEL will probably have to operate over a wide output tuning range. This can be achieved by variation of either electron beam energy or magnetic field; if the latter is employed then it would be wise to make provision for a value of $K \leq 3$.

For a helical magnet geometry the factor K^2 must be replaced by $2K^2$, and the corresponding optimum value becomes $K = \sqrt{2}$, a figure more readily achieved. Most undulator magnets have been designed to operate at lower levels, typically $K \sim 1$ (2-7).

MAGNETIC FIELD CALCULATIONS

In a plane periodic magnet the poles can be made wide enough to reduce the field to a two-dimensional form. For a magnet with period $\lambda 0$ and gap h, Maxwell's equations can be solved with appropriate boundary conditions to give the following field components (median plane y = o).

$$B_y = \sum_m B_m \sin(mkz) \cosh(mky)$$

$$B_z = \sum_m B_m \cos(mkz) \sinh(mky)$$

with

$$B_m = \frac{32\,\mu 0\,(NI)}{\pi\,\lambda 0} \; \frac{\sin(m\pi/4)}{m \sinh(mkh/2)}$$

$k = 2\pi/\lambda 0$ and (NI) = ampere turns per coil.

These solutions assume a square wave dependence of B_z with z at y = h/2 and include all the higher harmonics satisfying the symmetry conditions (m is odd). Accurate computer calculations[7] have confirmed that a satisfactory solution is given when the infinite series is truncated to only two terms, and this allows us to study the behavior of the component B_y. The field variation across the magnet gap results in a maximum field on axis, B_o, different from that at the pole tip, B_p.

$$\frac{B_o}{B_p} = \frac{1}{\cosh\,(kh/w)} \left(\frac{1 - \sinh\,(kh/2)/3\,\sinh\,(3kh/2)}{1 - \tanh\,(kh/2)/3\,\tanh\,(3kh/2)}\right)$$

and this expression is plotted in fig. 1. Clearly as $\lambda o/h$ is reduced the maximum on-axis field is also decreased for a specified value of pole tip field.

It will also be useful to examine the transverse field variation over the part of the gap occupied by the electron beam. At the centre of a magnet pole, the variation is

$$B_y = \frac{B_o}{1-f}\,[\cosh\,(ky) - f\,\cosh\,(3ky)]$$

$$\text{with } f = \frac{\sinh\,(kh/2)}{3\,\sinh\,(3kh/2)}$$

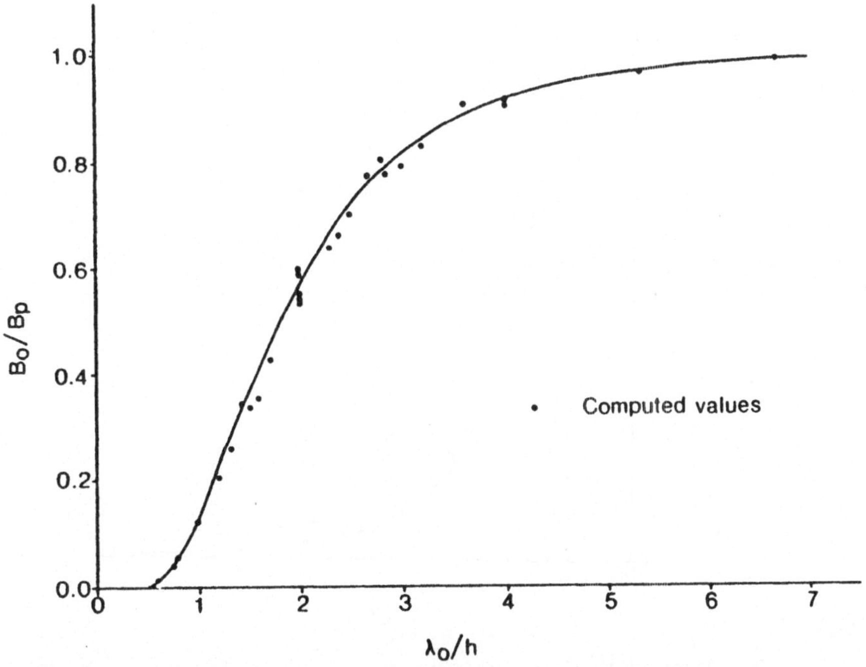

Fig. 1. Dependence of peak magnetic field on choice of magnet geometry.

For small y the field variation is approximately

$$B_y = B_o\ (1 + ay^2)$$

and

$$ah^2 = 2\pi^2 \quad (\lambda o/h)^{-2} \quad (1-9f) \quad (1-f)^{-1}$$

In fact for $\lambda o/h \lesssim 2$ the term f can be neglected and then

$$ah^2 = 2\pi^2 \quad (\lambda o/h)^{-2}$$

or

$$a = \frac{1}{2} \left(\frac{2\pi}{\lambda o} \right)^2$$

The expression ah^2 is plotted in fig. 2, showing the very rapid increase in field inhomogeneity for small values of $\lambda o/h$. It should be noted that the validity of the simpler expression for a implies that field variation near the magnet symmetry plane is dominated by choice of λo, with only a minor influence of the gap, h.

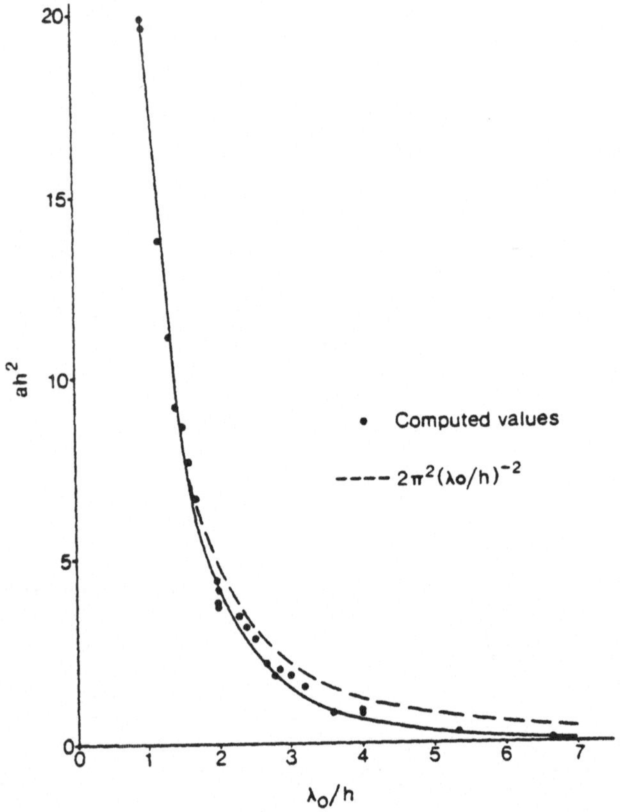

Fig. 2. Dependence of transverse field coefficient on magnet geometry.

Those results have assumed the use of a "square" magnet (pole width equal to pole separation), relatively low magnetic fields and large excitation coils within the coil window. If the pole width is increased for a fixed magnet period the field uniformity may be significantly improved, but the space for coils becomes less and severely limits any such improvement. Periodic magnets are much more vulnerable to saturation problems in the steel than more conventional magnet geometries, due to the field enhancement in the poles compared with the median plane and to the leakage of field between closely separated poles; saturation effects are likely to be relevant at fields as low as 0.5T in optimized magnet geometries. Small coils located near the pole corners can decrease the leakage fields, but tend to increase the field inhomogeneity in the gap; the inefficient use of the available coil window is also likely to make such coils impractical.

LINESHAPE

The transverse field inhomogeneities of the magnet in conjunction with the finite cross section of the electron beam will broaden the line by

$$\frac{\Delta\lambda}{\lambda} = \frac{K^2}{1 + K^2/2} \quad \frac{\Delta B}{B}$$

The magnitude of the effect may be estimated by use of the previously defined parameter a

$$\frac{\Delta\lambda}{\lambda} = \frac{K^2}{1 + K^2/2} \quad a\,(\Delta y)^2$$

$$\approx \frac{K^2}{1 + K^2/2} \quad \frac{1}{2}\left(\frac{2\pi}{\lambda 0}\right)^2 \quad (\Delta y)^2$$

If Δy is the radius of the electron beam then the contribution to the broadened homogeneous linewidth can be taken (pessimisticly) to be half the above full width. For a magnet length L the ratio, R, between the inhomogeneous and homogeneous parts of the total line shape is

$$R = \pi^2 \; L \; \frac{K^2}{1 + K^2/2} \quad \frac{(\Delta y)^2}{\lambda o^3}$$

For specified values of L, K and Δy there is thus a minimum required value of $\lambda 0$ if a permitted value of R is not to be exceeded. To illustrate the use of this equation, with K = 2, L = 5m and R = 0.5 (giving ~ 10% enhancement of the linewidth)

$$\lambda 0 \geq 50(\Delta y)^{2/3} \qquad \lambda 0,\ \Delta y \text{ in mm}$$

It can be seen that an electron beam radius of 5mm imposes a lower limit of $\lambda o \sim 150$mm. Such figures as this example are quite typical of single pass FEL schemes employing electron beam energies of 5-50 MeV.

The beam radius is related to the emittance, ε, and this leads to $\lambda o \propto \varepsilon^{1/3}$. For an electron storage ring, the emittance will be considerably smaller, at least in one plane, and this will allow a significant reduction in λo before line broadening becomes important.

It should also be realized that the above treatment assumes perfect alignment of the electron beam through the magnet and of the magnet poles to each other. In practice the ideal electron beam radius should be increased by some (hopefully small) factor to take account of such problems.

The effects discussed in this section apply to all types of periodic magnet, including permanent magnets and helical coils. For the circular geometry the field inhomogeneity occurs in both transverse planes, a distinct advantage for the ribbon beams and unequal beam apertures generally found in electron storage rings.

PERFORMANCE LIMITATIONS

Electromagnets

The maximum value of B_p is limited by the choice of technology (electromagnet, permanent magnet or superconducting magnet), and fig. 1 therefore illustrates that there is a maximum achievable value of K for a given $\lambda o/h$. Assuming $B_p = 1.5$T for a conventional electromagnet, and a minimum aperture h = 20mm, then the magnet period can be no less than $\sim$ 40mm ($\lambda o/h = 2$) if a value K = 3 is to be accessible; similar values of λo are found with other values of aperture.

A further limitation arises as λo is reduced for any given h, since there is a dramatic increase in the required excitation current to achieve the same K value, particularly below $\lambda o/h \sim 2$. This problem is exaggerated by the decreasing space available for coils, and the current density increases so rapidly that superconducting coils soon become necessary. The efficiency of the magnet design can be calculated using the previous analytic solutions, which give

$$\frac{NI}{K} = 1.18 \times 10^3 \sinh(kh/2) \left(1 - \frac{\sinh(kh/2)}{3\sinh(3kh/2)}\right)^{-1}$$

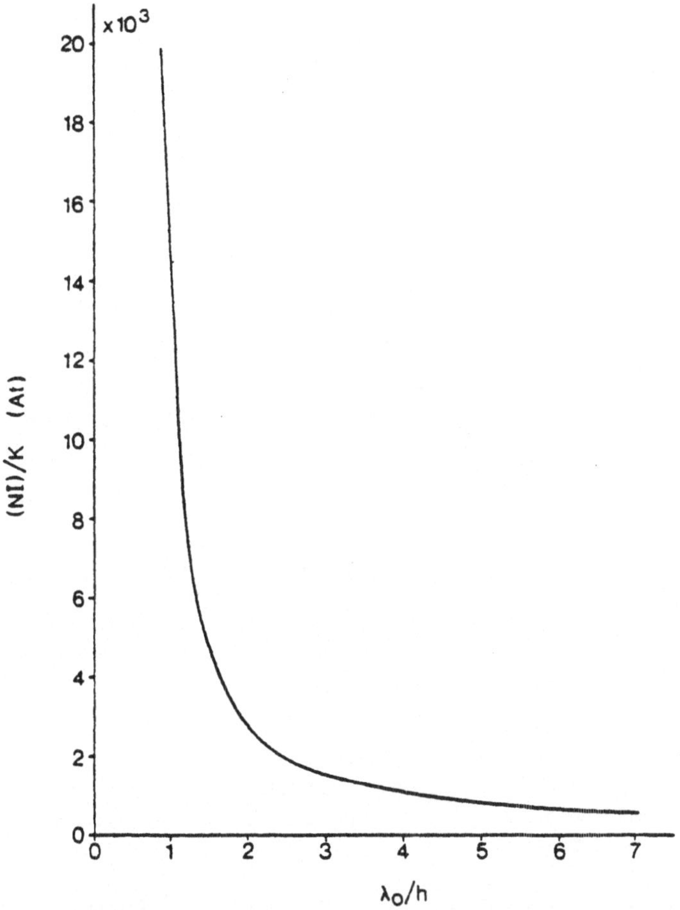

Fig. 3. Dependence of required magnet excitation on geometry.

and this function is shown in fig. 3, which demonstrates a sharp increase for $kh/2 > 1$ (small $\lambda 0/h$). As an example, for an aperture of 20mm and $\lambda 0/h = 3.3$, a current density of only 20 A/mm^2 is needed to reach $K = 2$, with coils filling a square space between poles. Reduction of $\lambda 0/h$ to 1.7 increases the required current density to 200 A/mm^2.

The analytic formulae are only valid for infinite steel permeability; saturation effects can be important at fields even as low as 0.3 - 0.4T in periodic magnets. As an example of this, the magnet installed in the ACO storage ring at Orsay requires ∿ 50% extra ampere-turns to overcome iron losses despite a maximum operating field of only 0.4T[6].

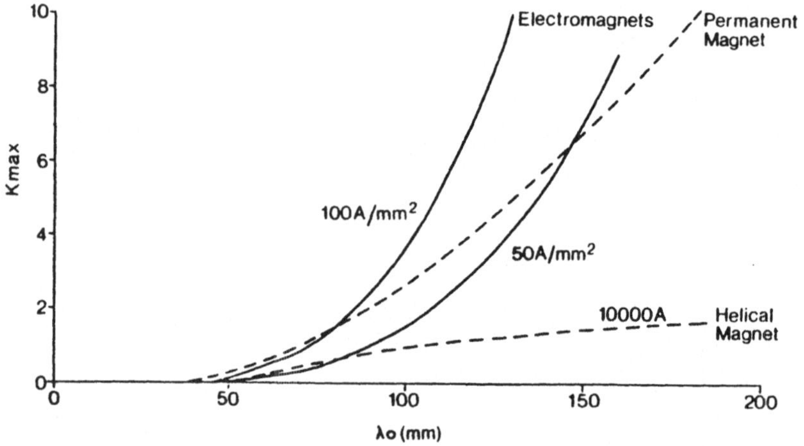

Fig. 4. Dependence of maximum achievable K value on magnet geometry.

As a demonstration of the above formula, the maximum K value of a practical magnet is plotted as a function of λ_0 in fig. 4 for two different allowed current densities, assuming square coils and an aperture h = 52mm.

For conventional magnets it is not easy to operate with a value $\lambda_0/h < 2$, and if this is essential then superconducting technology will probably have to be applied.

Permanent Magnets

The problem of high current density at small values of λ_0 is avoided by use of permanent magnets. For a linear array of magnetized blocks above and below the median plane, the maximum on-axis field is

$$B_o = 2 B_r \frac{\sin \pi/M}{\pi/M} (1 - e^{-kb}) e^{-kh/2}$$

with B_r = remanent field of material
M = number of material blocks per period
b = height of material block (width = λ_0/M)

Once again the reduction of B_o as $\lambda o/h$ is decreased limits the minimum permissible period. Fig. 4 includes the performance of a practical example with the same aperture (h = 52mm) as the electromagnets, for M = 4, B_r = 0.95T and b = $\lambda o/4$. This design competes very favorably with the alternative electromagnets, but is probably limited to Bo $\leq$ 0.3T at $\lambda o/h \leq 2$.

Helical Magnets

A circularly polarized transverse field is produced by a helical winding geometry; a double helix bifilar magnet with opposing currents allows cancellation of the axial field component, leaving only the spiralling transverse pattern. The amplitude of the transverse field is given by[5]

$$B_o = \frac{0.8\,\pi\, I}{\lambda o} \quad [ka \;\; K_o(ka) + K_1(ka)]$$

with

$$a = \text{bore radius}$$
$$K_o, K_1 = \text{modified Bessel functions}$$

This functional form results in an exponential decrease in Bo as $\lambda o/a$ reduces. It may be rewritten readily to show I/K as a function of ka, analogous to the expression previously quoted for a plane electromagnet. In fact it also exhibits a sharp increase for ka > 1, to be compared with a similar threshold at kh/2 $\sim$ 1 for the plane magnet.

In the region above ka $\sim$ 1, the expression can be stated in a simple form[4]

$$K \approx 4 \times 10^{-4} \;\; I \;\; e^{-0.9kh/2}$$

This expression has been included on fig. 4 for a representative current I = 10,000 A, and the inefficient generation of large values of K is clearly demonstrated.

The off-axis fields in such a helical magnet are represented by first-order Bessel functions. However the fields close to the axis can neglect all but the first term in each Bessel function series, and this leads to a simple sextupole error term analogous to that in the plane magnet geometry.

Of course the reason for the high current requirement in the helix is the lack of magnet steel. If the technical problems of including magnetic material could be overcome then improved designs could be adopted.

CONSTRUCTION FEATURES

Electromagnets

Periodic electromagnets have been designed at Orsay, Brookhaven, Frascati and Daresbury, but the choice of optimum parameters has varied greatly.

A superconducting design was adopted(6) for the magnet that has now been installed in the ACO electron storage ring. This allowed use of a very short period, λo = 40mm, and even smaller gap, h = 22mm (the magnet is moved around the beam some time after injection processes to the ring are completed). The required current density is $\sim$ 200 A/mm^2 and permits a maximum K $\sim$ 1.5; it is interesting to note that the superconducting magnet only provides up to 0.4T on the axis.

Several designs have been undertaken at Brookhaven.(2) The magnets have similar parameters to the Orsay device, except that it is assumed that a current density of as much as 150 A/mm^2 can be supported by a water cooled copper coil without the need for superconducting technology.

The Frascati magnet for the LELA project(8) adopts a more conservative design with a longer period ($\lambda o \sim$ 120mm) and a maximum current density of 30 A/mm^2. With $\lambda o/h \sim 3$ a field of nearly 0.5T should be attainable, giving access to values of $K \leq 5$.

The equally conservative FELIX magnet has $\lambda o/h = 4$ with λo = 200mm and a modest current density of 10 A/mm^2 in the water cooled coils; the possibility of air cooling has also been considered, in which case the current density would be lowered to $\sim$ 1.5 A/mm^2. This magnet would employ modular construction techniques to allow for future parameter changes, and like most plane periodic magnets, the top and bottom sets of poles would not be connected by any magnetic circuit.

Permanent Magnets

The design of these magnets has been revolutionized by the development of rare earth cobalt materials in recent years. This material is highly anisotropic, having an easy axis of magnetization. The magnetic characteristics give a permeability very close to unity and a remanent field in region 0.5 - 0.95T. With such a permeability blocks of material appear only as surface currents, allowing linear superposition exactly as if they were coils and greatly simplifying design procedures.

The material is usually obtained in sintered blocks that must be handled with care due to extreme brittleness. The most common example is samarium cobalt ($Sm\ Co_5$), which has a high remanent field $B_r \geq 0.9T$.

Such a magnet has been operated as part of the optical klystron on the VEPP3 storage ring at Novosibirsk. Magnet steel has been used to reduce the required quantity of material and also as pole caps to improve the field quality. The aperture at maximum field is 11mm and the period 110mm.

A detailed design for a $Sm\ Co_5$ magnet has been produced at Stanford, for inclusion in the SPEAR storage ring, with variable field up to $\sim$ 0.25T. A permanent magnet system with variable geometry and K $\sim$ 1 is also under consideration for the HERALD project at the TRW Research Laboratories.

Helical Magnets

The helical magnet for the Stanford FEL project(1) employs niobium-titanium conductor to achieve an on-axis field of up to 1.3T, although the FEL experimental results were obtained at a much lower field of 0.24T. A small bore diameter of 10mm was chosen, together with a period of only 32mm, close to the condition for maximum gain. Radial field variations were such that the electron beams radius had to be restricted to no more than $\sim$ 0.5mm; similar restrictions on beam alignment through the 5.2mm long magnet led to development of a novel survey technique. The helical magnet was surrounded by a solenoidal winding that was used to contain the orbited electron motion arising in the helical fields.

A helical magnet has been installed in the VEPP2 ring at Novosibirsk (for measurements of electron beam polarization), and employs water-cooled copper to achieve fields of up to 0.2T with a period of 24mm and bore diameter of 15mm; the field from the double helix is enhanced by a factor of 2 by a helical steel geometry added after the coil had been wound, and the device employs extremely high current densities $\sim$ 200 A/mm^2. However, no measurements of field homogeneity have been carried out.

CONCLUDING REMARKS

Considerable progress has been made in the design and construction of periodic magnets, and the factors influencing their performance have been studied in detail.

The minimum achievable period is likely to be set by line broadening in single pass FELs but by more fundamental technological limits in storage ring recirculation schemes, where high K values are also likely to be required.

Permanent magnet systems are likely to be preferred to the more complex superconducting technology, and it may even be possible to use them for a helical geometry. A major disadvantage of permanent magnets is the problem of varying the magnetic field (and also switching the magnet on or off). A typical solution is to vary the gap between opposite poles, but this may be an inconvenient exercise and will also affect the field homogeneity. Water cooled electromagnets will be used for their ease of control and cheapness if the magnet period is not too short.

Further technological problems will arise in the future with adoption of more unconventional magnet designs, employing gradient fields or tapering of parameters.

REFERENCES

1. L. R. Elias and J. M. Madey, Rev. Sci. Instr. 50 (11) 1335 (1979).
2. S. Krinsky, BNL 25698 (1979), Brookhaven Int. Report.
3. G. Chu, SLAC-PUB-1782, Stanford (1976).
4. J. P. Blewett and R. Chasman, J. Appl. Phys. 48(7), 2692 (1977).
5. Proc. Wiggler Workshop, 1977, SSRP 77/05, Stanford (1977).
6. Proc. Wiggler Meeting, 1978, Frascati (1978).
7. M. W. Poole and R. P. Walker, DL/SCI/P215A, Daresbury (1980). To be published in Nucl. Instr. Meth.
8. R. Barbini and G. Vignola, LNF 80/12(R), Frascati (1980).

TECHNICAL PROBLEMS IN THE CONSTRUCTION OF WIGGLERS AND UNDULATORS FOR FREE ELECTRON LASERS*

Alfredo Luccio

Brookhaven National Laboratory
Upton, New York 11973, U.S.A.

ABSTRACT

The main features of wigglers and undulators for free-electron lasers are reviewed, together with some of the technical problems involved in their construction and operation in conjunction with a particle accelerator, either in single-pass or in recirculated beam mode. A few examples are presented in some detail; helical undulator, flat undulator with standard electromagnet and with permanent magnets, and very high field superconduction wiggler.

I. INTRODUCTION

In a "classical" free electron laser, a beam of relativistic electrons from an accelerator is passed through a wiggler magnet or undulator. The electrons emit spontaneous radiation and, if an electromagnetic radiation field is present of the proper wavelength and polarization, amplification occurs of the latter at the expense of the energy of the electron beam.

Wigglers or undulators provide a static or, in some cases, a pulsed magnetic field along the electron beam path. The field is spatially modulated in the longitudinal direction and bends the electron trajectories into wiggling shapes.

Wigglers and undulators come in two main varieties: helical and flat (or transverse). In a helical structure, the magnetic field $\vec{B}$ is perpendicular to a cylindrical axis and rotates as we proceed

* Work supported by the U.S. Department of Energy.

along the helix. The electron trajectories are likewise helices wound around the axis. In a flat undulator, the field has only one "vertical" component on the symmetry plane, that inverts itself periodically in a sinelike way along the electron path. Here, the trajectories are wavy flat curves that lay in a plane perpendicular to $\vec{B}$.

We will call wigglers those magnets whose field is "high" and which therefore produce radiation with the typical continuous spectrum of synchrotron radiation, as discussed e.g. by A. Hofmann[1] or by D.F. Alferov et Al.[2] Undulators will instead denote "low" field wigglers, that produce quasi-monochromatic radiation.

The wiggler is generally inserted in a straight section of a storage ring, to produce more synchrotron radiation power and a harder spectrum than a normal accelerator bending section. The wiggler usually has only a few periods, since it does not need many to do this job.

Undulators are built with many periods, because the broadening of the spontaneous radiation spectrum is inversely proportional to the number N_w of periods, and a narrow linewidth is an important feature of a monochromatic radiation generator. The gain per pass of a free electron laser is moreover, as we will discuss later, proportional to the square of N_w.

Free electron lasers make use of undulators, and our discussion will be mostly on low-field undulators with many periods. However, we will recall also some of the features of high field wigglers, since the technical problems are the same for both kinds. We will also limit ourselves to static, i.e. not pulsed, undulators.

In the following, for the theory of the free electron laser (FEL), we will refer to the review paper by C. Pellegrini.[3]

II. UNDULATORS AND FELS

The wavelength λ of the spontaneous radiation from an undulator, in the forward direction, which equals the wavelength of the e.m. wave to be amplified, is related to the energy of the electrons and to the "undulator parameter" K by

$$\lambda = \frac{\lambda_w}{2\gamma_r^2} (1 + \kappa^2) \tag{1}$$

where γ_r is the "resonant" energy of the electrons in units of their rest energy mc^2 and κ is defined as follows

$$\kappa = 66.03\ B_w\ \lambda_w\ . \tag{2}$$

Here, B_w denotes the r.m.s. value of the undulator field on axis (in tesla units) and λ_w the undulator period (in cm). In a helical undulator it is

$$B_w = B_{wo} \qquad \text{(helical)}$$

and in a flat undulator

$$B_w = B_{wo}/\sqrt{2} \qquad \text{(flat)}$$

B_{wo} being the maximum value of the field on the unperturbed orbit.

The spontaneous radiation has a rather complicated spectrum, with harmonics of (1), and a complex angular distribution. The fundamental has a relative linewidth (homogeneous broadening) given by

$$\frac{\delta\lambda}{\lambda} \simeq \frac{1}{2N_w} \tag{3}$$

and is contained in a narrow cone of half-aperture

$$\delta\theta \simeq \frac{1}{\gamma\sqrt{2N_w}} \tag{4}$$

The presence of the number of periods N_w both in Eqs. (3) and (4) is due to interference of the radiation from all wiggles. In a high field wiggler, with a few poles, interference plays a minor role and the radiation is all contained in a narrow cone of aperture $\simeq 1/\gamma$ that beams tangentially out of the electron trajectory, and that accordingly describes a wider cone (for a helical wiggler), or sweeps a larger angle (for a flat structure).

The spectrum of the undulator radiation is particularly nice and clean when the parameter κ defined by (2) is of the order of one or less. This is certainly an important feature when we want to use the spontaneous radiation directly, but it is not so important when the undulator is being used for a FEL. In this case, the radiation spectrum is determined mainly by the time structure of the electron beam and by the characteristics of the optical cavity.[4] We will therefore feel free to use the best value of κ for amplification.

In the design of a FEL, the final parameters will be chosen on the basis of Eqs. (1) and (2), and of the expression of the gain per pass, defined as the relative enhancement of the laser radiation by a passage of the electron beam through the undulator.

This gain, for a constant period undulator, and for a monoenergetic beam, is given by

$$G_o = -32\sqrt{2}\,\pi^2 \frac{(\lambda^3\lambda_w)^{1/2}}{\Sigma} \frac{I_p}{I_A} N_w^3 f(x) \frac{\kappa^2}{(1+\kappa^2)^{3/2}} \tag{5}$$

where I_p is the electron beam peak current, and Σ the transverse cross section of the electron and photon beams, supposed equal. f(x) is the well known gain function

$$f(x) = \frac{1}{x^3}\left(\cos x - 1 + \frac{1}{2}\sin x\right), \quad x = 4\pi N_w \frac{\gamma-\gamma_r}{\gamma_r}$$

and I_A = 17 KA the so-called Alfven current.

The first two parameters to choose are the radiation wavelength and the gain. From the expressions above, it appears that our selection binds us to a given class of accelerators

Accelerator
- Energy γ,
- Peak current I_p,
- Beam size Σ,

and undulators

Undulator
- Type (helical or flat),
- Period λ_w,
- Field B_w,
- No. of periods N_w.

We have consequently several degrees of freedom, but also several constraints. However, in practice our freedom is limited, because almost invariably at this stage of FEL art the accelerator is given, with its energy γ, fixed within certain limits, its current I_p, the highest possible in order to achieve a high FEL gain, but nevertheless limited, and its beam size Σ and emittance. Thus, in most cases we are able to vary freely only the characteristics of the undulator.

Let us divide FEL's and accelerators in two broad categories:

i) Microwave-to-infrared FEL's fed by microtrons or linacs in single pass operation.

ii) Infrared-to-ultraviolet FEL's inserted in storage rings for recirculated beam operation.

Other combinations have been proposed, but we will not consider them, since they go beyond the limits of the present discussion.

Turning now more specifically to undulators, let us review some of the general considerations that help us in determining their design parameters. A few examples will aid.

III. HELICAL UNDULATORS

A helical undulator has a cylindrical symmetry that best matches the electron beam from a linac, an induction accelerator or a dc-machine. The undulator of the successful experiments of the Stanford Group[5] is helical.

The spontaneous radiation from a helical undulator is circularly polarized, and in an amplification experiment the electromagnetic wave to be amplified must also be circularly polarized and with the wavelength given by Eq. (1).

A helical undulator is made of a two-wire helical winding, the two wires carrying opposite currents. The field produced by this arrangement has been calculated by J. Blewett and R. Chasman[6], for a constant period undulator. They have obtained the following approximate expressions for the field near the axis:

$$\begin{cases} B_r = B_o \left(1 + \frac{3}{8} p^2 r^2\right) \sin(\theta - pz) \\ B_\theta = B_o \left(1 + \frac{1}{8} p^2 r^2\right) \cos(\theta - pz) \\ B_z = - p B_o r \left(1 + \frac{1}{8} p^2 r^2\right) \cos(\theta - pz) \end{cases} \qquad (6)$$

where r, θ, z, are cylindrical coordinates, and

$$p = \frac{2\pi}{\lambda_w} \qquad (7)$$

is the undulator wave number.

The expressions (6) exactly satisfy the curl condition

$$\nabla \times \vec{B} = 0$$

and, to order $(r/a)^2$, where a is the helix radius, the div condition

$$\nabla \cdot \vec{B} \simeq 0$$

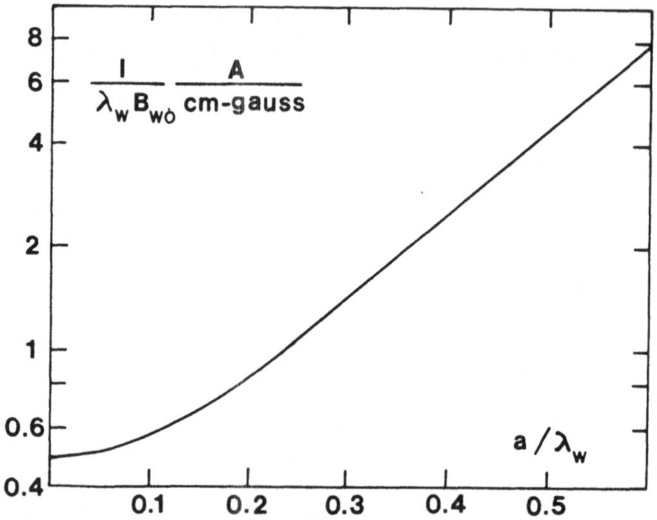

Fig. 1. Helical undulator. Current per turn required to produce an axial transverse field B_o in a helix of period λ_w and radius a. Modified after Blewett and Chasman.[6]

A curve for caluclating the current needed in a double helix to obtain the desired value of B_o is shown in Ref. 6 and reproduced here in Fig. 1. From the curve, it appears that to obtain a fraction of a tesla on the axis, with windings in air, only superconduction or pulsed operation of the undulator is possible.

Let us study an example, using data similar to that of the Stanford Group, as given in Table 1. There κ, and B_o are calculated from Eqs. (1) and (2), once λ, λ_w and γ are given. From the curve of Fig. 1, the high value of the current in the double helix is then found.

The main design features of a static superconducting helical undulator in air can be summarized as follows

i) Period of a few cm,
ii) Length of one meter or more, to allow many periods, since the gain is proportional to N_w^3, as it is shown by Eq. (5) ,
iii) Winding radius of the order of one cm, to minimize helix current (use Fig. 1),
iv) Vacuum beam pipe of a few mm inner radius, to allow for circulation of coolant (Helium) between conductor and vacuum vessel walls (cold bore).

Table 1. FEL with helical undulator. Example

Radiation wavelength	$\lambda = 10.6\ \mu m$
Undulator period	$\lambda_w = 3.2$ cm
Helix radius/period	$a/\lambda_w = 0.35$
Electron energy (Linac)	E = 24 MeV
Normalized energy	$\gamma = 47$
Undulator parameter	$\kappa = 0.67$
Magnetic field on axis	$B_o = 0.317$ T
Double helix current	$I_h = 20$ KA

These requirements make the construction of a helical undulator a rather difficult technical task. What makes the helical structure attractive resides perhaps in its geometrical simplicity, that gives to its field an almost perfect sinusoidal shape in both planes.

IV. FLAT UNDULATORS

The geometry of a flat undulator better matches the ribbon-like shape of the beam from a storage ring. The spontaneous radiation from a flat undulator is linearly polarized, with the electric vector lying in the plane of the wiggles, and the laser wave that is sent through the undulator together with the electron beam must be likewise polarized.

A flat undulator can be built either with a conventional or superconducting electromagnet, or also with permanent magnetic pole pieces. The shape of its field on the median symmetry plane is not as nearly sinusoidal as in its helical counterpart, unless the pole tips have a very sophisticated configuration.

The field of a flat idealized undulator has been given by following expression by J. Blewett and R. Chasman[7]

$$\begin{cases} B_x = 0 \\ B_y = B_o \cosh (py) \sin (pz) \\ B_z = B_o \sinh (py) \cos (pz) \end{cases} \tag{8}$$

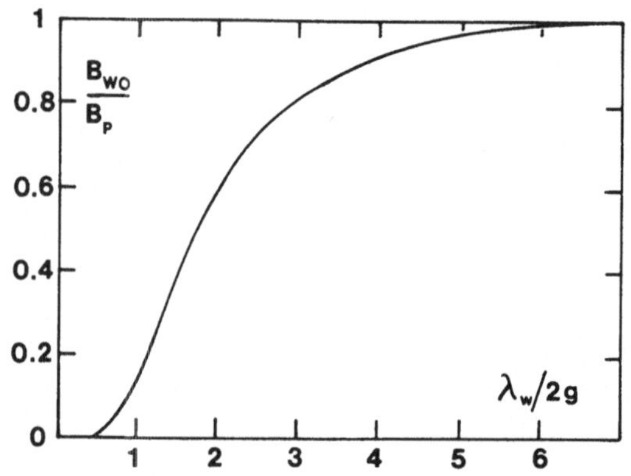

Fig. 2. Iron-cored flat undulator. Field on the median plane/field at pole tips, as a function of the undulator period/full gap. Modified after M.W. Poole and R.P. Walker.[8]

in cartesian coordinates, x (radical), y (vertical), and z (longitudinal, i.e. along the unperturbed orbit).

The maximum field B_o in the median plane, when generated by current layers in air, is related to the current density J_o (in A/m^2) by

$$B_o \simeq \frac{\mu_o}{2\pi} J_o \lambda_w e^{-pg} \qquad \text{(air)} \tag{9}$$

where g is the magnetic half-gap, and $\mu_o = 4\pi\ 10^{-7}$ henry/m is the magnetic permeability of vacuum.

If we build the magnet with a magnetic soft iron core, which is not easy for a helical undulator, the field pattern close to the symmetry plane can be still approximated by the expression (8), while for the maximum field B_o one has to add to the field of the currents in vacuum the non linear contribution of the magnetization of the core.

It is possible to relate the field on the median plane B_o to the field existing at the magnetic pole tips. A good formula is given by M.W. Poole and R.P. Walker[8]

$$\frac{B_o}{B_p} = \frac{1}{\cosh(pg)} \frac{1 - \dfrac{\sinh(pg)}{3\sinh(3pg)}}{1 - \dfrac{\tanh(pg)}{3\tanh(3pg)}} \qquad (10)$$

where B_p is the field at the pole tips. This is shown in Fig. 2.

Generally, for an iron-cored electromagnet, a detailed calculation with a magnetic computer code, either in two dimensions (codes of the POISSON or TRIM family), or in three dimensions (GFUN or analogous) is necessary. Computer results agree with Eq. (10) on the median plane of the undulator, and show that the distribution of the field off this plane is very sensitive to the shape of the steel pole pieces.

The spontaneous radiation lineshape from the undulator depends on the form of the field (inhomogeneous broadening) and the gain per pass of the FEL depends on the width of the line. The dependence of the inhomogeneous broadening on the detailed form of the field, in particular on the "vertical"* variation of the field B over the beam size in the undulator, can be appreciated by differentiating Eq. (1). With this, and with the definition (2), we obtain

$$\frac{d\lambda}{\lambda} = \frac{2\kappa^2}{1+\kappa^2}\frac{d\kappa}{\kappa} = \frac{2\kappa^2}{1+\kappa^2}\frac{dB_w}{B_w} \qquad (11)$$

Eq. (11) can be analyzed, as is common in accelerator physics, in terms of the relative importance of magnetic multipoles.[8]

As far as the radial variation of the field is concerned, it is assumed here that the poles of the undulator are so wide radially, that B is constant over beam horizontal size and wiggling trajectory.

As an example of a flat undulator, let us consider an undulator for a storage ring, similar to the one studied by S. Krinsky.[9] The machine parameters suggest operation in the visible region of the e.m. spectrum.

A consistent set of quantities is given in Table 2.

By using Eq. (9) in air, the field of Table 2 would require a current density

$J_o = 39\ \text{KA/cm}^2$ (air)

---

* For simplicity, we assume that the beam undulates in the "horizontal" plane, that is the most common situation in storage rings.

Table 2. FEL with flat undulator. Example

Radiation wavelength	$\lambda = 0.5\ \mu m$
Undulator period	$\lambda_w = 6$ cm
Half-gap	$g = 1.5$ cm
Electron energy (st. ring)	$E = 500$ MeV
Normalized energy	$\gamma = 980$
Undulator parameter	$\kappa = 3.87$
Max. magnetic field on m.p.	$B_o = 0.98$ T

which is a rather high figure even for superconducting coils.

By adding an iron core, and running POISSON for the same undulator parameters, the desired B_o is obtained with a current density

$$J_o = 22\ \text{KA/cm}^2 \qquad \text{(iron)}$$

This is shown in Fig. 3.

With the above value of the current, if we find that the half-gap selected is too small, we may increase g now, increasing the current again. This will ease the operation of the undulator inserted in a straight section of the storage ring, expecially at injection, when the internal clearance of the vacuum vessel must be kept as large as possible.

To examine this problem in some more detail, we use Eq. (10) or Fig. 2, and find that the undulator period-to-gap ratio should be equal at least to 4 or 5 to achieve a high field on the median plane, with reasonable values of the coil current.

Here also, as for the helical undulator, the problem of small magnetic gaps is difficult. Again, to allow adequate cooling of the superconductors, some extra space must be left for helium circulation between coils and vacuum vessel walls, which in turn must be built thick enough to resist pressure. The available space <u>inside</u> the beam pipe becomes generally very little. We will discuss this again later, when dealing with high field wigglers.

Other solutions have been proposed to the small gap problem, to allow the use of a short period undulator in a storage ring. For instance, J. Perot and M. Lemonnier[10] for the storage ring ACO, at

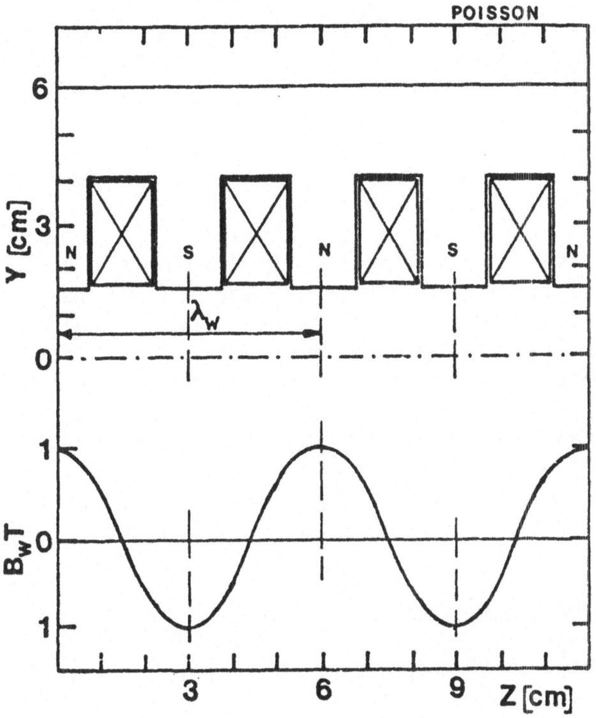

Fig. 3. Results of a POISSON run for a flat undulator with iron core, and J_o = 22 KA/cm^2. Data from Table 2.

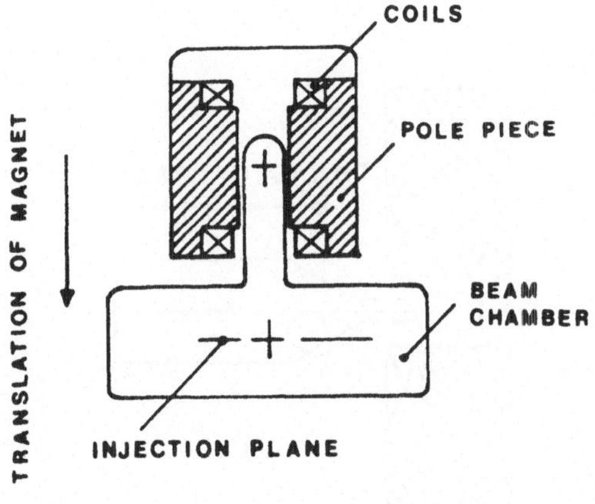

Fig. 4. Cross section of the vertical flat undulator for ACO with a "keyhole" beam chamber. Simplified sketch after J. Perot and M. Lemonnier.[10]

Orsay, France, in which the short straight sections made it desirable to decrease λ_w in order to obtain a large number of periods, have designed and built an undulator with the "keyhole" vacuum beam chamber shown in Fig. 4.

In this chamber we recognize two sections: a wider one to accommodate the beam at injection, and a narrow (23 mm magnet gap) for the beam stored and damped to its final size. To operate the undulator, the vessel is moved vertically down on the beam by about 7 cm, together with the superconducting magnet (this is a "vertical" flat undulator), such that the beam finds itself in the narrow gap. Elastic bellows are used to mantain the continuity of the vacuum in the accelerator donut.

A similar arrangement has been adopted for the undulator for the storage ring VEPP-3 at Novosibirsk, Soviet Union[11], sketched in Fig. 5. Here, the damped beam is displaced horizontally by means of magnetic fields into the narrower section of the accelerator chamber.

V. PERMANENT MAGNET UNDULATORS

The availability of permanent magnet materials with high remanent fields, such as Samarium-Cobalt, offers us other attractive design choices.

We can say in general that permanent magnet undulators of the flat species seem to show the following general features:

i) Simplicity of mechanical structure. The magnet is made of two ordinary steel beams with attached permanent pole pieces.

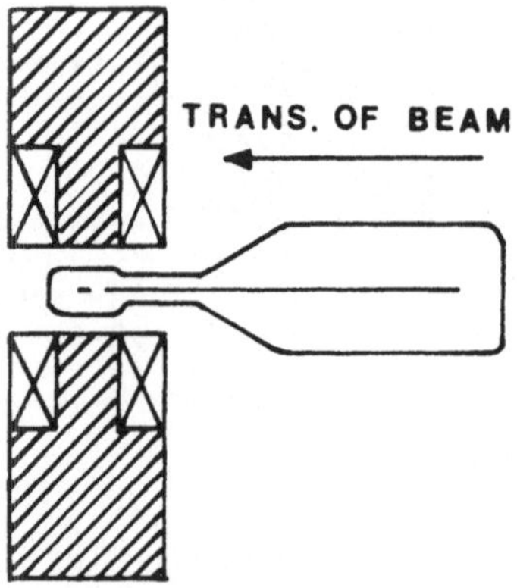

Fig. 5. Flat undulator with a "keyhole" beam chamber for VEPP-3. After L.M. Barkov et Al.[11]

ii) Very efficient use of the longitudinally available space (straight sections in a storage ring), since no dewar with cold-to-warm bore transition sections is needed.
iii) Ease in changing the magnetic field intensity on the median plane, thus tuning the laser, by simply varying mechanically the magnet gap.
iv) Possible changes in the basic structure of the undulator (making e.g. an "Optical Klystron", as proposed by N.A. Vinokurov and A.N. Skrinsky[12]), by simply rearranging the pole pieces.
v) Low construction cost and zero operation costs compared with superconducting undulators.

It seems that a good design is obtained by putting the whole magnet inside the vacuum enclosure, perhaps coating it with a good vacuum material, and allowing the two halves to move against each other to control field intensity. This arrangement would keep the gap wide and the field low when the electrons are injected (storage ring case), and furnish the right field for subsequent operation.

For an electron ring, a limit to the smallness of the gap would be presumably set by the longitudinal coupling impedance seen by the electron beam. The impedance, that determines the bunch length, could be lowered by lining the pole tips with two high conductivity thin metal ribbons.

The general appearence of an undulator with permanent magnets is shown in Fig. 6. Its preliminary design can be based on the following expression for the field on the median plane[13]

$$B_{wo} = 2\,B_r\,e^{-pg}\,(1 - e^{-p\ell})\,\frac{\sin(\pi/M')}{\pi/M'} \tag{12}$$

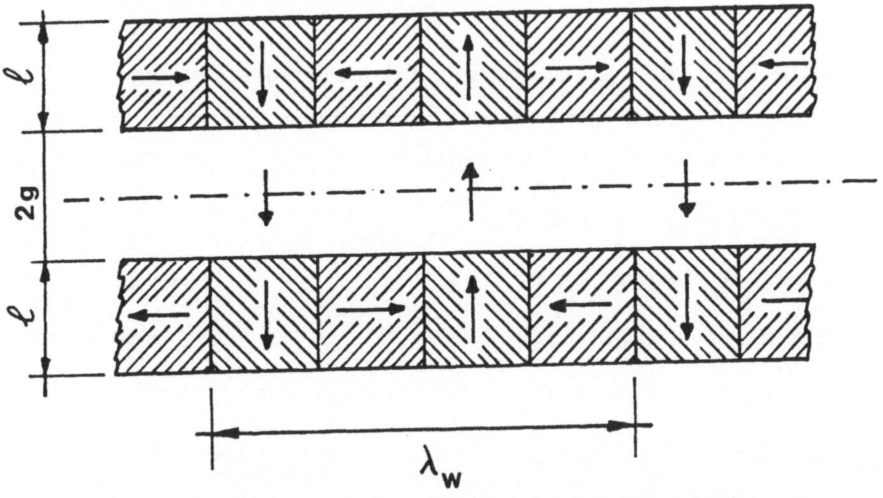

Fig. 6. Sketch of a permanent magnet undulator.

with the following meaning of the various quantities

B_r, remanent field in the magnetic material,

ℓ, transverse dimension of one magnet piece,

M', number of magnet pieces in a period.

With available Sm-Co_5 permanent magnets, take typically

$B_r = 0.85$ T, $\ell = \lambda_w/4$, $M' = 4$

and obtain

$$B_{wo} = 1.21\ e^{-pg} \tag{13}$$

i.e. an expression similar to (9).

It is interesting to study as an example the performance of a FEL with a permanent magnet undulator, to be used on a storage ring.[14]

Let us first re-write the expression (5) for the gain per pass in a manner which shows its dependence on the undulator period

$$G_{o,max} = -\frac{8}{\sqrt{2}}\frac{I_p}{I_A}\frac{f(x)_{max}}{\gamma^3}\frac{L_u}{\sigma^2\lambda_w}\frac{\kappa^2}{1+\frac{L_o\lambda}{8\pi\sigma^2}} \tag{14}$$

The values of all parameters used in (14) are summarized in Table 3a. λ and κ are related to λ_w and g by Eqs. (1), (2) and (13).

Table 3a. FEL with permanent magnet undulator. Example

Electron energy (storage ring)	$\gamma = 1000$
Electron peak current	$I_p = 108$ A
Undulator length	$L_u = 250$ cm
Optical cavity length	$L_o = 850$ cm
Beam transverse size	$\sigma_\perp^2 = 8.41\ 10^{-4} cm^2$
Gain function at maximum	$f(x)_{max} = 0.07$

Table 3b. FEL with permanent magnet undulator. Example

Undulator period	λ_w = 7 cm		
Magnetic gap	2g = 0.8	÷	1.6 cm
Gain (small signal)	G_o = 9.7	÷	7.15 %
Undulator parameter	κ = 3.9	÷	2.75
Magnetic field	B_o = 0.845	÷	0.590 T
Radiation wavelength	λ = 0.56	÷	0.30 μm

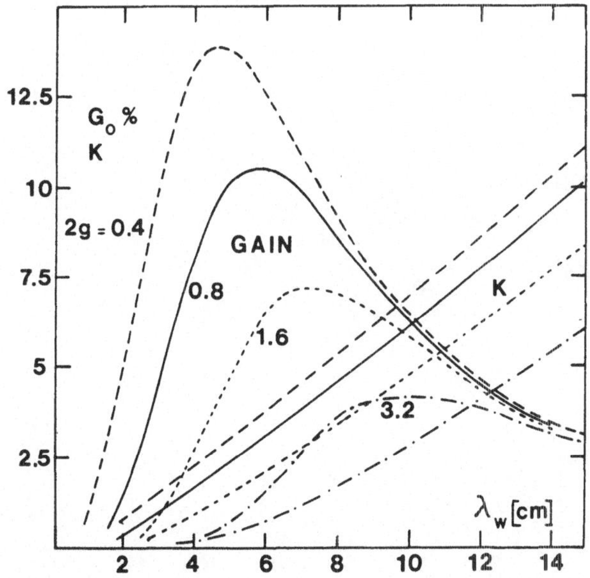

Fig. 7. Study of a permanent magnet undulator for the NSLS free-electron laser.[14] Gain and undulator parameter κ vs. period λ_w. Data of Tables 3a and 3b.

Table 3b. FEL with permanent magnet undulator. Example

Undulator period	$\lambda_w = 7$ cm		
Magnetic gap	$2g = 0.8$	÷	1.6 cm
Gain (small signal)	$G_o = 9.7$	÷	7.15 %
Undulator parameter	$\kappa = 3.9$	÷	2.75
Magnetic field	$B_o = 0.845$	÷	0.590 T
Radiation wavelength	$\lambda = 0.56$	÷	0.30 μm

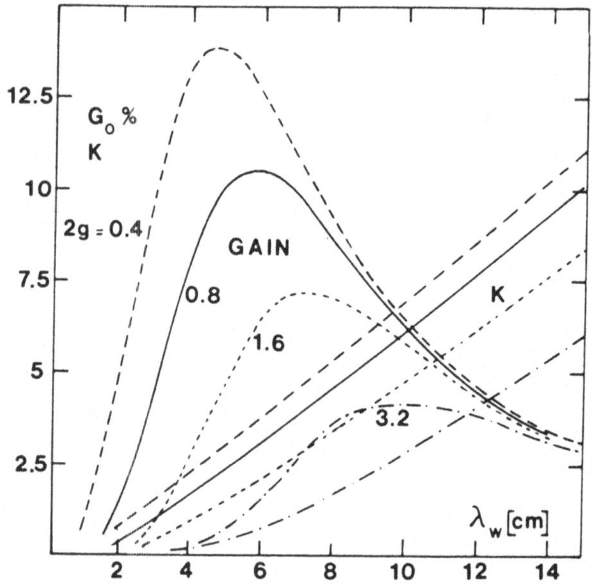

Fig. 7. Study of a permanent magnet undulator for the NSLS free-electron laser.[14] Gain and undulator parameter κ vs. period λ_w. Data of Tables 3a and 3b.

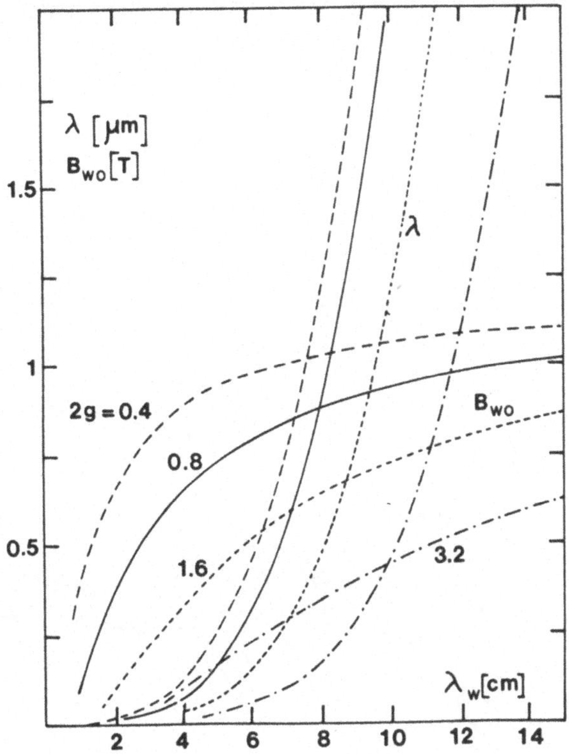

Fig. 8. Study of a permanent magnet undulator for the NSLS free-electron laser.[14] Radiation wavelength and magnetic field vs. period λ_w. Data of Tables 3a and 3b.

For every value of the magnetic half-gap g, the gain (14) exhibits a maximum for a given value of λ_w (Fig. 7). If we vary that g value, a wide variation of the laser wavelength, from UV to IR, is achieved (Fig. 8).

As an example, take an undulator period λ_w = 7 cm, and a gap between 0.8 and 1.6 cm. The values in Table 3b are obtained*.

---

* There are, at the present, better materials for permanent magnets than Sm-Co_5, with remanent fields in excess of 1 T.[13] With these, the values for the magnetic gap could be increased substantially.

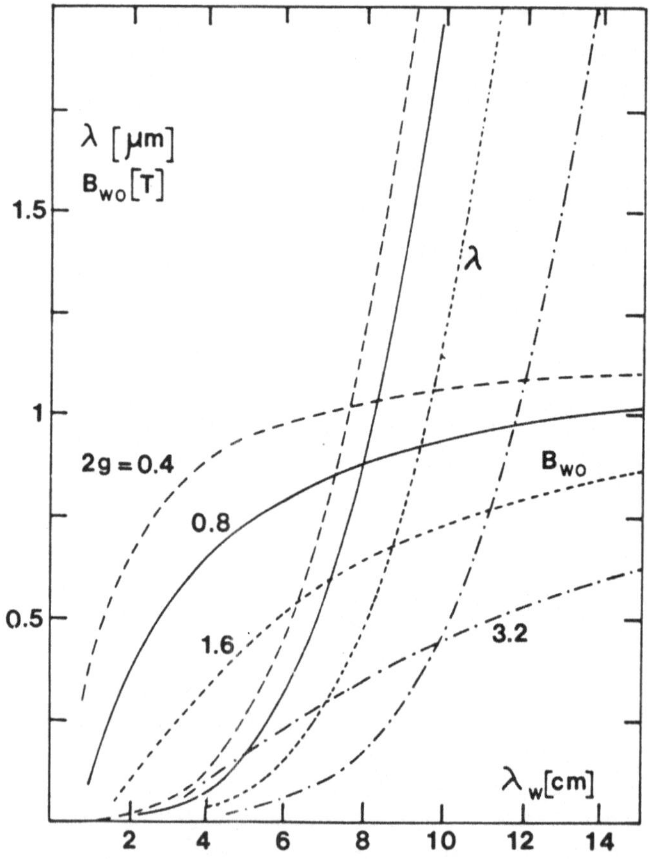

Fig. 8. Study of a permanent magnet undulator for the NSLS free-electron laser.[14] Radiation wavelength and magnetic field vs. period λ_w. Data of Tables 3a and 3b.

For every value of the magnetic half-gap g, the gain (14) exhibits a maximum for a given value of λ_w (Fig. 7). If we vary that g value, a wide variation of the laser wavelength, from UV to IR, is achieved (Fig. 8).

As an example, take an undulator period λ_w = 7 cm, and a gap between 0.8 and 1.6 cm. The values in Table 3b are obtained*.

* There are, at the present, better materials for permanent magnets than $Sm-Co_5$, with remanent fields in excess of 1 T.[13] With these, the values for the magnetic gap could be increased substantially.

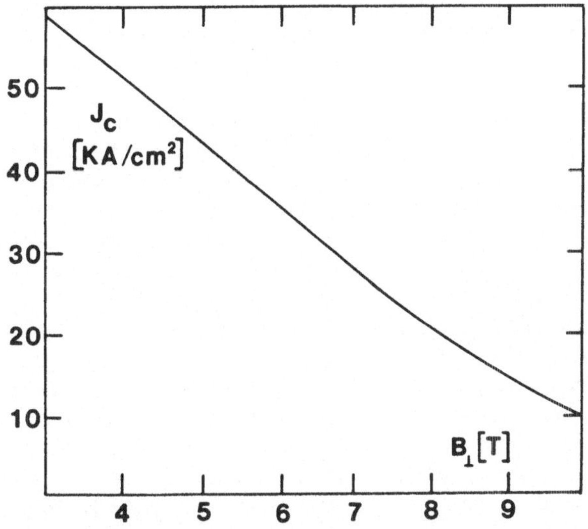

Fig. 9. Critical current density vs. transverse magnetic field in Nb-Ti multiwire conductor. 4.2 °K. Windings with 70% packing factor.

VI. HIGH FIELD WIGGLERS

For certain choices of parameters, reaching very high fields in undulators and wigglers may be desirable. The technical problems of obtaining fields in excess of some teslas are great, the costs of construction and operation of such devices are proportionally high.

The only choice is a superconducting design, either with Niobium-Titanium or Niobium-Tin windings. Some aspects to be mentioned are:

Magnet Coils

The technology of Nb-Ti is more advanced than that of Nb-Sn, and therefore the former material is preferred, unless very high fields, over 5 tesla, are needed. The conductor comes in flat ribbon or multi-strand round cable. Wire allows a better filling of the space available to the coils.

The coils in a superconducting magnet for very high fields are designed mainly according to the maximum field allowed inside the windings for a given current. The operating level depends also on the temperature that can be maintained. A curve of the critical current density vs. field in a typical commerical Nb-Ti multistrand cable is shown in Fig. 9, for operation at 4.2 °K.

At high field level, the iron core of the magnet is highly saturated and, even if the iron itself contributes to the field in the gap by its own magnetization, it does not funnel much of the flux, which simply spreads in the surrounding space. To avoid exceeding the critical field level, a good approach is to make windings in two or more layers, with different current densities, the lowest being closer to the poles, where the dispersed flux is more. Some kind of optimization in coil design may be thus achieved.

Fig. 10 shows the design of a 6 tesla wiggler for the X-ray ring of the National Synchrotron Light Source at Brookhaven U.S.A.[15] Fig. 11 shows the field pattern on the median plane and in the coils (maximum values), as calculated in two dimensions with POISSON. The figure shows also the critical levels of the field, B_c, not to be exceeded in the coils.

The Temperature

The temperature of the windings must be kept as low as possible, since the critical values decrease with increasing temperature. A typical value is 4.2 °K. Fig. 12, due to A.S. Carrol and G. Danby[16], shows a plot of the quantity $\sqrt{J_c B_{max}}$, that is approximately a linear decreasing function of the temperature. Even an increase in temperature from 4.2 to 4.5 °, of the helium bath is shown from the Figure to have a large effect. The coils must be then carefully engineered and surrounded by a good layer of coolant on every side, thus subtracting precious space from the vacuum beam chamber, as already pointed out.

The Mechanical Structure

A high field wiggler must be built strong enough to withstand the Lorentz forces exerted on the coils, and hence on the pole and yoke structure, due to the interaction of the currents in the coils with the magnetic field there.

In particular, we must avoid having the winding layers in the coils move with respect to each other under the stress. This would make thermal friction energy available locally, and the energy would spread throughout the coils, making them unstable. A discussion of this effect can be found e.g. in a paper by H. Brechna[17].

To counteract the forces on the coils, which are radially outward and would then bend outward and end pole pieces of a high field wiggler, a system of thick plates and longitudinal ties is needed.

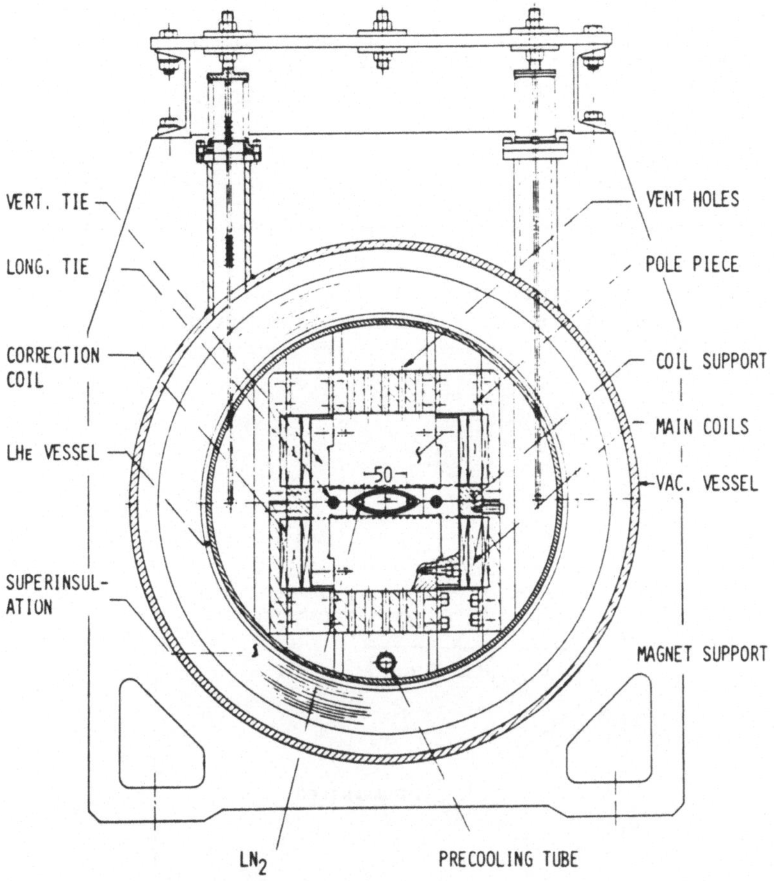

Fig. 10. Transverse cross section of the NSLS 6 tesla wiggler.[15]

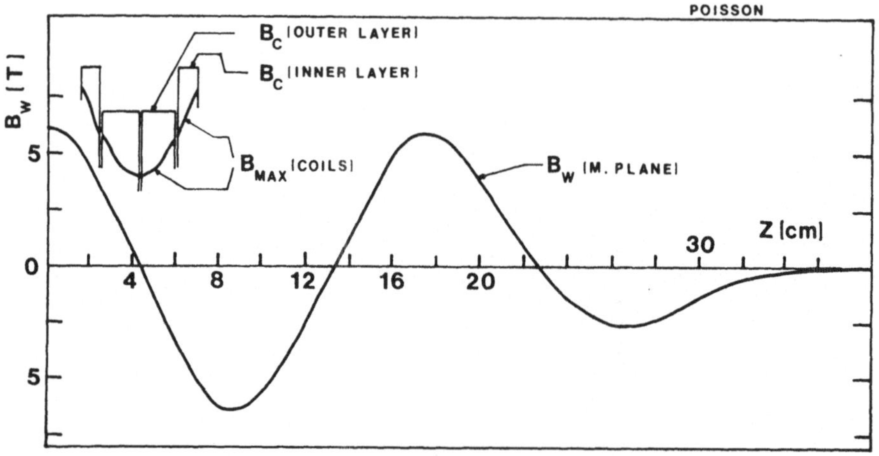

Figure 11. NSLS 6 tesla wiggler. POISSON field calculations on nedian plane and in coils. Inner layer 265 A X 695 turn, outer layer 265 A x 1690 turns.

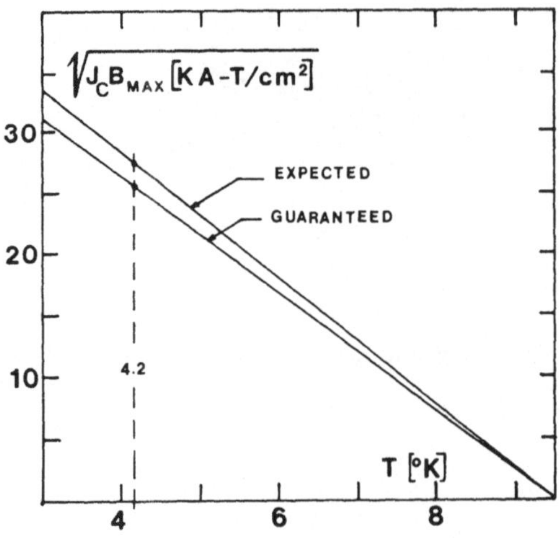

Figure 12. To show the temperature effect on a Nb-Ti multistrand wire. Modified with permission after A.S. Carrol and G. Danby[16].

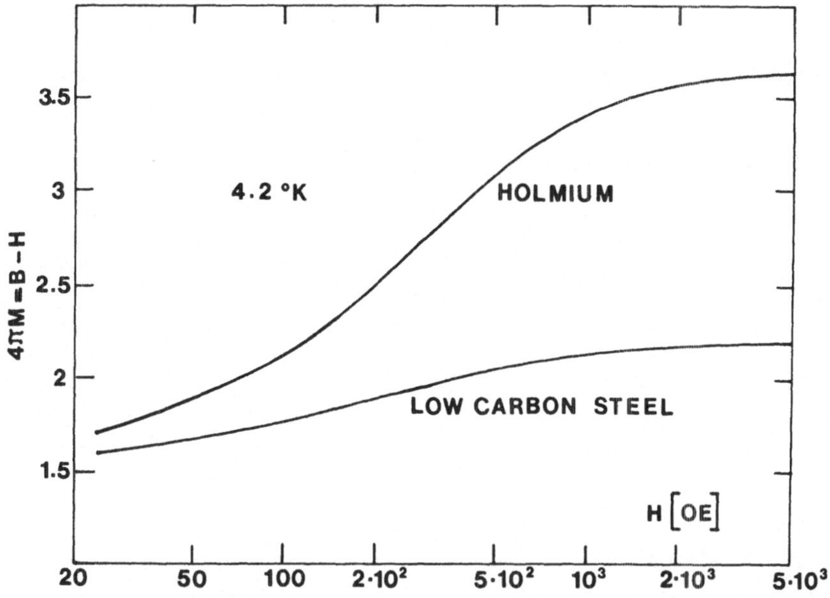

Figure 13 Magnetization curve of the metal holmium at 4.2 °K vs low carbon steel.

The Magnetic Core

The magnetic core helps to increase the field on the median plane, even at extreme saturation. To enhance this effect, one could make use of more exotic materials than soft iron, like rare earth metals (the best seems to be holmium), that at liquid helium temperature exhibit a large magnetization (fig. 13). With pole pieces of holmium, we would also gain the additional advantage that the magnetic flux is core contained in the core, away from the conductor coils.

Radiation Heating

A typical synchrotron radiation flux from a 5 to 6 tesla wiggler inserted in a high energy storage ring, such the one just described, amounts to about 1 Kwatt per pole. The power increases with the square of the electron energy and the square of the magnetic field. An approximate expression for the photon flux per unit beam current (in watt/mA) is[18]

$$\frac{dP}{dI} \simeq 0.828\ 10^{-7}\ \gamma^2\ (N_w + \frac{\sqrt{2}}{4})\ B_o^2\ \lambda_w \tag{15}$$

This power is absorbed by the side walls of the vacuum vessel, which must be provided by additional water and liquid nitrogen cooling.

REFERENCES

1. A. Hofmann, Nucl. Inst. and Methods, 152:17 (1978).

2. D.F. Alferov, Yu.A. Bashnakov, K.A. Belovintsev. E.G. Bessonov and P.A. Cherenkov, Particle Accelerators, 9:223 (1979).

3. C. Pellegrini, IEEE Trans. Nucl. Sci., NS-26:3791 (1979).

SOME POTENTIAL APPLICATIONS OF FREE ELECTRON LASERS TO PHOTOCHEMISTRY

K. L. Kompa

Projektgruppe für Laserforschung, Max-Planck-Gesellschaft, D-0846 Garching, FRG

It is very difficult to evaluate and predict applications of free electron lasers in the area of photochemistry, since this concerns the use of a source which does not exist in usable form, plus a research area which is still in its infancy and has not really taken shape either. In this situation different authors may give different judgements and the fairest approach may therefore be to look at this question from as many different directions as possible. This is the intention in the introduction to this paper. After that step, however, we will proceed to describe some specific applications areas.

1. Introduction

The laser offers a variety of radiation features, which - used alone or in combination - make this source superior to conventional radiation sources. These have often been listed as follows.

a. High spectral purity and small frequency bandwidth

b. High radiation powers and energies (pulsed or cw)

c. Many new wavelengths often combined with wavelength tunability

d. Spatial and temporal coherence

e. Availability of short (even ultrashort) pulse durations

To identify the photochemical application areas where these features could be most beneficial the above mentioned list shall be confronted with a catalogue of concepts, which have developed over the last few years in laser induced chemistry. This catalogue is arranged in such a way that the experiments where the highest degree of excitation specificity (state selectivity) is involved rank first and are followed by other applications in decreasing order of specificity in the excitation process.

a. Truly state-selective chemistry where a single individual quantum state is prepared and brought to react yielding a well controlled product distribution.

b. Bond- or group-selective chemistry where the excitation is still contained in a limited number of states, which couple to a specific product channel or - at least - yield a limited product distribution.

c. Species-selective chemistry where one compound out of a mixture is specifically converted to a product leaving the rest of the mixture unreacted. This species may be a rare isotope or an impurity.

d. Laser heating where the action of the laser is non-specific with regard to a given reaction mode but where the spatial and temporal laser characteristics may still be used advantageously.

Another way to look at the many diverse uses of lasers to control and probe chemical processes is shown in Figures 1 and 2. It becomes apparent at this point - looking at Figures 1 and 2, which are still limited to some of the more popular ideas in this field - that applications of lasers in photochemistry and molecular physics in general are practically endless. After all, chemistry is a very pluralistic discipline and every new reaction may call for different laser sources and excitation and probing concepts. We will therefore take the liberty of reducing the following discussion to one specific aspect of laser molecular interactions, which is related to the high radiation powers available with some present day lasers. It should be obvious from the beginning that the combination of high power and wavelength tunability is then the most important requirement, where the photochemist challenges the laser physicist. Discussions of the whole variety of existing laser sources with respect to chemical and spectroscopical applications have been given elsewhere[1]. These shall not be repeated here but just briefly summarized in Figure 3.

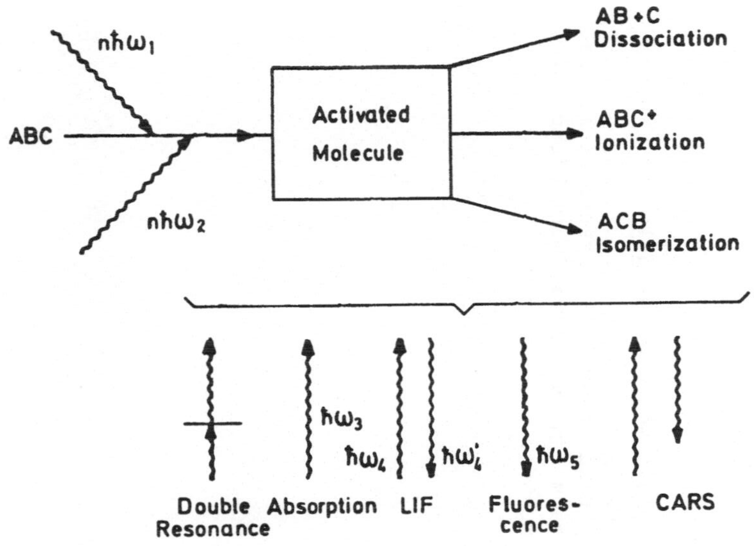

Fig. 1. Laser interaction with a polyatomic molecule leading to unimolecular processes like dissociation, ionization or isomerization. Excitation is achieved by single or multiple photon absorption using one laser, $h\omega_1$, or a combination of lasers $h\omega_1$, $h\omega_2$. The excited molecule and the possible products can be probed by observing spontaneous or laser-induced fluorescence (LIF). Likewise absorption (internal or external to the probe laser cavity) or coherent anti-Stokes Raman scattering (among other techniques not mentioned) may be used[2].

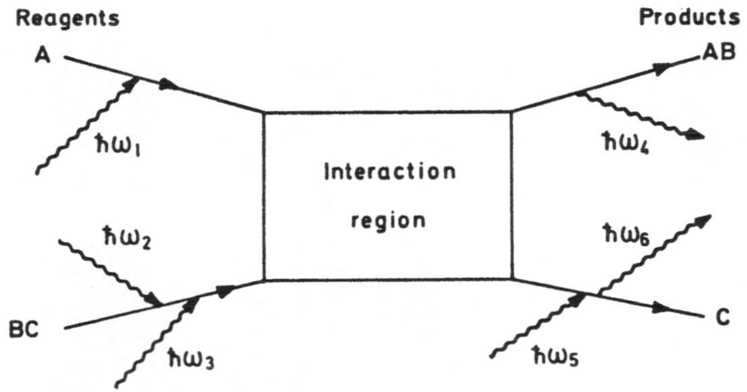

Fig. 2. Types of laser interaction with biomolecular reaction systems. Cases shown refer to pumping by a single photon $h\omega_1$ or by a combination of photons $h\omega_2 + h\omega_3$. Product state analysis can for instance be performed by observing $h\omega_4$ in spontaneous or stimulated emission (chemical lasers) or by laser-induced fluorescence, $h\omega_6$ after pumping by $h\omega_5$. Multiphoton excitation would correspond to $(h\omega_2)_n$. Not all types of laser probing are indicated here[2].

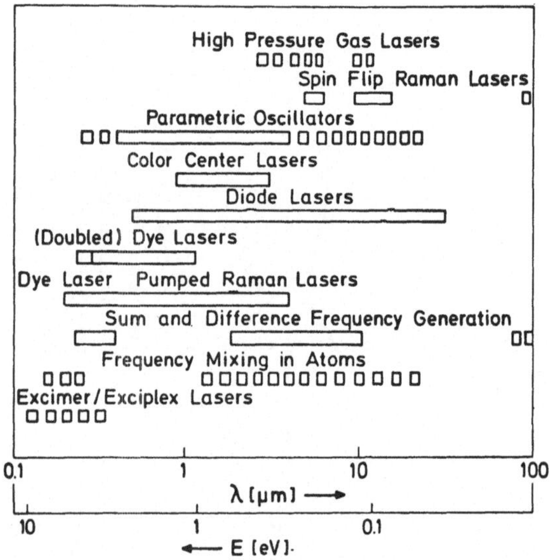

Figure 3. Survey of important laser sources suitable for experiments in chemistry and spectroscopy. ▭ Fully tunable, □ □ □ □ incompletely tunable or fixed frequency lasers. Many of the tuning gaps can in principle be closed by suitable technical means[2].

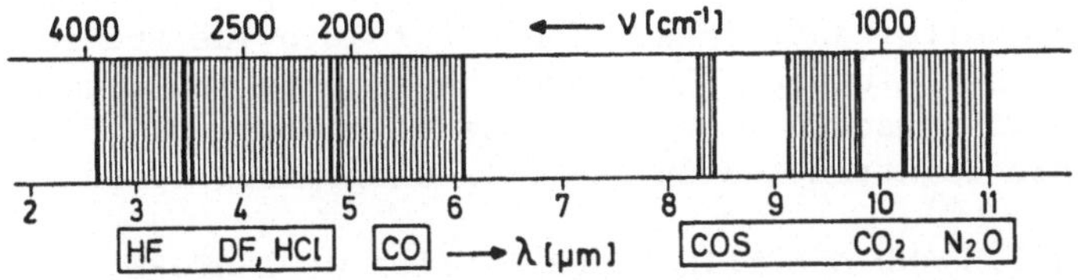

Figure 3a. Survey of the most popular high pressure infared gas lasers. The shaded regions denote the principal line tuning range.

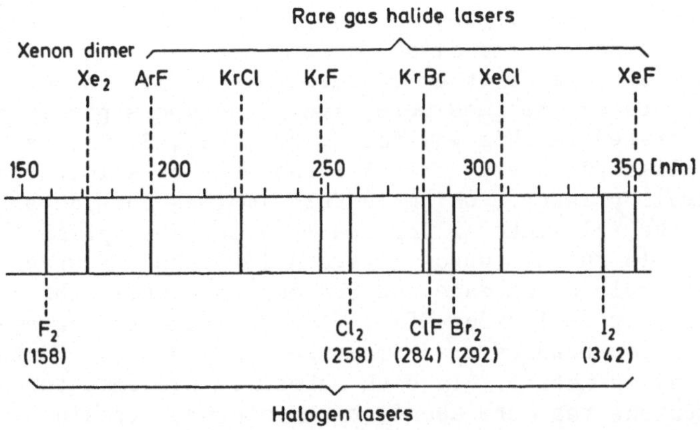

Figure 3b. High power ultraviolet pulse lasers: 3 classes of lasers are shown, namely the rare gas excimer, rare gas halogen and halogen dimer laser systems. In many cases broad fluorescence bandwidths point to the possibility of wavelength tuning. No systematic attempts have been reported, however, to investigate this possibility.

With regard to high laser pulse powers two groups of lasers primarily stand out in Figure 3: The high pressure infrared gas lasers (e.g. CO_2, CO, HF, etc.) and the new rare gas halide ultraviolet lasers (e.g. KrF, ArF, F_2, etc.). In addition, for reagent excitation as well as probing applications, dye lasers have turned out to be most valuable sources. Table 1 attempts to recapitulate some relevant performance data. These then provide the reference standards by which future FEL devices have to be measured. The requirements will become clearer if specific problems are considered. This is attempted in the following paragraphs.

2. PREPARATIVE AND ANALYTICAL APPLICATION OF ISOTOPE SELECTIVE INFRARED MULTIPHOTON DISSOCIATION

This section will summarize some numbers and figures for IR laser induced isotope separation involving the elements sulfur and carbon. Isotope separation by IR laser induced multiphoton dissociation was demonstrated on a laboratory scale (10 m/mole) with SF_6 and CF_3I. The question of prime interest concerns the degree of conversion per input photon relating to the effective energy absorption cross section. The experimental arrangement is shown in Figure 4. The central part of the set-up is a waveguide absorption cell ("gold finger"). A CO_2 pulsed laser is used to selectively induce dissociation of molecular gases. The photons from such a simple infrared laser cost only a few cents per mole. For dissociation, at least the bond energy (200-400 kJ/mole = 17-35 CO_2 laser quanta/molecule) must be expended. Taking into account additionally the quantum yield and coupling efficiency into the sample, $\gtrsim$100 photons/molecule may be needed. If, during the excitation and dissociation process, energy transfer to the other isotope or other mixture components can be suppressed, the process becomes selective. Isotope separation has been demonstrated in this way for H, B, C, Si, S, Cl, Se, Mo, Os and U[2]. Collisional energy transfer can only be eliminated by sufficiently small pressure, which in turn implies considerable absorption lengths (>10 meters) for the relevant absorption cross sections. In the set-up of Figure 4 1500 photons were needed to selectively dissociate one SF_6 molecule while the corresponding number is 100 for CF_3I. The experimental parameters and separation results are summarized in Table 2. Further details are found in Ref. 3. In both molecular systems, for experimental reasons the more abundant isotope was irradiated and dissociated. It would of course be more economic to dissociate the less abundant isotope in order to reduce the total photon input. This, however, is only possible if shorter pulses combined with smaller beam divergence can be used. Alternatively, "two-color" experiments may be conducted with two infrared lasers or one infrared plus one ultraviolet laser source. To make this a versatile scheme, which could be applicable to many different isotopes, tunable laser sources are needed.

Table 1. Laser Performance Data

Properties of some pulsed lasers

Laser	Wavelength m.	Pulse Length(s)	Pulse Energy Multimode (J)	Typical Rep. Rate (pps)
Discharge pumped KrF	0.248	10^{-8}	1	10
Discharge pumped ArF	0.193	10^{-8}	0.2	10
Amplified Nd:YAG	1.064	10^{-8}	1	10
doubled	0.53	10^{-8}	0.5	10
tripled	0.355	10^{-8}	0.3	10
quadrupled	0.265	10^{-8}	0.1	10
Nitrogen laser pumped dye	0.360-0.740	10^{-8}	10^{-3}	10
CO_2 - Q-switched	10.6	10^{-5}	0.1	100
CO_2 - TEA	10.6	10^{-7}	10	1-100
Discharge pumped HF	2.9	5×10^{-7}	1	10
Discharge pumped CO	5.6	10^{-5}	1	10
Optical Parametric Oscillators	2	10^{-5}	0.01	100
CF_4	16	10^{-7}	0.10	1

Characteristics of some cw lasers

Laser	Approximate Wavelength (μm)	Power (typical) TEM_{00}(W)	Beam Divergence (mrad)
Ar^+	0.514	15.	0.6
Ar^+ laser-pumped dye laser pumped	0.585	0.1	1.5
He-Ne	0.6328	0.005	1
Nd:YAG	1.064	20	2.5
CO_2	10.6/9.4	50	2.2
Didoe Lasers	IR	0.001	20*)
Chemical HF	2.9	10	1
Color Center	1.9	0.1	**)
Spin Flip Raman Lasers	6	1	20***)

*) Without external resonator several degrees in one direction, 20 mrad in the other.

**) Not definitively known.

***) Can be improved by external resonator.

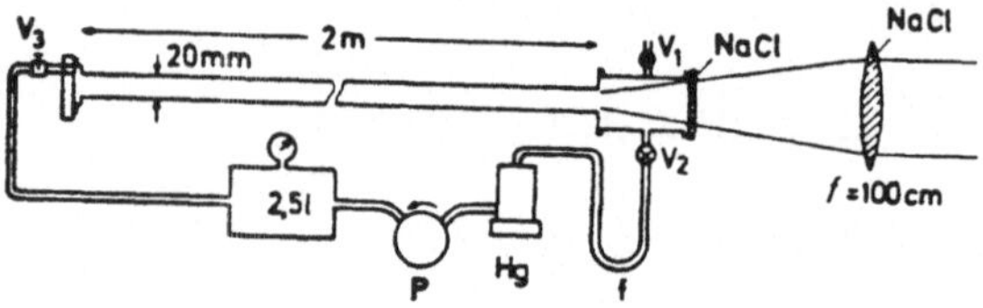

Figure 4. Experimental set-up for laboratory scale separation of light isotopes. $V_{1,2,3}$ valves, P membrane pump, Hg mercury pump, F cold trap with BaO suspension to capture SF_4 and HF or I_2, respectively[3].

Table 2. Isotope Separation of Sulfur and Carbon

	SF_6	CF_3I
Isotope ratio	$^{32}S:^{34}S$ =	$^{12}C:^{13}C$ =
Begin	95:4,2	98,9:1,1
End	20:80	30:70
CO_2 laser frequency	944,2 cm^{-1} (10 P 20)	1073,3 cm^{-1} (9 R 12)
Laser energy per pulse	20 J	8 J
Photons per pulse	10^{21}	$4 \cdot 10^{20}$
Number of shots	17000	2000
Conversion per shot	10% (^{32}S)	70% (^{12}C)
Pressure	0,4 mbar = 10^{16} cm^{-3}	0,4 mbar = 10^{16} cm^{-3}
Energy density	12 J/cm^2	4 J/cm^2
Waveguide losses	20%	20%
Absorption of Gas	20%	50%
Mean excitation	30 quanta/molecule	40 quanta/molecule
Photons expended	1500 quanta/molecule	100 quanta/molecule
Total conversion	10 mmol = 1,5g $^{32}SF_6$	10 mmol = 2g $^{12}CF_3I$

In general, isotope separation (or likewise purification) of gaseous mixtures are one area of successful laser photochemistry which would greatly benefit from improved laser sources.

3. UV LASER INDUCED MULTIPHOTON MASS SPECTROMETRY

Laser induced multiphoton spectroscopy including ionization spectroscopy is a new and active area of research, which has received a new impetus by the development of high power ultraviolet laser sources. The situation is somewhat similar to the infrared multiphoton excitation mentioned in the previous paragraph. In both cases, near resonant multistep photon absorption is often the process which is responsible for the generation of highly excited molecules with associated chemical consequences. In the case of UV excitation, however, not only dissociation but also ionization can occur with high probability. The two concepts may be compared in some of their features as is shown in Table 3.

Laser induced multiphoton ionization is not only a phenomenon of basic physical interest but provides also interesting prospects for mass spectrometry, and it is this practical aspect on which a few remarks will be offered. Total ion yields in multiphoton ionization can be very high (approaching 100 percent in favorable cases) if resonant or near-resonant intermediate states are available and if ionization requires not more than two photons. These conditions are satisfied for many polyatomic molecules (see below) and laser sources with photon energies $\gtrsim 5$ eV. In some cases this mass spectrometric concept allows for very sensitive (potentially single molecule) trace detection. In addition, the spectral dependence of the ionization probability makes very sleective detection of molecules possible. These features combine in the two-dimensional mass spectrometer which is described in Reference 6.

In order to exemplify the success of this ionization technique, multiphoton ionization of the benzene molecule is briefly discussed in the following. For orientation, reference 4 describes relevant states and the corresponding photon energies of some popular laser sources. Some experimental details are provided in Figure 5. A conventional time-of-flight mass spectrometer was employed here, modified to include a laser ionization source. Once the photo ions are formed they are collected by appropriate ion optics and are accelerated pulsewise by a series of grids into the field-free flight tube, where they separate in time according to their different m/e ratios. An entire mass spectrum at a given laser wavelength and intensity can thus be recorded in a single shot with excellent reproducibility and resolution. Without going into detail on the many and diversified experimental

Table 3. Parallel Aspects of Infrared Multiphoton Dissociation and Ultraviolet Multiphoton Ionization

IR-MPD	UV-MPI
1) Study spectroscopy and chemistry of highly excited molecules	
2) Study radiationless transitions (constraints restricting the number of participating levels)	
3) Dissociate	ionize
selectively isotopes or trace compounds	
4) Observe secondary product	
fragmentation	ionization
5) Generate interesting	
radicals	ions
6) Study collisional effects (energy transfer, photochemistry)	
7) Collisions between excited molecules can lead to	
"Treanor Pumping"	Triplet-Triplet annihilation

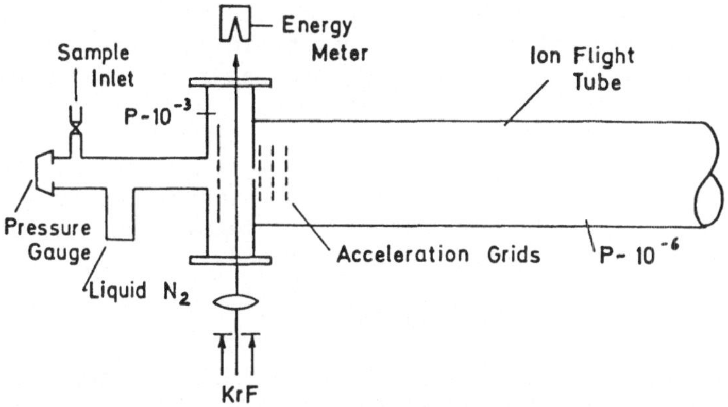

Figure 5a. Experimental arrangement for photo ionization mass spectrometry. The ions are formed in the interaction region between the UV laser beam and the molecular gas and are subsequently drawn out into the time-of-flight tube[4].

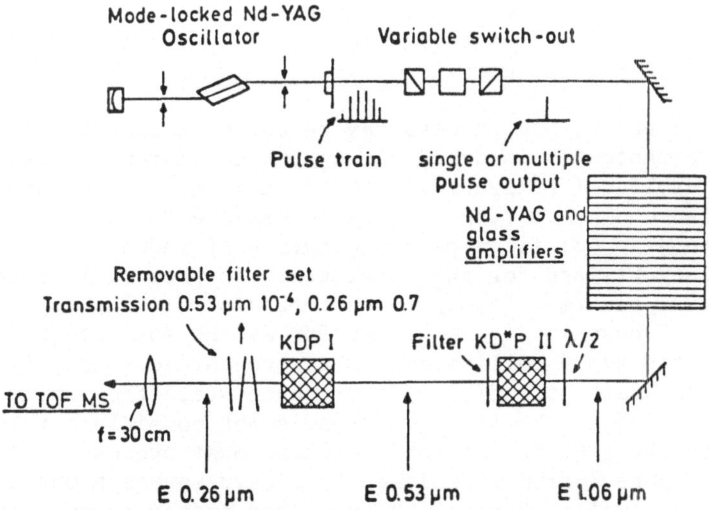

Figure 5b. Picosecond Nd-YAG laser with provision for frequency doubling and quadrupling. Single pulses or pulse trains can be gated from a mode-locked oscillator and after amplification may either be used at one frequency or at a combination of the fundamental and harmonic frequencies to ionize molecular gases[5].

and theoretical results, it should be mentioned that more than 10 percent of all the benzene molecules present in the interaction volume between the molecular sample and the laser beam have been ionized and detected[5]. Recent results[6] extend this concept to many other molecules including species of biological interest. Again progress is hampered by the nonexisting or at most limited wavelength tunability of the available high power UV lasers. A breakthrough in photo ionization mass spectrometry might well be expected if better laser sources could be found.

4. MISCELLANEOUS FUTURE APPLICATIONS

As mentioned in the beginning, there are quite different philosophies amont laser photochemists. Some practitioners in the field are interested in applications where the laser replaces a conventional excitation source in an existing reaction scheme. The laser then must compete with other sources which are usually much more developed and established. Correspondingly, this often becomes a difficult task. A second philosophy is trying to locate problems where existing excitation sources do not perform well or where results can be obtained with a laser which are impossible to obtain otherwise. The two application areas remarked upon in Sections 2 and 3 are of this kind. A third philosophy might look for applications which have only become conceivable with the very special characteristics of laser sources. In a pictorial way, one might call this "inventing the questions to which the laser could provide answers". Laser assisted collisions (or bimolecular reactions in a braoder sense) fall into this latter category.

In general terms, any process may be called a radiative collision if a photon is absorbed during the collision process, in this way enabling ("triggering") the formation of a product which otherwise would not form. A simple example for such a concept is found in collision pair excitation of xenon. This experiment also utilized for the first time the potential of the 1580 A molecular fluorine laser[7] in photochemistry. The experimental setup is shown in Figure 6. As far as the spectroscopic parameters of the xenon collision pair are concerned there, is a shallow van der Waals potential in the ground state from which F_2 laser excitation to the $^{1,3}\Sigma^+_u$ states is not possible for energetic reasons. The excitation therefore must proceed according to Figure 7 from a position of closer approach outside the van der Waals well. Part of the absorbed energy is emitted by the xenon excimer, which fluoresces at 1720 A. The pressure dependence of the fluorescence intensity shows a quadratic behaviour as expected. Besides pure xenon, xenon-halogen mixtures were also investigated[8] and provided information which is

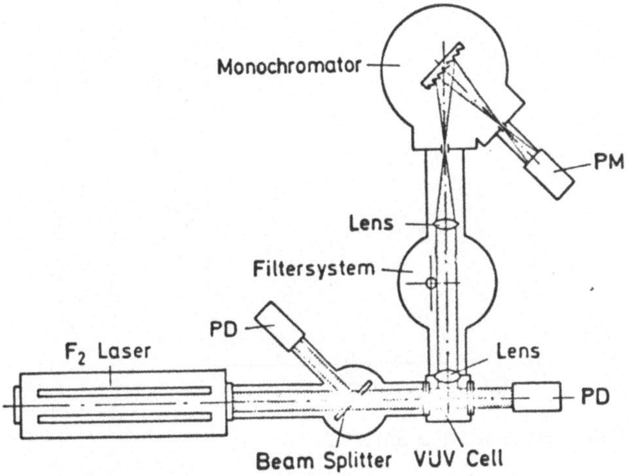

Figure 6. Experimental apparatus for radiative collision studies[8] employing a rapid TEA discharge F_2-laser[7] whose emission wavelength is at 158 nm. The laser beam is guided in vacuum into the high pressure absorption cell. The fluorescence is collected at right angles.

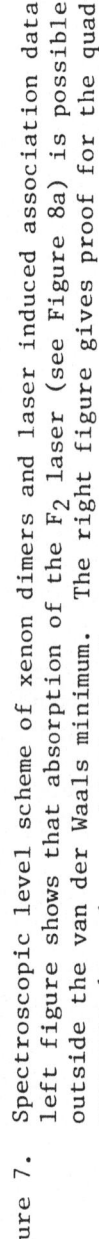

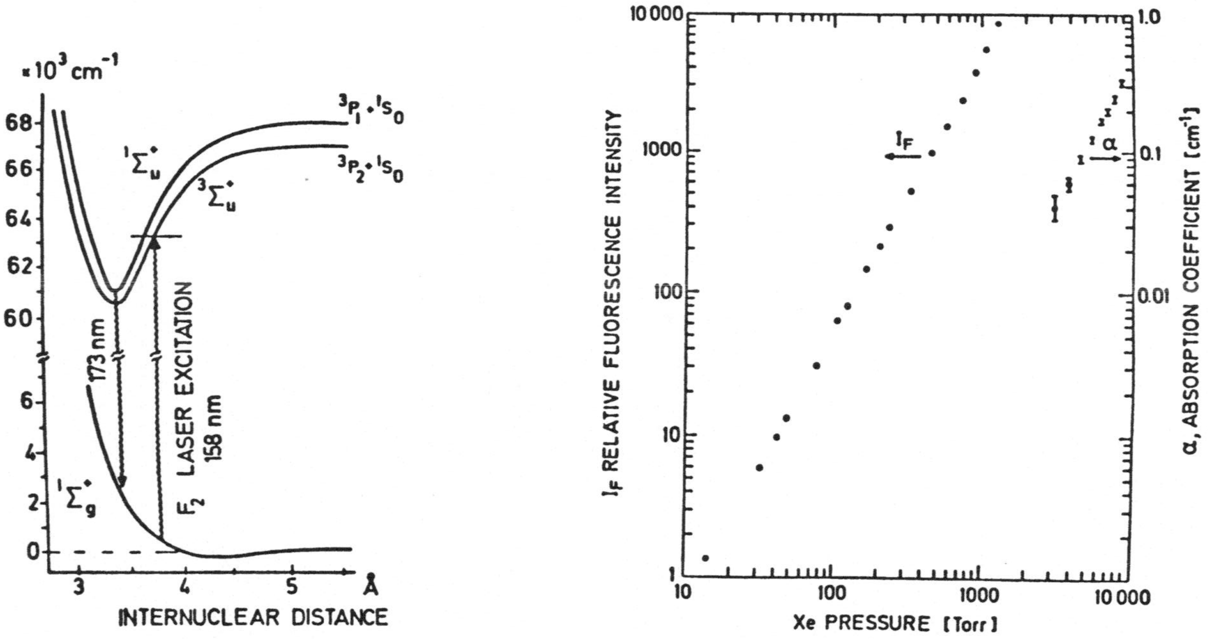

Figure 7. Spectroscopic level scheme of xenon dimers and laser induced association data. The left figure shows that absorption of the F_2 laser (see Figure 8a) is possible only outside the van der Waals minimum. The right figure gives proof for the quadratic pressure dependence of both the fluorescence intensity and the absorption cross section. For further details consult Ref. 8.

relevant to the studies of kinetic processes in rare gas halide lasers. Obviously, it would be more interesting to do such experiments with tunable high power lasers in order to be able to probe a larger range of the interatomic potential. This again examplifies a case where better lasers would open up new research possibilities.

While the above mentioned xenon collision pair absorption is on the border line of radiative collisions, various analyses have been given for energy transfer and reactive collisions induced by laser photons. Consider for instance radiative collisions of the type

$$A(i) + BC\ (v_i) + \quad \rightarrow \quad A(f) + BC\ (v_f)$$

$$\rightarrow \quad A(i) + BC\ (v_f)$$

where A is an alkali atom and BC a diatomic molecule (N_2, CO, NO, O_2 etc.)[9]. The initial and final electronic (vibrational) states of the atom A (molecule BC) are denoted by $i(v_i)$ and $f(v_f)$, respectively. The excitation frequency is far detuned from the absorption lines of the isolated particles (no absorption at infinite separation). The analysis[9] shows that cross sections large enough to be experimentally observable may be expected in such processes. Experiments are in progress now to demonstrate the validity and the magnitude of the laser control. Eventually, this may lead to a new type of laser chemistry where initial state relaxation is eliminated because the laser energy is only applied at the point of close encounter of the reagents. In addition, one may hope for some specificity of product formation and energy distribution in the products. Needless to say, such concepts again call for new and better laser sources which combine high radiation powers with frequency control.

REFERENCES

1. For a chemistry oriented laser comparison see for instance the report of the workshop held in Riva del Garda (Trento) Italy in June 1979 on the possible impact on free electron lasers on spectroscopy and chemistry, G. Scoles, ed.

Additional references are for instance M. J. Berry, in L. E. St.-Pierre, ed., Future Sources of Organic Raw Materials, Pergamon Press, Elmsford, NY, 1979 or A. Ben-Shaul, Y. Haas, R. D. Levine, K. L. Kompa, Lasers and Chemical Change, Springer series in Chem. Phys., Vol. 9, Springer-Verlag Berlin, Heidelberg, New York, 1980.

2. R. V. Ambartzumian and V. S. Letokhov, in Chemical and Biochemical Applications of Lasers, Vol. 3, C. B. Moore, ed., Academic Press, New York, 1977.

R. V. Ambartzumian, V. S. Letokhov, Acc. Chem. Res. 10, 61 (1977).

N. Bloembergen, E. Yablonovitch, Physics Today, May 1968, p. 23.

C. D. Cantrell, S. M. Freund, J. L. Lyman, in Laser Handbook, Vol. 3, M. L. Stitch, ed. North-Holland, Amsterdam, 1978.

3. W. FuB, W. E. Schmid, Ber. Bunsenges. Phys. Chem. 83, 1148 (1979).

4. M. Seaver, J. W. Hudgens, J. J. Decorpo, Int. J. Mass Spectr. Ion. Phys. 34 159 (1980).

5. J. P. Reilly, K. L. Kompa, J. Chem. Phys. in print.

J. P. Reilly, K. L. Kompa, to be published in Advances in Mass Spectrometry, Vol. 8, (1980), Heyden & Sons, London.

6. V. S. Antonov, V. S. Letokhov, Appl. Phys., in print.

7. H. Pummer, K. Hohla, M. Diegelmann, J. P. Reilly, Opt. Commun. 28, 104 (1979).

8. H. P. Grieneisen, K. Hohla, K. L. Kompa, to be published in Int. J. Mass Spectr. Ion Phys.

9. P. Hering, Y. Rabin, to be published in Chem. Phys. Lett.

INDEX